Lecture Notes in Electrical Engineering

Volume 768

The book series *Lecture Notes in Electrical Engineering* (LNEE) publishes the latest developments in Electrical Engineering - quickly, informally and in high quality. While original research reported in proceedings and monographs has traditionally formed the core of LNEE, we also encourage authors to submit books devoted to supporting student education and professional training in the various fields and applications areas of electrical engineering. The series cover classical and emerging topics concerning:

- Communication Engineering, Information Theory and Networks
- Electronics Engineering and Microelectronics
- Signal, Image and Speech Processing
- Wireless and Mobile Communication
- Circuits and Systems
- Energy Systems, Power Electronics and Electrical Machines
- Electro-optical Engineering
- Instrumentation Engineering
- Avionics Engineering
- Control Systems
- Internet-of-Things and Cybersecurity
- Biomedical Devices, MEMS and NEMS

For general information about this book series, comments or suggestions, please contact leontina.dicecco@springer.com.

To submit a proposal or request further information, please contact the Publishing Editor in your country:

China

Jasmine Dou, Editor (jasmine.dou@springer.com)

India, Japan, Rest of Asia

Swati Meherishi, Editorial Director (Swati.Meherishi@springer.com)

Southeast Asia, Australia, New Zealand

Ramesh Nath Premnath, Editor (ramesh.premnath@springernature.com)

USA, Canada:

Michael Luby, Senior Editor (michael.luby@springer.com)

All other Countries:

Leontina Di Cecco, Senior Editor (leontina.dicecco@springer.com)

**** This series is indexed by EI Compendex and Scopus databases. ****

More information about this series at http://www.springer.com/series/7818

Anuradha Tomar · Hasmat Malik · Pramod Kumar ·
Atif Iqbal
Editors

Machine Learning, Advances in Computing, Renewable Energy and Communication

Proceedings of MARC 2020

Springer

Editors
Anuradha Tomar
Electrical Engineering Group
Eindhoven University of Technology
Eindhoven, The Netherlands

Hasmat Malik
BEARS
University Town
NUS Campus
Singapore, Singapore

Pramod Kumar
Department of Computer Science
and Engineering
Krishna Engineering College
Ghaziabad, India

Atif Iqbal
Department of Electrical Engineering
Qatar University
Doha, Qatar

ISSN 1876-1100 ISSN 1876-1119 (electronic)
Lecture Notes in Electrical Engineering
ISBN 978-981-16-2353-0 ISBN 978-981-16-2354-7 (eBook)
https://doi.org/10.1007/978-981-16-2354-7

This Springer imprint is published by the registered company Springer Nature Singapore Pte Ltd.
The registered company address is: 152 Beach Road, #21-01/04 Gateway East, Singapore 189721, Singapore

Preface

The papers presented at the 2nd International Conference on Machine Learning, Advances in Computing, Renewable Energy and communication (MARC 2020) held at Krishna Engineering College in Ghaziabad, Uttar Pradesh, India, on December 17 and 18, 2020, are compiled in this volume. The International Conference on Machine Learning, Advances in Computing, Renewable Energy and communication focuses on advanced research in the area of electrical and computer science engineering and will provide a forum for sharing insights, experiences and interaction on various facts of evolving technologies and patterns related to these areas. The objective of MARC 2020 is to provide a platform for leading academic scientists, researchers, scholars and students to get together to share their results and compare notes on their research discovery in the development of electrical engineering and high-performance computing. Numerous participants attended the conference, made technical presentations and indulged in various technical discussions. The number of papers published in this volume and the number of unpublished presentations at the conference indicate the evidence of growing interest among students, researchers and teachers in manufacturing and advanced computing. More than 300 research papers were submitted, out of which 57 were accepted and presented.

We would like to extend our sincere gratitude to Springer LNEE for giving Krishna Engineering College the opportunity and the platform to organize this conference which helped in reaching out to eminent scholars and fellow researchers in the field of electrical and computer science engineering and helping them in widening the areas of the subject.

We thank all the contributors of this book for their valuable effort in producing high-class literature for research community. We are sincerely thankful to the reviewers to provide all the reviews/comments/suggestions in a short period of time.

We express sincere gratitude to our patrons, Dr. Sandeep Tiwari, Krishna Engineering College, Ghaziabad, and Dr. Sukumar Mishra, IIT Delhi, for their motivation and support in hosting MARC 2020. Our sincere thanks and appreciation to our General Chairs, Dr. Bijaya Ketan Panigrahi, IIT Delhi, and Dr. Dugesh Pant, USAC & USERC, Dehradun, India, for their solid support blended with encouragement and incomparable motivations to achieve the remarkable milestone. We wish to acknowledge our gratitude to Program Chairs, Dr. Yog Raj Sood, NIT Hamirpur, and Dr. Sudeep Tanwar, Nirma University, Ahmedabad, to set the tone of the conference at the higher launch. We sincerely acknowledge all the keynote speakers for disseminating your knowledge, experience and thoughts. We express our sincere gratitude to the Management of Krishna Engineering College, Conference Executive Chair, Publication Chair and Technical Committee Members for their kind support and motivation.

We wish to thank our colleagues and friends for their insight and helpful discussion during the production of this book. We would like to highlight the contribution, suggestion and motivation of Prof. Imtiaz Ashraf, Aligarh Muslim University, India; Prof. M. S. Jamil Asghar, Aligarh Muslim University, India; Prof. Salman Hameed, Aligarh Muslim University, India; Prof. A. H. Bhat, NIT Srinagar, India; Prof. Kouzou Abdellah, Djelfa University, Algeria; Prof. Jaroslaw Guzinski, Gdansk University of Technology; Prof. Akhtar Kalam, Victoria University of Technology, Australia; Prof. Mairaj Ud Din Mufti, NIT Srinagar, India; Prof. Majid Jamil, JMI, India; Prof. A. P. Mittal, NSUT Delhi, India; Prof R.K Jarial, NIT Hamirpur (HP), India; Prof. Rajesh Kumar, GGSIPU, India; Prof. Anand Parey, IIT Indore, India; and Prof Yogesh Pandya, PIEMR Indore, India.

We would like to express our gratitude to our family members for their love and affection and for their intense feeling of deep affection.

Woodlands, Singapore/Delhi, India — Dr. Hasmat Malik
JSS, Noida India — Dr. Anuradha Tomar
KEC, Ghaziabad, India — Dr. Pramod Kumar
Doha, Qatar — Prof. Atif Iqbal

Contents

About the Editors

Dr. Anuradha Tomar has 12 years of experience in research and academics. She is currently working as Postdoctoral researcher in Electrical Energy Systems Group, Eindhoven University of Technology (TU/e), the Netherlands and is responsible for project management and research as a team member in European Commission's Horizon 2020, UNITED GRID and UNICORN projects. She is also associated as an Associate Professor with the Electrical Engineering Department at JSS Academy of Technical Education, Noida, India. She has received her B.E Degree in Electronics Instrumentation & Control with Honours in the year 2007 from University of Rajasthan, India. In the year 2009, she has completed her M.Tech Degree with Honours in Power System from National Institute of Technology Hamirpur. She has received her Ph.D in Electrical Engineering, from Indian Institute of Technology Delhi (IITD). Dr. Anuradha Tomar has committed her research work efforts towards the development of sustainable, energy efficient solutions for the empowerment of society, humankind. Her areas of research interest are Photovoltaic systems, Microgrid, Energy conservation and Automation. She has authored or co-authored 69 research/review papers in various reputed International, National Journals, and Conferences. She is an Editor for books with International Publication like Springer, Elsevier. Her research interests include photovoltaic systems, microgrids, energy conservation, and automation. She has also filled seven Indian patents on her name. Dr. Tomar is Senior member of IEEE, Life member of ISTE, IETE, IEI, and IAENG.

Dr. Hasmat Malik (M'16, SM'20) received M.Tech degree in electrical engineering from National Institute of Technology (NIT) Hamirpur, Himachal Pradesh, India, and the Ph.D. degree in Electrical Engineering from Indian Institute of Technology (IIT), Delhi. He is a Chartered Engineer (CEng) and Professional Engineer (PEng). He is currently a Research Fellow at BEARS, University-Town, NUS Campus, Singapore since Jan., 2019 and served as an Assistant Professor for 5+ years at Division of Instrumentation and Control Engineering, Netaji Subhas Institute of Technology (NSIT) Delhi, India. Dr Hasmat organized five international conferences, and proceedings have been published by Springer. He is a Senior Member of the IEEE, USA, Life Member of ISTE (Indian Society for Technical Education), Life

Member of IETE (Institution of Electronics and Telecommunication Engineering), Life Member of ISRD (International Society for Research and Development) London and Member of CSTA (Computer Science Teachers Association) USA, Member of ACM (Association for Computing Machinery) EIG, and Member of Mir Labs, Asia. He has published widely in International Journals and Conferences his research findings related to Intelligent Data Analytic, Artificial Intelligence, and Machine Learning applications in Power system, Power apparatus, Smart building & automation, Smart grid, Forecasting, Prediction and Renewable Energy Sources. Dr. Hasmat has authored/co-authored more than 100 research papers, eight books and thirteen chapters, published by IEEE, Springer and Elsevier. His principle area of research interests is artificial intelligence, machine learning and big-data analytics for renewable energy, smart building & automation, condition monitoring and online fault detection & diagnosis (FDD).

Prof. Dr. Pramod Kumar is working as Professor, Head (CSE) & Dean (CSE & IT) in Krishna Engineering College, Ghaziabad since January 2018. He also served as Director of the Tula's Institute, Dehradun, Uttarakhand. He has more than 22 years of experience in academics. He completed his Ph.D in Computer Science & Engineering in 2011 and M.Tech (CSE) in 2006. He is a senior member of IEEE (SMIEEE) and Joint Secretary of IEEE U.P Section. He has published widely in International Journals and Conferences his research finding related to Computer Network, Internet of things(IoT) and Machine Learning.He has Authored/Co-Authored more than 70 research papers and several Chapters in edited books. He has supervised and co-supervised several M.Tech and Ph.D students. He has organized more than 10 IEEE International Conferences and all the research paper of these conferences is now available in IEEE Explore. He is the editor of two books. He has conducted more than 15 Faculty Development Program in the collaboration of EICT, IIT Roorkee, EICT, IIT Kanpur and AKTU. He has organised IEEE National workshop on Research Paper writing on 27 March 2017 and IEEE Women Symposium on 21 Feb 2016.

Prof. Atif Iqbal Fellow IET (UK), Fellow IE (India) and Senior Member IEEE, Vice-Chair, IEEE Qatar section, DSc (Poland), Ph.D. (UK)- Associate Editor, IEEE Trans. On Industrial Electronics, IEEE ACCESS, Editor-in-Chief, I'manager Journal of Electrical Engineering, Former Associate Editor IEEE Trans. On Industry Application, Former Guest Associate Editor IEEE Trans. On Power Electronics. Full Professor at the Dept. of Electrical Engineering, Qatar University and Former Full Professor at the Dept. of Electrical Engineering, Aligarh Muslim University (AMU), Aligarh, India. Recipient of Outstanding Faculty Merit Award academic year 2014-2015 and Research excellence awards 2015 and 2019 at Qatar University, Doha, Qatar. He received his B.Sc. (Gold Medal) and M.Sc. Engineering (Power System & Drives) degrees in 1991 and 1996, respectively, from the Aligarh Muslim University (AMU), Aligarh, India and Ph.D. in 2006 from Liverpool John Moores University, Liverpool, UK. He obtained DSc (Habilitation) from Gdansk University of Technology in Control, Informatics and Electrical Engineering in 2019. He has been employed as a Lecturer in the Department of Electrical Engineering, AMU, Aligarh since 1991 where he served as Full Professor until Aug. 2016. He is recipient of Maulana Tufail Ahmad Gold Medal for standing first at B.Sc. Engg. (Electrical) Exams in 1991 from AMU. He has received several best research papers awards e.g. at IEEE ICIT-2013, IET-SEISCON-2013, SIGMA 2018, IEEE CENCON 2019, IEEE ICIOT 2020 and Springer ICRP 2020. He has published widely in International Journals and Conferences his research findings related to Power Electronics, Variable Speed Drives and Renewable Energy Sources. Dr. Iqbal has authored/co-authored more than 440 research papers and four books and several chapters in edited books. He has supervised several large R&D projects worth more than multi million USD. He has supervised and co-supervised several Ph.D. students. His principal area of research interest is Smart Grid, Complex Energy Transition, Active Distribution Network, Electric Vehicles drivetrain, Sustainable Development and Energy Security, Distributed Energy Generation and multiphase motor drive system.

Control Plane Systems Tracing and Debugging—Existing Problems and Proposed Solution

Gleb Peregud, Maria Ganzha, and Marcin Paprzycki

Abstract Hierarchical control plane systems are hard to debug and reason about, among others, because of the prevalence of intent-driven actuation. Moreover, an industry-adopted distributed systems tracing model, called OpenTracing, does not handle activity tracing in presence of coalescing effects, materializing in control plane systems, e.g., cloud platforms and build systems. The goal of this contribution is to outline a solution for reasoning about such systems, by creating a novel distributed systems tracing mechanism, based on an extension of the OpenTracing model.

Keywords Distributed systems tracking · Provenance tracking · Hierarchical control plane systems · Build systems

1 Introduction

The complexity of computer systems is growing [1] and results in individuals' losing ability to fully comprehend systems they create/use. Today, programs may be developed by tens of thousands of developers [2, 3]. Moreover, there exist services with millions of lines of code, with thousands of Remote Procedure Call interfaces [4]. The growth of system complexity is followed by the increasing complexity of software automation. Today, it ranges from simple build systems (Make [5]), through imperative automation systems (Puppet [6]), to complex multi-cloud systems (Terraform [7]), which become too complex for individual developers. For instance,

G. Peregud · M. Ganzha
Warsaw University of Technology, Warsaw, Poland
e-mail: gleb.peregud@gmail.com

M. Ganzha
e-mail: M.Ganzha@mini.pw.edu.pl

M. Paprzycki (✉)
Systems Research Institute Polish Academy of Sciences, Warsaw, Poland
e-mail: marcin.paprzycki@ibspan.waw.pl

A. Tomar et al. (eds.), *Machine Learning, Advances in Computing, Renewable Energy and Communication*, Lecture Notes in Electrical Engineering 768,
https://doi.org/10.1007/978-981-16-2354-7_1

understanding a complex service deployment would require expertise in Terraform, Docker [8], Linux kernel, Kubernetes [9], cloud provider API(s), and deployment requirements of the servers. Among the most complex automation systems, are the cloud services [10, 11]. They serve public APIs to cloud customers and drive the cloud infrastructure.

1.1 Intent-Driven Actuation

Deployment systems are shifting from an imperative execution model (like Ansible), to a declarative-first model (like Terraform). They follow a scheduling policy to reach an intended state, executing a sequence of, usually small/restricted, imperative steps. We call these *intent-based systems*, and their execution an *intent-driven actuation*. For example, Kubernetes is an intent-based system: its API objects [12] describe the desired state (user intent) and the system match updates the cluster to match it. Furthermore, deployment systems (NixOps [13]) and modern build systems [14, 15] employ intent-driven actuation. They build a new instance of a target object from the scratch, based on the specification, and replace the current object with the new one (imperative operation).

1.2 Coalescing Effects

Systems that employ intent-driven actuation exhibit the *coalescing effects*. In general, coalescing occurs when work units (from multiple requests) are batched, before joint execution [16]. In build systems, coalescing effect occurs in a "diamond dependency graph", since the shared dependency is built in response to both incoming requests. Here, no single attribution of causality is possible. Overall, coalescing effects change the relationship between requests and activities in the system from one-to-many to many-to-many.

In intent-driven actuation, coalescing effects appear as actuation aggregates "multiple intents". Here, actuation can take an "arbitrary path" between the current state and the desired state. The system can choose to batch user requests and satisfy them all at once, to execute multiple actions based on a single request, or do something in-between. Let us illustrate this with a few examples:

- If two consecutive requests desire the same state, only one is acted upon.
- If two consecutive requests modify the same property, in quick succession, only the latter actuation is likely to occur, due to the, so-called, debouncing.
- If three requests modify the state in quick succession from A to B, to C, and back to B, typically only the actuation from A to B will be executed.

- If a request considerably changes the desired state, the automation system will perform a series of actions over a long period of time. For example to deliver 10,000 VMs, the cloud will not start them instantaneously, but over time.

Observe that a combination of such behaviors results in *arbitrary relationships between requests and actuations*. This makes the majority of debugging tools inaccurate. In this context, a widely adopted industry approach to reasoning is *distributed systems tracing*, and the most popular tools follow OpenTracing model [17] based on Google's Dapper. However, this model does not handle coalescing effects (see [16]).

1.3 Control Plane Systems

Coalescing effects are present in *control plane systems*, in the upper layers of cloud service stacks. Cloud APIs can be imperative, declarative, or both. Imperative API allows direct actions on cloud resources. Declarative API allows users to declare an arbitrarily complex intent with a single call. An example of such desired intent is a shape of service deployment, e.g., size and regional distribution of a group of virtual machines. Typically, declarative API calls trigger asynchronous work changing the deployment to match the desired state.

Additionally, in Platform-as-a-Service (PaaS) and Infrastructure-as-a-Service (IaaS) solutions, deployment systems may be provided as services. For example, any PaaS offering Kubernetes API exposes an intent-based solution. It can be assumed that PaaS offerings from Amazon Web Services, Google, or Azure combine their systems scale with complex execution models of intent-based systems.

1.4 Hierarchical Control Plane Systems

In industry practice, some cloud service APIs are implemented on top of other cloud services. For example, Google Cloud Functions (GCF) runs on top of Google Cloud Run (GCR), which runs on top of Google Kubernetes Engine (GKE). In turn, GKE runs on top of Google Cloud Engine (GCE). Hence, the control plane system may be *hierarchical*, and the hierarchy can have quite a complex structure of dependencies making it hard to be reasoned about.

1.5 Build Systems in Control Plane Systems

Let us now consider that control plane systems, with declarative APIs, need to be provided with the desired state. Build system outputs are often used for this purpose. For example, Infrastructure-as-Code (IasC) model recommends treating service deployment as any other code—to be built and tested using *build systems*. Here, the Bazel build system can define the desired shape of Kubernetes deployments [18]. Immutable infrastructure [19] model often prescribes the use of the Docker containers and Docker's primitive build system [15].

Furthermore, *build/build-like systems* are often parts of control plane systems. For example, Terraform Cloud [20] and EPAM Cloud [21] execute Terraform planning process as part of their API. Tools like NixOps [13] and Disnix [22] define deployments end-to-end, starting with individual binaries, through the VM/container images, to the "shape of the cloud".

2 Problem Statement

Hierarchical control plane systems employing intent-driven actuation (HCPS) are hard to reason about. These are multi-layered systems, with a large number of servers. Hence, reasoning about them requires tracing both their activities and mutations of the state, managed by any layer of the system. A control plane system in a cloud service is inevitably multi-tenant. The behavior of individual tenants can affect how other tenants are being served and can affect the health of the system. Since such interactions are difficult to reproduce, tools providing information about the control plane system must be always on. Understanding system behavior requires observing related activities and associated state changes across many services and machines. Given that these systems are developed by large organizations with multiple teams, it may not be obvious which services are used for a given request.

The aim of undertaken work is to propose a debugging solution, applicable to a hierarchical control plane system employing intent-driven actuation.

3 Related Work

To address this problem, it is necessary to collect data about the state of a HCPS. Here, four related areas can be identified: *distributed systems tracing*, *monitoring and logging*, *provenance tracking*, and *build systems*.

3.1 Distributed Systems Tracing

The state-of-the-art solutions for *distributed tracing* closely follow the OpenTracing model. The OpenTracing approach uses "tainting", where an incoming request is "tainted" with a unique identifier. It is propagated across all activities participating in handling the request. This has been standardized by W3C as the Trace Context recommendation [23].

Tainting propagates a single identifier only, hence it cannot deal with coalescing effects. None of OpenTainting open-source implementations solve this problem. Let us note that intent-driven actuation cannot be faithfully recorded using tree-shaped execution traces, where an actuation may be linked with more than one incoming request. Hence, intent-based systems **necessitate a more flexible model**. Note also that the tainting model cannot be efficiently extended by propagating a set of trace identifiers instead of a single one, due to super-linear storage requirements and duplication of spans.

3.2 *Monitoring and Logging*

Monitoring and logging mechanisms allow capturing metrics and unstructured log entries. Metrics are most suitable to capture aggregated statistics of the system. They can be labeled with additional information, allowing finer-grained view, but they still deliver aggregated numbers [24]. Logging mechanisms are typically based on unstructured log entries. Neither mechanism is suitable to track relationships between incoming requests and individual work units.

3.3 *Provenance Tracking*

Coalescing effects are associated with shared mutable states. Hence, provenance tracking becomes relevant, as it allows tracking state objects, their mutations, and relations. It is related to database provenance (data lineage tracking). Provenance traces [25] track provenance at query time. This is a bottom-up approach using the proposed formal language. The solution generates very large traces, as all transformations are recorded. If applied naively, this approach is not feasible for a large-scale control plane system, since it (1) requires writing from scratch and (2) does not allow precise specification of which tracking data is to be gathered.

It has been recognized [26] that provenance tracking in clouds requires aggregation across multiple layers of the environment. Provenance tracking has been applied in intent-based control plane systems in networking. Intent-based networking (IBN) concerns automated and policy-aware network management [27]. ProvIntent [28] is a framework extension for the SDN control plane that accounts for intent semantics. It extends the ProvSDN [29] to explicitly incorporate intent evolution in provenance tracking. It uses the W3C PROV data model.

W3C PROV expands on OpenProvenance model (OPM) [30] as data model for provenance on the Web. It is sufficiently expressive to represent coalescing effects—it can represent non-tree-shaped activities and it can record objects as triggers for activities. However, the PROV model is not suitable since: (1) Most activities in HCPSs do have a tree-shaped control flow. Here, PROV Activities have to be represented using *wasStartedBy* and *wasEndedBy* relations, causing substantial storage overhead. (2) While [31] proposes the use of OPM to represent message passing in distributed systems, the proposed representation is verbose compared to OpenTracing. (3) PROV model is expressive and complex much beyond OpenTracing. This complexity could hinder its adoption in the industry. Systems like SPADE [32] do use the PROV model to capture provenance data in distributed settings. SPADE is focused on gathering low-level information from OS audit logs, network artifacts, etc. Therefore, SPADE's syscall and library call level instrumentation would not scale for a large production system.

Provenance tracking has been researched from the perspective of *formal systems*. Here, Souilah et.al. [33] presented a formal provenance, based on the π-calculus. The approach is based on enriching exchanged data with provenance information, similar to the tainting approach. Hence, this approach is not suitable for our problem. Why-across-time provenance [34] provides state machines-based mechanisms to track data provenance in time-varying stateful distributed systems. However, wat-provenance requires determinism, while the large-scale distributed systems involve a fair share of non-determinism in load balancing, bin-packing, resource allocation, load shedding, etc. Additionally, both this and π-calculus-based formalism depart from the well-established OpenTracing.

Provenance tracking in security is focused on threat detection. This puts constraints on available approaches. ProTracer [35] uses a mix of logging and tainting and is focused on Advanced Threat Protection. Hence, it uses kernel-level audit logging and syscall interception, as a black-box, zero-trust approach. ProTracer does not use a general data model, making it impossible to represent abstract entities, e.g., cloud resources. CamFlow [36] automates provenance capture, as a Linux Security Module (LSM), designed for single-machine system auditing. It outputs results in W3C PROV format.

Security-focused provenance tracking addresses different issues than these in HCPSs. Security requires zero trust, while the HCPSs case allows for full trust. It restricts security provenance solutions to use *observed* provenance. However, provenance tracking in large-scale control plane systems cannot be based solely on observed provenance, due to the use of abstract entities and their scale [37]. Provenance tracking in HCPSs should rather apply *disclosed* provenance.

Provenance has been approached in the context of workflow management systems and data transformation pipelines. RAMP is a data-intensive scalable computing (DISC) provenance framework [38]. It is restricted to data-intensive computations over static data. Therefore, it cannot be applied to components, like storage systems, coordination services, load balancers, etc., where control flow evolves with the system, and data passes via a large set of mechanisms (RPCs, databases, pub/sub systems, etc.). LogProv is a provenance logging system, implemented for Apache Pig and Apache Hadoop [39]. It supports dynamically shaped big data workflows and pipelines. It uses structured logging and the ElasticSearch. Overall, RAMP is a very specialized framework for DISC provenance tracking, while LogProv is less specialized and more flexible. Here, we believe that the LogProv approach can be generalized.

3.4 Build Systems

As discussed, build systems play important role in HCPSs, while reasoning about build systems combines tracing and provenance tracking. Build systems are inherently intent-based, since a build target is described declaratively. Moreover, memorization and incremental recomputation are coalescing effects, used to achieve a minimality property [15]. Here, build systems, like Bazel [40] or Nix [41], provide mechanisms to reason about, debug, and optimize operations. These mechanisms are implementation-specific, informal, and not interoperable.

Separately, tracing build processes have been used to uncover license compliance inconsistencies [42], which is a form of provenance tracing. This approach does not address the hierarchical nature of control plane systems and captures only two specialized levels of provenance for build tasks and files used in builds.

Finally, build system dependency tracking can be seen as provenance tracking, for build targets based on their inputs (disclosed provenance in [43]). However, at best, it works at the level of the build graph, which is too high level of granularity. For instance, when build systems operate within a control plane system, the ability to track the relationship between build system inputs and actions of control plane components taken based on build system outputs is necessary.

4 Solution Outline

We will now outline the requirements for a system that can solve the posed problem. Here, we believe that the solution should extend OpenTracing (to facilitate adoption). Hence, such a solution lies between the OpenTracing and the OpenProvenance in the design space of debuggability tools. Hence, based on analysis of literature and existing industrial tools, the proposed solution should have the following properties:

1. *Coalescing effects support*—to support tracing of intent-driven actuation;
2. *Support for abstract entities*—essential to align with cloud APIs;
3. *Support for composite entities*—to support objects like archives, VM images, container images, etc. which are prevalent in cloud APIs;
4. *Low storage overhead*—necessary for large-scale systems;
5. *Full coverage*—for all tracked activities and resource state mutations;
6. *Gradual fidelity execution tracing*—to selectively apply execution tracing;
7. *Gradual fidelity provenance tracking*—to selectively track provenance;
8. *Minimal mental burden*—to support adoption in industry;
9. *Cross-host tracking*—for distributed systems tracing support;
10. *Multi-layer systems support*—to allow tracking across hierarchical layers;
11. *Asynchronous data intake*—to support data ingestion in presence of unreliable network, unpredictable latencies, and lack of ordering guarantees;
12. *Event-based data production*—to deal with faults and to avoid buffering;
13. *Flexible control flow support*—to deal with pre-existing control flows.

We believe that a system, which satisfies these requirements, will be able to trace (1) non-ultra-large online serving systems and (2) hierarchical control plane systems employing intent-driven actuation. Note that existing solutions, like Dapper or CamFlow, satisfy only a subset of these requirements.

4.1 Provenance-Enhanced Distributed Systems Tracing Model

We propose Provenance-Enhanced Distributed Systems Tracing (PEDST) model, as a foundation for the solution. PEDST extends the OpenTracing model to record interactions between actions in the system and objects the system interacts with. Hence, the following concepts are used in PEDST model. *Execution* (similar to the OpenTracing "span") tracks *operations* on *entities*, which are recorded. *Entity* is "anything" that is "important-enough" to track. Read and write *operations*, performed by an *execution*, allow tracking provenance of objects they operate on. Each write *operation*, on an entity, gives rise to a new *incarnation* (immutable *entity* revision). *Entities* are mutable. *Executions* of the same logic are grouped, by association with a *process*, which describes a procedure. A pair of *executions* can interact. Recorded *interaction* consists of *messages*. A *message* may carry an *incarnation* as a payload.

The PEDST model supports both *activity tracing* and *provenance tracking*. The vehicles of tracing are *executions* and their parent–child relationships [16]. Correspondingly, the vehicle for provenance tracking is read and write operations performed on *incarnations*. This puts the proposed approach in the "data provenance" class [44]. The capability to dynamically track the provenance of objects (e.g., files, configs, resources, etc.) contributes to "support for abstract entities".

Concepts of *executions*, *incarnations*, and *operations* are sufficient to perform both execution tracing and provenance tracking. Concepts of *process* and *entity* allow improving the usability of envisioned tools. *Annotations* are necessary for the model to be a superset of OpenTracing. *Interactions* and *messages* are necessary to track provenance propagation through RPCs and other message passing mechanisms found in distributed systems.

Furthermore, the use of OpenTracing as the foundation allows to partially satisfy the "minimal mental burden" requirement, due to familiarity with this model. Nevertheless, PEDST is more complex than OpenTracing. However, this complexity is inherent to the problem. Here, the *entities* concept and versioning with *incarnations* and *operations* as a link between activity tracing and provenance tracking are the minimal extension of the OpenTracing model, to make it comply with other requirements.

4.2 *Architecture*

Finally, to solve the problem, the model needs to be implemented as part of a software system. We propose to use an architecture akin to Dapper's or LogProv's, but extended accordingly. The data model described above is used to represent the final processed data, while data collection from individual servers is done with a distributed structured logging mechanism, using a logging data model. The logging data model needs to support aggregation into the PEDST model, by log-tailing processors. Processor ingests data from the logging mechanism and stores it into the PEDST storage. Optionally, data can be transformed into a graph-based representation, for a more flexible querying support. Additionally, a visualization tool needs to expose the processed data to a user of the system.

5 Conclusions

In this work, we have identified characteristics of hierarchical control plane systems, with intent-driven actuation, which makes their debuggability an unsolved problem. Next, we have identified *provenance* as a possible solution. We have introduced main aspects of a provenance-enhanced distributed systems tracing model; an extension of the Dapper tracing model, with elements of provenance tracking. Finally, an architecture of a software implementation was suggested. Further work on research, implementation, and application of the proposed model is underway and will be reported in subsequent publications.

References

1. Dvorak D (2009) Nasa study on flight software complexity. In: AIAA infotech@ aerospace conference and AIAA unmanned... unlimited conference, p 1882
2. Facebook Engineering: 9.9 million lines of code and still moving fast-facebook open source in (2014). https://engineering.fb.com/core-data/9-9-million-lines-of-code-and-still-moving-fast-facebook-open-source-in-2014/
3. Potvin R, Levenberg J (2016) Why google stores billions of lines of code in a single repository. Commun ACM 59(7):78–87. https://doi.org/10.1145/2854146
4. Barroso LA, Ranganathan P (2010) Datacenter-scale computing. IEEE Micro 30, 6–7. http://www.computer.org/portal/web/csdl/doi/10.1109/MM.2010.63, special issue of the IEEE Micro Magazine
5. Fowler G (1990) A case for make. Software: practice and experience 20(S1), S35–S46. https://doi.org/10.1002/spe.4380201305, https://onlinelibrary.wiley.com/doi/abs/10.1002/spe.4380201305
6. Loope J (2011) Managing infrastructure with puppet: configuration management at scale. O'Reilly Media, Inc
7. Brikman Y (2019) Terraform: up & running: writing infrastructure as code. O'Reilly Media
8. Merkel D (2014) Docker: lightweight linux containers for consistent development and deployment. Linux J 2014(239):2

9. Burns B, Grant B, Oppenheimer D, Brewer E, Wilkes J (2016) Borg, omega, and kubernetes. Queue 14(1):70–93
10. Low C, Chen Y, Wu M (2011) Understanding the determinants of cloud computing adoption. Ind Manag Data Syst
11. Wood K, Anderson M (2011) Understanding the complexity surrounding multitenancy in cloud computing. In: 2011 IEEE 8th international conference on e-business engineering. IEEE, pp 119–124
12. Understanding kubernetes objects (2020) https://kubernetes.io/docs/concepts/overview/working-with-objects/kubernetes-objects/. Accessed 13 May 2020
13. Dolstra E, Vermaas R, Levy S (2013) Charon: declarative provisioning and deployment. In: 2013 1st international workshop on release engineering (RELENG). IEEE, pp 17–20
14. McNerney PJ (2020) Beginning bazel. Apress, Berkeley, CA. https://doi.org/10.1007/978-1-4842-5194-2
15. Mokhov A, Mitchell N, Peyton Jones S (2020) Build systems à la carte: theory and practice. J Funct Program 30(1). https://doi.org/10.1017/s0956796820000088. https://oadoi.org/10.1017/s0956796820000088
16. Sigelman BH, Barroso LA, Burrows M, Stephenson P, Plakal M, Beaver D, Jaspan S, Shanbhag C (2010) Dapper, a large-scale distributed systems tracing infrastructure. Technical report, Google, Inc. https://research.google.com/archive/papers/dapper-2010-1.pdf
17. The Open Tracing Project (2020) (1) https://opentracing.io. Accessed 7 June 2020
18. This repository contains rules for interacting with kubernetes configurations/clusters (2020). https://github.com/bazelbuild/rules_k8s. Accessed 30 May 2020
19. Mikkelsen A, Grønli TM, Kazman R (2019) Immutable infrastructure calls for immutable architecture. In: Proceedings of the 52nd hawaii international conference on system sciences
20. HashiCorp: Terraform (2020). https://www.terraform.io/docs/cloud/index.html. Accessed 1 Sept 2020
21. Terraform as a service (2020) https://cloud.epam.com/site/competency_center/e=p=c_services/terraform_as_a_service(=t=a=s). Accessed 30 May 2020
22. Van Der Burg S, Dolstra E (2014) Disnix: a toolset for distributed deployment. Sci Comput Program 79:52–69
23. Kanzhelev S, McLean M, Reitbauer A, Drutu B, Molnar N, Shkuro Y (2020) Trace context (2). https://www.w3.org/TR/trace-context. Accessed 1 Sept 2020
24. Turnbull J (2018) Monitoring with Prometheus. Turnbull Press
25. Cheney J, Acar U, Ahmed A (2008) Provenance traces. arXiv:0812.0564
26. Imran M, Hlavacs H, Khan FA, Jabeen S, Khan FG, Shah S, Alharbi M (2018) Aggregated provenance and its implications in clouds. Futur Gener Comput Syst 81:348–358
27. Sivakumar K, Chandramouli M (2017) Concepts of network intent. Internet Research Task Force, Internet Draft
28. Ujcich BE, Bates A, Sanders WH (2020) Provenance for intent-based networking. In: Proceedings of the IEEE conference on network softwarization
29. Ujcich BE, Jero S, Edmundson A, Wang Q, Skowyra R, Landry J, Bates A, Sanders WH, Nita-Rotaru C, Okhravi H (2018) Cross-app poisoning in software-defined networking. In: Proceedings of the 2018 ACM SIGSAC conference on computer and communications security, pp 648–663
30. Moreau L, Clifford B, Freire J, Futrelle J, Gil Y, Groth P, Kwasnikowska N, Miles S, Missier P, Myers J et al (2011) The open provenance model core specification (v1. 1). Futur Gener Comput Syst 27(6):743–756
31. Groth P, Moreau L (2011) Representing distributed systems using the open provenance model. Future Gener Comput Syst 27(6):757–765. https://doi.org/10.1016/j.future.2010.10.001
32. Gehani A, Tariq D (2012) Spade: support for provenance auditing in distributed environments. In: ACM/IFIP/USENIX international conference on distributed systems platforms and open distributed processing. Springer, pp 101–120
33. Souilah I, Francalanza A, Sassone V (2009) A formal model of provenance in distributed systems. In: Workshop on the theory and practice of provenance, pp 1–11

34. Whittaker M, Teodoropol C, Alvaro P, Hellerstein JM (2018) Debugging distributed systems with why-across-time provenance. In: Proceedings of the ACM symposium on cloud computing, pp 333–346
35. Shiqing M, Zhang X, Xu D (2016) Protracer: towards practical provenance tracing by alternating between logging and tainting. In: NDSS (01). https://doi.org/10.14722/ndss.2016.23350
36. Pasquier T, Han X, Goldstein M, Moyer T, Eyers D, Seltzer M, Bacon J (2017) Practical whole-system provenance capture. In: Symposium on cloud computing (SoCC'17). ACM
37. Raju B, Elsethagen T, Stephan E, Van Dam KK (2016) A scientific data provenance api for distributed applications. In: 2016 international conference on collaboration technologies and systems (CTS). IEEE, pp 104–111
38. Ikeda R, Park H, Widom J (2011) Provenance for generalized map and reduce workflows. CIDR 273–283
39. Wang R, Sun D, Li G, Atif M, Nepal S (2016) Logprov: logging events as provenance of big data analytics pipelines with trustworthiness. In: 2016 IEEE international conference on big data (Big Data). IEEE, pp 1402–1411
40. Optimizing performance (2020). https://docs.bazel.build/versions/master/skylark/performance.html. Accessed 30 May 2020
41. Dolstra E (2006) The purely functional software deployment model. Utrecht Uni
42. van der Burg S, Dolstra E, McIntosh S, Davies J, German DM, Hemel A (2014) Tracing software build processes to uncover license compliance inconsistencies. In: Proceedings of the 29th ACM/IEEE international conference on Automated software engineering, pp 731–742. https://doi.org/10.1145/2642937.2643013
43. Braun U, Garfinkel S, Holland DA, Muniswamy-Reddy KK, Seltzer MI (2006) Issues in automatic provenance collection. In: International provenance and annotation workshop. Springer, pp 171–183
44. Herschel M, Diestelkämper R, Lahmar HB (2017) A survey on provenance: what for? what form? what from? VLDB J 26(6):881–906

Distribution Expansion Planning in a Deregulated Environment

Abhilasha Pawar, R. K. Viral, and Anuprita Mishra

Abstract Electricity is the most widely used and versatile utility of modern times and this fact makes its planning extremely important. Proper planning and detailed knowledge of resources will lead to fetch better prices for this utility. This paper discusses the transition of the power market structure from a conventional to a modern deregulated structure. It further investigates the inclusion of distributed generation and its expansion planning along with the description of various methods, their related advantages and limitations. Moreover, various techno-economic impacts of distributed generation planning are also detailed. The related optimization methods which are generally used for the planning of these distributed generators are discussed. A brief classification of various methods available for optimization of planning, economic sustainability, loss minimization and enhancement of voltage profile in the distribution system is also presented by the authors.

Keywords Distributed generation · Deregulation · Optimization methods

1 Introduction

Electrical power is considered as a utility. Its control, planning and operation were earlier limited to a central single utility. The electrical power industry is witnessing a transformation in its operational and business model where the vertically integrated structure of the utilities is unfolding and unlocked for competition with private players. The main outcome of this development is to open up markets for private players and to provide a lesser price to the consumer [1]. The deregulated power market has opened up opportunities for providing electricity at competitive prices, encouraged competition in the generation sector, supply of electric power and also

A. Pawar (✉) · R. K. Viral
Department of Electrical & Electronics Engineering, Amity School of Engineering and Technology, Amity University, Noida, Uttar Pradesh, India

A. Mishra
Department of Electrical & Electronics Engineering, Oriental Institute of Science and Technology, Bhopal, Madhya Pradesh, India

A. Tomar et al. (eds.), *Machine Learning, Advances in Computing, Renewable Energy and Communication*, Lecture Notes in Electrical Engineering 768,
https://doi.org/10.1007/978-981-16-2354-7_2

improve the steadiness and economy of the power system. Distributed generation (DG) has powered the deregulated market through its advantages and versatility. The expansion planning of distribution system with DG integration has opened up new avenues of research and innovation in the electrical power industry. Today's emphasis is to find out optimum ways to empower the end-users in terms of cost and maintenance. Various methods have been used for optimization and still, it is a matter of debate so as to find an approximate method to reduce the losses and improve efficiency while providing minimum cost to consumers.

2 Power Markets and Structure

Electricity markets have witnessed many changes in the last 30 years due to restructuring and reforms which were aimed at increasing efficiency and providing an open structure to the facility. This led to the advancement and modernization of the conventional power system into a new deregulated power system which is also referred to as restructuring or reforms [2]. Power market refers to buying and selling of electric power between players of the energy industry.

2.1 Conventional Power Market

The conventional power market followed a vertical integrated approach. The term vertical integrated utility came from the fact that there is a vertical integration of all tasks. So, the earlier setup of the power system was controlled by a single utility for all functions that is, generation, transmission and distribution. Figure 1 depicts the conventional model.

2.2 Modern Deregulated Power Market

Modern-day deregulated power market offers a choice to consumers by opening up the vertical integrated model and allowing more private participation in the various sections, i.e. generation, transmission and distribution. The deregulation has pushed the industry to endorse distributed generation and look for advanced technologies for curbing transmission congestion. The modern deregulated power market is depicted in Fig. 2.

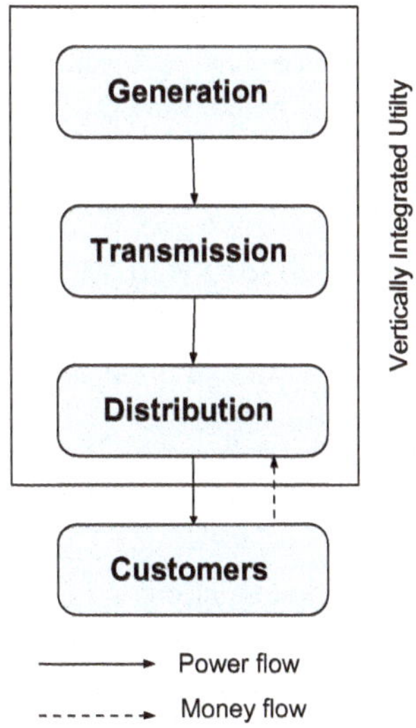

Fig. 1 Conventional power market

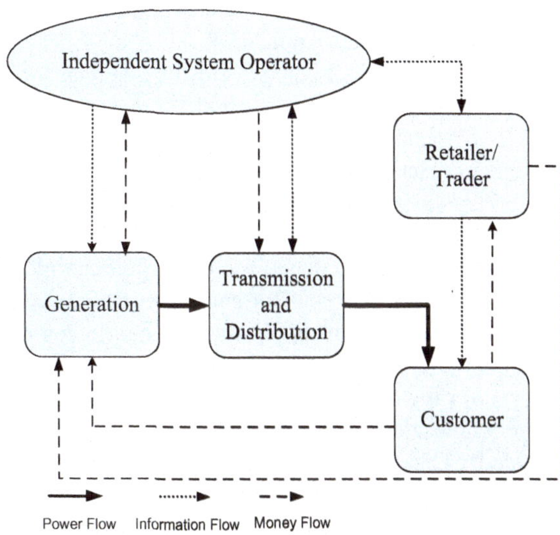

Fig. 2 Modern deregulated power market

2.3 Deregulation Process

Regulation means a set of rules that have been framed and imposed by the government to ensure a smooth operation of the power system. This process of re-defining rules and regulations which govern the electrical industry and offer consumers a choice of electricity suppliers by allowing competition is called deregulation. The deregulation process has introduced some new entities to the market like generation companies (GenCo), transmission companies (TransCo), distribution companies

(DisCom), independent power producer (IPP), independent system operator (ISO), power exchange (PX) and retail energy service companies (ResCo) which play their own part in the process.

2.4 Various Market Models

The basic market models which are founded on predetermined provisions are classified as monopoly model, single buyer model, wholesale competition model and retail competition model [3]. While in the monopoly model, a single entity, mostly the government, is accountable for the generation, transmission and distribution of electric power to the consumers. This model was followed throughout the world before deregulation.

The single-buyer model views competition in the generation sector. In this model, the single-buyer agency purchases power from independent power producers (IPPs) along with its own generation which sells the power to distribution companies in the chosen area. The advantages of this model include the participation of private players in power generation, and thus the introduction of competition in the existing model. This model also suffers from disadvantages like problems associated with long-term contracts, and price not being decided by demand–supply interaction.

The wholesale competition model is nearer toward competition. In this model, there is an organized market where generators sell their produce at competitive rates. This market can be organized either by a separate entity or by the system operator itself. Very little choice is left for the end-user here. Still, the end-user remains affiliated to the DisCom or retailer operating in that geographical area. Here, the bulk customers are privileged to choose their energy provider. This model facilitates a competitive environment for generators because wholesale pricing is decided by the interaction between supply and demand. In contrast, the retail price of electrical energy remains structured or controlled because the small consumers still do not have a choice for their supplier. The distribution companies are then exposed to vagaries of the wholesale price of the commodity. Merits of this model include freehand provided to DisComs and bulk consumers to choose the seller and competitive pricing as pricing is decided by an interaction between demand and supply.

But this model also suffers from disadvantages like no choice to consumers because they have to buy power from affiliated DisCom and also the rates for end consumers are regulated rather than competitive.

The retail competition model provides customers to have admission to compete for generators either directly or indirectly through their own choice of retailer. In this model, open access is provided to both transmission and distribution.

This model is a multi-buyer, multi-seller model and here the power pool acts like an auctioneer. This model has a facility to bid into the spot market. The pool considers the supply and demand while determining the spot price hourly on daily basis. The pool gathers money from buyers and dispenses it to the producers. The main advantage of this model over the monopoly model is that it introduces competition in the area of both wholesale and the retail system. This model is regarded to be a truly deregulated market model. The retail price is not regulated due to the fact that small consumers can alter their retailers regarding better price options. This model proves to be economically efficient as the interaction between the demand and supply is considered for price setting. The wholesale competition model takes care of relatively few customers which are regulated by Discoms and here a spot market is not essential.

However, in the retail competition model, spot markets are essential, since a third party owns a network that facilitates contractual arrangements between customers and producers. Metering is the major problem of the retail competition model. If proper metering is not facilitated, it may create disputes and logistic problems for the increasing number of customers. This model is a fully deregulated model as every customer has a buying choice. Moreover, the demand–supply interaction determines the pricing. This feature was lacking in the monopoly model. In addition to this, there is no regulation in energy pricing. But this method also suffers disadvantages like the requirement of constitutional and structural changes at both, wholesale and retail level, complex settlement system due to a large number of participants and requirement of additional infrastructural support. A comparison of the above models is shown in Table 1.

3 Distributed Generation and Expansion Planning

The world today encounters a lot of problems due to the use of fossil fuels which are a compulsory component of conventional power plants. But these fossil fuels present a lot of problems and have hazardous environmental effects which can be seen in the form of climate change, higher variations in temperatures, increase in pollution levels etc. Also, fossil fuels are costly and their use leads to high generation costs and as the conventional power plants are situated in far-off places, the transmission costs are also increased. Today's focus is on small-scale power production at the utility end to find an answer to this problem. This leads to the integration of distributed generation.

3.1 Distributed Generation

According to the definition given by Ackermann [4], "Distributed Generation is an electric power source which is connected directly to the distribution network or on the customer site of the meter". So, distributed generation (DG) is the use of small generating units in the load side, i.e. the distribution side of the power system. In the papers [5, 6], distribution generation technology has been classified into three groups, namely (a) renewable distributed generation, (b) non-renewable distributed generation and (c) for energy storage. Figure 3 shows the classification of DG.

The various problems which were encountered with distributed generation are:

Table 1 Comparison of various market models

Sl. No.	Type of model	Features	Advantages	Disadvantages/Problems
1	Monopoly model	• Single entity is responsible for Generation, Transmission and Distribution of power • Monopoly is with the Government	All control with central utility i.e. Government	No competition, consumer has to take the price fixed by the central utility
2	Single buyer model	• The single buyer agency purchases power from Independent Power Producers (IPPs) • Sells the power to distribution companies in the designated service area	Allows participation of private players and hence introduces competition in the existing model	Electricity price not being decided by demand-supply interaction
3	Wholesale competition model	• Wholesale pricing is determined by the interaction between supply and demand	Facilitates competitive environment for generators	Rates for end consumers are regulated rather than competitive
4	Retail competition model	• Both, transmission and distribution are provided open access in this model	Competition is introduced in both wholesale and retail areas of the system	Metering

- Reverse power flow
- Reactive power
- System frequency
- Voltage levels
- Protection schemes
- Harmonics injection
- Stability issues
- Optimal placement of DGs
- Islanding
- Amplified fault currents based on DG location.
- High monetary cost per KW of generation with renewable energy.

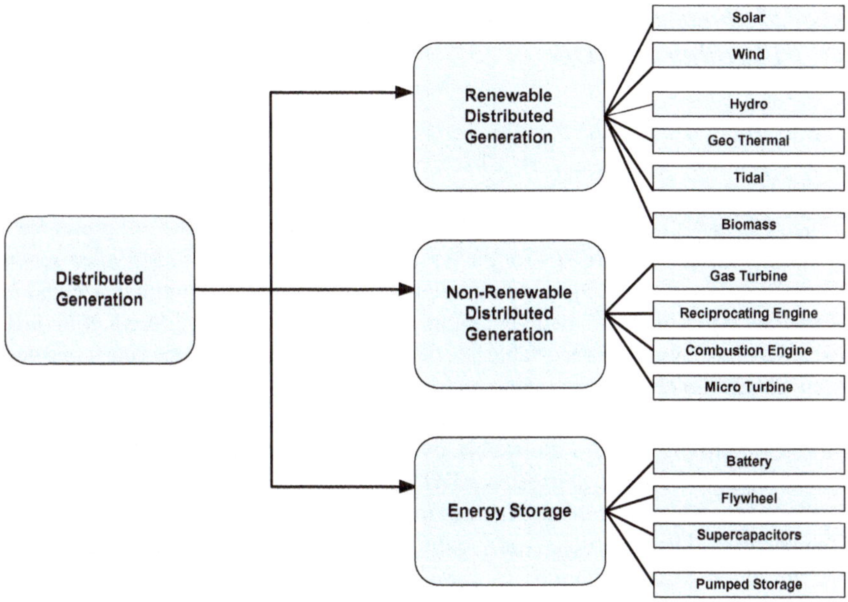

Fig. 3 Classification of distributed generation technology

Adding to this, there are some non-technical issues also such as scarcity of skilled workers and exclusion of distributed generation sources from the race which discourages the setting up of new power plants for reserve purposes.

3.2 Distribution System Expansion with DG

Deregulation in the power system has given way to revolution and modernization in distribution system planning (DSP). New energy policies are inspiring distributed energy resources which in turn is causing RESs planning and an increase in the number of DG installations as well. DG has evolved as an answer to ever-increasing load demand. However, un-coordinated management of these alternative energy resources can exert pressure on the power grid. Network configuration, storage systems, location and size of distributed generations (DGs) units play a major role in the optimal performance of distribution networks and their impact has to be properly considered in the planning process. The uncertainties with respect to the irregular nature of renewable DGs, load demand and price of energy should be considered in calculating the cost components [7].

3.3 Methods and Techniques for Distribution Expansion Planning with DG

The distribution expansion planning can be studied by considering various methodologies followed by researchers. Chandreja et al. [8] calculated the advantages of DG in the radial system for the reduction of losses. Keane et al. [5] paid attention to methods and strategy for DG planning and integration. Cao et al. [6] presented a profit-driven planning strategy for the regulation of integration of distributed generation (DG). Hung et al. [9] found out methods for deciding the optimal size and pf of various types of distributed generators for minimizing losses. Zhang et al. [10] discussed the planning of renewable DG resources taking uncertainties into consideration. Shaaban et al. [11] presented a multi-objective optimization approach, using the genetic algorithm (GA) for the optimal allocation of DG in the distribution system. Ganguly et al. [12] used a multi-objective optimization algorithm for DG expansion planning. They also suggested a DG allocation strategy for radial distribution networks considering load uncertainties using an adaptive genetic algorithm (GA). Liang et al. [13] used particle swarm optimization (PSO) for the location and sizing of a substation. Scarlatache and Grigoras [14] used a fuzzy approach for DG planning and optimization. Grisales et al. [15] used a hybrid methodology of genetic algorithm and particle swarm optimization algorithm for allocating the devices.

The DG planning methodologies can be categorized into four broad groups on the basis of the above literature:

(A) Traditional
(B) Heuristic
(C) Hybrid
(D) Others

The traditional approach is again divided into (i) optimal flow method, which basically consists of optimal flow or load flow study using algorithms like Gauss-Seidel, Newton-Raphson method, etc. It numerically analyzes approximate power flow in the numerous branches of an interconnected system; (ii) analytical method which analyzes the functioning of the system using algebraic expression and hence can be used for reaching an optimum solution; and (iii) two-thirds rule based on zero-point analysis.

Heuristic or artificial intelligence (AI) search method is an intelligent search-based method. These algorithms may have exponential time and space complications as they store complete information of the path also including the explored intermediate nodes. This method envelops a wide range of artificial intelligence (AI) techniques, such as evolutionary algorithm, particle swarm optimization, genetic algorithm, simulated annealing, fuzzy systems, tabu search, Hereford ranch, bat algorithm, ant colony system, artificial immune system, firefly algorithm, cuckoo search algorithm, intelligent water drop algorithm etc.

Hybrid methods are the blending of existing methods to reach a better solution to the optimization problem. The basic idea here is to merge two or more search methods and find a solution that will reduce the limitations of individual methods and incorporate the advantages of the considered methods. Examples of such hybrid systems as available in the literatures include genetic-fuzzy (GA-FZ), Monte-Carlo simulation genetic algorithm, genetic tabu search (GA-TS), genetic particle swarm optimization (GA-PSO), tabu-fuzzy (TS-FZ), genetic algorithm optimal power flow (GA-OPF), particle swarm optimization-load flow, etc.

Other methods comprise mixed-integer linear programming (MILP), mixed-integer nonlinear programming (MINLP), Monte-Carlo simulation, etc.

3.4 Major Advantages of DGs Integration with Expansion Planning

The advantages of distributed generation include modularity, i.e. it can be made in small modules. This particular feature allows resiliency in installation and expansion planning as it takes less time as compared to conventional methods. Also, DG is less site-specific, i.e. it can be installed anywhere with less chances of failure and also provides faster replacements and easy maintenance as well. Other advantages include better voltage profile, enhancement of reliability, eases congestion reduces pollution, improvement of power quality and reduction in system losses. Common DG applications include its usage as baseload plant, peak load plant, act as energy storage and hence provide backup to the distribution system.

Competitive electricity markets, the rapid growth of electricity demand, reliability issues of electric power supply, technological advancement in power generation resources, utility supporting devices and increased assimilation of telecommunication and information technology in distribution networks (smart grid) completely transform their behavior and operation. Further, the integration of dispatchable or non-dispatchable distributed energy resources near the load centers results in the enhancement of distribution network performance.

4 Distributed System Expansion Planning with DG Under Deregulated Market

The deregulated market provides an opportunity for DG integration as the market players are trying out methods to reduce overall cost and provide better pricing for power. Distribution system expansion planning incorporates DG to enhance reliability and overcome other issues of conventional power markets.

4.1 Impacts of Deregulation in DG Expansion Planning

Distributed generation is regarded as a substitute to provide power to new customers owing to its fast response time and minimum risk in investment as it is constructed in modules and can take care of load variations more efficiently, and being a comparatively new technology, the scope of improvement is always there. As deregulation has opened up markets for private players, the integration of DG in the distribution system is being emphasized due to its technical, economic and non-polluting nature.

4.1.1 Technical Impacts

Technical impacts include feasibility, reliability, better voltage profile, reduces system losses and congestion, less time-consuming, the requirement of less specialized skills and easy maintenance and replacement. This will also include the capacity of feeder, substation transformer capacity and system voltage profile while considering an increase in load. The problem of reverse power flow also has to be dealt with.

4.1.2 Economic Impacts

Economic impacts mean the effect of incorporating DG into the distribution system in terms of money [16–20]. This includes competition in electricity price and hence advantage to the consumers. This would further include the cost of installation and operation of DG. As capacitors are required, their installation cost will also be considered. New feeder lines and transformers will also be required for this purpose. The irregular nature of DG and the cost of energy that is lost in the process also have to be considered.

4.2 Best Practices in DG Expansion Planning Under Deregulated Market

Many researchers have tried various methods for DG expansion planning and have used numerous techniques and algorithms. But the best practices are still limited to fund a basic cause, i.e. to minimize the cost of DG installation and operation. As discussed in the literature above, every method has its own advantages and limitations but some scope is still there in reaching an optimum solution by adding two or more techniques or hybridization of techniques to reach the desired goal.

5 Recommendations

The authors recommend the following points on the basis of the study of various aspects regarding deregulation and expansion planning:

- Deregulation has helped to ease out stress on the central utility which was conventionally vertical and provided options for customers to chose their service providers keeping cost and efficiency in mind. This has changed the whole scenario of the power market in modern times.
- Distribution system expansion with DGs has opened up a new arena in technical and economic aspects of power system distribution. Better opportunities are being presented for DG integration by the government in the present scenario owing to its advantages over conventional sources which further motivates research in this field.

6 Conclusions

From the above work, it was observed that there is a basic struggle between accurateness, consistency and computational time in the commonly used methods. It is often difficult to reach at a solution that can optimize all objectives without any compromise. Researchers have tried and tested many ways to reach an optimal solution but most efforts depend upon certain considerations. None of the prevailing approaches are found to take care of both operational as well as feasible aspects. Few researchers tried to provide a guaranteed solution but it turned out to be lengthy and arduous trials, whereas those methods which have modest and interesting techniques lack accurateness. The uncertainties involved in distribution system expansion planning and operation may lead to difficulties along with the restructured environment of the distribution system with the usage of distributed generation.

References

1. Trehan NK, Saran R (2004) Electric utility deregulation: failure or success. In: Proceedings of IEEE symposium conference record nuclear science, Rome, pp 4614–4617
2. Jain S, Kalambe S, Agnihotri G, Mishra A (2017) Distributed generation deployment: state-of-the-art of distribution system planning in sustainable era. Renew Sustain Energy Rev 77:363–385
3. Acharya N, Mahat P, Mithulananthan N (2006) An analytical approach for DG allocation in primary distribution network. Int J Electr Power Syst 28(10):669–678
4. Ackermann T, Anderson G, Soder L (2001) Distributed generation: a definition. Electr Power Syst Res 57:195–204
5. Keane A et al (2013) State-of-the-art techniques and challenges ahead for distributed generation planning and optimization. IEEE Trans Power Syst 28(2):1493–1502
6. Cao X, Wang J, Zeng B (2018) Distributed generation planning guidance through feasibility and profit analysis. IEEE Trans Smart Grid 9(5):5473–5475

7. Ganguly S, Samajpati D (2015) Distributed generation allocation on radial distribution networks under uncertainties of load and generation using genetic algorithm. IEEE Trans Sustain Energy 6(3):688–697
8. Chiradeja P (2005) Benefit of distributed generation: a line loss reduction analysis. In: 2005 IEEE/PES transmission and distribution conference and exposition: Asia and Pacific, Dalian, pp 1–5
9. Hung DQ, Mithulananthan N, Bansal RC (2010) Analytical expressions for DG allocation in primary distribution networks. IEEE Trans Energy Convers 25(3):814–820
10. Zhang C, Li J, Zhang YJ, Xu Z (2018) Optimal location planning of renewable distributed generation units in distribution networks: an analytical approach. IEEE Trans Power Syst. 33(3):2742–2753
11. Shaaban MF, Atwa YM, El-Saadany EF (2011) A multi-objective approach for optimal DG allocation. In: Proceedings of 2nd international conference on electric power and energy conversion systems (EPECS), Sharjah, pp 1–7
12. Ganguly S, Sahoo NC, Das D (2014) Multi-objective planning for reactive power compensation of radial distribution networks with unified power quality conditioner allocation using particle swarm optimization. IEEE Trans Power Syst 29(4):18011810
13. Liang H, Liu S, Su J (2012) Optimal planning of distribution system considering distributed generators. In: Proceedings of CIRED workshop: integration of renewables into the distribution grid, Lisbon, pp 1–4
14. Scarlatache F, Grigoras G (2016) A fuzzy approach in optimal DG planning. In: Proceedings of international conference and exposition on electrical and power engineering (EPE), Iasi, pp 738–742
15. Grisales LF, Montoya OD, Grajales AR, Hincapie A, Granada M (2018) Optimal planning and operation of distribution systems considering distributed energy resources and automatic reclosers. IEEE Latin Am Trans 16(1):126134
16. Nandan NK et al (2018) Solving nonconvex economic thermal power dispatch problem with multiple fuel system and valve point loading effect using fuzzy reinforcement learning. J Intell Fuzzy Syst 35(5):49214931. https://doi.org/10.3233/JIFS-169776
17. Jadoun VK et al (2021) Optimal scheduling of non-convex cogeneration units using exponentially varying whale optimization algorithm. Energies 14(4):1–30. https://doi.org/10.3390/en14041008
18. Mahto T et al (2018) Load frequency control of a solar-diesel based isolated hybrid power system by fractional order control using particle swarm optimization. J Intell Fuzzy Syst 35(5):5055–5061. https://doi.org/10.3233/JIFS-169789
19. Vigya T et al (2021) Renewable generation based hybrid power system control using fractional order-fuzzy controller. Energy Rep 7C:641–653. https://doi.org/10.1016/j.egyr.2021.01.022
20. Bajaj M et al (2021) Optimal design of passive power filter using multi-objective Pareto-based firefly algorithm and analysis under background and load-side's nonlinearity. IEEE Access. https://doi.org/10.1109/ACCESS.2021.3055774

Density-Based Remote Override Traffic Control System

Gunjan Varshney, Anshika Jaiswal, Udit Mittal, Abhilasha Pawar, and Satyajeet

Abstract Vehicular traffic is growing almost everywhere in the world which is the prime cause of congestion, especially at road intersections. Due to unreliable and unpredictable road travel conditions, the emergency and essential services may get trapped in traffic congestion. The traffic lights traditionally have a fixed time-scheduling and this further blocks the smooth flow of traffic. In this paper, a density-based remote override traffic control system has been proposed which provides a solution to smartly manage the dynamic traffic conditions. In this test device, the design of the junction density calculation would be centered on infrared (IR) sensors, installed on each lane and interfaced with the microcontroller. The IR will be enabled as vehicles travel through the path. This may be achieved by changing the sequential order of traffic lights generated by the autonomous road surveillance network by utilizing the automated remote sensing-based system.

Keywords Traffic light signals · Traffic congestion · Traffic monitoring · Traffic management · IR transmitter and receiver · 8051 microcontroller · LCD display

1 Introduction

Street traffic management is a significant problem all around the globe. In India, the problem is deeply felt in almost all big urban areas such as Bengaluru, Pune, Hyderabad and Delhi-NCR. This is mostly attributed to rapid development in the IT industry and also to a rise in population contributing to the need for transport. The everyday life of the commuters is disrupted by traffic congestion. The major source of anger, anxiety and other physiological issues for the typical Indian youth is the everyday challenge and determination to stop traffic, noise and reckless drivers. On average, a person spends his or her day driving anywhere from 30 min to 2 h. It is, therefore, essential to provide an economic and effective traffic control solution. Traffic management can be improved with density-based regulation. Here we suggest

G. Varshney · A. Jaiswal (✉) · U. Mittal · A. Pawar · Satyajeet
Department of Electrical Engineering Department, JSSATE, Noida, Uttar Pradesh, India

A. Tomar et al. (eds.), *Machine Learning, Advances in Computing, Renewable Energy and Communication*, Lecture Notes in Electrical Engineering 768,
https://doi.org/10.1007/978-981-16-2354-7_3

an efficient traffic management scheme by evaluating road congestion. In the present situation, one of the main issues in metropolitan centers around countries is traffic congestion, especially during rush hours. Typically, it has been found that the green traffic light is always in 'ON Mode' while there is almost no vehicle on that road. As usual, it can be noticed that long vehicle queues exist in one lane while the adjacent lane lies vacant. This is mainly due to improper traffic management and can be tackled utilizing the current advancements in the field of information technology.

8051 is a device that is interfaced with IR technologies and requires a microcontroller. Three IR sensors, i.e. transmitters and a receiver are mounted on every path. Traffic density regulation is calculated as low, medium and large as the car travels, the IR sensors are enabled, and the microcontroller inputs and the time delay of the green signal correlate with the density value. LCD monitor is used to show the period. Each sensor is interfaced to the microcontroller, which in fact regulates the traffic signal network in compliance with the density sensed by the sensors. When the density on the distinctive side is extreme, a lot of attention should be given to that side. Sensors constantly maintain sensing density in all ways such that the experimental signal is sent to the side on a high priority to enable the sensors to sense high density on that particular side. The first priority point leads the route with a corresponding priority stage.

1.1 Long Traffic Queues

With a rise in the number of automobiles on the road, serious traffic congestion has risen dramatically in big cities. Typically, this occurs in the morning and evening, during office hours at all major junctions.

1.2 Poor Traffic Management

People have to wait due to the immovability of traffic at road junctions. Traffic lights are so designed that it stays red for a predetermined timeframe and the commuters have to wait for the light to turn green. The key focus is on managing traffic in a limited period of time and creating an optimal solution for traffic jams.

2 Literature Review

In the last few years, monitoring and video surveillance technologies have been commonly utilized for traffic management [1–3]. Historically, vehicle detectors like radars and ultrasonic and microwave detectors have been used, but due to small sensors, with reduced resources and challenging to manage, deployment and development problems increased repair costs. Metal barriers in the area of the road impact radar sensors. Vehicle mathematical simulation parameters are planned. Mathematically, the spatial location of the object, the sunlight and the car are related to the values recorded by the sensor.

Manual intersection dependence contributes to a complex derivation of fault parameters. Parameters are dangerous because many of the issues are related to the time variations of these parameters. The techniques are built to predict traffic congestion on the basis of sensing. Another way is to calculate the traffic intensity dependent on the number of traffic fractions controlled by the RF signals on the roadside. This approach was inefficient since substantial manual labor was needed on various roads [4–6].

A number of advances have been created to estimate traffic intensity dependent on image analysis. Although these techniques need good photographs, the quality of which depends on the environment, particularly rain and fog. Algorithms to model the different traffic situations, such as fuzzy logic, have been used [7, 8].

Traffic light signals that work on set signal timing delays earlier, and this comprises uncontrolled traffic congestion in the current scenario. Where traffic intensity exceeds higher than the maximum on a specific route, a longer green light period is required to minimize traffic density. The key issue with the current traffic light network is that the fixed timings are unchanged in technology and excessive processing period as if there are no cars on the opposing road. Because the car has been standing in proper rows, there is a lot of flow [9–12]. Some of the solutions available in the literature are inductive loop tracking, passive infrared sensors, wireless network tracker and recognition in radio frequencies [13–15]. Moreover, some examples are available in [16–19].

Our program uses the Arduino microcontroller which is interfaced with the IR sensors. IR transmitters and receiver are mounted on every structured path. As the vehicle moves through the IR sensors, the photodiode is triggered and the entity identified by the sensor is decreased. Collective details on the traffic intensity of the through path of the '+' junction is measured and it dynamically adjusts the latency of the green signal. Traffic intensity is calculated as small, medium and high based on these details, and the length of the traffic signal differs along different routes.

3 Methodology and System Design

3.1 Design of Density-Based Traffic Light Control System

The way to deal with this plan is acknowledged through the structure and execution of its input subsystem, control unit (control program) and yield subsystem. The information subsystem is made of sensors, modified and executed utilizing some previously existing standards to accomplish ideal execution. The control unit is acknowledged by a microcontroller-based control program, which deciphers the information and qualifies it to create an ideal yield.

These include:

1. Mains supply
2. DC force supply
3. Sensors clusters
4. Controller
5. Traffic lights.

The principle supply gives 230 VAC force which is changed over to 5 VDC (V_{DD}) by the DC power supply used to control the sensor exhibits, the controller, the observation camera and traffic lights. The sensors give a contribution to the controller which at that point plays out some coherent tasks to control the traffic lights as yield utilized for controlling traffic at street crossing points. Also for the proposed reconnaissance framework, the camera is interfaced with the controller to catch permit number plates of traffic defaulters for capacity and law authorization purposes. In picking the sensors, the accompanying highlights were thought about: precision. In spite of the fact that the infrared (IR) sensors are generally upset by a commotion in the encompassing, for example, radiations, surrounding light and so forth, they were utilized for this plan since they are modest and promptly accessible in the market.

3.2 *Choice of Microcontroller and Peripherals*

Despite the fact that the microcontroller PIC 8051 has a couple of impediments, yet it is picked on account of the accompanying highlights which have been mentioned below and the pin diagram is as shown in Fig. 1.

- Pins 1–8: It has no other function. Port one may be a domestic force up, similar metal directional input/output port.
- Pin 9: The RESET pin is used to reconnect the 8051 microcontroller to its primary values. The RESET pin must be removed for 2 rotations of the unit.
- Pins 10–17: In addition, these pins provide many alternative functions such as timer input, interrupts, and serial communication indicators.
- Pins 18–19: This area unit is used to interface the external system to relinquish the clock pulses.
- Pin 20: Indicated as V_{ss}-symbolizes land (0 V) connection.
- Pins-pair 1–28: recognized as Port pair (P 2.0—P 2.7)—apart from acting as input/output port, senior order address bus field indications multiplexed with this identical metal directional port.
- Pin-29: PSEN is used to view the output of the external program memory.
- Pin-30: It is used to require or prohibit external memory interfaces. If no external memory is required, this pin is dragged high by linking it to V_{cc}.

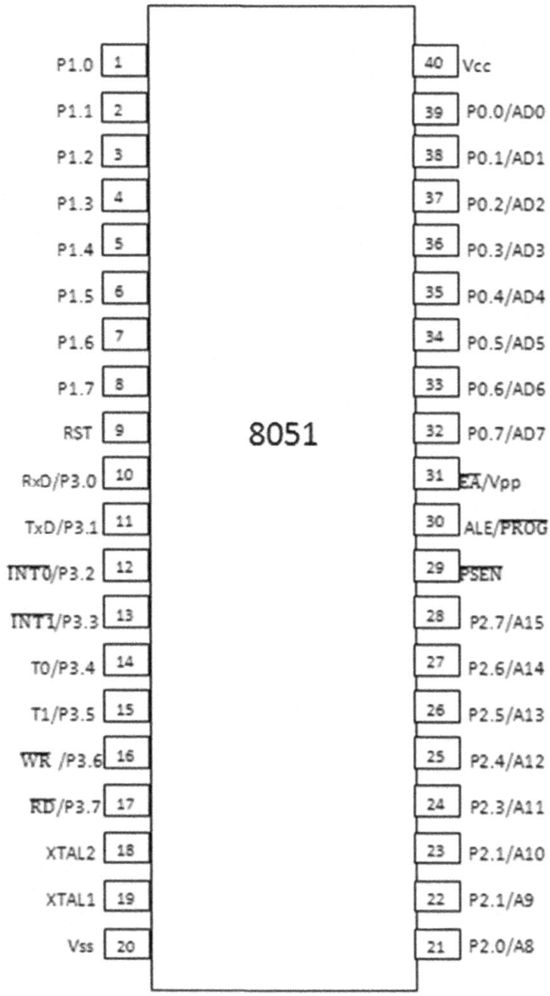

Fig. 1 Microcontroller 8051

- Pin-31: Address Latch alters is used to de-multiplex the address information for port zero.
- Pins 32–39: apart from being used as an input/output port, low order information and address bus signal multiplexed with this port (to use external interface).

LCD Display

Liquid crystal display (LCD) is a low-power, low-volume, flat-panel display. This can be conveniently programmed and is used in a broad variety of optical and electronic applications. It uses a complex layout wherein the active agent containing the pixel cell is at the junction of multiple electrode buses. In the proposed system we have specifically used a 16 × 2 LCD module which is capable of displaying data over 2 lines having 16 characters each. In addition, two register sets are needed to set up the LCD; a command registry is required for LCD activation, screen clearing, cursor location selection and monitor control, as long as the data registry retains the ASCII.

IR Sensor

IR sensor is an electronic system used to identify barriers or to distinguish between items based on their characteristics. It is usually used to calculate the heat of the body. The IR sensor produces or absorbs infrared radiation (430 THz–300 GHz) and is therefore invisible to the human eye. The light emitting diode (LED) will serve as an IR emitter, while the IR detector is a photodiode portion that is responsive to the IR light of the same frequency as the radiation emitted. As the IR radiation from the LED enters the photodiode, the output voltage varies based on the magnitude of the IR light.

3.3 Choice of Traffic Light Indicators

Three light emanating diodes, 'GREEN', 'YELLOW' and 'RED', each have their typical significance of 'GO', 'READY' and 'STOP' separately as shown in Fig. 2. They are constrained by the control transports of the microcontroller relying upon the sensible choices taken by the controller to control the paths of traffic as indicated by their densities.

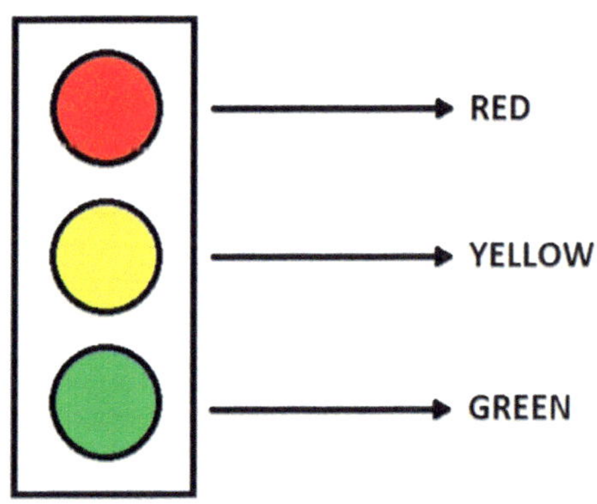

Fig. 2 Traffic light pointers

4 Implementation

4.1 Arrangement and Implementation of Infrared Sensors

Specification

This design primarily focuses on the traffic scale detection on each road. Depending on the density within each lane using infrared sensors, and the IR sensors, the road map is designed to perform the function efficiently.

Simulation of microcontrollers

The Proteus microcontroller simulation operates by adding either a hex file or a debug file to the schematic component of the microcontroller. This is then co-simulated with both analog and digital circuitry related to it. It allows its usage in a large variety of model prototyping.

4.2 System Testing

The density-based traffic management system is a mechanism that will count the vehicles on either side of the junction road when the vehicles enter the junction. When the circuit is attached and the code is recorded, test it by sensing the word IR sensor used to identify the optoelectronic means of detecting it, most usually some sort of photodetector. The unit can be tested with Proteus. This one we use to create 8051 controller programs. Upon writing scripts, we will dump code to the controller using 8051 programmers. Just use an IR signal, and the device can now be built utilizing photodiode and phototransistors. Next, write a program to allow LCD to check the board next. To trigger the LCD, submit the appropriate commands and customize the serial ports, parity and number of bits once all of the instruments attached to the controller have been configured.

4.3 Algorithm

The entire system's algorithm is represented in the flowchart. The system starts at the start of the flow diagram, and so do the microcontroller and RAM. It clears and initializes the microcontroller's all flags. Initially, all stop flags are set, which means that all traffic light indicators for traffic control on all four lanes display a RED light which stops all traffic at the beginning of its operation. The state of all the sensor arrays on each traffic lane is then read and given to the microcontroller as input for logical operations. The system then goes further to assign operating serial number to each lane based on their densities and assigns lane one to the lane with the most density.

Step 1: Start.
Step 2: Variables initialization.
Step 3: Set the variables to zero.
Step 4: Begin the while loop.

4.1: Loop for the red light.
4.2: Loop for the yellow light.
4.3: Loop for the green light.

Step 5: Check for the emergency signals.
Step 6: Apply the time delay signal.
Step 7: Repeat till the conditions are obtained.
Step 8: Stop.

5 Result

The fixed scheduling of traffic lights creates a serious problem. The effect of a low-efficiency traditional traffic network makes a significant impact on the cultural, safety and environmental sectors. The trouble with the transport network and inadequate control will contribute to vehicle crashes, traffic delays and road pollution, which bring heavy loads on companies and employment. Advances in technology and small control units, equipment and sensors have made it possible to create complex and intuitive integrated systems to solve challenges faced by humans and create a simpler lifestyle. This traffic light control aims to support society to develop traffic lighting structures and regulate the movement of cars at intersections by introducing new architecture. The planned smart traffic network consists of a traffic light controller that controls the traffic lights at the '+' intersection of highways. On the basis of this knowledge, the period assigned to green light should be increased to accommodate a wide movement of cars in the case of a traffic jam or shortened to avoid excessive standing period where there are no cars. For the above setup, a microcontroller-based traffic light controller has been built and programming has been developed. The results are shown in Fig. 3 in ON and OFF conditions. The shown outputs validate the results related to density-based traffic light controllers.

6 Conclusion

The proposed device is complemented by a handheld computer for emergency services trapped in traffic. By way of safe contact utilizing the wireless network, the portable device moves the traffic controller in emergency mode and offers a

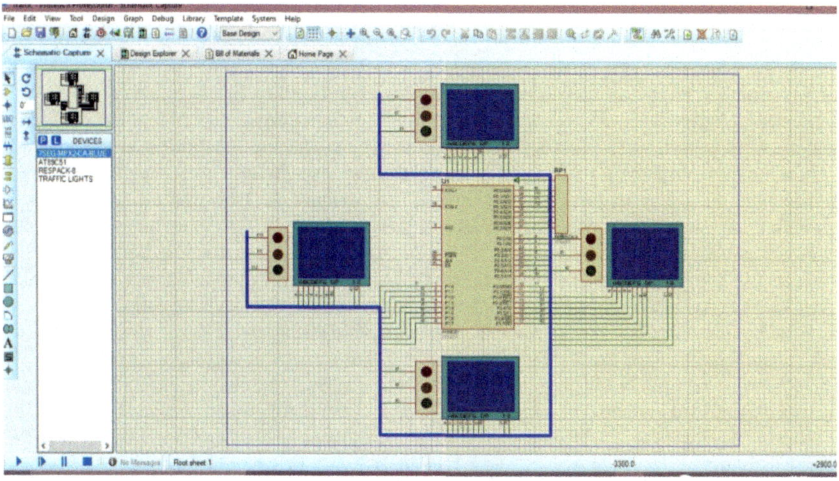

(a) OFF Mode

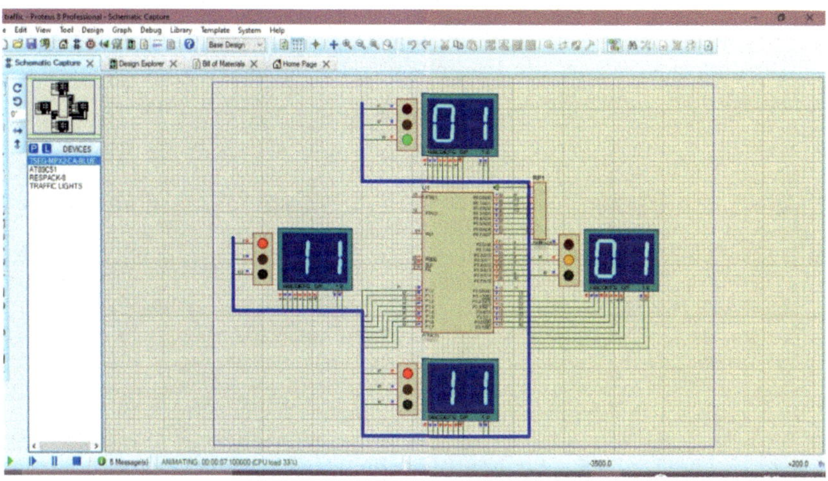

(b) ON Mode

Fig. 3 Simulation results

clear route before the trapped emergency vehicle passes through the intersection. The developed framework is applied, realized electronically and reviewed to certify the complete validity of its activities and functions. The existing architecture can be encouraged by tracking and regulating the intersection of dual roads. Future enhancements, such as a pedestrian crossing icon, time limit screens and a traffic crash may be added to the system.

References

1. Bhilawade V, Ragha LK (2018) Intelligent traffic control system. Int J Scientific Res Publ 8(2)
2. Kaur G, Sharma S (2017) Traffic management using digital image processing. IJCST 8(2)
3. Shirazi MS, Morris BT (2016) Vision-based turning movement monitoring: count, speed & waiting time estimation. IEEE Intell Transp Syst Mag 8(1):23–34
4. Mehboob F, Abbas M, Jiang R, Al-Maadeed S, Bouridane A, Tahir MA (2016) Automated vehicle density estimation from raw surveillance videos. In: SAI computing conference, London, UK
5. Tejashri G, Priyanka C, Komal C, Togrikar PS (2016) Implementing intelligent traffic control system for congestion control, ambulance clearance and stolen vehicle detection
6. Lan C, Chang G (2015) A traffic signal optimization model for intersection experiencing heavy scooter vehicle mixed traffic flows. 16(4):1771–1783
7. Monika G, Kalpana N, Gnanasundari P (2015) An intelligent automatic traffic light controller using embedded systems. Int J Innov Res Sci Eng Technol 4(4)
8. Kavya G, Saranya B (2015) Density based intelligent traffic signal system using PIC microcontroller. Int J Res Appl Sci Eng Technol (IJRASET) 3(1):205–209
9. Kham N, Nwe C (2014) Implementation of modern traffic light control system. Int J Sci Res Publ 4(6)
10. Isa I, Shaari N, Fayeez A, Azlin N (2014) Portable wireless traffic light system (PWTLS). Int J Res Eng Technol 3(2):242–247
11. Ali MDH, Kurokawa S et al (2013) Autonomous road surveillance system proposed model for vehicle detection and traffic signal control. Procedia Comput Sci 19
12. Chandrasekhar M, Saikrishna C, Chakradhar B, Kumar PP, Sasanka C (2013) Traffic control using, digital image processing. IJAEEE 2(5). ISSN (Print): 2278-8948
13. Jaiswal S, Agarwal T, Singh A, Lakshita (2013) Intelligent traffic control unit. Int J Electri Electron Comput Eng 2(2):66–72
14. Tyagi V, Kalyanaraman S, Krishnapuram R (2012) Vehicular traffic density state estimation based on cumulative road acoustics. IEEE Trans Intell Transp Syst 23(3)
15. Niksaz P (2020) Automatic traffic estimation using image processing. In: 2012 international conference on image, vision and computing
16. Aggarwal S et al (2020) Meta heuristic and evolutionary computation: algorithms and applications. Springer Nature, Berlin, p 849. https://doi.org/10.1007/978-981-15-7571-6. ISBN 978-981-15-7571-6
17. Jafar A et al (2021) AI and machine learning paradigms for health monitoring system: intelligent data analytics study in big data, vol 86. Springer Nature, Berlin. ISBN 978-981-334-411-2
18. Gopal et al (2021) Digital transformation through advances in artificial intelligence and machine learning. J Intell Fuzzy Syst 1–8. https://doi.org/10.3233/JIFS-189787
19. Mahto T et al (2020) Traffic signal control to optimize run time for energy saving: a smart city paradigm, metaheuristic and evolutionary computation: algorithms and applications. Studies in computational intelligence. Springer Nature, pp 491–497. https://doi.org/10.1007/978-981-15-7571-6_21

Reactive Power Pricing Framework in Maharashtra

Shefali Tripathi, D. Saxena, Rajeev Kumar Chauhan, and Anant Sant

Abstract High penetration of renewable generation into the grid has introduced many uncertainties and technical challenges like voltage variation, chances of reverse power flow and increased fault level. One of the crucial challenges in the Indian power system is to consider the high variability and unpredictability of generation from renewable, productive and economical grid service. High renewable energy (RE) penetration results in a serious impact on the stability of the grid, thus making it imperative to maintain adequate reserves of reactive power to ensure secure and reliable operation of the grid. Voltage-control and reactive-power management are two features of a single activity that ensure reliable and secure operation of the grid. Thus, for effective voltage control, reactive-power management in the grid is necessary. This paper proposes an implementable reactive energy accounting and settlement framework for state entities in the Maharashtra state, which involves payments from and into the reactive energy account depending on the drawl and injection by respective state entities and system voltages.

Keywords State transmission utility (STU) · Energy accounting · Availability-based tariff (ABT) · Regional entities · Generators · Regional reactive charges (RRC) · State reactive charges (SRC) · Reactive reserve amount (RRA) · Reactive power · Voltage control

1 Introduction

The energy trends are shifting from a major dependency on fossil fuels to renewable energy sources. This can be attributed to reasons like sharp decline in RE prices,

---

S. Tripathi · D. Saxena
Department of Electrical Engineering, MNIT Jaipur, Jaipur, India

R. K. Chauhan (✉)
IMS Engineering College, Ghaziabad, India

A. Sant
Idam Infrastructure Advisory Pvt. Ltd., Mumbai, India

A. Tomar et al. (eds.), *Machine Learning, Advances in Computing, Renewable Energy and Communication*, Lecture Notes in Electrical Engineering 768,
https://doi.org/10.1007/978-981-16-2354-7_4

global temperature concerns and environment degradation. Indian power system witnessed a transition from five independent regional grids to a unified national grid in December 2013. The power sector in India is gradually changing its characteristics by synchronizing the national grid to a single frequency and increasing the availability of power from inadequate to surplus to meet the demand for the system. Earlier renewable energy generators could be disconnected without the significant impact on grid stability but with India's dedication to a greener future where substantial energy demand from renewable sources in the system is met, their increased penetration will pose a threat to grid stability and reliability, thus making it impossible to connect/disconnect renewable energy generators (REGs) at system operator's discretion. More penetration distributed generators (DGs) have also resulted in power quality problems like harmonics, voltage sags and swells. Variation in power generated from REGs causes voltage fluctuations, thus changing voltage profile along the network depending on how much power is consumed and produced at that system level. Voltage transients also appear as a result of the connection and disconnection of generators [1]. Background flow of energy in an AC system which arises from the production of magnetic and electric field affects constitutes reactive energy which has the potential to affect system voltages. Similar to frequency, which is consistent across the network, reactive power cannot be made to flow throughout the system. It is important to provide local reactive power support to restore the voltage at its nominal, while simultaneously ensuring that reactive power does not travel far in the system. Reactive power if made to flow through the system disturbs the voltage profile throughout, thus making it necessary to provide local reactive power support. To restore system voltages, compensation devices are required to be installed wherever required. The devices which are capable of storing the energy by virtue of electric and magnetic field are capable of providing reactive power support.

Hence to improve voltage profile, reactive power support has to be provided locally. The reason for providing local power support is if reactive power is made to flow throughout the system or if the reactive power support is provided far from the source, then it will lead to disturbing the entire voltage profile. Voltage is a local problem and hence local support has to be provided, unlike the frequency which is a global problem. Moreover, several examples related to the reactive power pricing framework are represented in the digital domains [2–8].

2 Impact of RE Penetration on the Grid

A large number of converter-based distributed generators (DGs) are continuously being integrated into the system. One of the main power system parameters, namely reactive-power is affected by the high levels of penetration of renewable power, causing steady-state voltage and dynamic/transient stability problems [9].

3 Grid Stability

Power grids are complex, dynamic structures having a large number of systems connected to them. Optimal and secure operation of the power grid requires multiple small control areas which constitute the electrical grid to maintain an optimal balance between the power generated and the demand while simultaneously ensuring congestion and overloading the free system. Grid stability is directly linked to the balance between the generation and demand while simultaneously system voltage as close to the nominal value and between the permissible limits as mentioned in the grid code. Proper reactive and active power balance is essential for grid stability. Although reactive power is vital for secure grid operation, it is essential to have it in right amount. Low system voltage results in higher currents for the same amount of power produced, thus resulting in high system losses (transmission power losses). Voltage should be maintained well within the limits because even a small deviation in the system voltage can be fatal enough for the working of the equipment causing wear and tear. Compensation for reactive power improves the amount of active power that can be transmitted effectively, thereby improving the stability of the system.

4 Need for Reactive Power Management

Reactive power flow that arises from the energy refers to the background energy movement. It is the energy that helps the transformers to transform voltage levels, enables generators to generate electricity and also enables motors to rotate. In other words, it can be called the energy required for setting up a magnetic field essential for the production and smooth flow of active power (Fig. 1).

Complex Power $= \sqrt{}$(True_power2 + Reactive_power2).

Complex Power $= \sqrt{}(P2 + Q2)$.

With the increase in penetration of REGs, there is an increasing need for maintaining system security and reliability. Reactive power is required to maintain voltage within the permissible limits to deliver active power efficiently via transmission lines.

Fig. 1 Power triangle

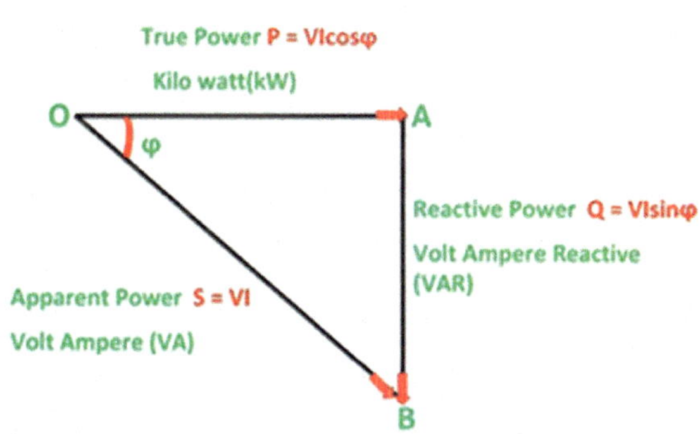

Power system equipment and loads are designed to operate at a fixed voltage with a small permissible deviation. At low voltages, the performance of the equipment is poor and can make equipment underperform and can also cause overheating.

5 Reactive Power Compensation Methods

Reactive power can be supplied using various sources, like static VAR compensators, static capacitors, shunt reactors, generators and synchronous condensers. Reactive power if made to travel for long distances affects the voltage profile wherever it flows. Thus, it is always recommended to provide support near to the point where needed.

5.1 Synchronous Condenser

The synchronous machine's ability to function as a condenser enhances the power factor and increases the reliability of the power network. To operate the synchronous machine in condenser mode, it is first started as a synchronous generator from standstill till it is synchronized with the grid after this changeover of synchronous machine condenser mode takes place. During the changeover, the excitation is introduced which in turn reduces the induced e.m.f. making it lower than the terminal voltage. Owing to this, power flow is reversed with the synchronous machine absorbing active power from the grid and simultaneously the flow of water to the turbine is stopped. Thus synchronous machine stops generating power and is considered as a pure reactor or a condenser (capable of supplying/absorbing reactive power). When a machine is operating in synchronous condenser mode that is under excited or overexcited, varying excitation current controls reactive power. Active power (P) negative indicates that active power is being absorbed from the grid to keep the machine under synchronous condenser mode and to supply losses. P positive indicates that the synchronous machine working as a generator and supplying active power to the grid. Reactive power (Q) negative implies that the synchronous machine is absorbing reactive power and Q positive means that the reactive power is being supplied by the synchronous machine [10].

5.2 Induction Generator

The induction machine absorbs reactive power irrespective of the mode in which it is operating. As the rotor speed is elevated above the synchronous speed, the induction machine functions as a generator. Now to operate the induction machine as a generator, armature current has to be supplied by giving an initial excitation. This excitation can be given in any of the three ways. One is through the residual magnetic field present in the induction machine; the second way is installing the capacitors which get charged during the operation of the machine by the residual magnetism and the third by directly absorbing it from the grid. Sometimes when the load is large, the excitation provided by the residual magnetism present in the machine is not sufficient to operate it as a generator. For this reason, induction machines are installed with capacitors which provide the dual function of providing excitation and power factor improvement. For charging the capacitors a DC excitation can be provided, thus making them capable of providing starting current to the machine.

5.3 *Reactive Power Capability of Generators*

The generator's reactive power capacity curve decides the reactive power that a generator can produce without suffering any heating due to losses. A generator's capacity curve depends on the three limits, which are the current limit of the armature, the current limit of the field and the heating limit of the end component. The flow of armature current in the stator winding results in ohmic losses which cause heat. Thus, during the design of the generator, a cooling system is provided which puts a limit on the maximum armature current that can flow in the winding and beyond which armature will suffer damage. The field current limit is determined by the rotor winding current, which is again determined by the amount of ohmic losses it can absorb or deliver. The end part heating refers to the heating of the end region of the stator core of the generator due to the flow of eddy currents. When it is under excitation, the system operates at the leading power factor, so the field current increases to generate the necessary flux, leading to further flux concentration in the end area (Fig. 2).

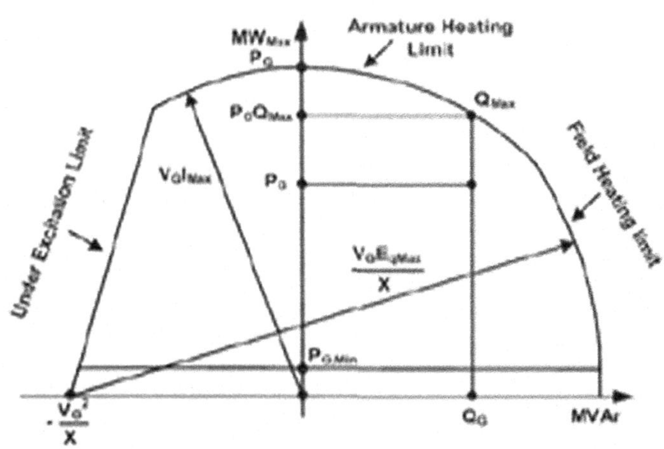

Fig. 2 Capability curve of generator

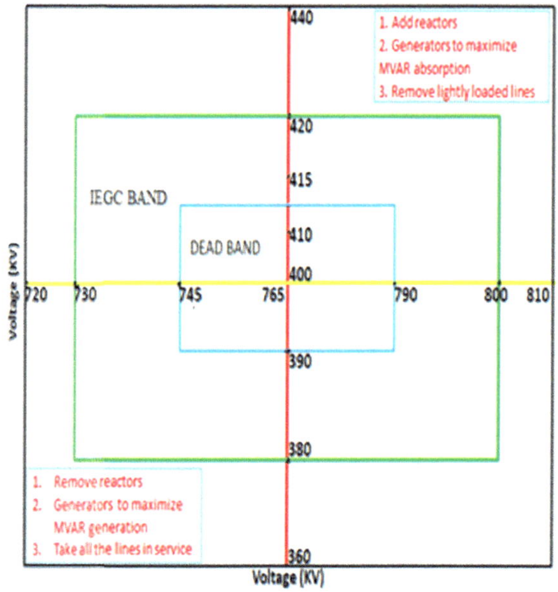

Fig. 3 765/400 kV transformers

5.4 OLTC Transformer

This technology is used for controlling voltage in the power system. By adjusting the transformer's turn ratio under loading condition, it regulates network voltage. Taps are provided along the primary or secondary winding. It is capable of automatically regulating the voltage under loading condition. The transformers which are equipped with the OLTC can change the number of turns of either the primary or secondary in order to change the turns-ratio under loading condition. As per the draft of Maharashtra Electricity Grid Code (MEGC), transformer tap changing criteria is as shown below [11]:

Figures 3 and 4 show the voltage values for which transformer tap settings have to be changed automatically. The transformer taps are expected to be changed when the voltage levels fall outside the IEGC band, i.e., in the dead band.

5.5 STATCOM

It is a shunt-connected FACTS device consisting of a single-voltage source converter (VSC) and its related shunt-connected transformer, connected by minimizing feeder power losses to enhance the power factor. STATCOM acts as a controllable voltage source. By regulating the amount of reactive power that is pumped into or absorbed from the grid, STATCOM also increases the voltage at the point of common coupling. Since it is capable of absorbing or producing reactive power, it is connected near to

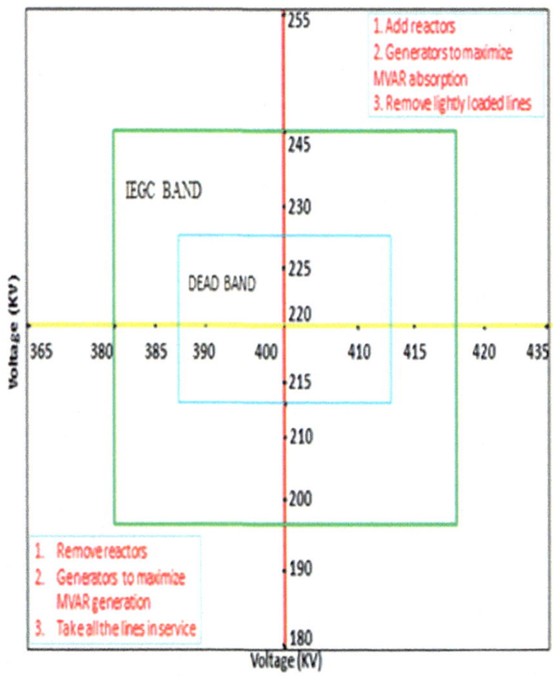

Fig. 4 400/220 kV transformers

the near to loads where voltage profile is to be improved. It is similar to a synchronous condenser with the only difference that it has no inertia because of the absence of any moving part. It is a flexible fast response reactive power compensating device. In the case of steady-state operation, when the system voltage is high, STATCOM consumes reactive power and when the system voltage is low injects reactive power. In the case of STATCOM, the reactive power that it can inject or absorb is directly proportional to the voltage at the point of common coupling. These are capable of providing a much faster response in comparison to SVC. STATCOM has the capability to feed the grid with the maximum available reactive current.

5.6 Static VAR Compensators (SVC)

It is a set of electrical devices belonging to the FACTS device family having internal switchgear as the only moving part. It is also referred to as the dynamic reactive power compensating device. It provides dynamic fast response reactive power support. It is capable of controlling voltages under normal, steady-state as well as contingency conditions. Thyristor-controlled reactor/thyristor switched capacitor SVC acts as a controllable reactance that is connected in parallel.

5.7 Dynamic Voltage Regulator (DVR)

It is connected to the power network in series and can be formed by three voltage source converters, where via an LC filter and a transformer, each VSC is connected to the power network.

6 Reactive Power Practices in India

Substation voltages are controlled using two schemes. One is the switching of the shunt capacitors (SCs) and the shunt reactors (ShR). The other scheme is the tap operation of the transformer. Maintaining system voltage within limits is the responsibility of the system operator (SLDC). When the system voltage goes out of the lower or upper limit as specified in the grid code, the system operator turns the SCs and ShRs on and off to bring the system voltage within the limits. The operation of SC is controlled by the system operator using SCADA. SCs with the potential to increase system voltage are usually turned on during morning when the demand is at its peak in order to increase injection of reactive power into the grid and turned off during evening peak time when the demand is reduced in order to reduce the reactive power injection into the grid. To control the voltage, transformer tap control is used by adjusting the tap location when the transformer is underloaded. The tap changing operation is carried out by the substation engineer by the manual pressing of a button. The tap changing frequency is different for different transformers. For transformers with the voltage level of 765/400 KV, 400/220 KV and 220/132 KV, the frequency of change of tap settings is less in comparison to the transformers with the voltages 220/66 KV, 132/66 KV and 132/11 KV. In order to maintain terminal voltages within the limits as defined in the Indian Electricity Grid Code 2010, generators are equipped with an automatic voltage regulator (AVR). Usually, the terminal voltage of the transformer is fixed. When the system voltage is outside the limit or is suspected to go outside the limits, the system operator instructs the plant owner to absorb reactive power and reduce the system voltage. Similarly, when the system voltage is low, the SLDC instructs the plant owner to inject reactive power into the system by operating the conventional plants in condenser mode.

7 Energy Accounting: Overview and Importance

Energy accounting takes into account energy flowing through transformers, transmission network, energy injected by the generating stations, energy drawn by the beneficiary and the difference as the energy loss of the transformer. This accounting involves active energy and reactive energy accounting. SLDC has the responsibility of regulating generation, load and the intrastate interchange between the entities.

7.1 Active Energy Accounting

Active energy accounting is essential to keep track of deviation between the actual drawl and the schedule which accounts for unscheduled interchanges (UIs). This UI settlement is done by SLDC by apportioning charges amongst DISCOM and other entities connected to the state transmission utility (STU) network. Availability-based tariff ABT helps in the segregation of charges for UI among various STU users on the basis of accounted deviations from schedules by entities, thus forming the basis for active energy accounting. Very few states like Maharashtra, Delhi, Madhya Pradesh Gujarat, West Bengal and Chhattisgarh have energy accounting systems [12].

7.2 Reactive Energy Accounting

Reactive energy accounting monitors the reactive power flows within the grid. It measures and records reactive power injection and withdrawals by various entities connected to the grid. However, the reactive energy accounting framework is not in place at the state level which makes it difficult to effectively manage reactive power in the grid at the state level. In inter-state energy accounting, each of the states is considered as a single entity and drawl/injection at the state periphery is monitored for accounting respective charges.

8 Regulatory Framework for Reactive Power Management in India

With the view of 22 GW of power based on solar photovoltaic and wind by 2022 [13], it is necessary to develop a reactive power management framework. Presently in India, reactive power management is incentive-based. VArh is paid at the rate of 10 paise/KVArh with an annual escalation of 0.5 paise/KVArh. As per IEGC 2010, incentives for reactive power compensation are given in Table 1 based on the following guidelines [14]:

Table 1 Guidelines for reactive power compensation incentives

Activity	Beneficiary/Utility	Voltage
Q Injection	Gets paid @10 paisa/KVArh from DSM pool account	$V < 97\%$
Q Drawl	Pays @10 paisa/KVArh to DSM pool account	
Q Injection	Pays @10 paisa/KVArh to DSM pool account	$V > 103\%$
Q Drawl	Gets paid@10 paisa/KVArh from DSM pool account	

Table 2 Voltage limits as per (CEA) Central Electricity Authority manual on transmission planning criteria

Nominal (KV)	Maximum (KV)	Minimum (KV)
765	800	728
400	420	380
220	245	198
132	145	122
110	121	99
66	72	60
33	36	30

(i) With the exception of the generating stations, the regional body pays for VAr drawing and VAr injection when the voltage at the metering point is below 97% and above 103%, respectively.
(ii) Except for the generating stations, regional bodies are charged for VAr drawl and VAr injection when the voltage at the metering point is above 103% and below 97%, respectively.
(iii) Regional bodies, with the exception of generating stations, shall strive to reduce the VAr drawl at an interchange point if the voltage at that point is below 95% of the rated voltage, and if the voltage is above 105% shall not return VAr.
(iv) The ISGS and other generating stations connected to the regional grid shall, in compliance with the instructions of the RLDC, generate/absorb reactive power within the capacity limits of the respective generating units, without compromising the active generation needed at that time. No charges for such VAr generation/absorption shall be made to the generating companies.

Voltage limits as per (CEA) Central Electricity Authority manual on transmission planning criteria are shown in Table 2 [15].

9 Issues of Reactive Power Management in India

9.1 PPA and Reactive Power Charges

Utilities that enter into PPA with the generation utility do not compensate the latter well for the reactive power support. They are called upon to supply power as and when required. When the generator is called upon to supply or absorb reactive power, its real power generation is reduced and is asked to absorb real power from the grid for which it is billed by the grid, i.e., to run the synchronous generator in synchronous mode, the generating company is asked to pay for the input energy consumed. This energy is roughly 1–1.5% of the machine rating which is small when considered for a short duration but becomes substantial when the machine generator operates in condenser mode for a long duration. There is no provision of incentives for the generators for reactive power support in the grid.

9.2 Lack of Incentives

Since many of the states in India lack an energy accounting framework, reactive energy quantum is not accounted for and no reactive power incentives are given for reactive-power support to the participating generating stations. The technical grid connectivity guidelines specify the power factor ranges within which each generator is required to operate, thus making reactive power support more obligatory.

9.3 Absence of Reactive Energy Accounting Framework

In India, due to a lack of reactive energy accounting, it becomes very difficult to keep a track of actual and scheduled VAR drawl/injection.

10 Suggested Reactive Power Pricing Framework for Maharashtra

The present system does not compensate generators for the reactive power support that they provide. Hence reactive power pricing framework for India should focus on compensating generators for the reactive power support that they provide. All generators should provide the obligatory as well as the enhanced reactive power support services as and when required.

10.1 Methodology for Accounting and Settlement

For discouraging VAR over drawl or over injection by transmission system utilities and generating sources, VAR exchanges should be priced as per the voltage levels as shown in Table 3.

The charge of 13.00 paise/kVArh with an escalation 0.50 paise/kVArh annually shall be levied. In the case security of the grid is endangered, SLDC may direct generating stations and TSUs for curtailing VAR injection/ drawl.

Table 3 Voltage levels at which VAR exchanges should be priced

Voltage/condition of TSU and generating unit	VAR drawl from InSTS	VAR injection into InSTS
Metered V < 97% of Bus V	Pay into the pool	Get paid from the pool
Metered V > 103% of Bus V	Get paid from the pool	Pay into the pool

10.2 Accounting and Settlement of Reactive Energy

STU shall be vested with the responsibility of smart energy meters at all the G <> T and T <> D in the InSTS along with AMR facility. Energy accounting and settlement shall be done in accordance with FBSM mechanism initially on monthly basis and later on weekly basis. The responsibility of providing meter data of reactive energy to SLDC is with STU in accordance with the state ABT order within 10 days for accounting and computation of the successive next week.

10.3 Settlement Procedure

Case I—"RRC is payable (P) (+) by the State and [RRC + SRC (Receivable (R)) < SRC (P)": Balance sum shall be retained as a reserve (RRA) after paying out RRC and SRC(P).

Case II—"RRC is payable (+) by the State and [RRC + SRC (R)] > SRC (P)": The excess amount available in reserve (RRA) shall be removed if any in order to comply with [RRC + SRC(R)] and SRC (P). SRC(P) and SRC(R) shall be apportioned to balance the total payables and total receivables if there is no reserve or if it is insufficient to fulfill the difference.

Case III—"RRC is payable (−) by the State and [RRC + SRC (P)] > SRC (R)": After paying out RRC and SRC (P), the balance sum is held as a reserve (RRA).

Case IV—"RRC is payable (+) by the State and [RRC + SRC (P)] > SRC (R)": The excess amount available in reserve (RRA) shall be removed if any in order to comply with [RRC + SRC(R)] and SRC (P). SRC(P) and SRC(R) shall be apportioned to balance the total payables and total receivables if there is no reserve or if it is insufficient to fulfill the difference.

Case V—No RRC for the State, No SRC (P), only SRC (R) and no RRA: No reactive power charges shall be paid to the TSUs.

Case VI—RRC is payable (+) by the State, No SRC (P), SRC (R) and no RRA: The amount available in the state UI pool account may be utilized for payment to the regional reactive pool purely on a temporary basis. In such instances, once the balance is made in the state reactive pool account, the same shall be transferred to the state UI pool account.

11 Conclusion

This paper proposes an implementable framework of energy accounting and reactive power in the state of Maharashtra. The specialty of this framework is that it compensates for all the state utilities including the generators. The study of the reactive power compensation frameworks in different states reveals that it is only the generator that was not compensated for providing the reactive power support within its reactive power capability. The proposed framework provides a fixed charge/KVArh compensation for the reactive power accounted for in the energy accounting framework.

12 Way Forward

At present, active energy accounting takes place effectively, and due to this, we are able to develop a competitive bid-based active power market. But with the reactive energy accounting framework in place in different cities in India, it will be possible to develop a competitive market for reactive power as well. The auction shall be based on the reactive power required to support voltage at a point in the grid so as to provide reactive power service at the least cost. The auction mechanism is to invite DERs to enter into the market, prevent collusion and make sure that no one dictates the market. On the basis of the requirement, reactive power can be procured through an auction in which the potential energy sources which are capable of delivering reactive power at the required point compete. The potential DERs need to quote the price along with the quantum of reactive power that they are willing to supply.

Auctions can become highly transparent and predictable, thus giving an opportunity to incumbents to learn to win an auction. The auction process can be in two ways:

- **Auction process-1**: All the reactive power suppliers bid simultaneously and compete against each other for the supply of reactive power to the locations as identified by the system operator.
- **Auction process-2**: In this, all the suppliers of reactive power of a particular location are entered into the system keeping in mind the key factor of the effectiveness of respective reactive power service.

Some incumbents can show predatory behavior by quoting a very low price initially for the service to deter the entrants and then coordinate among themselves to offer prices much higher in the long run. The final objective of the auction mechanism should be the maximization of economic welfare rather than the minimization of the price to be paid in the auction. Suppliers of reactive power offering services at lower quality and lower cost shall be given less priority over those who are offering high-quality reactive power service but at a higher cost. More effective reactive power should be valued more in comparison to the less effective reactive power [16].

References

1. Dulau LI, Abrudean M, Bica D (2014) Effects of distributed generation on electric power systems. In: The 7th international conference interdisciplinarity in engineering (INTER-ENG 2013), vol 12, no 1, pp 681–686
2. Fatema N et al (2021) Intelligent data-analytics for condition monitoring: smart grid applications, Elsevier, 268 pp. ISBN 9780323855112
3. Aggarwal S et al (2020) Meta heuristic and evolutionary computation: algorithms and applications. Springer Nature, Berlin, 949 pp. https://doi.org/10.1007/978-981-15-7571-6. ISBN 978-981-15-7571-6
4. Yadav AK et al (2020) Soft computing in condition monitoring and diagnostics of electrical and mechanical systems. Springer Nature, Berlin, 496 pp. https://doi.org/10.1007/978-981-15-1532-3. ISBN 978-981-15-1532-3
5. Smriti S et al (2019) Applications of artificial intelligence techniques in engineering, vol 1, Springer Nature, 643 pp. https://doi.org/10.1007/978-981-13-1819-1. ISBN 978-981-13-1819-1
6. Gopal et al (2021) Digital transformation through advances in artificial intelligence and machine learning. J Intell Fuzzy Syst (Pre-press) 1–8. https://doi.org/10.3233/JIFS-189787
7. Jafar A et al (2021) AI and machine learning paradigms for health monitoring system. In: Proceedings of intelligent data analytics. Springer Nature, Berlin, 496 pp. https://doi.org/10.1007/978-981-33-4412-9. ISBN 978-981-33-4412-9
8. Smriti S et al (2018) Special issue on intelligent tools and techniques for signals, machines and automation. J Intell Fuzzy Syst 35(5):4895–4899. https://doi.org/10.3233/JIFS-169773
9. Dugan RC, McGranaghan MF, Santoso S, Beaty HW (2012) Electrical power systems quality
10. Vaidya GA, Gopalakrishnan N, Nerkar Y (2011) Reactive power pricing structure for hydroelectric power station in condenser mode operation. Int J Electr Power Energy Syst 33(8):1420–1428
11. Draft Maharashtra Electricity Regulatory Commission (Grid Interactive Rooftop Renewable Energy Generating Systems) Regulations (2019)
12. Report on Scheduling, Accounting, Metering and Settlement of Transactions in Electricity (SAMAST) (2016)
13. Ministry of New and Renewable Energy (MNRE) (2018-19)
14. Central Electricity Regulatory Commission Indian Electricity Grid Code (2010)
15. Central Electricity Authority (CEA) Manual on Transmission Planning Criteria (2013)
16. Principles for Efficient and Reliable Reactive Power Supply and Consumption (2005)

Design Optimization of Solar Thermal Energy Storage Tank: Using the Stratification Coefficient

Jasmeet Kalra, Rajesh Pant, Pankaj Negi, Vijay kumar, Shivani Pant, and Sandeep Tiwari

Abstract Thermal stratification is a technique for maintaining separate layers of fluid having different temperatures. It plays a significant role in creating a large thermal gradient which in turn helps in storing more thermal energy in a solar thermal energy storage system. This paper investigates the effect of storage tank variables in terms of aspect ratio, equivalent diameter and its relationship with average stratification coefficient by varying them to different ranges to propose the optimized models.

Keywords Renewable energy · Solar thermal storage · Thermal stratification · Optimization · Equivalent diameter

1 Introduction

Today, the energy demands of the world are dependent on fossil fuels which definitely will create a crisis in the coming future. This problem has motivated mankind to search for an alternative form of non-degradable energy. Naturally available energies like wind, tidal, hydro and solar energy are the best resources for bringing ecological balance and fulfilling the futuristic energy demands of the world [1]. Designing and building the cheapest and feasible storage system based on the above-mentioned renewable energies is a solar thermal storage system. Thermal energy storage (TES) system is a technique of storing heat energy by increasing and decreasing the temperature of a medium, stored in a reservoir which can be later used for further applications [2]. The most common applications of TES are in industries, central air conditioning

J. Kalra (✉) · R. Pant · P. Negi
Department of Mechanical Engineering, Graphic Era Hill University, Dehradun, Himachal Pradesh, India

V. kumar · S. Pant
Department of Physics, Graphic Era Hill University, Dehradun, Himachal Pradesh, India

S. Tiwari
Krishna Engineering College, Ghaziabad, Uttar Pradesh, India

A. Tomar et al. (eds.), *Machine Learning, Advances in Computing, Renewable Energy and Communication*, Lecture Notes in Electrical Engineering 768,
https://doi.org/10.1007/978-981-16-2354-7_5

[3] and green buildings [4]. The advantages of TES comprise overall energy saving and less pollution at minimum expenditure. However, varying intensity of sunlight and different incident angle, at different places, pose a challenge in meeting load demand [5]. In TES, thermal stratification plays a significant role as it increases the overall efficiency of the system. Stratification occurs due to the property of a liquid that gets lighter on heating and forms different fluid layers of varying temperature with the hottest fluid layer at the top and coldest at the bottom. This creates a thermal gradient which provides an opportunity for large thermal energy storage [6] ease of use. Moreover, several examples are available in the digital domain [7–13].

1.1 Methods of Thermal Storage

The most common methods used in TES are storing sensible heat, latent heat and thermo-chemical storage reservoirs. The latter two methods have given promising results in storing and extracting the maximum amount of heat with minimum losses.

1.2 Sensible Heat of Storage

In this method, a solid or liquid medium increases its temperature by absorbing thermal energy. Quantity of energy stored is a function of specific heat of medium, temperature gradient and percentage of storing medium present [14].

$$Q = mCp\left(T_i - T_f\right) \tag{1}$$

where Q is stored heat in Joules; m denotes the mass of thermal storage medium in kg; Cp is specific heat in J/(kg K); T_i and T_f are initial and final temperatures in degree centigrade. Water being easily available, non-toxic and having high heat capacity (about 4180 kJ m^{-3} K^{-1}) is best suited as a medium for sensible heat storage method below 100 °C. Above it, other mediums like molten salts, oils and liquid metals are used.

1.3 Latent Heat of Storage

It requires a phase change material (PCM) which delivers or accumulates energy by undergoing from one physical state to another and the sensible heat as per Eq. 1. Phase change in PCM can take place as solid–liquid, liquid–gas, solid–gas and solid–solid. Many researchers have presented review papers [15–18] of available latent thermal energy storage systems and exploring different kinds of PCMs and their thermal properties.

Thermo-chemical energy storage. It consists of thermo-chemical medium which absorbs and releases energy by splitting and recombining of molecular bond in a fully reversible process. During charging, thermo-chemical material A absorbs heat and breaks into B and C, which can be easily stored separately. Whenever energy is needed, these two products (B and C) can mix together under suitable conditions of pressure and temperature releasing a huge quantity of energy. The complexity of the storage system increases, with the management of two materials with their separate storage capacities but it is one of the best techniques to store thermal energy due to high overall efficiency and low losses. Metal oxides have also been considered for a thermo-chemical medium that evolves oxygen that can be further used or discarded into the atmosphere [19]. It is still under investigation.

2 Methodology

As per the objective function for optimization of design parameters, we had considered the equivalent diameter of the storage material, aspect ratio of the storage tank and void fraction for variation within a range of values. The effect of changing the values was plotted on graphs along with its effect on Wu and Bannerot stratification coefficient [20].

Equivalent Diameter. The size or dimension of the bed element is called equivalent diameter and is represented by D_e. It may not necessarily be a spherical shape. Mathematically, it may be represented as:

$$D_e = \left(\frac{6}{\pi}V\right)^{\frac{1}{3}} \tag{2}$$

where D_e is the equivalent diameter, and V is the storage tank volume.

Void fraction is the term that represents the volumetric air gaps between the bed elements inside the storage tank. It is the ratio of volumetric air gaps to the total volume of the bed. With the rise in the volume of bed elements within the storage tank, void fraction decreases, and vice versa. Void fraction affects heat transfer within the storage tank as the material changes and in turn the heat transfer rate. Mathematically, it is calculated from the relation.

$$Ep = \frac{V - V_s}{V} \tag{3}$$

where E_p is the void fraction,

V denotes the volume of the bed material,
V_s denotes the volume of storage material.

Table 1 Range of selected variable parameters

Parameter	Range
Void fraction, E_p	0.3–0.6
Equivalent diameter, D_e (m)	0.05–0.20
Aspect ratio, *L/D*	1–10

Aspect ratio also known as L/D ratio. It is the ratio of either height or length of the storage tank to the diameter of the storage tank and is given by (Table 1):

$$\text{Aspect ratio} = L/D \tag{4}$$

Stratification coefficient is defined as the index showing the degree of stratification achieved [20]. It is based on the deviation of mean storage temperature from the mean square temperature of storage (Table 2).

$$ST_{WU} = \frac{1}{m_{store}} \sum_n m_n [T_{bn} - T_{avg}]^2 \tag{5}$$

where m_{store} is the mass of storage,

m_n is the mass of one element of bed,
T_{bn} is the bed element temperature, and
T_{bm} is the bed mean temperature.

Sizing of packed bed: The sizing of the packed bed is established on the general principle of energy to be stored within a specified interval of time. The bed size is considered on the basis that it should be able to absorb the maximum quantity of

Table 2 Range of selected fixed parameters

S. no.	Description	Parameter	Value
1	The volume of packed bed (m^3)	V_b	15
3	Number of bed element	N	60
4	Initial bed temperature (°C)	T_{bi}	25
5	Density of air (kg/m^3)	ρ_a	1.1
6	Dynamic viscosity of air (kg/s-m)	μ_a	1.865×10^{-5}
7	Inlet air temperature to bed (°C)	T_{ai} or T_{ib}	40
8	Ambient temperature (°C)	T_∞	25
9	Density of storage material (kg/m^3)	ρ_s	1920
10	Specific heat of air (J/kg °C)	C_{pa}	1008
11	Specific heat of storage material (J/kg °C)	C_{ps}	835
15	Time interval (min)	Δt	5

Table 3 Different values of aspect ratio, the length (L) and diameter (D) of the bed is working out to be as given below

Aspect ratio (L/D)	Length of bed (L in m)	The diameter of bed (D in m)
1	2.65	2.65
3	5.53	1.84
7	9.75	1.40
10	12.40	1.24

energy from incoming hot air during the charging and the average bed temperature becomes nearly equal to inlet air temperature at the end of the charging.

Assuming a uniform system load of 5 kN, implying that the collectors would collect at a rate higher than the load because it must supply an energy equivalent of 5 kN to be supplied over 16 h which is to be stored in the storage bed.

Energy to be stored works out to be 2.88×10^5 kJ using the following equation for the mass of the storage:

$$MCp\Delta T = 2.88 \times 10^5$$

Using a maximum value of temperature rise of 15 °C, (the difference between temp. of air at the inlet of 40 °C and starting temperature of 25 °C), the mass of the storage bed works out to be 19,000 kg. For the density of bed material of 1920 kg/m^3 and void fraction of 0.30, the volume of the storage bed works out to be 15 m^3 (Table 3).

3 Mass Flow Rate of Air (Heat Transfer Fluid)

It is defined such that the equivalent amount of energy fed by the hot air during 8 h is the energy that is needed for charging the bed. This mass flow rate can be determined from the following relationship:

$$(\dot{m}c_p)_{air}(T_{ai} - T_{bi})t_{ch} = V(\rho_{cp})_s(1 - E_p)(T_{bm} - T_{bi})$$

$\dot{m}$ stands for the required mass flow rate, kg/s.
C_{pa} stands for specified heat of air, J/kg K.
T_{ai} stands for the temperature of the incoming air, °C.
T_{bi} denotes the initial temperature of the bed, °C.
T_{ch} is the charging time, (s).
T_{bm} is the maximum bed temperature, °C.
V is the bed volume, m^3.
ρ_b is the density of bed material, k/m^3.
Cp_s stands for the specific heat of bed material, J/kg K.
Ep is the void fraction of the bed.

Using the bed volume determined in the earlier section and the values of other parameters listed in Table 2, the rate of flow for air works out to be 0.065 kg/s.

A computer program has been developed in C++ language for the determination of bed temperature Tb, air temperature Ta and stratification coefficient developed by Wu and Bannerot [20] ST_{WU} is in relation to system parameters, namely the equivalent diameter of bed element (D_e), the void fraction of bed (E_p) and aspect ratio of bed (*L/D*) in the given range for fixed values of other parameters.

4 Results and Discussion

The time-averaged value of the stratification coefficient has been plotted in Figs. 1, 2, 3 and 4. A designer can hence use these optimized values directly to make the

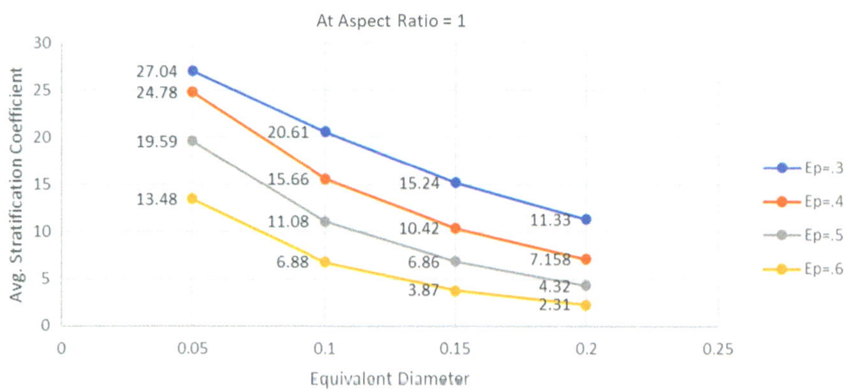

Fig. 1 Variation of average stratification coefficient with an equivalent diameter at aspect ratio 1

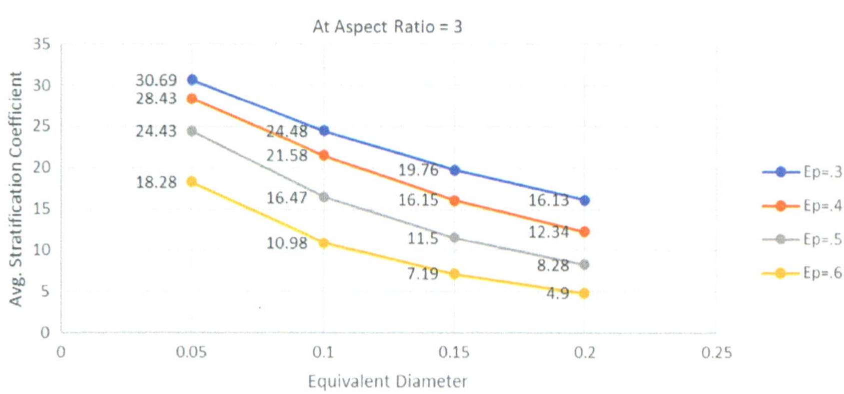

Fig. 2 Variation of average stratification coefficient with an equivalent diameter at aspect ratio 3

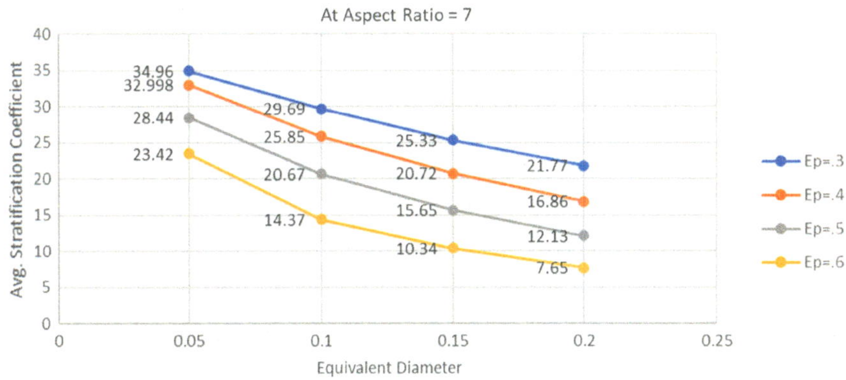

Fig. 3 Variation of average stratification coefficient with an equivalent diameter at aspect ratio 7

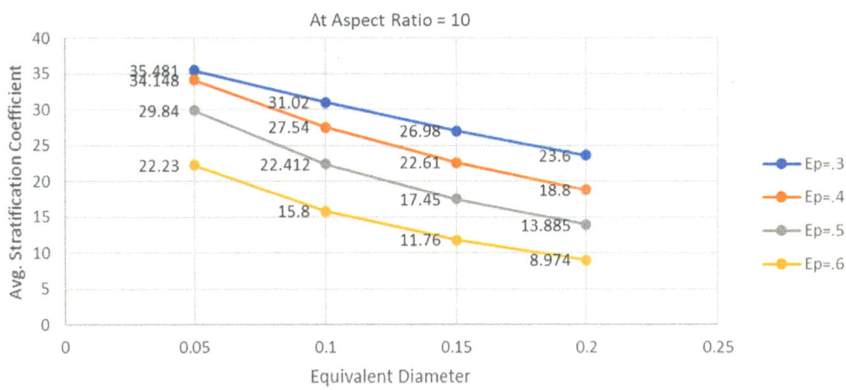

Fig. 4 Variation of average stratification coefficient with an equivalent diameter at aspect ratio 10

best solar collector storage system. Figure 1 shows that for a fixed value of aspect ratio (L/D = 1) of the storage tank, the best possible set of equivalent diameter and void fraction can be seen to be 0.05 and 0.3, respectively. Figures 2, 3 and 4 show the average value of the stratification coefficient corresponding to aspect ratios of 3, 7 and 10, respectively, where the maximum value of this coefficient can be 30.69, 34.96 and 35.48, respectively.

These plots can be used by the designer to arrive at suitable optimum values of system parameter sets for any fixed value and other variable values. For instance, in case the void fraction value of, say 0.5, has been fixed as a system constraint, then Fig. 4 shows that the optimum set of equivalent diameter and aspect ratio will be 0.05 m, 10 and the maximum value of stratification coefficient can be seen as 29.84.

References

1. Twidell J, Weir T (2015) Renewable energy resources. Routledge, London, UK
2. Sarbu I, Sebarchievici C (2016) Solar heating and cooling: fundamentals, experiments and applications. Elsevier, Oxford, UK
3. Saito A (2002) Recent advances in research on cold thermal energy storage. Int J Refrig 25:177–189
4. Hasnain SM (1998) Review on sustainable thermal energy storage technologies, part II: cool thermal storage. Energy Convers Manage 39:1139–1153
5. Elhab BR, Sopian K, Mat S, Lim C, Sulaiman MY, Ruslan MH (2012) Optimizing tilt angles and orientations of solar panels for Kuala Lumpur, Malaysia. Science
6. Kalra J, Raghav G, Nagpal M (2016) Parametric Study of Satratification in packed bed sensible heat, solar energy storage system. Appl Solar Energy 52(4):259–264
7. Fatema N et al (2021) Intelligent data-analytics for condition monitoring: smart grid applications. Elsevier, 268 pp. ISBN: 9780323855112
8. Aggarwal S et al (2020) Meta heuristic and evolutionary computation: algorithms and applications. Springer Nature, Berlin, 949 pp. https://doi.org/10.1007/978-981-15-7571-6. ISBN 978-981-15-7571-6
9. Yadav AK et al (2020) Soft computing in condition monitoring and diagnostics of electrical and mechanical systems. Springer Nature, Berlin, 496 pp. https://doi.org/10.1007/978-981-15-1532-3. ISBN 978-981-15-1532-3
10. Smriti S et al (2019) Applications of artificial intelligence techniques in engineering, vol 1. Springer Nature, 643 pp. https://doi.org/10.1007/978-981-13-1819-1. ISBN 978-981-13-1819-1
11. Gopal et al (2021) Digital transformation through advances in artificial intelligence and machine learning. J Intell Fuzzy Syst. Pre-press 1–8. https://doi.org/10.3233/JIFS-189787
12. Jafar A et al (2021) AI and machine learning paradigms for health monitoring system: intelligent data analytics. Springer Nature, Berlin, 496 pp. https://doi.org/10.1007/978-981-33-4412-9. ISBN 978-981-33-4412-9
13. Smriti S et al (2018) Special issue on intelligent tools and techniques for signals, machines and automation. J Intell Fuzzy Syst 35(5):4895–4899. https://doi.org/10.3233/JIFS-169773
14. Kumar A, Shukla SK (2015) A review on thermal energy storage unit for solar thermal power plant applications. Energy Procedia 74:462–469
15. Xu B, Li P, Chan C (2015) Application of phase change materials for thermal energy storage in concentrated solar thermal power plants: a review to recent developments. Appl Energy 160:286–307
16. Zhang H, Baeyens J, Cáceres G, Degrève J, Lv Y (2016) Thermal energy storage: recent developments and practical aspects. Prog Energy Combust Sci 53:1–40
17. Cárdenas B, León N (2013) High temperature latent heat thermal energy storage: phase change materials, design considerations and performance enhance-ment techniques. Renew Sustain Energy Rev 27:724–737
18. Liu M, Saman W, Bruno F (2012) Review on storage materials and thermal performance enhancement techniques for high temperature phase change thermal storage systems. Renew Sustain Energy Rev 16:2118–2132
19. Brumleve TD (1974) Sensible heat storage in liquids, solar energy technology division. Sandia Laboratories Report SLL-73-0263. Livermore, CA 94550, United States
20. Wu L, Bannerot RB (1987) Experimental study of the effect of water extraction on thermal stratification

Investigation of Hydro Energy Potential and Its Challenges in Himalayan State: Uttarakhand

Rajesh Pant, Jasmeet kalra, Pankaj Negi, Shivani Pant, and Sandeep Tiwari

Abstract Indian Himalayan region is rich in biodiversity of flora and fauna as well as it is the source of numerous perennial rivers. These rivers not only flourish the ecosystem of Indian plains but are also the source of immense renewable water energy. Indian Himalayas accounts for 79% of the total hydro potential of the nation, but currently, only 12.3% is being extracted from it. This paper studies the possible causes of the scarce utilization of waterpower in the Himalayan state of Uttarakhand. It also highlights the strength and flaws of hydropower projects on the environment and its local residents.

Keywords Renewable energy · Sustainable energy · Hydropower energy · Uttarakhand

1 Introduction

Hydropower is one of the cleanest and non-polluting sources of renewable energy. Its adaptability for peak and baseload variation makes it reliable for power grids. India has a total potential of 148,700 MW [1] and pumped storage of about 94,000 MW [2]. India ranks fifth in the world hydropower generation [3]. It utilizes 12.3% of its total utility power generation capacity with an installed capacity of 46,000 MW as of 31 March 2020 [4]. Most of the untapped hydropower potential lies in the northern and northeastern regions of India [1]. The four Himalayan states Jammu-Kashmir, Himanchal Pradesh, Uttarakhand and Arunachal Pradesh have almost 85% of total hydropower potential [5]. Uttarakhand alone is considered to have an approximate potential of 20,000 MW with large, medium, small, mini and micro hydropower plants [6]. Moreover, some examples related to hydro energy potential and its challenges are listed in [7–13].

R. Pant (✉) · J. kalra · P. Negi · S. Pant
Department of Mechanical Engineering, Graphic Era Hill University, Dehradun, India

S. Tiwari
Krishna Engineering College, Ghaziabad, Uttar Pradesh, India

A. Tomar et al. (eds.), *Machine Learning, Advances in Computing, Renewable Energy and Communication*, Lecture Notes in Electrical Engineering 768,
https://doi.org/10.1007/978-981-16-2354-7_6

Table 1 Types of HPPs

S. no.	Name	Capacity
1	Large HPP	Above 30 MW
2	Medium HPP	10–30 MW
3	Small HPP	Less than 10 MW
4	Micro HPP	Up to 100 KW

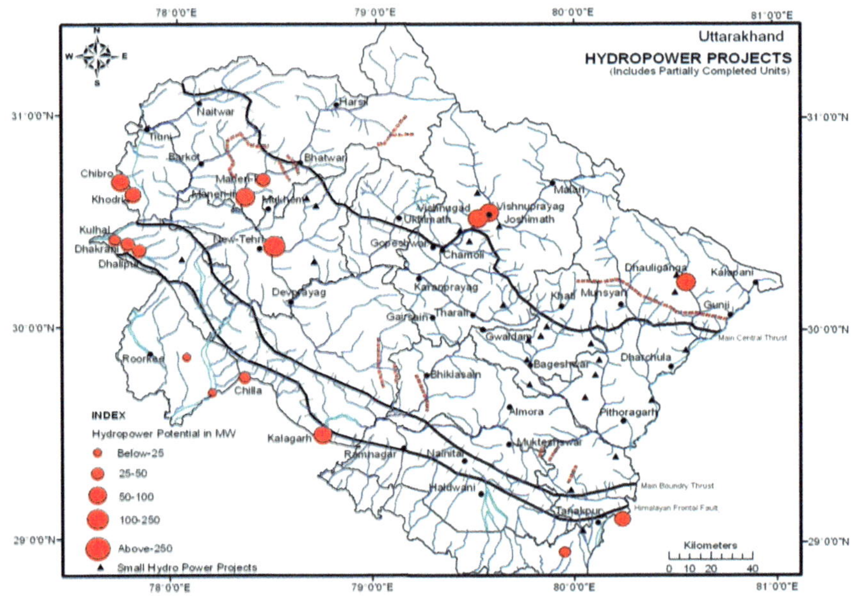

Fig. 1 Hydropower projects running in Uttarakhand [6]

Hydropower plant classification (HPP): As per the department of energy, the HPP can be classified as large, medium, small and micro HPP [14] on the basis of power generated as shown in Table 1.

Small hydropower plants give an added benefit of a longer lifespan and help in environment conservation (Fig. 1).

2 Development of Hydropower Resources in Uttarakhand

Uttarakhand was formed on 9 November 2000. It is located between 28° 4″ to 30° 25″ North latitude and 77° 35″ to 81° 02″ East longitudes. It has 53,483 square kilometers spread, of which 86% is a hilly area. It shares international boundaries with Nepal, Tibet and China. It is having numerous sources of water which are being provided by snow-fed glaciers. The state is endowed with 17 rivers and 31 lakes. The unsystematic distribution of water also creates water scarcity in a few of its districts. Alaknanda, Bhagirathi, Sub Ganga, Yamuna, Ramganga and Sharda are the principal rivers basins of the state.

Uttarakhand completes 97% of its energy demands. As per the Uttarakhand-DRE-Plan-Report data, the energy demand and energy availability of the state in 2017–2018 were 13,457 MU and 13,426 MU, respectively, with 31 MU deficits [15]. The projected energy demand of Uttarakhand in 2021 will be 19,406 MU which is 33% more than 2017–2018 for achieving the aim of 24 × 7 power supply. This need can be fulfilled by harnessing maximum water energy. Tables 1 and 2 show the working hydropower project and Table 3 shows the identified hydropower projects of Uttarakhand (Table 4).

Table 2 List of operational hydropower project [16]

Project	River	Design head (mts)	Installed capacity (MW)
Chibro	Tons	110	240
Khodri	Tons	57.9	120
Dhakrani	East Yamuna	19.8	33.75
Chila	Sub Ganga	32.5	144
Pathri	Sub Ganga	9.75	20.4
Mohammadpur	Sub Ganga	5.7	93
Bhilangana II	Bhagirathi	218	24
Bhilagna	Bhilagna	168.4	22.5
Maneri Bhali 2	Bhagirathi	247.6	304
Rishiganga	Alaknanda	52	13.2
Urgam	Alaknanda	196	3
Birahi Ganga	Birahi Ganga	54.5	7.2
Kaliganga	Kali Ganga	166	4
Rajwaqti	Nandakini	46.5	3.6
Vanala	Nandakini	135.98	15
Srinagar	Alaknanda	65.97	330
Kanchauti	Kanchautigad	400	2
Relagad	Relagad	264.83	3
Khatima	Sharda	17.98	41.4
		Total capacity	**1340.35**

Table 3 List of pumps storage-based projects [16]

Project	River	Design head (m)	Installed capacity (MW)
Dhalipur	Yamuna	30.48	51
Kulhal	Yamuna	18	30
Maneri Bhali I	Bhagirathi	147.5	90
Tehri I	Bhagirathi	228	1000
Koteshwer	Bhagirathi	81	400
Ramganga	Ramganga	84.4	198
		Total capacity	**1769**

Table 4 List of proposed hydro projects with estimated cost [17]

Project	River	Installed capacity (MW)	Approximate cost (Crores)/Base year
Deora Mori	Tons	**27**	**112 (1995)**
Mori Hanol	Tons	27	215 (1995)
Hanuman-Chatti—Syana-Chatti	Yamuna	33	149 (2000)
Barnigad-Naiangaon	Yamuna	34	234 (2000)
Chunni-Semi	Mandakini	26	110 (1996)
Markura Lata	Dhauliganga	45	154 (1996)
Lata Tapovan	Dhauliganga	108	247 (1996)
Vishnugad Pipalkoti	Alaknanda	340	1048 (1996)
Utyasu Dam	Alaknanda	1000	1985 (1996)
	Total capacity	**1640**	

3 Challenges in Development of HPPs in Uttarakhand

Harnessing hydro energy will bring numerous benefits to Uttarakhand but there are many challenges that need to be addressed beforehand. The construction and installation of an HPP is a mammoth task, requiring a huge initial investment with sound planning. Once the project is started, delay in construction costs higher interest rates. Preproject time-consuming activities involve site identification with a feasibility study, seismic investigation, acquiring forest/land clearance, preparing and sanctioning of environmental impact assessment and addressing rehabilitation and resettlement of local community. HPP requires blasting and tunneling which makes these plants hazardous to flora and fauna.

Uttarakhand falls in the highly seismic zone of range IV or V which poses a great challenge in the development of such projects. Another threat to such projects is flash floods and landslides which is a common site in Uttarakhand during the rainy season. In June 2013, severe rain and multiple cloud burst was a reason for flash flood in the Kedarnath valley which was named as the Himalayan tsunami.

As per the state government report, 169 people died, 4021 people reported missing (presumed to be dead), 11,091 livestock lost, 4200 villages affected, 70,000 tourists and 1 lakh locals were stranded in mountains with heavy damage to private and public property [18]. The Supreme Court of India constituted a 17-member committee to investigate this disaster, a report of which is still not made public. But an informal network 'SANDRP' (South Asia Network on Dams, Rivers and People) and eminent *geologist Prof. K. S. Valdiya* suggested that the building of a hydropower plant in the seismic and landslide-prone area was the reason for this disaster [19, 20]. With this report, it becomes a challenge for Uttarakhand to utilize its water potential.

4 Strength and Opportunities of HPPs

Continuous energy is needed for the welfare and unhindered growth of any state/country for which fossil fuel cannot be a good choice due to its unavailability, economic cost and, moreover, its negative ecological impact. Renewable energy is looked upon as a continuous source of green energy, especially in developing countries due to its economical, technical and environmental benefits [21]. Among renewable energy, waterpower is widely accepted due to its all-year-round availability, non-polluting nature, high density, easy power predictability and fast adaptability to load variations. It can be directly used or can be stored in the form of pumped storage or in charging batteries for small industries with many other direct and indirect advantages [22]. Nepal is a country that is generating huge revenue by selling HPP power to India and Bangladesh [21]. The low running and maintenance cost of hydropower plants is an added advantage that assists in ease to meet a low breakeven point of HPP despite its high construction and installation cost. The micro and small HPP are regarded as highly eco-friendly and have a large lifespan with little maintenance. Micro HPPs have an efficiency of 60–90% when compared to their solar counterparts [23]. Hydroelectricity can yield prosperity and self-sustenance to developing countries by meeting their current and future energy demands and producing more jobs in the service sector and the Uttarakhand government is motivating entrepreneurship by encouraging investment in micro and small HPP projects.

5 Flaws of HPPs

The HPP brings great risk to socio-economic conditions and the ecological system of an area. The power production in HPP requires a large amount of water to be stored with a high head (height from datum) forming a big lake in a catchment area. Heavy rainfall brings logs and sediments which have a negative impact on the working and lifespan of turbines, penstock and other equipments of HPP, hence the water is released during such times bringing floods in lower plains [24]. Building a dam in a river obstructs the natural flow of water, sediments [25] and even the migration of marine life like fish and other aquatic animal breeding [26]. A large-scale structure causes deforestation in the catchment area, riverbank erosion, flooding of agricultural land, water scarcity during low rains, water-borne diseases and significant deviation of an intricate ecosystem [27]. One important concern in HPP is that it emits a large amount of greenhouse gases which likely reduces snowlines in the Himalayas, making the lower plains venerable to floods as happened in the Kedarnath valley, Uttarakhand in June 2013 [19, 20]. Reports suggest that the total emission of greenhouse gases from an HPP may be more than its thermal counterpart working with carbon, natural gases and oils over a period of 25 years of operations [28]. Another important criterion for an HPP is meeting project deadlines as it greatly affects the cost of the project. Tehri project missed many deadlines, and the revised cost for the project was estimated INR 8392 crores during commissioning instead of the initial cost estimate of INR 3000 crores. Similarly, the estimated initial cost for the Alaknanda project in Srinagar was INR 20.69 bn in the year 2007 but it is revised to INR 36.75 bn in 2011 due to time runs [29]. There are many other examples that show a huge difference in estimation and commissioning costs of the hydro project all across India.

6 Conclusions

As per our investigation, there is a huge potential for hydro energy in Uttarakhand. The large difference between installed and available capacity gives us insight into many other factors which play a crucial role in developing Uttarakhand as a hydro-energy-rich Himalayan state. But there are challenges that hinder the growth of this sector in Uttarakhand and other Himalayan states such as heavy rainfall leading to flood and sedimentation negatively affecting turbine life. Large catchment area, heavy deforestation, soil erosion, rehabilitation and a large increase in greenhouse gases are some of the challenges which cannot be ignored. To overcome these challenges, an environmental impact assessment (EIA) study should be done in-depth for any proposed project. EIA is a powerful tool that investigates the impact of any hydro project on the environment, local residents and gives an optimum solution acceptable to all, such as project coordinators, investors, local residents and government agencies involved. Therefore, it is necessary to increase research in developing technology and management skills to overcome these challenges and finding solutions for utilizing the maximum hydro potential of Uttarakhand and other Himalayan states.

References

1. Central Electricity Authority (2016) Review of performance of hydro power stations 2015–16. New Delhi, India
2. Ministry of Power (2008) Hydropower policy. New Delhi, India
3. IHA Report, India overtakes Japan with fifth-largest hydropower capacity in the world. Accessed 30 May 2020
4. All India installed capacity of power stations, March 2020 (PDF). Accessed 25 May 2020
5. Sharma AK, Thakur NS (2016) Analyze the factors effecting the development of hydro power projects in hydro rich regions of India. Perspect Sci 8:406–408. ISSN 2213-0209. https://doi.org/10.1016/j.pisc.2016.04.090
6. Kamlesh K, Ashutosh S, Singh D, Pursotam K, Purushottam GK (2013) Uttarakhand Himalayas: hydropower developments and its impact on environmental system. J Environ 2
7. Fatema N et al (2021) Intelligent data-analytics for condition monitoring: smart grid applications. Elsevier, 268 pp. ISBN: 9780323855112
8. Aggarwal S et al (2020) Meta heuristic and evolutionary computation: algorithms and applications. Springer Nature, Berlin, 949 pp. https://doi.org/10.1007/978-981-15-7571-6 (ISBN 978-981-15-7571-6)
9. Yadav AK et al (2020) Soft computing in condition monitoring and diagnostics of electrical and mechanical systems. Springer Nature, Berlin, 496 pp. https://doi.org/10.1007/978-981-15-1532-3. ISBN 978-981-15-1532-3
10. Smriti S et al (2019) Applications of artificial intelligence techniques in engineering, vol 1. Springer Nature, 643 pp. https://doi.org/10.1007/978-981-13-1819-1 (ISBN 978-981-13-1819-1)
11. Gopal et al (2021) Digital transformation through advances in artificial intelligence and machine learning. J Intell Fuzzy Syst, Pre-press, pp 1–8. https://doi.org/10.3233/JIFS-189787
12. Jafar A et al (2021) AI and machine learning paradigms for health monitoring system: intelligent data analytics. Springer Nature, Berlin, 496 pp. https://doi.org/10.1007/978-981-33-4412-9. ISBN 978-981-33-4412-9
13. Smriti S et al (2018) Special issue on intelligent tools and techniques for signals, machines and automation. J Intell Fuzzy Syst 35(5):4895–4899. https://doi.org/10.3233/JIFS-169773
14. EERE homepage. https://www.energy.gov/eere/water/types-hydropower-plants. Last accessed 21 Nov 2020
15. Uttarakhand DRE plan report on Uttarakhand Decentralized Renewable Energy Plan 2018–19, p 15. https://www.thecleannetwork.org/wp-content/uploads/2018/10/Uttarakhand-DRE-Plan-Report.pdf
16. Agarwal S, Kansal M, Issues of hydropower development in Uttarakhand region of Indian Himalaya. Water Energy Int 59
17. Uttarakhand irrigation homepage. https://uttarakhandirrigation.com/hydropower-projects. Last accessed 21 Nov 2020
18. NIDM, India disaster report 2013. https://nidm.gov.in/PDF/pubs/India%20Disaster%20Report%202013.pdf
19. SANDRP homepage. https://sandrp.wordpress.com/2014/04/29/report-of-expert-committee-on-uttarakhand-flood-disaster-role-of-heps-welcome-recommendations/. Last accessed 21 Nov 2020
20. Valdiya KS (2014) Damming rivers in the tectonically resurgent Uttarakhand Himalaya. Curr Sci 106(12):25
21. Manzano-Agugliaro F, Taher M, Zapata-Sierra A, Juaidi A, Montoya FG (2017) An overview of research and energy evolution for small hydropower in Europe. Renew Sustain Energy Rev 75:476–489. https://doi.org/10.1016/j.rser.2016.11.013
22. Haidar AMA, Senan MFM, Noman A, Radman T (2012) Utilization of pico hydro generation in domestic and commercial loads. Renew Sustain Energy Rev 16(1):518–524. https://doi.org/10.1016/j.rser.2011.08.017

23. Behrouzi F, Nakisa M, Maimun A, Ahmed YM (2016) Global renewable energy and its potential in Malaysia: a review of Hydrokinetic turbine technology. Renew Sustain Energy Rev 62:1270–1281. https://doi.org/10.1016/j.rser.2016.05.020
24. Vassoney E, Mochet AM, Comoglio C (2017) Use of multicriteria analysis (MCA) for sustainable hydropower planning and management. J Environ Manage 196:48–55. https://doi.org/10.1016/j.jenvman.2017.02.067
25. Kocaman AS, Modi V (2017) Value of pumped hydro storage in a hybrid energy generation and allocation system. Appl Energy 205:1202–1215. https://doi.org/10.1016/j.apenergy.2017.08.129
26. Ioannidou C, O'Hanley JR (2018) Eco-friendly location of small hydropower. Eur J Oper Res 264(3):907–918. https://doi.org/10.1016/j.ejor.2016.06.067
27. Sivongxay A, Greiner R, Garnett ST (2017) Livelihood impacts of hydropower projects on downstream communities in central Laos and mitigation measures. Water Res Rural Develop 9:46–55. https://doi.org/10.1016/j.wrr.2017.03.001
28. Hidrovo B, Andrei JU, Amaya M-G (2017) Accounting for GHG net reservoir emissions of hydropower in Ecuador. Renew Energy 112:209–221. https://doi.org/10.1016/j.renene.2017.05.047
29. AHPCL homepage. https://www.power-technology.com/projects/alaknanda-hydroelectric-project-uttarakhand/. Last accessed 21 Nov 2020/

Big Data Preprocessing Phase in Engendering Quality Data

Bina Kotiyal and Heman Pathak

Abstract The change in the behavior of humans in the past decade has shown a tremendous generation in the data. The various researchers have given various definitions and discussed the different characteristics of big data. In the present study, we emphasize on the less focused areas of big data. One such zone is big data preprocessing. Extracting valuable information from big data has broadly three phases: first is acquisition and storage, second is data preprocessing, third is applying data mining and, at last, analysis of data. The contribution of this paper is that it shows generating the valuable information from big data not dependent on opting an advanced algorithm or novel algorithm but more than that it depends on acquisition of relevant data and preprocessing phase. The preprocessing phase plays a significant role in generating valuable data which serves as a great input in decision-making. At last, this paper gives a brief survey and analysis on big data preprocessing techniques used to handle imperfect data, reduction of data size and imbalanced data. It also theoretically discusses the different problems associated with the various phases and gives future directions where the researchers can work.

Keywords Feature selection · Preprocessing · Big data analytics process · Big data analytic techniques

1 Introduction

The rising popularity of smartphones and the tremendous change in human behavior over the internet leads retailers to deal with different kinds of data sources (internal data or external data). These data sources demand high analytics to be performed. However, before performing the analytics to the data, another issue arises from the rapid generation of data and variety (data present in a different format) which makes the traditional machines to be failed in processing and extracting some valuable

B. Kotiyal (✉) · H. Pathak
Gurukul Kangri Vishwavidyalaya, 60, Rajpur Road, Dehradun, Uttarakhand, India

A. Tomar et al. (eds.), *Machine Learning, Advances in Computing, Renewable Energy and Communication*, Lecture Notes in Electrical Engineering 768,
https://doi.org/10.1007/978-981-16-2354-7_7

information that plays a significant role in real-time applications for making decisions, such as for business intelligence, health sector etc. Big data technology plays a pivotal function where the conventional data management systems failed to bear, hoard and process this mountainous data immediately. Big data has taken the attention of not only the academicians, but it is a point of concern to government offices and corporate sectors too. The volume of data, one of the leading concerns of big data, has been addressed like "what size of data is called as big data?", therefore justifying the size of macro data is a need according to the author [1].

This paper contributes primarily to focus on some of the neglected zones of big data. One such zone is big data preprocessing. Many researchers have conducted research in big data but very few of them have discussed the preprocessing stage. The researchers have focused on all the phases but very few works have been done in preprocessing phase. The data preprocessing phase involves data cleaning, data integration and data transformation that confiscates the noise from the data [2]. Data quality has a lead role in the generation of valuable insights from data. It is among the most significant phases in big data analytics and it prepares the raw data for further processing and hence results in more efficient processing [3]. Moreover, several recent examples and applications of big data in different domains are listed in [4–10].

The storage phase is also an area of concern on account of exponential growth in data size because of the drastic change in human behavior over the past years continuously contributing to the ocean of data by means of online shopping, sharing of views on social network sites and many more. In near future, the storage and processing of big data will bring new heights of challenges to practitioners and researchers because of the various characteristics of big data. However, preprocessing phase plays a crucial role in generating high-quality data and rich performance in results, and this phase has a significant impact on decision-making; therefore, this area needs to be focused.

2 Characteristics of Big Data

Volume. Volume indicates the extensive data generated in seconds by each one of us. The time and type of data affect the definition of big data; therefore, it is not feasible to sketch a well-accepted threshold for big data volume that makes a "big dataset" [1]. Big dimensions can be in terabytes or in petabytes.

Variety. The data collected from diverse sources is a kind of heterogeneous data such as data collected from social media, weather data, geographical data etc. The data in spreadsheet or relational databases format is known as structured data. This data is only 5% of all remaining data [1]. The structure of data only creates one-fourth of the actual data according to statistics [11]. Semi-structured data looks like unstructured data. The unstructured data consists of text, images, videos, audio etc. Unstructured data does not have any format and forms 95% of the actual data.

Velocity. Velocity adverts to the speed at which the data generation takes place and it should be analyzed. It encompasses the speed (determined through batch data or real-time data) with which the data is treated and must comply with the swiftness with which the data is generated [12].

Veracity. The term veracity is invented as the fourth V, which represents the unpredictable (uncertain, incompleteness and inconsistency) characteristics present in some sources of data, such as the human behavior sentiments with respect to social media that are not certain in nature. However, they contain valuable information. Therefore, handling unpredictable data is one more important feature of big data technology built for the management and removal of uncertain data.

Variability. Variability and complexity are two more vital values of the big data familiarized by SAS. Variability shows the changes in the data flow rate, whereas complexity comes when the data is collected from several different sources.

Value. Value points to the usefulness of data for decision-making; it indicates the worthiness of the data extracted from the substantial volume of data. The organization can be rich in data but can be poor in valuable information until it is processed or if that valuable information cannot be used by an organization.

3 Big Data Analysis Process

The big data analysis process has five phases. Figure 1 shows the analysis process.

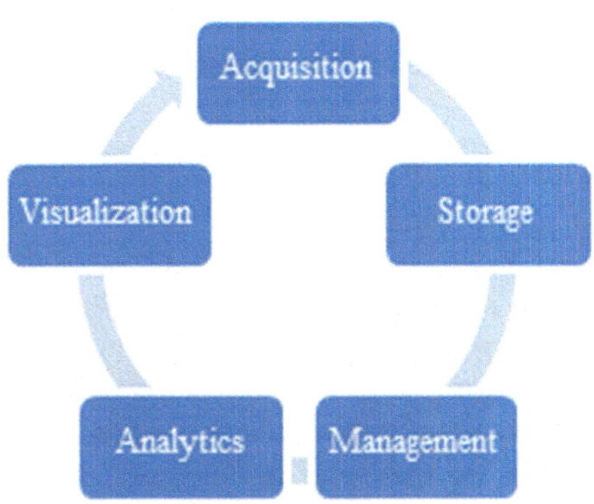

Fig. 1 Big data analysis process. The first step in the analysis process is data acquisition (data collected through different mediums and in different forms); the second step is to store the data; the third step is data management; and the fourth step is applying the analytic method to the data for extracting the meaningful information and at last data visualization is used for pictorial graphical representation

3.1 Data Acquisition

Data acquisition addresses a wide range of data ingesting processes, like collecting, filtering and cleaning in a databank. In this step, the data from the plurality of sources are collected and stored for a value creation drive. It has two sub-parts: data collection and data staging. Data collection is the way to collect the unrefined data from a factual environment and build it skillfully such as data collected through sensors or log files, whereas data staging is defined by the author [13]. In the stream processing model, the data is continuously generated at an amazingly fast speed and therefore needs to be analyzed immediately to extract its consequences, whereas in the batch processing model the data is saved and then processed. The two sub-steps of staging in batch processing are: data exploration and preprocessing.

3.2 Data Storage

A significant role is played by the data storage, as the size of the data is continuously increasing, and it results in the need for large and efficient storage. Hadoop ecosystem is used to store the data. However, it is not worth storing the entire data for processing, therefore storage optimization needs to be focused. This can be achieved by some techniques, like principal component analysis, random forest, and feature selection, thus compressing the data and reducing the storage space that will result in high performance [14].

3.3 Data Management

Data management ensures the effectiveness of big data storage. It can be broadly classified into two categories: the first is data preprocessing and the second is data analytics. Big data preprocessing is a formidable phase as existing tactics are not useful due to its size and heterogeneity of data, data generation through internal and external sources, the speed with which the data is generated, the complexity of data, noise in data, missing values, inconsistency and many more. The high volume of data with the aforementioned characteristics needs more sophisticated tools to process it [15, 16]. Apache spark technology can be used as a better solution than Hadoop as it can alleviate the challenges of data preprocessing. Apache spark processes the data through a directed acyclic graph (DAG), which automatically dispenses the data through the clusters and does the essential actions in parallel [17]. The working of apache spark is comparatively better than Hadoop as it does in-memory processing of large datasets. Sparks implements MLib with ten learning algorithms which encourage the integration of novel preprocessing methods in the future. Apache Flink can be used in the future as it fills the gap between Hadoop and

Spark. The Spark technology works on micro-batches, whereas Flink technology works on batches. Flink uses the FlinkML, a machine learning library [18].

3.4 *Data Analytics*

Big data gives a better decision and thus results in business intelligence. This can be achieved when we apply the analytical methods for generating meaningful data from the raw data. The four analytical methods are: prescriptive analytics, predictive analytics, diagnostic analytics and descriptive analytics.

3.5 *Data Visualization*

Visualization is related to the graphical representation of a design through images, tables, diagrams in a way to make a clear picture of the data. The prior visualization tools are not capable of handling the mammoth of data as the data is growing at a continuous pace. Latency is the challenge that comes with high volume of data with its continuous generation that needs to be tackled. Its job is to identify the patterns and correlations as per the need of organizations. However, much of the result is dependent on the beginning phases such as the collection of information, data preprocessing and using the analytics techniques according to the domain. Reducing the dimensions for the purpose of good storage and not making use of efficient techniques in the processing phase land up in losing the interesting patterns, whereas considering more dimensions can result in dense visualizations [19]. An example of visualization tools is Pentaho and Tableau. Pentaho generates reports from unstructured and structured data. It helps in making sound decisions. It is easy to use and gives a detailed visualization but makes use of less advanced analytics techniques than Tableau [20]. On the other hand, Tableau can process a huge number of datasets, and it is the fastest-growing tool used in the business industry. It has the proficiency to convert large and complex datasets into untaught depictions [21].

4 Big Data Analytics Techniques

Extracting the knowledge from the massive data is dependent on data analytics. The different types of analytics are:

Data Mining. It is the process of extracting hidden patterns and correlations from the datasets [22]. It includes the techniques such as association rule mining, classification, regression and clustering. Data mining serves as a base for machine learning and artificial intelligence [23]. The research studies show that we can also employ the existing data mining algorithms.

Web Mining. Web mining technique discovers the pattern from huge web data. It can be used to find the effectiveness of a website. Web mining can be classified into three parts [24].

Machine Learning. It is the ability of a system to learn from historical data without being programmed. Machine learning is a significant application of artificial intelligence. Making the automatic vital decisions and discovery of knowledge are the aim of machine learning [15] and it can also do predictions [23]. The size of data brings challenges to the on-hand machine learning techniques. Another challenge is machine learning works on horizontal scaling that uses MapReduce, streaming or graph-based solutions that show improvement in performance but do not handle other challenges such as on-feature engineering, the curse of dimensionality etc. Therefore, the amalgamation of a new learning paradigm along with processing manipulations with algorithms provides research opportunities.

5 Survey and Analysis

Big data preprocessing has many challenges associated with the characteristics of big data. To name a few are imperfect data (missing value, noisy [25]), feature extraction, feature section, heterogeneity, scalability and class imbalance. The research studies show that the quality of data depends on performing the data preprocessing phase properly with the domain-specific data. Therefore, it is very crucial to focus on this face. This paper has focused on three parts: imperfect data, feature extraction and class imbalanced and performed the theoretical analysis.

This paper analyses some of the techniques that are used to handle the imperfect data, reduction of data size and imbalanced data. Imperfect data deals with the techniques associated with missing value or noise. Missing values can result in bad decision-making and can be a potential reason for the loss of efficiency in the extraction of knowledge; therefore, it is very necessary to deal with it [26]. Discarding the missing values and computing the value to the missing data by means of mean, median, mode and likelihood are the traditional means for handling the missing data. Little and Rubin gave the solution of discarding the missing values. This approach can lead to biases in the learning process and significant information can be vanished. However, using the likelihood probability can be used to fill the missing values. The second approach to handle missing data is by computing the likelihood based on each use case. It does not impute the data. The maximum likelihood assessment of a factor is in the assessment of the factor that is most likely to have resulted in the experimental data. For the missing data, the maximum likelihood is calculated individually for cases with whole data on some variables and others with whole data on all variables. After that, they are maximized together to find the assessments. The advantage of the maximum likelihood is it does not need a careful selection of variables for impute values and gives unbiased factor assessments. The disadvantages were that it is restricted to linear models and the model for the dataset was not known in advance. However, using machine learning techniques can be fruitful in computing the missing values as it does not require any prior knowledge.

The data often contains noise that can affect the input or output or both of them, and thus leads to the poor quality of data [27]. Noise related to the input is called attribute noise and the noise related to the output variables is called class noise. The class noise shows that the data is more biased. The approaches used to handle the class noise are data polishing and noise filters [28]. Data polishing is a challenging task and can be applied to a small portion of data, whereas a noise filter can be used to remove the noisy instances without changing the existing data mining algorithms. Some of the conventional methods used for handling the noise are binning, regression and clustering. The binning method takes the neighboring data to smooth the value of a bin. It works on organized data values arranged into several bins. The smoothing is done by bin means, bin medians and bin boundaries [29]. When the values are not sorted then this method cannot perform. Another approach to handling the noisy data is through regression. It is a method that adapts the attributes of a dataset to a function. Regression can be linear regression and multiple regression. The linear regression is able to predict the value of one attribute with respect to the other attribute, whereas multiple regression is the extension of linear regression in which more than two attributes are considered. The clustering technique can be used to handle the noisy data. When the dataset has minimum and maximum variations in the values then clustering can be used to find the outliers. It makes a group of similar values and discards the dissimilar one. Clustering forms the group for similar data values and identified the irregular pattern in the group. The value that lies outside of the group boundary is outliers with an unusual pattern. This method is very efficient in handling the homogeneous sampled data [30]. The author [14] has given two approaches to cope with noisy data in big data classification problems. But these approaches are limited to the classification problem. An advanced algorithm needs to be developed in handling the noise and generating clean and high-quality data known as smart data.

The major hurdle in social network analysis is its size. It is computationally costly to process such a massive network. Dimension reduction is another technique to deal with the data size. It is a challenging process whose job is to reduce the high dimensions to low dimensions also known as the space transformation technique. It can lead to a better performance of the system by employing dimension reduction and compression techniques [31]. The three main characteristics of big data need to be addressed, and attributes are reduced to take out the valuable information. Dimensionality reduction could be achieved by feature selection and principal component analysis [24]. Feature selection (FS) is considered to be a more promising technique as it deletes the redundant data from the feature set. The feature selection method is classified into supervised, unsupervised and semi-supervised. Based on their learning it can be of three types that are filter, wrapper and embedded [16] which are employed to increase the processing performance of a system. In the filter method, the features are opted on the basis of their scores with the outcome variables using statistical tests

[32]. It has less computation compared to the other methods; however, the evaluation criteria are critical. It uses the chi-square test, correlation coefficient etc. Some of the filter methods consider the relationship between the features and class labels. Pertaining to big data the FS faces some challenges; it takes time to learn the data and have difficulty to handle velocity characteristic. The presence of noise makes it difficult to select the correct features of data. The existing methods are impotent in managing big data properly. For handling the unreliable data, incomplete data researchers have proposed feature-based heuristic algorithms. To name a few are the Fisher score algorithm, genetic algorithm etc. These algorithms are helpful in generating better feature sets with high speed. The survey shows that these methods are good in increasing the performance of the system, reducing the processing time and the cost incurred in overall phases and at last providing rich data. The wrapper method uses a subset of features and builds a model. The inferences are drawn from the previous model and the features are added or removed from the subset. Recursive feature elimination, forward feature selection and backward feature elimination are some of the examples of wrapper methods. Computationally, it is an expensive method and suffers from an overfitting problem. The embedded method is the combination of filter and wrapper method. LASSO regression [33] and decision tree are used in the embedded method. The embedded method has more accuracy and less error rate over filter and wrapper method but it has high computation and shows the problem of overfitting in high-dimensional data.

Imbalanced data or class imbalance is the challenge associated with the large growing of data, also known as the volume of data, and assuming that the data are not correctly classified across the distribution. Solving the problem of class imbalanced, big data preprocessing is potentially one of the main attention of researchers. If the class imbalance problem is not handled properly it can deteriorate the performance of a system because it has varying probability of existence. It has been a topic of research for more than a decade.

6 Conclusion

The paper emphasizes on the preprocessing phase. It shows the importance of preprocessing phase for generating great insight from large datasets. The paper also conducted a brief survey on big data preprocessing methods and presented some of the methods for handling the imperfect data, reduction of data size and imbalanced data. The implementation of the methods used for preprocessing the data is good in increasing the performance of the system, reducing the processing time and the cost incurred in overall phases and at last providing rich data. However, preprocessing is still in its infancy.

References

1. Gandomi A, Haider M (2015) Beyond the hype: big data concepts, methods, and analytics. Int J Inf Manag 35(2):137–144. https://doi.org/10.1016/j.ijinfomgt.2014.10.007
2. Han J, Pei J, Kamber M (2011) Data mining: concepts and techniques. Elsevier
3. Yogish HK, Raju GT, Manjunath TN (2011) The descriptive study of knowledge discovery from web usage mining. Int J Comput Sci Issues (IJCSI) 8(5):225
4. Fatema N et al (2021) Intelligent data-analytics for condition monitoring: smart grid applications. Elsevier, p 268. ISBN: 9780323855112
5. Aggarwal S et al (2020) Meta heuristic and evolutionary computation: algorithms and applications, Springer Nature, Berlin, p 949. https://doi.org/10.1007/978-981-15-7571-6. ISBN: 978-981-15-7571-6
6. Yadav AK et al (2020) Soft computing in condition monitoring and diagnostics of electrical and mechanical systems. Springer Nature, Berlin, p 496. https://doi.org/10.1007/978-981-15-1532-3. ISBN: 978-981-15-1532-3
7. Smriti S et al (2019) Applications of artificial intelligence techniques in engineering, vol 1. Springer Nature, p 643. https://doi.org/10.1007/978-981-13-1819-1. ISBN: 978-981-13-1819-1
8. Gopal et al (2021) Digital transformation through advances in artificial intelligence and machine learning. J Intell Fuzzy Syst (Pre-press) 1–8. https://doi.org/10.3233/JIFS-189787
9. Jafar A et al (2021) AI and machine learning paradigms for health monitoring system: intelligent data analytics. Springer Nature, Berlin, p 496. https://doi.org/10.1007/978-981-33-4412-9. ISBN: 978-981-33-4412-9
10. Smriti S et al (2018) Special issue on intelligent tools and techniques for signals, machines and automation. J Intell Fuzzy Syst 35(5):4895–4899. https://doi.org/10.3233/JIFS-169773
11. Chahal H, Jyoti J, Wirtz J (2018) Understanding the role of business analytics: some applications. https://doi.org/10.1007/978-981-13-1334-9
12. Sivarajah U, Kamal MM, Irani Z, Weerakkody V (2017) Critical analysis of big data challenges and analytical methods. J Bus Res 70:263–286
13. Alabdullah B, Beloff N, White M (2018) Rise of big data—issues and challenges. In: 2018 21st Saudi computer society national computer conference (NCC). IEEE, pp 1–6
14. Chakravarthy SK, Sudhakar N, Reddy ES, Subramanian DV, Shankar P (2019) Dimension reduction and storage optimization techniques for distributed and big data cluster environment. In: Soft computing and medical bioinformatics. Springer, Singapore, pp 47–54
15. Chen CLP, Zhang C (2014) Data-intensive applications, challenges, techniques and technologies: a survey on big data. Inf Sci 134
16. Cai J, Luo J, Wang S, Yang S (2018) Feature selection in machine learning: a new perspective. Neurocomputing 300:70–79
17. Salloum S, Dautov R, Chen X, Peng PX, Huang JZ (2016) Big data analytics on apache spark. Int J Data Sci Anal 1(3–4):145–164
18. https://data-flair.training/blogs/hadoop-vs-spark-vs-flink
19. Ali SM, Gupta N, Nayak GK, Lenka RK (2016) Big data visualization: tools and challenges. In: 2016 2nd international conference on contemporary computing and informatics (IC3I). IEEE, pp 656–660
20. Yaqoob I, Hashem IAT, Gani A, Mokhtar S, Ahmed E, Anuar NB, Vasilakos AV (2016) Big data: from beginning to future. Int J Inf Manag 36(6):1231–1247
21. García S, Ramírez-Gallego S, Luengo J, Benítez JM, Herrera F (2016) Big data preprocessing: methods and prospects. Big Data Anal 1(1):9
22. Bhandari B, Goudar RH, Kumar K (2018) Quine-mccluskey: a novel concept for mining the frequency patterns from web data. Int J Educ Manag Eng 8(1):40
23. L'heureux A, Grolinger K, Elyamany HF, Capretz MA (2017) Machine learning with big data: challenges and approaches. IEEE Access 5:7776–7797

24. Kotiyal B, Kumar A, Pant B, Goudar RH (2014) Classification technique for improving user access on web log data. In: Intelligent computing, networking, and informatics. Springer, New Delhi, pp 1089–1097
25. Little RJ, Rubin DB (2019) Statistical analysis with missing data, vol 793. John Wiley & Sons
26. Nguyen G, Dlugolinsky S, Bobák M, Tran V, García ÁL, Heredia I, ... Hluchý L (2019) Machine learning and deep learning frameworks and libraries for large-scale data mining: a survey. ArtifIntell Rev 52(1):77–124.
27. Miller JA, Bowman C, Harish VG, Quinn S (2016) Open source big data analytics frameworks written in scala. In: 2016 IEEE international congress on big data (BigData Congress). IEEE, pp 389–393
28. Pandey M, Litoriya R, Pandey P (2016) Mobile applications in context of big data: a survey. In: 2016 symposium on colossal data analysis and networking (CDAN). IEEE, pp 1–5
29. Hariharakrishnan J, Mohanavalli S, Kumar KS (2017) Survey of pre-processing techniques for mining big data. In: 2017 international conference on computer, communication and signal processing (ICCCSP). IEEE, pp 1–5
30. García-Gil D, Luengo J, García S, Herrera F (2019) Enabling smart data: noise filtering in big data classification. Inf Sci 479:135–152
31. Rong M, Gong D, Gao X (2019) Feature selection and its use in big data: challenges, methods, and trends. IEEE Access 7:19709–19725
32. Frénay B, Verleysen M (2013) Classification in the presence of label noise: a survey. IEEE Trans Neural Netw Learn Syst 25(5):845–869
33. Mudiyanselage TB, Zhang Y (2019) Feature selection with graph mining technology. Big Data Min Anal 2(2):73–82

Performance Evaluation of Sliding Mode Control for Underactuated Systems Based on Decoupling Algorithm

Ajit Kumar Sharma and Bharat Bhushan

Abstract The present work investigates an approach of sliding mode control with a decoupling algorithm to stabilize a class of underactuated systems. The decoupled method offers an unexacting mean to attain asymptotic stability for nth-order nonlinear systems. The translator oscillator rotational actuator (TORA) is chosen here to test the efficacy of the proposed control scheme. The detailed mathematical framework of underactuated systems and the proposed SMC are presented in the article. The underactuated systems are provided with different control inputs. The simulation results of the TORA system with the proposed control scheme demonstrate robust performance in a wide range of operations and disturbances. The system's stability, accuracy and transient performance measures like peak overshoot and response time improved with increment in the type of control input to the system.

Keywords Sliding mode control (SMC) · Decoupling algorithm · TORA system

1 Introduction

In the past decade, underactuated systems received sustained focus in the research field. The underactuated systems are employed in several applications, including locomotive systems, robotics, marine systems, underwater robots, and aerospace systems [1]. In an underactuated system, its control input numbers are lesser than its degree of freedom. The nonlinearity of the system increases the complexity between the directly actuated states and un-actuated states of control design. Few underactuated systems are not able to complete the requirement of the existence of stability by feedback law [2]. Some studies have proposed that stabilization through continuous feedback may remove the complications confronted by smooth feedback stabilization [3]. The other proposed methods include back-stepping control, adaptive non-smooth control, energy-based or mechanism based on passivity, intelligent control [4], hybrid control, sliding mode control, and decoupling algorithm [5] which also give better

A. K. Sharma (✉) · B. Bhushan
Electrical Engineering Department, Delhi Technological University, Delhi, India

A. Tomar et al. (eds.), *Machine Learning, Advances in Computing, Renewable Energy and Communication*, Lecture Notes in Electrical Engineering 768,
https://doi.org/10.1007/978-981-16-2354-7_8

results in system stabilization. The use of a decoupling algorithm to design the SMC-based underactuated systems will give satisfactory results in system complexity and better stability. SMC is a prominent robust control approach and a ubiquitous control strategy for nonlinear systems [6]. Its popularity in the research domain can be attributed to the following factors: (i) The system's behaviors are independent of plant parameter's variations in some cases, (ii) Reliability of desired dynamical properties of the systems in their sliding mode regime, (iii) Realization simplicity of compensator of the systems, and (iv) Sliding mode regime invariance to disturbances. The design of SMC is divided into two stages [7]. In the first stage, an appropriate surface is chosen and the second stage gives a control law which is formulated to guide the state of the system to desired states on the sliding surface. SMC affects a reduction in the order of the system, thereby increasing the system scope to minimize the impact of disturbances and uncertainties [8]. The discrete SMC strategy is appealing due to the prospect of undemanding implementation in digital controllers. Several researchers have explored discrete SMC design [9]. Two approaches are seen in the literature for discrete SMC. The first approach is focused on the mapping of continuous-time SMC to discrete-time [10–11]. The second approach is modeled on the disturbance observer and equivalent control design [12–13]. Higher-order SMC techniques are also discussed by some researchers for different types of machines, [14–19]. Higher-order sliding mode controllers demonstrate additional advantages like subdued chattering and high precision.

The motivation of this work is to formulate the SMC strategy based on a decoupling algorithm, which stabilizes the underactuated systems globally to its all degree of freedom. An SMC decoupling model has been designed and implemented in an underactuated system in a MATLAB environment. Moreover, different recent examples and applications using recent algorithms are represented in the digital domain [18–24].

This paper is formulated in different sections. Section 1 briefs the background and purpose of the study. Section 2 explains the mathematical framework of underactuated systems along with the decoupling algorithm. Section 3 expounds on decoupled sliding mode control design scheme. Section 4 contains the simulation findings from the work. Section 5 embodies the conclusion followed by references.

2 System Description

A translational oscillator rotational actuator (TORA) model is taken here as an underactuated system and shown in Fig. 1. A wall mass (M) is joined by a spring. The spring stiffness is k. The cart can move in only one dimension. L is a distance from its center, at which point the mass m rotates. Since movement is only in the horizontal plane, the effect of gravitational force is neglected.

In Fig. 1 u represents the control torque applied to m. The disturbance force on the cart is x. θ is the angular position and the rotational actuator of mass m is denoted by α.

Fig. 1 A translational operational rotational actuator system

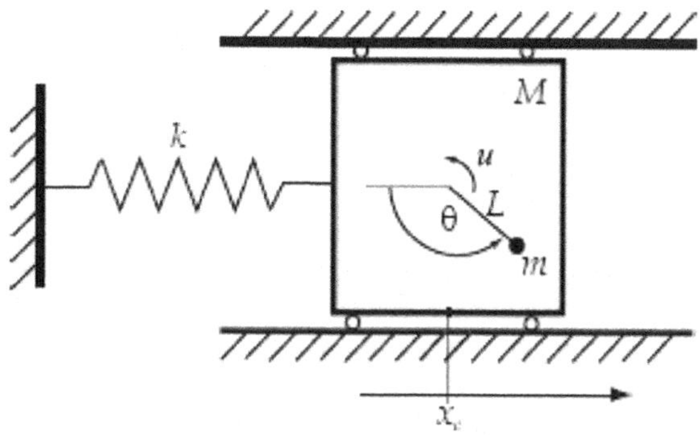

A standard system to test the efficacy of the proposed controllers on a TORA is developed. The proposed system is further improved and gives the general dynamics [13] for a TORA system as:

$$\dot{z_1} = z_2$$

$$\dot{z_2} = \frac{-z_1 + \varepsilon\theta_2^2 \sin\theta_1}{1 - \varepsilon^2 \cos^2\theta_1} - \frac{\varepsilon\cos\theta_1}{1 - \varepsilon^2\cos^2\theta_1} v$$

$$\dot{\theta_1} = \theta_2$$

$$\dot{\theta_2} = \frac{\varepsilon\cos\theta_1 (z_1 - \varepsilon\theta_2^2 \sin\theta_1)}{1 - \varepsilon^2\cos^2\theta_1} + \frac{1}{1 - \varepsilon^2\cos^2\theta_1} v$$

where v denotes the control input, z_1 is platform's normalized displacement from the equilibrium position $z_2 = \dot{z_1}$. The objectives of control are

$$z_1, \dot{z_1}, \theta_1, \dot{\theta_1} \to 0, \text{ for, } t \to \infty$$

Apply the decoupling algorithm to the above equation:

$$f_1 = \frac{-z_1 + \varepsilon\theta_2^2 \sin\theta_1}{1 - \varepsilon^2\cos^2\theta_1}, g_1 = \frac{-\varepsilon\cos\theta_1}{1 - \varepsilon^2\cos^2\theta_1}$$

$$f_2 = \frac{\varepsilon\cos\theta_1 (z_1 - \varepsilon\theta_2^2 \sin\theta_1)}{1 - \varepsilon^2\cos^2\theta_1},$$

$$g_2 = \frac{1}{1 - \varepsilon^2\cos^2\theta_1},$$

and

$$\left.\begin{aligned} x_1 &= z_1 + \varepsilon \sin\theta_1 \\ x_2 &= z_2 + \varepsilon\theta_2 \cos\theta_1 \\ x_3 &= \theta_1 \\ x_4 &= \theta_2 \end{aligned}\right\} \tag{1}$$

We have, $\frac{g_1}{g_2} = \varepsilon\cos\theta_1$

From Eq. 1, the control goals $z_1, \dot{z_1}, \theta_1, \dot{\theta_1} \to 0$ are equivalent to $x_i \to 0$, i = 1, 2, 3, 4.

Since,

$$\dot{x_2} = \dot{z_2} + \varepsilon\dot{\theta}_2\cos\theta_1 - \varepsilon\theta_2^2\sin\theta_1$$

$$\begin{aligned} &= \frac{-z_1 + \varepsilon\theta_2^2\sin\theta_1}{1-\varepsilon^2\cos^2\theta_1} - \frac{\varepsilon\cos\theta_1}{1-\varepsilon^2\cos^2\theta_1}v \\ &+ \varepsilon\left(\frac{\varepsilon\cos\theta_1(z_1 - \varepsilon\theta_2^2\sin\theta_1)}{1-\varepsilon^2\cos^2\theta_1} + \frac{1}{1-\varepsilon^2\cos^2\theta_1}v\right)\cos\theta_1 - \varepsilon\theta_2^2\sin\theta_1 \\ &= \frac{-z_1 + \varepsilon\theta_2^2\sin\theta_1}{1-\varepsilon^2\cos^2\theta_1} + \frac{\varepsilon^2\cos\theta_1(z_1 - \varepsilon\theta_2^2\sin\theta_1)}{1-\varepsilon^2\cos^2\theta_1} - \varepsilon\theta_2^2\sin\theta_1 \\ &= \frac{-z_1(1-\varepsilon^2\cos\theta_1) + \varepsilon\theta_2^2\sin\theta_1(1-\varepsilon^2\cos\theta_1)}{1-\varepsilon^2\cos^2\theta_1} - \varepsilon\theta_2^2\sin\theta_1 \end{aligned} \tag{2}$$

$$z_1 = -x_1 + \varepsilon\sin x_3 \tag{3}$$

$$\text{u} - \frac{\varepsilon\cos\theta_1(z_1 - \varepsilon\theta_2^2\sin\theta_1)}{1-\varepsilon^2\cos^2\theta_1} + \frac{1}{1-\varepsilon^2\cos^2\theta_1}v,$$

also, from Eq. 2

$$-z_1 = -x_1 + \varepsilon\sin x_3 \tag{4}$$

Combining both Eqs. 3 and 4

$$\text{v} = \varepsilon\cos x_3\left(x_1 - \left(1 + x_4^2\right)\varepsilon\sin x_3\right) - \left(1 - \varepsilon^2\cos^2 x_3\right) \tag{5}$$

From the above analysis, Eq. 1 can be decoupled as

$$\left.\begin{aligned} \dot{x_1} &= x_2 \\ \dot{x_2} &= f_1(x_1, x_3) = -x_1 + \varepsilon\sin x_3 \\ \dot{x_3} &= x_4 \\ \dot{x_4} &= u \end{aligned}\right\} \tag{6}$$

The above equation must be satisfied with three assumptions as follows:
Assumption 1: $f_1(0,0) \rightarrow 0$;
Assumption 2: $\frac{df_1}{dx_3} is invertible$;
Assumption 3: If $f_1(0, x_3) \rightarrow 0$, then $x_3 \rightarrow 0$.
If Eq. 6 is not satisfying assumption 2, we can redefine $f_1(x_1, x_3)$ as:

$$f_1(x_1, x_3) = -x_1 + \varepsilon \sin x_3 + 11\varepsilon x_3$$

Then, $\frac{df_1}{dx_3} = \varepsilon \cos x_3 + 11\varepsilon$ and Eq. 6 becomes

$$\left.\begin{aligned} \dot{x_1} &= x_2 \\ \dot{x_2} = f_1(x_1, x_3) &= -x_1 + \varepsilon \sin x_3 + 11\varepsilon x_3 \\ \dot{x_3} &= x_4 \\ \dot{x_4} &= u \end{aligned}\right\} \tag{7}$$

3 Controller Design

To realize $x_i \rightarrow 0$, define the error equation as

$$\left.\begin{aligned} \dot{e_1} &= x_2 \\ e_2 = \dot{e_1} &= x_2 \\ e_3 &= f_1(x_1, x_3) \\ e_4 = \dot{e_3} = \dot{f_1}(x_1, x_3) &= \frac{df_1}{dx_1}x_2 + \frac{df_1}{dx_3}x_4 \end{aligned}\right\} \tag{8}$$

The sliding mode function would be

$$s = c_1 e_1 + c_2 e_2 + c_3 e_3 + e_4 \tag{9}$$

where c_1, c_2, c_3 are positive constants.

From $\frac{d}{dt}\left(\frac{df_1}{dx_1}\right) = 0$, we have

$$\dot{s} = c_1 \dot{e_1} + c_2 \dot{e_2} + c_3 \dot{e_3} + \dot{e_4}$$

$$= c_1 x_2 + c_2 (f_1 - 11\varepsilon x_3) + c_3 e_4 \frac{d}{dt}\left(\frac{df_1}{dx_1}x_2 + \frac{df_1}{dx_3}x_4\right) \tag{10}$$

where $\frac{d}{dt}\left(\frac{df_1}{dx_1}x_2 + \frac{df_1}{dx_3}x_4\right)$

$$= \frac{df_1}{dx_1}(f_1 - 11\varepsilon x_3) + \frac{d}{dt}\left(\frac{df_1}{dx_3}x_4\right) + \frac{df_1}{dx_3}u \tag{11}$$

From Eqs. 10 and 11.

$$\dot{s} = c_1x_2 + c_2(f_1 - 11\varepsilon x_3) + c_3e_4 + \frac{df_1}{dx_1}(f_1 - 11\varepsilon x_3) + \frac{d}{dt}\left(\frac{df_1}{dx_3}x_4\right)\frac{df_1}{dx_3}u \tag{12}$$

$$\text{Let M} = c_1x_2 + c_2(f_1 - 11\varepsilon x_3) + c_3e_4 + \frac{df_1}{dx_1}(f_1 - 11\varepsilon x_3) + \frac{d}{dt}\left(\frac{df_1}{dx_3}x_4\right) \tag{13}$$

Then, the design of the SMC is given as

$$\text{U} = \left[\frac{df_1}{df_3}\right]^{-1}(-M - nsgns - ks) \tag{14}$$

where n and k are positive constants.

So far, we have discussed the model parameters of the underactuated system (TORA) and SMC controller. To get stable and distortionless performance, the systems and the SMC controller must obey similar parameter functions in the algorithm.

To achieve this, the SMC must follow the Lyapunov function.

To get this Lyapunov function, $\dot{s} = -nsgn(s) - ks$

The Lyapunov function: V $= \frac{1}{2}s^2$

then $\dot{V} = s\dot{s} = -n|s| - ks^2$ must be negative.

4 Simulation and Results

Simulation is performed on the MATLAB/Simulink environment to evaluate the results under various conditions (i.e., k = 10, 50, and 100).

The initial states are $\left[\,1\ 0\ \pi\ 0\,\right]$

To satisfy $\lambda_{ref}(-A) > \boldsymbol{\gamma}$, choose a = 5, η = 0.50, and switch function Δ = 0.10.

Figure 2 shows the simulation model of SMC and TORA.

Various cases are discussed below:

Case (a): For k = 10, the location of the pole is closer to the imaginary axis. The damping coefficient has a higher value, which increases disturbance in system response. The response time is slower which makes the system less stable, as we can see in Figs. 3, 4 and 5.

Case (b): For k = 50, the pole location moves away from the imaginary axis. The damping is decayed out as shown in Fig. 6. The improved control input makes the actuator position and displacement, as shown in Figs. 7 and 8, arrive at the rest position. Hence the system becomes more stable.

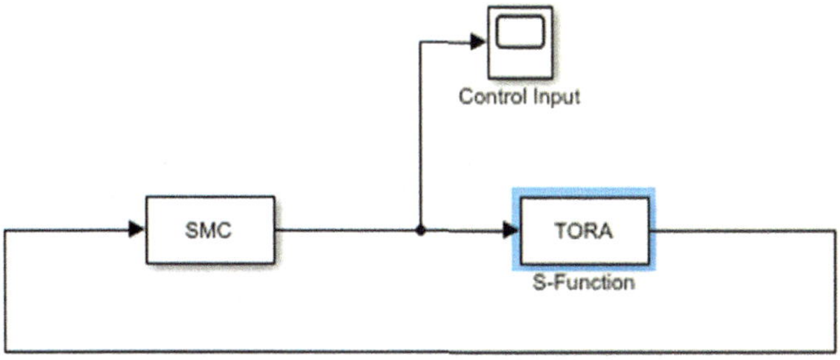

Fig. 2 Simulink model of the TORA system

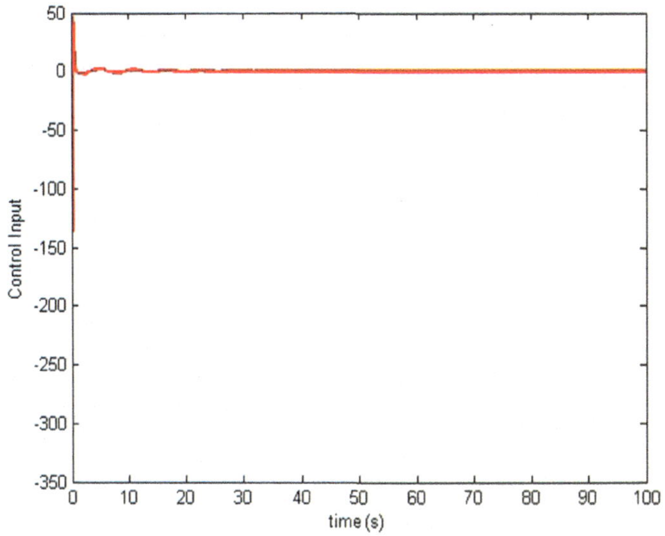

Fig. 3 Control input for TORA system (k = 10)

Case (c): For k = 100, the poles are located in the dominant pole region. The damping coefficient and transient response time are reduced to zero as shown in Fig. 9. In the dominant pole region actuator, the position and displacement become more stable, as shown in Figs. 10 and 11.

The distance from the imaginary axis is called the dominant pole and measure for stability. The nearer to the imaginary axis, the less stable is the system due to its possible move to the right side. This is also represented in Table 1 where like poles moving far from the imaginary axis system become more stable (k = 100).

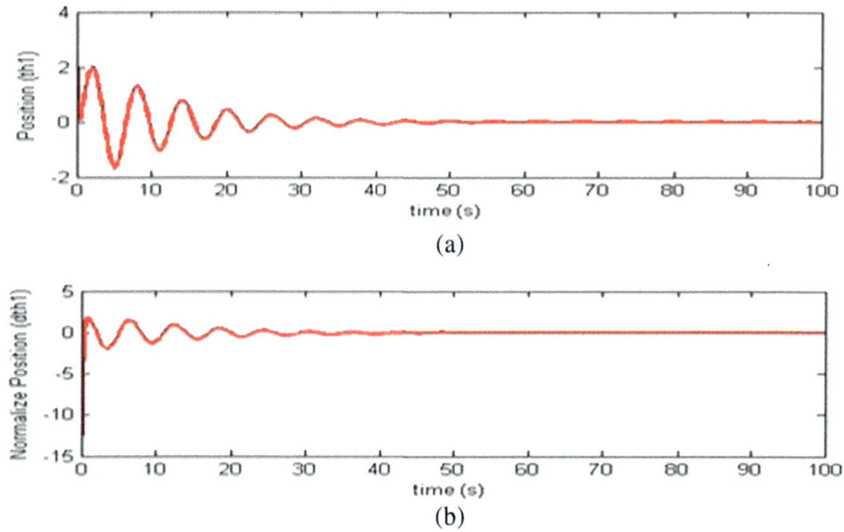

Fig. 4 **a** An actuator angle, **b** derivative of actuator angle (k = 10)

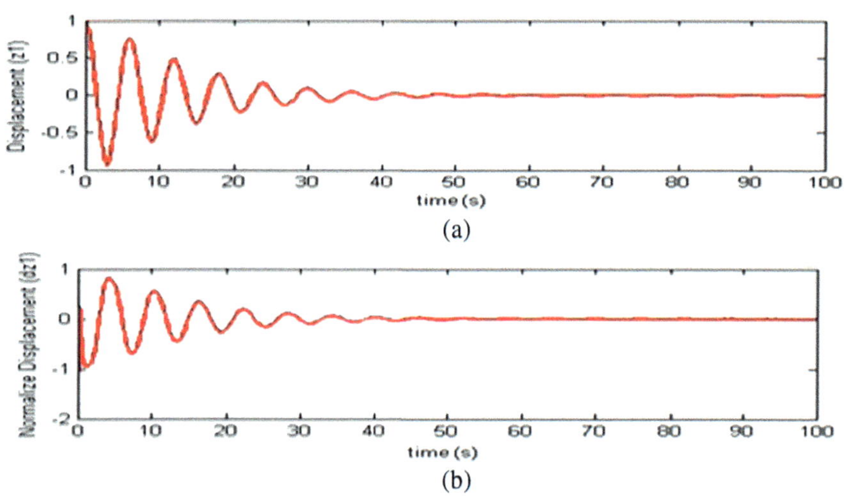

Fig. 5 **a** An actuator displacement, **b** derivative of actuator displacement (k = 10)

5 Conclusions

Here, an SMC with a decoupled algorithm is investigated to stabilize underactuated systems. Inverted pendulum and TORA system were taken as test cases. The performance of the control scheme was evaluated based on transient performance, stability, overshoot, and settling response. Simulation results showed that the proposed SMC

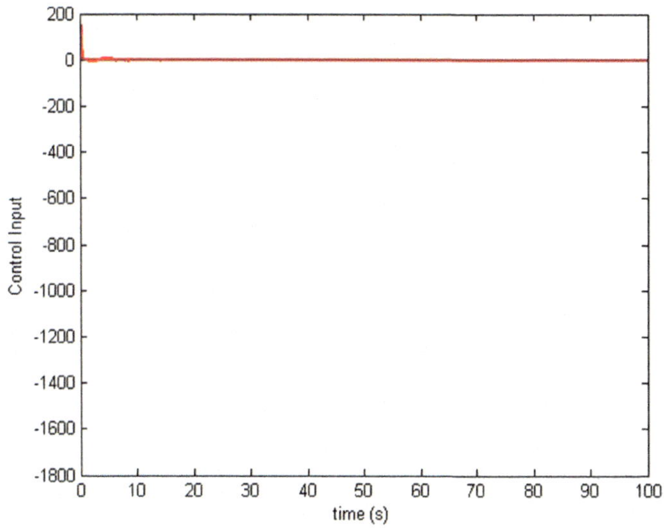

Fig. 6 Control input for TORA system (k = 50)

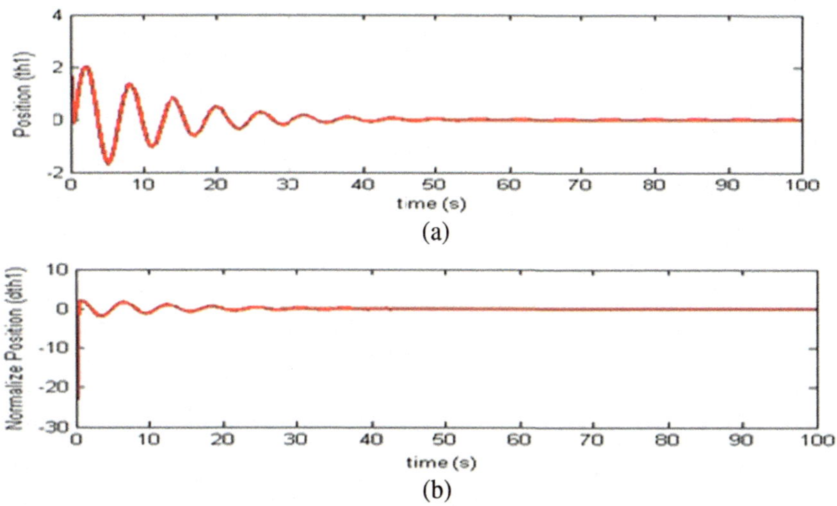

Fig. 7 **a** An actuator angle, **b** derivative of actuator angle (k = 50)

was able to give a robust performance in a wide range of operations and disturbances. The system's stability, accuracy, and transient performance measures like peak overshoot and response time improved with increment type control input to the system.

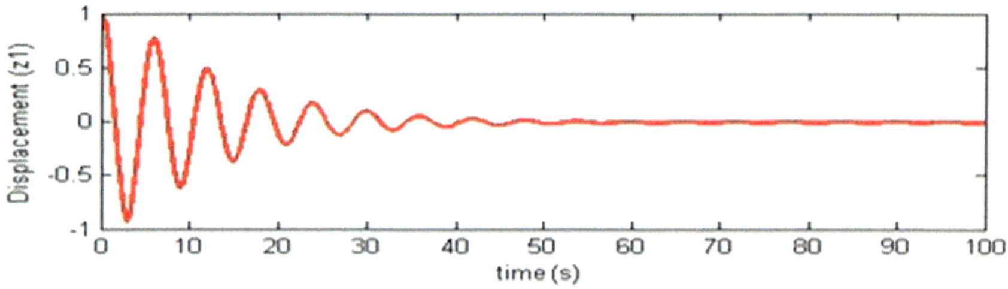

(a)

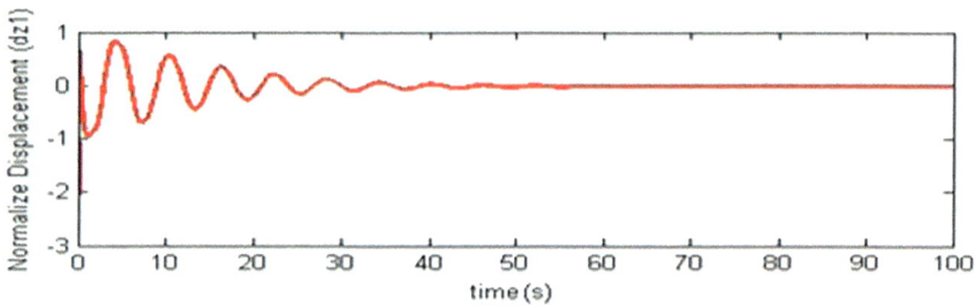

(b)

Fig. 8 **a** An actuator displacement, **b** derivative of actuator displacement (k = 50)

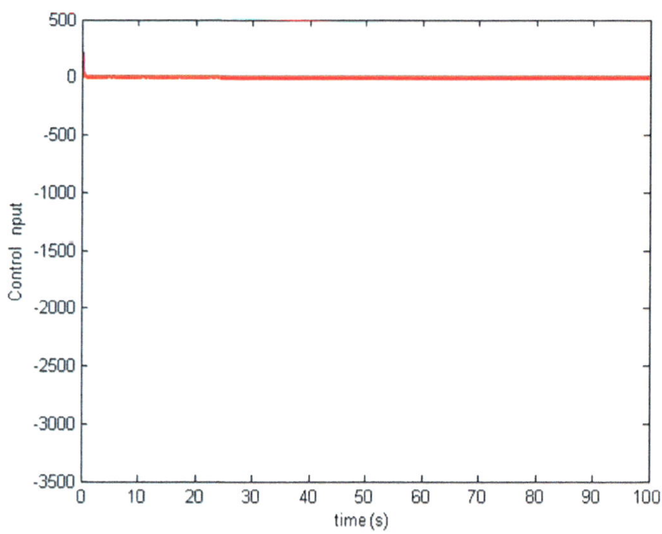

Fig. 9 Control input for TORA system (k = 100)

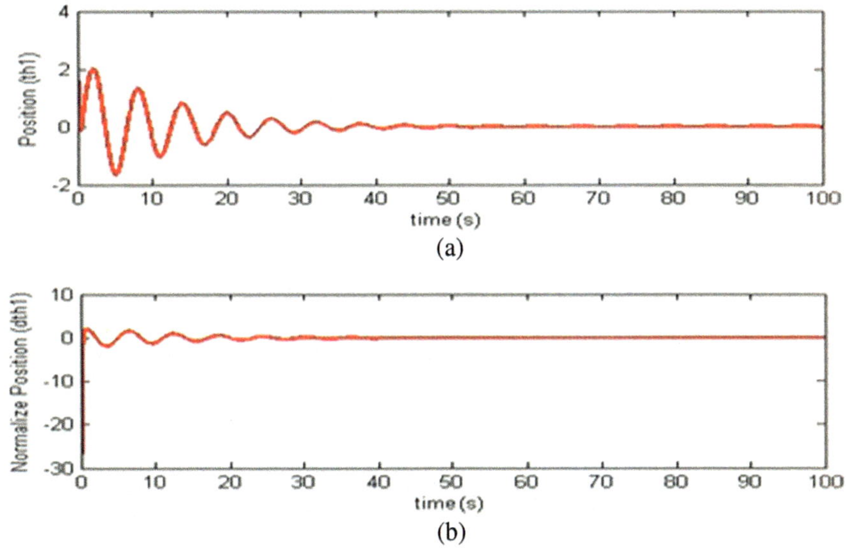

Fig. 10 **a** An actuator angle, **b** derivative of actuator angle (k = 100)

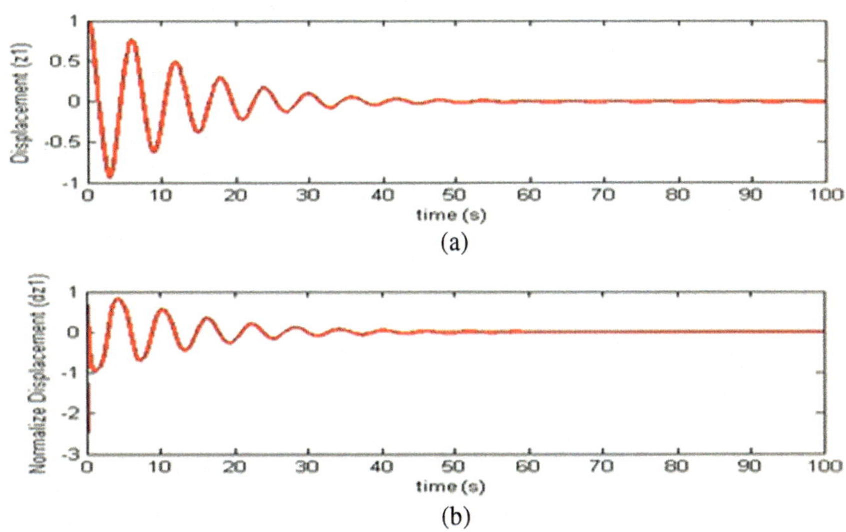

Fig. 11 **a** An actuator displacement, **b** derivative of actuator displacement (k = 100)

Table 1 Performance analysis of TORA system for different values of “k”

Analysis	TORA system		
	k = 10	k = 50	k = 100
Peak overshoot	50	150	250
Settling time	25 s	5 s	0.2 s
Stability	Less	Less	More

References

1. Wang W, Yi J, Zhao D, Liu D (2004) Design of a stable sliding-mode controller for a class of second-order underactuated systems. IEE Proc Control Theory Appl 151(6):683–690
2. Brockett RW (1983) Asymptotic stability and feedback stabilization. Differ Geom Control Theory 27(1):181–191
3. Kawski M (1989) Stabilization of nonlinear systems in the plane. Syst Control Lett 12(2):169–175
4. Seto D, Baillieul J (1994) Control problems in super-articulated mechanical systems. IEEE Trans Autom Control 39(12):2442–2453
5. Lozano R, Fantoni I, Block DJ (2000) Stabilization of the inverted pendulum around its homoclinic orbit. Syst Control Lett 40(3):197–204
6. Angulo MT, Carrillo-Serrano RV (2014) Estimating rotor parameters in induction motors using high-order sliding mode algorithms. IET Control Theory Appl 9(4):573–578
7. Traoré D, Plestan F, Glumineau A, De Leon J (2008) Sensorless induction motor: High-order sliding-mode controller and adaptive interconnected observer. IEEE Trans Ind Electron 55(11):3818–3827
8. Fierro R, Lewis FL, Lowe A (1991) Hybrid control for a class of underactuated mechanical systems. IEEE Trans Syst Man Cybern Part A Syst Hum 29(6):649–654
9. Utkin V (1997) Variable structure systems with sliding modes. IEEE Trans Autom Control 22(2):212–222
10. Bupp RT, Bernstein DS, Coppola VT (1998) A benchmark problem for nonlinear control design. Int J Robust Nonlinear Control. IFAC Aff J 8(4–5):307–310
11. Olfati-Saber R (2002) Normal forms for underactuated mechanical systems with symmetry. IEEE Trans Autom Control 47(2):305–308
12. Olfati-Saber R (2002) Global configuration stabilization for the VTOL aircraft with strong input coupling. IEEE Trans Autom Control 47(11):1949–1952
13. Liu J (2017) Sliding mode control using MATLAB. Beihang University, Beijing, China. Academic Press, Elsevier, pp 317–318
14. Gao W, Wang Y, Homaifa A (1995) Discrete-time variable structure control systems. IEEE Trans Ind Electron 42(2):117–122
15. Bartoszewicz A (1998) Discrete-time quasi-sliding-mode control strategies. IEEE Trans Ind Electron 45(4):633–637
16. Zhang J, Feng G, Xia Y (2014) Design of estimator-based sliding-mode output-feedback controllers for discrete-time systems. IEEE Trans Ind Electron 61(5):2432–2440
17. Zhao L, Huang J, Liu H, Li B, Kong W (2014) Second order sliding mode observer with online parameter indentification for sensorless induction motor drives. IEEE Trans Ind Electron 61(10):5280–5289
18. Smriti S et al (2018) Special issue on intelligent tools and techniques for signals, machines and automation. J Intell Fuzzy Syst 35(5):4895-4899. https://doi.org/10.3233/JIFS-169773
19. Fatema N et al (2021) Intelligent data-analytics for condition monitoring: smart grid applications. Elsevier, p 268. ISBN 9780323855112
20. Yadav AK et al (2020) Soft computing in condition monitoring and diagnostics of electrical and mechanical systems. Springer Nature, Berlin, p 496. https://doi.org/10.1007/978-981-15-1532-3. ISBN 978-981-15-1532-3
21. Smriti S et al (2019) Applications of artificial intelligence techniques in engineering, vol 1. Springer Nature, p 643. https://doi.org/10.1007/978-981-13-1819-1. ISBN 978-981-13-1819-1
22. Gopal et al (2021) Digital transformation through advances in artificial intelligence and machine learning. J Intell Fuzzy Syst Pre-press 1–8. https://doi.org/10.3233/JIFS-189787
23. Jafar A et al (2021) AI and machine learning paradigms for health monitoring system: intelligent data analytics. Springer Nature, Berlin, p 496. https://doi.org/10.1007/978-981-33-4412-9. ISBN 978-981-33-4412-9

24. Aggarwal S et al (2020) Meta Heuristic and evolutionary computation: algorithms and applications. Springer Nature, Berlin, p 949. https://doi.org/10.1007/978-981-15-7571-6. ISBN 978-981-15-7571-6

Fuzzy Logic-Based Range-Free Localization in WSN

Jigyasa Chadha and Aarti Jain

Abstract Received signal strength (RSS)-based location estimation is one of the most acceptable methods for localization in wireless sensor networks. However, am imperfection in the value of RSS due to various fading affects error in the estimated location results. Here, we have proposed to use fuzzy logic to rectify the uncertainty in the value of received signal strength which in turn results in more localization accuracy. The proposed localization algorithm first finds the interval for the maximal possible value of RSS received by an unknown node from the anchor node and then uses a similarity index to find out the best RSS value for an unknown node. The simulation results show that the proposed method leads to better per node localization accuracy and overall network localization accuracy at different node and anchor node densities as compared to the weighted centroid method.

Keywords Fuzzy set · Fuzzy weighted graph · Membership function · Similarity index · Localization · Wireless sensor networks

1 Introduction

In today's era, the localization of sensor nodes in wireless sensor networks (WSNs) is one of the emerging fields of research. In WSNs the main objective of deploying the network is to collect and communicate the sensed information to the intended user or base station for further action. The wireless sensor nodes in WSNs are mostly randomly deployed and are programmed to work autonomously. The location of nodes is a very important parameter in WSNs for the required use of collected information at the user side, thus nodes are programmed to send their location information along with the collected data for effective use at the receiver end. However, mostly the nodes deployed for creating a sensor network are location unaware and they use localization algorithms for estimating their location coordinates [1, 2].

J. Chadha (✉) · A. Jain
Ambedkar Institute of Advanced Communication Technologies and Research, Delhi 110031, India

A. Tomar et al. (eds.), *Machine Learning, Advances in Computing, Renewable Energy and Communication*, Lecture Notes in Electrical Engineering 768,
https://doi.org/10.1007/978-981-16-2354-7_9

The received signal strength-based location algorithms are very popular in WSNs for localizing the unknown nodes. The most popular received strength-based localization algorithms are the methods proposed by Dil et al. [3], Al-Homidan and Fletcher [4], Cheng et al. [5], and Peng and Sichitiu [6]. These methods have mainly used angle-of-arrival, time-of-difference, or RSS-based Euclidean distance calculations for estimating the position of an unknown sensor node.

In this paper, we have proposed to use fuzzy logic to rectify the imprecise value of received signal strength which in turn results in more localization accuracy. The outcome shows that the new proposed fuzzy-aided method leads to better individual node localization accuracy and average localization accuracy of the network at different nodes and anchors densities.

The paper is categorized into different sections. Section 1 presents the introduction to the proposed method in received signal strength-based localization algorithms, and the main factor which adds localization error is the imprecise value of the RSS itself. This imprecise value of the RSS results due to fading effects of the channel, terrain constraints, or environmental issues; Sect. 2 presents the groundwork; Sect. 3 presents the proposed method in detail; Sect. 4 depicts the simulation results, and at last, Sect. 5 concludes the conclusion followed by references.

2 Preliminaries

A. Fuzzy Set [7]

For a universe of discourse X which contains objects denoted by x, the fuzzy set $\tilde{X}$ is defined as: $\tilde{X} = \{(x, \mu_{\tilde{X}}(x))|x \in X\}$. $\mu_{\tilde{X}}(x)$ is the degree of mapping of object x in X and is usually termed as membership function. The range of membership function for a normalized fuzzy set is defined as$\mu_{\tilde{X}} : X \to [0, 1]$. In general, the set theory feature of the fuzzy set is the set whose boundaries are not crisp or sharply defined.

B. Fuzzy Interval-Valued Graph [8]

A fuzzy edge graph $G(V, E, m)$ is defined as the graph consisting of set V with n number of nodes and set of E edges between the nodes and given by $e_{ij} = (n_i, n_j)\ \ E \subseteq V \times V$. Both the vertices and edges are crisp identities; however, for every defined edge, an interval-valued fuzzy number is linked. For an edge (n_i, n_j), the linked interval-valued fuzzy number represents the range of uncertainty in RSS value received by the node n_j from the node n_i at different instances of time t.

C. The Fuzzy Number and Membership Function [7]

The fuzzy number is associated with the fuzzy set as a numerical number associated with the crisp set. However, similar to a fuzzy set, the boundaries of fuzzy numbers are not precise. For example, for a fuzzy set of real numbers, a fuzzy number near to "5" can be represented as follows in Fig. 1.

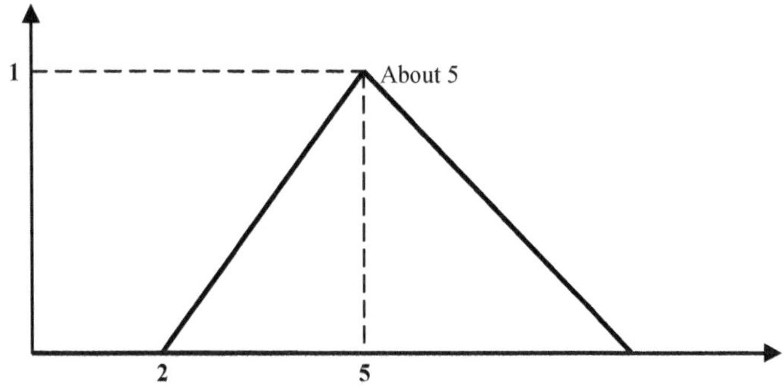

Fig. 1 Fuzzy numbers near to "5"

In this paper, a triangular fuzzy number is considered to represent the uncertainty in the RSS value received by a node from an anchor node at different instances of times. For a triangular fuzzy number, a fuzzy set may be represented by a triangular membership function. For example, to implement a fuzzy set near to 100, by a triangular fuzzy number, the fuzzy membership function can be defined as $A(96, 100, 104)$. In general, the fuzzy membership function for a fuzzy set is defined as triplet $A(a, b, c)$, where c and a represent the upper and lower bound, respectively, for a defined fuzzy number. A triangular fuzzy number $A(a, b, c)$ is defined by the following function:

$$triangle(x : a, b, c) = \begin{cases} 0 & x < a \\ (x - a)/(b - a) & a \leq x \leq b \\ (c - x)/(c - b) & b \leq x \leq c \\ 0 & x < c \end{cases} \quad (1)$$

The appearance of the defined triangular function depends on the chosen parameters "*a, b, c*". Figure 2 shows the triangular function $triangle(x : a, b, c)$. Moreover, the reader may refer to [9–17] for the detailed and advanced-level analysis of the fuzzy logic application and implementation for better understanding.

3 Proposed Localization Algorithm

In an RSS-based localization algorithm, anchor nodes transmit a number of beacon packets containing their own location information to all other nodes at maximum allowable transmission power. After receiving the beacon packets, each node decodes the location information of sender anchor nodes and uses RSS indicator I (inbuilt in sensor node) to estimate its own distance from the respective anchor node. Finally, after receiving beacon (location packets) from three or more anchor nodes, each

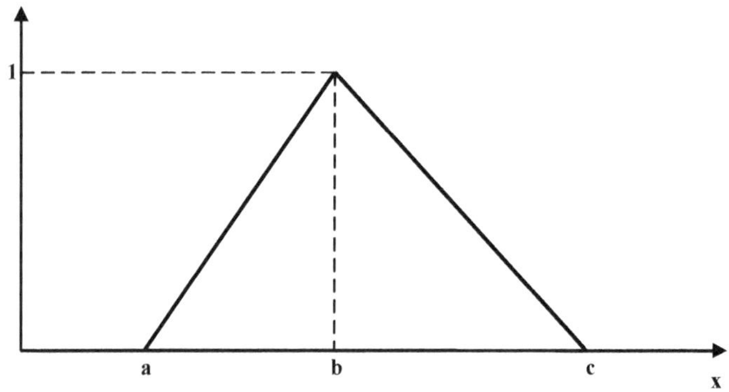

Fig. 2 Fuzzy triangular membership function

unknown sensor node uses the proposed localization algorithm to estimate its own position.

In this proposed method, the RSS values received from anchor nodes at a different instance of time have been modeled as a triangular fuzzy number, $triangle(x : a, b, c)$, where "a" represents the minimum received RSS value, "c" represents the maximum received RSS value, and "b" represents the average of all received RSS values at a different instance of time.

The value of RSS value received by the node n_i from a node n_j at a different instance of time is imprecise in nature due to various fading and shadowing effects. This imprecision is represented by the collection of RSS values by using a fuzzy triangular number.

The choice of anchor nodes for the estimation of position is an important parameter for reducing the localization error in WSNs. In this paper, first of all, on the basis of RSS values received from all anchor nodes, the best possible anchor node position $\overline{R_{\max}}$ for localization is calculated. After that, a similarity index has been used to select the three best anchor nodes for position estimation. The same has been done as follows:

Let two triangular fuzzy numbers (R_1 and R_2) representing the RSS value from anchor node A_1 and A_2 and given by $\overline{R}_1 = (a_1, b_1, c_1)$ and $\overline{R}_2 = (a_2, b_2, c_2)$, then $\overline{R_{\max}}$ is defined as follows:

$$\overline{R_{\max}} = \sup\{R_p \big| R_p = \max(R_p);\ p = 1, 2, \ldots n$$

Here p is the number of anchor nodes.

Hence For $p = 2$,

$$\overline{R_{\max}}(a, b, c) = (\max(a_1, a_2), \max(b_1, b_2), \min(c_1, c_2)) \quad (2)$$

Finally, a similarity index between the maximum RSS value $\overline{R_{\max}}$ and $\overline{R_p}$ is calculated as:

$$S\left(\overline{R_p}, \overline{R_{\max}}\right) = \begin{cases} 0, & \overline{R_p} \cap \overline{R_{\max}} = \emptyset \\ \dfrac{100\left(c - a_p\right)^2}{2\left(c_p - a_p\right)\left[(c - b) + \left(b_p - a_p\right) + \left(c_p - b_p\right)\right]} & \overline{R_p} \cap \overline{R_{\max}} \neq \emptyset \end{cases} \quad (3)$$

The step-wise details of the proposed localization algorithm are as follows:

Step 1: Record the RSS value for p anchor nodes at t different instance of time and compute the fuzzy number values $\overline{R}_p$, $p = 1, 2, ..., n$ corresponding to the possible n RSS values.
Step 2: Find the fuzzy maximum RSS value by using Equation no.
Step 3: Use a fuzzy similarity index as defined in the equation. To give the similarity value $S(\overline{R}_p, \overline{R_{\max}})$ between $\overline{R}_p$ and $\overline{R_{\max}}$ for $p = 1, 2, ..., n$ is computed.
Step 4: Obtain the three anchor nodes with optimum RSS value with top three similarity values $S(\overline{R}_p, \overline{R_{\max}})$.
Step 5: Use the weighted triangulation method to estimate the position of the unknown node.

4 Simulation Results

The proposed fuzzy number-based localization method has been simulated using MATLAB. For simulations, in the proposed method, 100 unknown sensor nodes are deployed randomly in the field of size 100 × 100 units. The noisy distance has been formulated by adding random Gaussian noise $\pm N_i$ in the actual distance between the unknown nodes and anchor nodes. The performance of the proposed method has been compared with the performance of the weighted centroid method. In the proposed method, the weights corresponding to the three selected anchor nodes have been assigned in accordance with the similarity index value. The simulation has been performed with 10, 20, and 30% of anchor nodes to the deployed unknown nodes, i.e., for 10, 20, and 30 anchor nodes.

I. **Per Node Localization Error**

Figures 3, 4, and 5 show the resulted per node localization error for anchor density 10, 20, and 30 anchor nodes for 100 unknown nodes. It can be observed from the results that for the proposed method, the maximum per node error has been lower than the weighted centroid method. This reduction in maximum error can be endorsed to fuzzy number-based modeling of the RSS value.

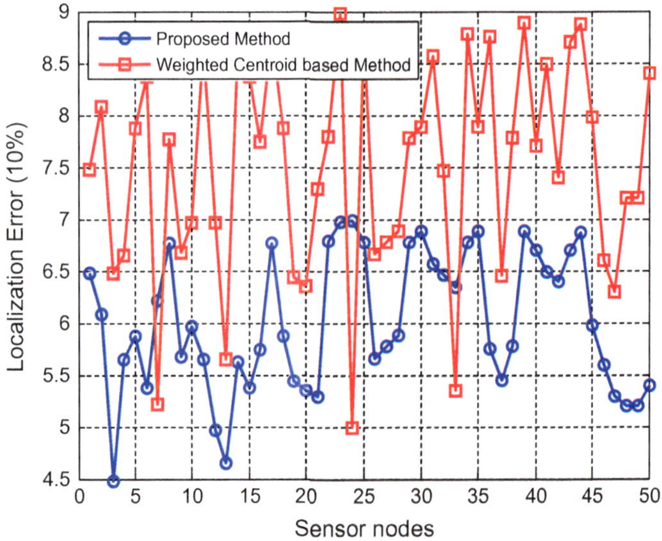

Fig. 3 Localization error at anchor density 10%

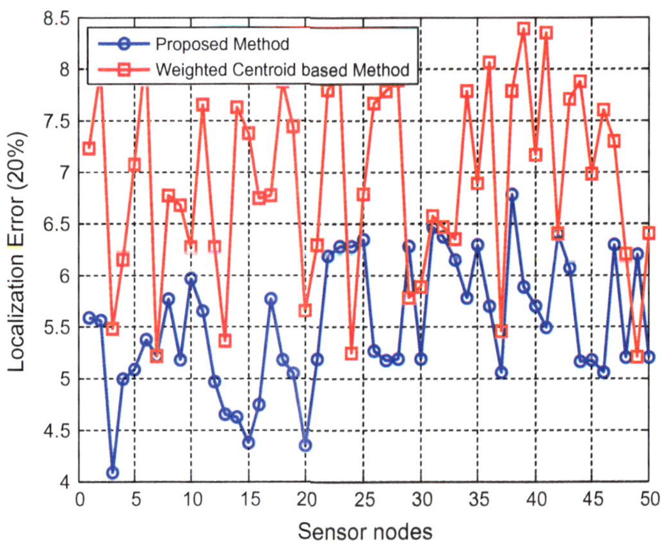

Fig. 4 Localization error at anchor density 20%

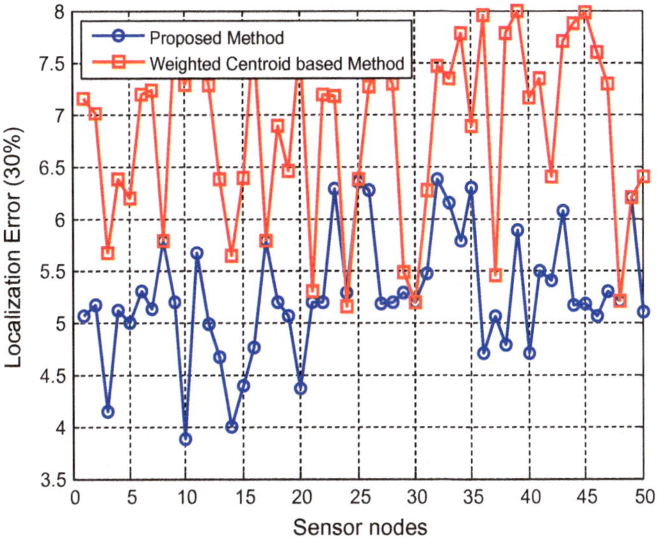

Fig. 5 Localization error at anchor density 30%

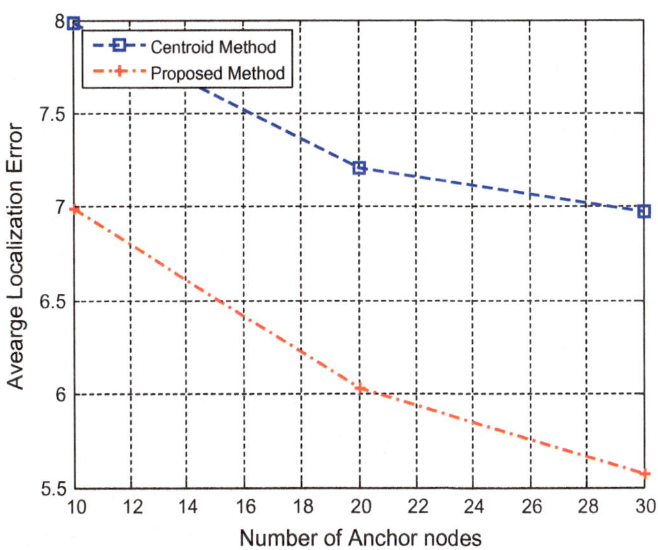

Fig. 6 Overall localization error

II. **Overall Localization Error**

Figure 6 depicts the average value of localization error for the proposed method and the centroid method. It can be observed that the proposed method results in overall lesser localization error for different anchor node densities.

5 Conclusion

It can be concluded that the fuzzy logic-based approximation of RSS value has led to decreased localization error in WSNs. Moreover, the computational complexity of the proposed method is comparable to the centroid method. The proposed method resulted in better per node localization accuracy and overall network localization accuracy at different node and anchor node densities. This work can be extended further for localization using soft techniques like neural networks and particle swarm optimization.

References

1. Xiao J, Zhou Z, Yi Y, Ni LM (2016) A survey on wireless indoor localization from the device perspective. ACM Comput Surv (CSUR) 49(2):25
2. Han G, Xu H, Duong TQ, Jiang J, Hara T (2013) Localization algorithms of wireless sensor networks: a survey. Telecommun Syst 52(4):2419–2436
3. Dil B, Dulman S, Havinga P (2006) Range-based localization in mobile sensor networks. In: European workshop on wireless sensor networks. Springer, Berlin, Heidelberg, pp 164–179
4. Al-Homidan S, Fletcher R (1995) Hybrid methods for finding the nearest Euclidean distance matrix. Recent Adv Nonsmooth Optimizat 1–17
5. Cheng X, Thaeler A, Xue G, Chen D (2004) TPS: a time-based positioning scheme for outdoor wireless sensor networks. In: INFOCOM 2004. Twenty-third annual joint conference of the IEEE computer and communications societies, vol 4, pp 2685–2696. IEEE
6. Peng R, Sichitiu ML (2006) Angle of arrival localization for wireless sensor networks. In: 2006 3rd annual IEEE communications society on sensor and ad hoc communications and networks, vol 1, pp 374–382. IEEE
7. Yen J, Langari R (1998) Fuzzy Logic intelligence, control and information. Pearson. ISBN 978-81-317-05346
8. Akram M (2012) Interval-valued fuzzy line graphs. Neural Comput Appl 21(1):145–150
9. Sharma R et al (2017) Transmission line fault classification using modified fuzzy Q Learning. IET Generat Trans Distribut 11(16):4041–4050. https://doi.org/10.1049/iet-gtd.2017.0331
10. Yadav AK et al (2013) Application of neuro-fuzzy scheme to investigate the winding insulation paper deterioration in oil-immersed power transformer. Electri Power Energy Syst 53:256–271. https://doi.org/10.1016/j.ijepes.2013.04.023
11. Sharma R et al (2017) EMD and ANN based intelligent fault diagnosis model for transmission line. J Intell Fuzzy Syst 32(4):3043–3050. https://doi.org/10.3233/JIFS-169247.
12. Kukker A et al (in Press) Reinforcement learning based genetic fuzzy classifier for transformer faults. IETE J Res. https://doi.org/10.1080/03772063.2020.1732844

13. Mishra S, Malik H (2016) Application of Fuzzy Q Learning (FQL) technique to wind turbine imbalance fault identification using generator current signals. In: Proceedings of IEEE PIICON-2016, pp 1–6, 25–27 Nov. https://doi.org/10.1109/POWERI.2016.8077283
14. Sharma R et al (2020) Fuzzy reinforcement learning based intelligent classifier for power transformer faults. ISA Trans (in Press). https://doi.org/10.1016/j.isatra.2020.01.016
15. Yadav AK et al (2020) A novel hybrid approach based on relief algorithm and fuzzy reinforcement learning approach for predicting wind speed. Sustain Energy Technol Assess 43. https://doi.org/10.1016/j.seta.2020.100920
16. Vigya et al (2021) Renewable generation based hybrid power system control using fractional order-fuzzy controller. Energy Rep 7C:641–653. https://doi.org/10.1016/j.egyr.2021.01.022
17. Jarial RK et al (2012) UV/VIS response based fuzzy logic for health assessment of transformer oil. Elsevier Procedia Eng 30:905–912. https://doi.org/10.1016/j.proeng.2012.01.944. ISSN: 1877–7058

Analysis of Algorithms in Medical Image Processing

Tina, Sanjay Kumar Dubey, Ashutosh Kumar Bhatt, and Mamta Mittal

Abstract Biomedical imaging has gained tremendous attention over the years by aiming to attain better visualization of the abnormalities of structures or disorders in the images of the organs of the body. The motive behind the ensuing procedures of image processing in the medical domain is to improvise the decision-making in the clinical diagnosis and to resolve the issues faced by surgeons to operate the organ by providing the higher dimensional view of internal parts of that organ. The image processing techniques include various steps, including image acquisition, image reconstruction enhancement and analysis. This paper will summarize the imaging (medical) processing techniques used in healthcare informatics along with the challenges faced by the image construal with diverse algorithms, such as image segmentation by edge detection method, k-means etc. The interpretation formed by the physicians after visualizing medical images will improve remote healthcare and help in the treatment of patients in a better manner. The challenges faced by researchers while using a high-technology medical probe to generate better visualization promote their interest in the medical field domain.

Keywords Image processing · Visualization · Image segmentation · Medical imaging modalities · Edge detection

The original version of this chapter was revised: The author name "Mamata Mittal" has been changed to "Mamta Mittal". The correction to this chapter is available at https://doi.org/10.1007/978-981-16-2354-7_58

Tina (✉) · S. K. Dubey
Amity University, Noida, Uttar Pradesh, India

A. K. Bhatt
Birla Institute of Applied Science, Bhimtal, India

M. Mittal
GB Pant Government Engineering College, New Delhi, India

A. Tomar et al. (eds.), *Machine Learning, Advances in Computing, Renewable Energy and Communication*, Lecture Notes in Electrical Engineering 768,
https://doi.org/10.1007/978-981-16-2354-7_10

1 Introduction

Medical image processing is the technique of creating imageries of the internal structure of organs for clinical research, research in medicines or surgical guidance. The images formed by digital imaging will provide a better view of the abnormalities of structures, disorders, approximation closeness of adjacent structures, the area covered by the abnormal structures, density or depth in case of higher dimensional view. The medical imaging modalities include computed tomography (CT), X-ray, Ultrasound, sonography, radiography etc. This procedure incorporates radiological imaging which utilized electromagnetic energies, sonography and thermal imaging. Around several different advances castoff to record data approximately the area and capacity of the body. These procedures take restrictions equated with modulating the produced imageries. Biomedical imaging crops the imageries of the structures of the body deprived of obtrusive techniques [1]. The medical images were delivered using fast processors and because of transformation of the vitalities numerically and reasonably to signals. These signals are then transformed into digitized images. The signal illustrates various kinds of nerves in the body [2]. The role of medical image is very prominent in the field of healthcare. The dispensation of medical images is done over the series of steps involved and implemented by tools used in digital imaging and medical probes. This kind of medical image handling incorporates several kinds of methods and operations such as acquiring an image, storing, enhancing and then visuals correspondingly. The medical image has dimensions that imply measures, like color, illumination, texture, entropy etc. in organs. Medical image processing has numerous advantages, such as the result can be attained fast, better quality, storage and the processing is not so complex. The hindrances of digitized medical images are its misuse for different purposes, failure to sizing again with saving the eminence, the necessity of massive memory and the need for a quicker mainframe for control.

Medical image is the utilization of techniques involved with medical equipment to control the digitized medical view of an organ [3]. This strategy has numerous advantages such as flexibility, versatility, storing the data of image and communiqué. Through the advent of various resizing (image) algorithms or methods, digitized images can be reserved proficiently and it takes numerous processes to achieve digitized images side by side. The dimensional view, either two-dimensional or three-dimensional, can be treated in multiple dimensions. Digital imaging has two primary sorts of images. First is the Raster-image which is portrayed as four-sided prearrangement for frequently experimented qualities called pixels. Digitized imaging has static resolution because of its pixel size. The digitized images may lose their quality while resizing procedure as some information might get missed in between. These are mainly used in photography due to their color and shades. These digitized images include tiff, png and bmp formats. Secondly, vector images can be depicted as a crumpled and determined object that is characterized unambiguously by the system. The vector partakes numerous characteristics, including the width of the line, measurement and shade. Vectors images stand effectively scalable besides these can be recreated in various scales deprived of losing their eminence. Vectors remain appropriate aimed at the configuration, painting and outlines. Image processing in the medical domain commonly works on uniform sampling data with standard x, y and z spatial space, which can be either in 2D or 3D images. At each sampled point, information is generally articulated to vital structure, either signed or unsigned short. The specific importance of the data at the sample point relies upon the type of mode; for instance, a computed tomography scanner gathers radio compactness data points, while brain MRI acquires might fold weight image. Similarly, ultrasound which is a curved array can collect Fan-like images and additionally require differential algorithmic methods to process. Other modalities include different data forms incorporating sheared image because of scaffold tilt, while image acquisition and amorphous mesh.

Various differential equations and mathematical models are the foundation of medical image processing [4]. These representations with esteem to evidence separated after images retains on presence a major procedure for realizing logical development in quantifiable, medicine then other related exploration. Biomedical images are acquired by a scope of strategies over every biological scale. In the present scenario, clinical images might be viewed as a geometrically organized array of samples of data which measure such different physical phenomenon. The broadening degree of medicinal imaging act as a slant to compose our acuities of the biophysical world that has provoked an expressive augmentation in our dimensions to smear novel handling methods via consolidating frequencies of evidence to advanced and composite scientific models of biological dimensions and dysfunction. The mathematical foundations of the biomedical domain provide a platform to develop software procedures that can be combined in the therapies or other relevant systems. Such frameworks encourage the effective transference of imaging methods trivially, invasive health techniques etc. Even though the main reason for medical processing is a clinical diagnosis, but it has importance in other fields also such as medicine, forensics etc. In the case of medicine, it is widely used for research purposes, and numerous processes are utilized; for example, image segmentation and analysis of texture which is utilized for the identification of disorders [5]. In the arena of medicine and healthcare, informatics on the firmness systems of an image can be utilized for communication remotely. However, in the field of forensics, the basic procedures utilized in this field include denoising, edge detection and safety. It depends on the database about the persons and also coordinates the information with the database to characterize the individual personality. Moreover, the reader may refer to [6–12] for other applications of the algorithms in image processing and their analysis.

2 Applications of Medical Imaging

2.1 Computed Tomography (CT)

Among various medical image modalities available, CT scanning is done through X-ray beams which will produce the medical image showing the tissues of the body in multiple slices in different directions. It is used to assess the circulation of body fluids, parenchyma for organs and the musculoskeletal system. In this procedure, the subject's part is placed in the aperture and then the part is scanned with the help of an X-ray tube which rotates in every direction. Neck, head, cardiac, lungs etc. can be scanned using CT scanners.

2.2 Ultrasound Imaging

In this imaging modality, high-frequency sound waves will be used to illustrate the density of the tissues in the organs. Its applications include breast ultrasound, obstetric ultrasound, ultrasound molecular imaging etc. These are done using a medical probe which will send the pulses of the ultrasound into tissue and then these pulses reverberation off tissues with various reflection properties and are recorded and shown as an image.

2.3 Nuclear Medicine

It utilizes radio-isotopes to create a digitized image for elements of various edifices. The pharmaceutical materials label the radioisotopes to guide specific organs. The transmitted photons of patients are gotten in the indicators and then converted into a signal. The individual signal is then changed to an understandable medical image. Numerous sorts of medicine (nuclear) examining are performed, for example, positron, planar, and tomographic emanations.

2.4 X-Ray Imaging

X-ray modality slog on frequency and wavelength that penetrates through the surface of the organ and absorb that in different amount depending upon the density it passes through. Its application includes imaging the structure of bones and is also used to detect cancer via mammography.

2.5 Magnetic Resonance Imaging

This imaging uses radio waves and strong magnetic field that align with the protons to spin in order to analyze abnormalities. Through this modality, joints or ligament structures can be shown with a better visualization. It examines the body structures, especially injuries in the spinal cord, brain, tumors, and can also detect strokes.

3 Steps in Medical Image Processing

Medical image processing is the process of implementing image processing techniques on medical imaging modalities that transform a medical image to a digital form [13]. Operations can be performed on digitized data which will be helpful in enhancing the image and extracting desired information from it. It is a kind of signal indulgence wherein inputs an image, analogous to videos or else photos and attributes associated with that biomedical image. Classically, the framework for image dispensation integrates imageries as 2D signals, while smearing efficiently signal-formulating practices to them. A medical image is a collection of estimations in 2D or 3D space. In these modalities, these medical image intensity works as radioactivity absorption for X-ray and the aural pressure in ultrasound imaging. It is further categorized as a scalar image where each location has a single measured value. Medical imaging has been undergoing an upheaval in the former decade through the approach of more rapidly, increasingly precise, and then fewer intrusive medical equipment. This has determined the prerequisite for concerning software development advancement given a significant force to newer scheming in signals and image processing techniques [14].

The imaging modalities are grouped by various measurements, like image contrast, signal-to-noise (SNR), entropy and illumination. The most common image processing technique histogram, and among them, grayscale histograms are considered vital for improving and evaluating the images. The grayscale histogram is a method to demonstrate the pixel qualities rather than location. This will show whether a medical image either bright or not. The mean of the pixel values is assimilated in the histogram by adding previously formed pixels besides steady bin-altitude, in addition to separating by the whole figure of pixel [15]. Histogram adjustment utilized to compare numerous imageries gained on certain centers. Given the method of work through fluctuating histogram toward developed, even indistinguishable, then the mean estimation of centered pixel intensities is adjusted and assigned to perfect brightness. Somewhat force above or else beneath varieties the imageries are darker or even brighter. The SNR of a medical modality is utilized to relate the degree of the anticipation to contextual signals. Signal-to-noise (SNR) is characterized by way of a proportion of its intensity.

$$\text{Signal to Noise ratio (SNR)} = \text{Power of signal/Power of Noise} \quad (1)$$

The image processing practices are implemented following the steps which are building blocks of the imaging structure, where the main components are image acquisition, pre-processing of an image and visualization which are further categorized into different steps. Figure 1 presents the steps trailed in medical imaging. The first phase of medical image processing includes image acquisition which will help in acquiring the digital image through the tools and techniques using a computer and medical probe [16]. The obtained digital data needs to be converted into desired

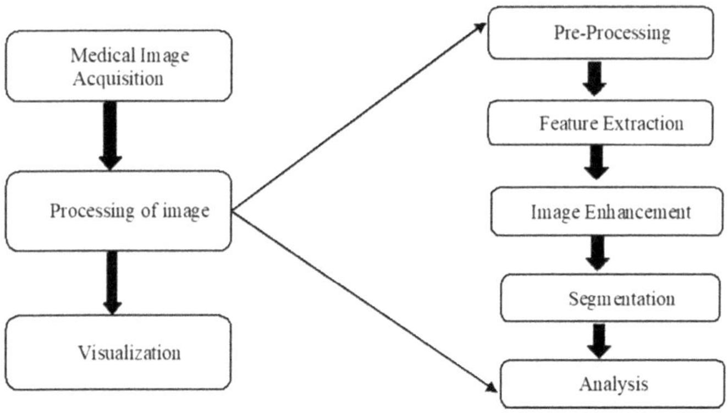

Fig. 1 Steps involved in medical image processing

output as the raw data has inadequacies. To get over such blemishes, data needs to knowledge diverse phases of processing.

In the second phase, image pre-processing carries the steps which will improvise the view of the medical images and helpful in the treatment of patients. With the implementation of algorithms and models on the input of data, the distortions from image features will be suppressed. The medical image enhancement procedure is used to progress the excellence of medical image and then measurable quality by using software aids [3]. This strategy incorporates both goal and abstract improvements. This method incorporates local operations and points also. The neighborhood activities rely upon that region's input pixel values. It is categorized into two parts: first, a spatial technique that right away works on the pixel's value, and secondly, a transformation technique is performed on the Fourier transformation.

Image segmentation is the most important stage as it is defined as a technique of separating the image into numerous segments (Fig. 2). These sections resemble numerous organs, tissue modules or organically related structures [17]. The essential fact of this separation is to make the image meek to analyze and protecting the quality. Noise, low contrast and other imaging ambiguities have often arisen in medical imaging modalities. This process is utilized to trace the border of structure within the image and label the pixel as per its attributes and intensities. This is utilized to make a three-dimensional delineation for experimental determinations. It is utilized in machine discernment, functional as well as anatomical investigations, computer-generated reality perception and abnormality examination [15]. Medical image analysis integrates phases of processing, which are utilized for quantifiable approximations impartial as well as exclusive interpretations for medical images.

The last phase of medical image processing includes image visualization which performs operations for manipulation of matrices, hence ensuing an enhanced output of the medical image. It will manage all the methods and techniques that give the effective communication, storage, transmission, and recovery of imaging data. In a nutshell, it can be summarized as visualization techniques generally used to view the procedures engendered from the algorithms applied.

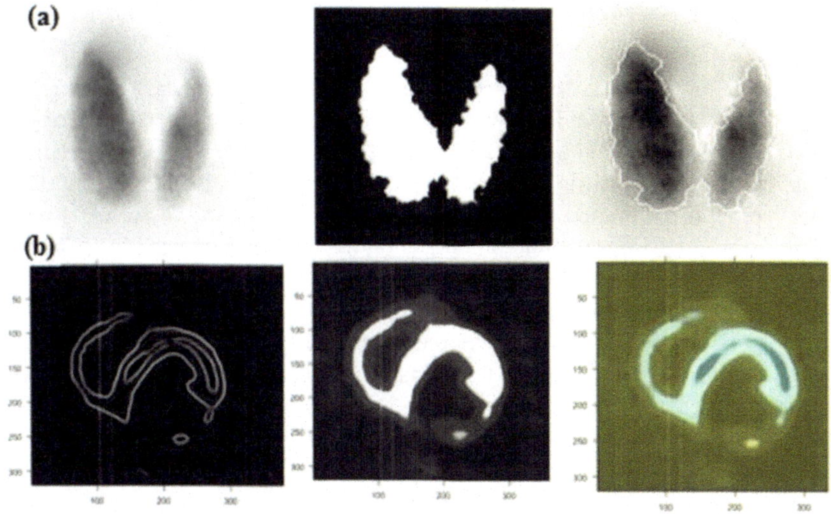

Fig. 2 Image segmentation. **a** Thyroid gland. **b** Heart

4 Algorithms in Medical Image Processing

Imaging modalities have multiple coordinate systems, either two-dimensional or three-dimensional coordinate system, and image registration is one of the main techniques used to transform a set of data points of an image into one coordinate system which can be further utilized for analysis [18]. The dataset points may have different time stamps, depth, intensities or viewpoints etc. Registration methods are castoff to compare or integrate the data attained from the stated dimensions. It is solitary of the furthermost extensively known calculations in imageries modalities and through the utmost GPU executions. Linear interpolation is the main purpose behind the graphics processing unit medical equipment's sustenance for direct addition, in which it is plausible to transmute the image and then the dimensions proficiently [16]. A distinctive procedure is to tenancy a graphic processing unit figure as a resemblance measurement, with most frequent mutual evidence over the medical modalities in equivalence, while the processing unit runs a serial optimization method to discover the boundaries (scaling, rotation, translation) that bounce the top match among the two modalities. Mutual information among two discrete variables (X, Y) can be demarcated as

$$I(X; Y) = \int\int f(x, y) \log f(x, y) f(x) f(y) \, dx \, dy \tag{2}$$

4.1 Feature Extraction

Feature extraction is a mainstream technique utilized for processing medical imaging modalities as it is more viable for stimulating fragments more perceptible. Contrast limited adaptive histogram equalization (CLAHE) methods can be used for feature extraction technique as it is a variant of adaptive histogram equalization (AHE), where there will be limited contrast amplification, which someway reduces noise amplification. CLAHE will limit amplification by cutting or clip the histogram in a predefined range prior tp computation of the distribution function. The imaging modality is part of disconnected regions, and in every region nearby histogram equalization out is functional. At that point, the limits among the sections are dispensed through a bilinear exclamation, and the fundamental target of the technique is to illustrate the points changed inside a neighborhood window by means of the suspicion that a strength inducement inside it is an aloof interpretation of adjacent dissemination of power approximation of the whole image. The neighborhood window is supposed to be unpretentious by the steady variability of intensities among the edges and image centers. In this scenario point distribution transformation is circumscribed about the mean intensity for the neighborhood and it conceals the whole range of strength of the image. Suppose a medical image having dimensions $n * n(n^2)$ engrossed on pixel $P(i, j)$ and the image is detached to fetch alternative sub-image P of $(n * n)$ pixel by way of specified by the ailment stated as:

$$pn = 255 \cdot \left[(\emptyset_w(p) - \emptyset_w(\min))/(\emptyset_w(\max) - \emptyset_w(\min)\right] \tag{3}$$

Based on the global values of the histogram, Poisson distribution (Eq. 3) can be applied, which will enhance the extraction process for gray-level medical image modality defined as:

$$\emptyset_w = inv\left[1 + \frac{\exp(\mu_w - p)}{\sigma_w} \right] \tag{4}$$

where μ_w shows the mean of the local window and variance of the local window.

4.2 SLIC Algorithm

A simple linear iterative cluster (SLIC) algorithm is used especially for high dimensions, such as 4D or 5D spaces, and it performs local clustering between color space and the x and y coordinate system of the pixels. It will enable the regularity in superpixel (defined as the cluster of pixels with the same characteristics) shapes and compactness. This algorithm can be implemented with either grayscale or color images such that the generation of superpixels will depend upon proximity and color similarity of the image plane. When compared with other superpixel strategies, SLIC is rapid, proficient and takes incredible bound observance. The number of superpixel stands the primary boundary, exploiting SLIC forthright also because of the boundary. Another superpixel computation of SLIC operates k-means consortium slant for a group of superpixels [5]. The deliver usually evaluation super-pixel, the matrix S equal to N/k. For color images, CIELAB (a color space), clustering flinches with an instatement where the k primary pixel centers L_i, x_i, y_i, z_i, a_i and T are inspected on a standard framework dispersed S pixel detached. This is a complete desist from fixating a superpixel on edge in addition to moderating the prospect of scattering a superpixel through an uproarious pixel. Distance measure (D_i) in Eq. 4 utilized in the 5D space in normalized function is given as:

$$D_i = d_{\alpha ß} + (m/S) * d_{xy} \tag{5}$$

where $d_{\alpha\beta} = \surd((l_k - l_i)^2 + (a_k - a_i)^2 + (b_k - b_i)^2) d_{xy} = \surd((x_k - x_i)^2 + (y_k - y_i)^2)$.

$d_{\alpha\beta}$ is the sum of lab distances and d_{xy} is the normalized grid interval.

Variable m is utilized to regulate the compactness in the grid for superpixels. The range of m can be between 1 and 20.

4.3 Edge Detection

Medical image processing implements the edge detection method for the analysis of edges or boundaries within medical images. It will be able to identify the disjointedness of the entity in the organ with respect to brightness. Edge detection techniques are primarily utilized for the analysis of the region of interest where massive dissimilarity of intensities occurs in the medical image to find disorders or abnormality. Contemplate the perfect instance of a positive object dark foundation. A physical article stands articulated through projection arranged medical imaging I. IO is a characteristic function of object customs segmentation. Subsequently, the object O is distinguished on the background, and diversities of intensities I are massive on boundaries ∂O. Although it is consequently normal to describe ∂O by way of locus points wherever the middling gradient $|\nabla I|$ remains huge, this technique proposed somewhat extraordinary discrete convolution masks to coarse the gradient of medical image. To avoid noise edges are not able to localize smoothly, so for this, a leveling pre-processing step and post-processing stage are included to assure that the edges are predominantly localized. The heuristic is defined by accepting that the edges are level bends and all added unconditionally, expect that close to an edge the digital images for the structure wherever S is smooth capacity $|\nabla S| = O(1)$, which vanishes on edge; ϕ is defined as $R \to [0, 1]$ a smooth function and ε is the parameter relative to edge width. This technique incorporates Sobel kernels, Prewitt kernel and Roberts kernels which are described underneath.

4.3.1 Sobel Kernels

This edge detection technique incorporates the first derivative of the Gauss kernel with a 3×3 matrix and will be dependent on the mean of central pixels. The derivative for Gaussian kernel stated as G_x and G_y are demarcated in Eqs. 6 and 7, respectively:

$$\begin{bmatrix} a_0 & a_1 & a_2 \\ a_3 & a_4 & a_5 \\ a_6 & a_7 & a_1 \end{bmatrix}$$

$$G_x = (a_2 + 2a_3 + a_4) - (a_0 + 2a_7 + a_6) \tag{6}$$

$$G_y = (a_6 + 2a_5 + a_4) - (a_0 + 2a_1 + a_2) \tag{7}$$

Sobel Kernels reduces the noise of modalities in medical imaging such as in the case of red blood cells digital modality where the neighboring cells are difficult to analyze because of the noise.

4.3.2 Robert Kernels

Robert Kernel edge detection technique regulates the forward difference of two pixels with respect to its closeness. It utilizes gradient operator and first-order derivative (fractional) in Eqs. 8 and 9 to find the high noises in images.

$$\partial f/\partial x = f(i, j) - f(i + 1, j + 1) \tag{8}$$

$$\partial f/\partial x = f(i + 1, j) - f(i, j + 1) \tag{9}$$

4.3.3 Prewitt Kernel

Prewitt Kernel technique is dependent on the central difference between pixels and implemented through gradient magnitude and gradient direction technique (Fig. 3). The matrix has pixels arranged in order where the attributes are defining nearby pixels of the image, with c as constant, and the fractional derivative so obtained for this technique for G_x (Eq. 9) for matrix:

$$\begin{bmatrix} a_0 & a_1 & a_2 \\ a_7 & [i, j] & a_3 \\ a_6 & a_5 & a_4 \end{bmatrix}$$

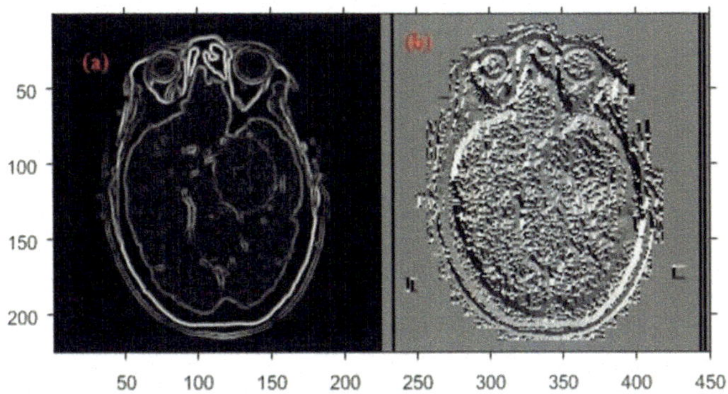

Fig. 3 Prewitt kernel edge detection. **a** Gradient magnitude. **b** Gradient direction

$$G_x = (a_2 + ca_3 + a_4) - (a_0 + ca_7 + a_6) \tag{10}$$

5 Conclusion and Future Scope

In this paper, the concept for medical imaging modalities with respect to the image processing techniques that are applied is given. Various medical imaging modalities are defined with the view of the basic parameter required to analyze the subject's part. Medical image processing consists of steps that are substantial while performing any illustrations in modalities by dividing it into three phases, i.e., image acquisition, processing of medical image and visualization or reporting. The objective of image modality to divide into phases is such that the image can be divided into segments through which the targeted region can be chosen based on the level of pixel intensity in that region. A review of algorithms is depicted which perform image segmentation such that region of interest can be demonstrated based on the level of intensities in an image. Edge detection, feature extraction and SLIC algorithm are the common algorithms implemented with respect to the output of imaging modality required.

The future of medical image processing will include the scanning of every possible higher dimensional visualization for better treatment of patients. Because of advances in healthcare informatics and related innovations, there will be many algorithms in this domain impending time, changing the way the image visualization of medical images is overseen. The future inclination of advanced image processing techniques using deep learning models will increase the scope of the medical domain.

References

1. Hesamian MH et al, Deep learning techniques for medical image segmentation: achievements and challenges. J Dig Imag 32(4):582–596
2. Mohamed Y et al (2019) Research in medical imaging using image processing techniques. intechopen.com/books (June)
3. Haque IRI, Neubert J (2020) Deep learning approaches to biomedical image segmentation. Inf Med 18:100297
4. Liu S, Wang Y et al (2019) Deep learning in medical ultrasound analysis: a review. Engineering 5(2):261–275
5. Ghimire M et al (2020) A-SLIC: acceleration of SLIC superpixel segmentation algorithm in a co-design framework. Adv Intell Syst Comput (May)
6. Fatema N et al (2021) Intelligent data-analytics for condition monitoring: smart grid applications. Elsevier, 268 pp. ISBN: 9780323855112
7. Aggarwal S et al (2020) Meta heuristic and evolutionary computation: algorithms and applications. Springer Nature, Berlin, 949 pp. https://doi.org/10.1007/978-981-15-7571-6. ISBN 978-981-15-7571-6
8. Yadav AK et al (2020) Soft computing in condition monitoring and diagnostics of electrical and mechanical systems. Springer Nature, Berlin, 496 pp. https://doi.org/10.1007/978-981-15-1532-3. ISBN 978-981-15-1532-3
9. Smriti S et al (2019) Applications of artificial intelligence techniques in engineering, vol 1. Springer Nature, 643 pp. https://doi.org/10.1007/978-981-13-1819-1. ISBN 978-981-13-1819-1)
10. Gopal et al (2021) Digital transformation through advances in artificial intelligence and machine learning. J Intell Fuzzy Syst, Pre-press, 1–8. https://doi.org/10.3233/JIFS-189787
11. Jafar A et al (2021) AI and machine learning paradigms for health monitoring system: intelligent data analytics. Springer Nature, Berlin, 2021, 496 pp. https://doi.org/10.1007/978-981-33-4412-9. ISBN 978-981-33-4412-9
12. Smriti S et al (2018) Special issue on intelligent tools and techniques for signals, machines and automation. J Intell Fuzzy Syst 35(5):4895–4899. https://doi.org/10.3233/JIFS-169773
13. Alvarez L, Lions PL, Morel JM, Image selective smoothing and edge detection by nonlinear diffusion. SIAM J Numer Anal 29:845–866
14. Dash S, Acharya BR, Mittal M, Abraham A (2020) In: Kelemen A (ed) Deep learning techniques for biomedical and health informatics. Springer Nature
15. Haque RI, Neubert I (2020) Deep learning approaches to biomedical image segmentation. Inf Med Unlock 18
16. Shruthishree SH, Tiwari H (2017) A review paper on medical image processing. Int J Res Granthaalayah 4(4) (April)
17. Kaur B, Mittal M et al (2018) An improved salient object detection algorithm combining background and foreground connectivity for brain image analysis. Comput Electr Eng 71:692–703
18. Haskins G, Kruger U, Yan P (2020) Deep learning in medical image registration: a survey. Mach Vis Appl 31. Article number: 8
19. Gao F, Yoon H, Wu T, Chu X (2020) A feature transfer enabled multi-task deep learning model on medical imaging. Expert Syst Appl 143:112957 (April)
20. Gandhi M et al (2020) Preprocessing of non-symmetrical images for edge detection. Augment Human Res 5. Article number: 10
21. Zhou T, Ruan S, Canu S (2019) A review: deep learning for medical image segmentation using multi-modality fusion. Array 3–4:100004 (September–December)

22. Kamnitsas K et al (2017) Efficient multi-scale 3D CNN with fully connected CRF for accurate brain lesion segmentation. Med Image Anal 36:61–78 (February)
23. Pandey P et al (2019) Pragmatic medical image analysis and deep learning: an emerging trend. In: Advancement of machine intelligence in interactive medical image analysis, pp 1–18
24. Cao X, Fan J, Dong P, Ahmad S (2020) Image registration using machine and deep learning. In: Handbook of medical image computing and computer assisted intervention 2020, pp 319–342

Technoeconomic Feasibility and Sensitivity Analysis of Off-Grid Hybrid Energy System

Sumit Sharma, Yog Raj Sood, and Ankur Maheshwari

Abstract This paper aims to analyze and configure the optimal configuration of a hybrid renewable energy system to fulfill the electric load requirement of unelectrified rural areas in Chamarajanagar district, Karnataka (India). The renewable energy sources available at the location are solar, wind, and biomass with pumped hydro storage used for the energy-storing purpose. This research paper identified the best-suited design to satisfy the village load demand of a hybrid renewable system in a variety of combinations. Different case studies are compared and evaluated based on the cost of energy (COE), total net present value (NPV), initial capital value (ICV), and operating cost. Also, the behavior of different cost patterns is studied with the variable inputs; therefore, sensitivity analysis is presented in this study.

Keywords Renewable energy sources (RES) · Solar photovoltaic (SPV) · Pumped hydro storage · Diesel generator

Nomenclature

$Cost_{TNP}$	Total net present value
$Cost_{Tann}$	Total annualized value
CRF	Capital recovery factor
rate	Annual interest rate (%)
life	Plant life (years)
T_{Load}	Total load served (kWh/year)
P_S	Solar rated capacity (kW)
D_S	Solar derating factor (%)
I_r	Solar radiation in present time (kW/m^2)
$I_{r,STC}$	Solar radiations at standard conditions (1 kW/m^2)
α_T	Temperature coefficient (%/°C)

S. Sharma (✉) · Y. R. Sood · A. Maheshwari
Electrical Engineering Department, National Institute of Technology, Hamirpur 177005, India
e-mail: sumitsharma8882@nith.ac.in

A. Tomar et al. (eds.), *Machine Learning, Advances in Computing, Renewable Energy and Communication*, Lecture Notes in Electrical Engineering 768,
https://doi.org/10.1007/978-981-16-2354-7_11

T_S	Solar cell temperature in present time (°C)
$T_{S,STC}$	Solar temperature under standard conditions (25 °C)

1 Introduction

The usage of renewable energy sources (RES) has been increasing in recent times because of the electricity problem that arises in remote communities, where grid-connected electricity is not possible. A hybrid energy system is characterized as an independent and organized cluster of various RES. In modern society, the energy is put next to the fundamental humanities necessities, such as food, clothes, and shelter. The electricity produced from conventional energy sources is costly and creates a lot of environmental problems. Many researchers across the world studied the working of the hybrid energy system (HES) in rural communities and observed that HES would be the best solution for electricity supply in remote areas. Kumar S. et al. presented the cost assessment and sizing estimation of the HES in the western Himalaya region in Himachal Pradesh, India [1]. Different types of control methodologies used in the evaluation of HES are discussed by Subho Upadhyay [2]. As RES are available for electricity production, but mainly the solar, wind, hydro, and biomass are available in abundance and generally used for electricity production. There are different storage technologies like batteries, pumped hydro storage, hydrogen fuel cell, etc. R. Singh et al. discussed the different storage technologies and system architecture of the HES [3]. Among all the storage technologies, batteries are mostly used because of their availability, but there is a lot of drawbacks of batteries. In this research paper, the technology used for storing the electrical energy is pumped hydro storage. Sambeet Mishra et al. focused on the benefits and advantages of HES in comparison with an islanded system [4]. Due to the dependency on fossil fuels by the majority of the population, a stage arises of higher electricity costs and maximum environmental problems. Therefore, there will be a need to go on renewable energy sources for electrical energy production. Bagul, A. D et al. discussed the benefits of HES in electricity generation in rural areas [5]. In [6], the author discussed the sensitivity analysis using the various types of storage systems. Sharma S., et al. focused on the technoeconomic assessment of green microgrid systems in the simulation atmosphere [7]. Moreover, the reader may refer to [8–15] for other applications of technoeconomic feasibility and sensitivity analysis of off-grid hybrid energy system.

This paper presents the cost assessment and sizing analysis of a hybrid system with the available resources, i.e., solar, wind, biomass, and converter with a pumped storage system for storing purposes. Section 2 presents the input parameters and resource variables data used in the study. Sections 3 and 4 present the economic modeling and results of this research paper.

2 Input Parameters and Resource Variables

In view of the electricity requirements of the village cluster, the electrical load demand of the selected location has been calculated. The location chosen for the study is a cluster of three unelectrified villages situated in district Chamarajanagar, Karnataka (India) [16]. The majority of the population is living in the hilly areas and forests where the electricity supply is not feasible. RES available at the location are solar, wind, biomass, and pumped hydro storage. The daily load demand for the selected location is present in Fig. 1. The electrical load in kWh per day is approximated as 724.80 kWh/day, with a peak load demand of 108.60 kW.

The electrical energy produced from the solar photovoltaic is due to solar rays. The solar system generated the DC power output but the load is AC electrical; therefore, a bidirectional converter is used in the system for the conversion of DC electrical output into AC electrical power. A good amount of wind speed is necessary at the site for electrical energy generation. To avoid emission problems diesel generator is replaced by the biomass generator. Table 1 shows the value of resource components, i.e., solar radiations (kWh/m^2/day), wind speed (m/s) and available biomass (tonnes/day). Figures 2 and 3 present the block diagram of different components and flowchart describing the purposed analogy in HES, respectively.

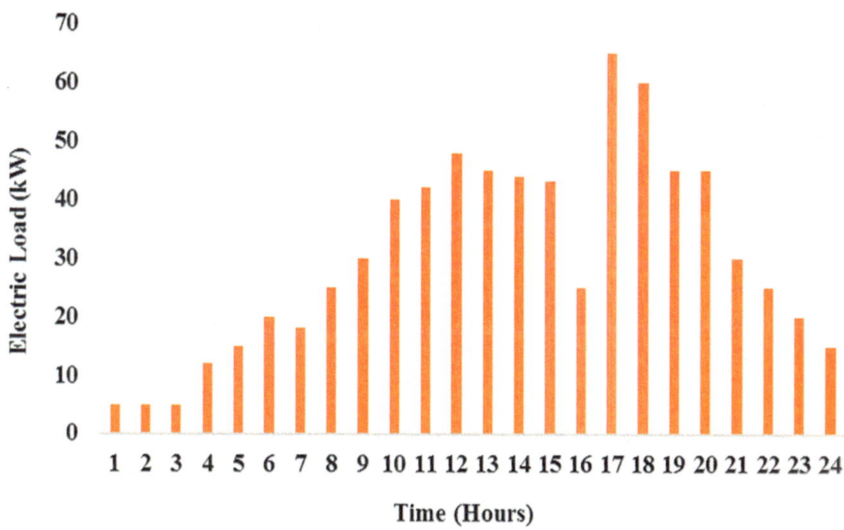

Fig. 1 Electric load profile for 24 h

Table 1 Resource parameters values of solar, wind, and biomass

Months	Solar radiations ($kWh/m^2/day$)	Wind speed (m/s)	Available biomass (tonnes/day)
Jan	5	2.8	3
Feb	5.8	2.7	4
Mar	6.5	2.6	5
Apr	6.4	2.5	6
May	6.3	2.7	7
Jun	5	3.4	8
Jul	4.8	3.2	8
Aug	4.9	3	7
Sep	5	2.4	6
Oct	4.9	2.2	5
Nov	4.8	2.3	4
Dec	4.9	2.8	3
Annual average value	5.36	2.72	5.50

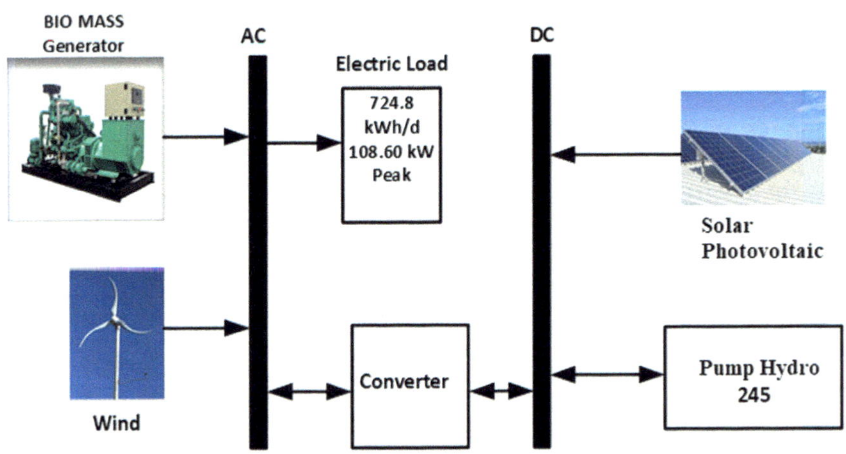

Fig. 2 Block diagram of hybrid energy system containing solar, wind, biomass generator, pumped storage, and converter

3 Economic Modeling

The simulation model is constructed using the net present value which corresponds to the total installation and operation costs of the system over its lifetime. The total NPV contains many costs like replacement, maintenance, fuel, and capital costs. The net present cost is expressed as [17]

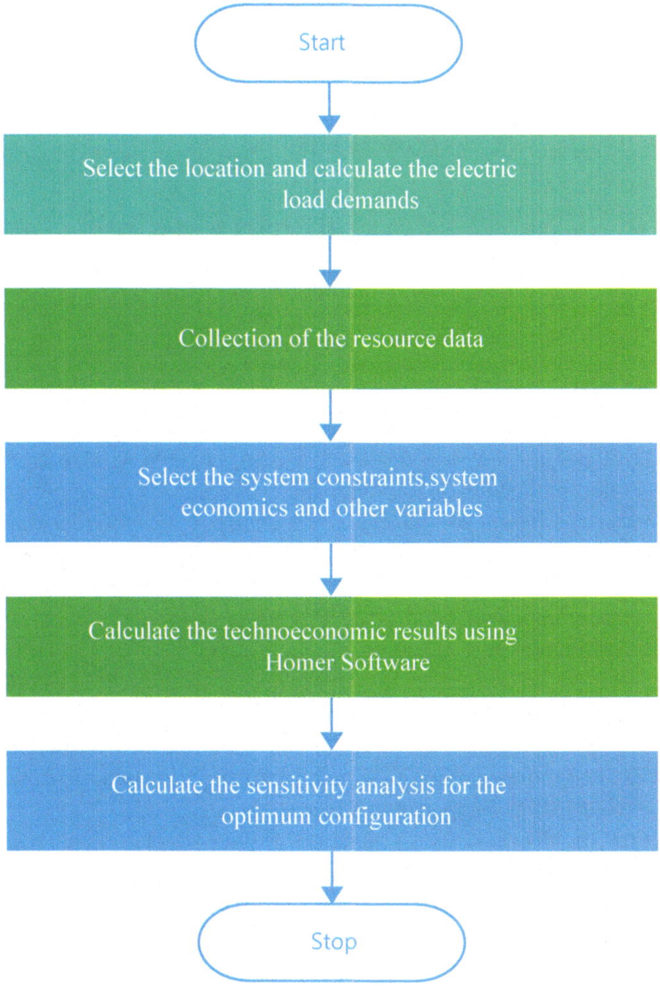

Fig. 3 Flowchart describing the purposed analogy

$$\mathrm{Cost}_{\mathrm{TNP}} = \frac{\mathrm{Cost}_{\mathrm{Tann}}}{\mathrm{CRF(rate,\ life)}} \tag{1}$$

Levelized COE is calculated as the average cost per kWh of energy generated by the system. CRF is dependent on the project life and annual interest rate and is expressed as [18]

$$\mathrm{COE} = \frac{\mathrm{Cost}_{\mathrm{Tann}}}{\mathrm{T}_{\mathrm{Load}}} \tag{2}$$

$$\text{CRF} = \frac{\text{rate}(1 + \text{rate})^{\text{life}}}{\text{rate}(1 + \text{rate})^{\text{life}} - 1} \tag{3}$$

SPV is generating electrical energy from the sun's rays. The mathematical equation used for the solar power output is [19]

$$P_{SPV} = P_S D_S \left(\frac{I_r}{I_{r,STC}}\right)\left[1 + \alpha_T \left(T_s - T_{S,STC}\right)\right] \tag{4}$$

4 Results and Discussion

The HES developed for the chosen site consists of the SPV, wind turbine, biomass gasifier generator, pumped hydro storage, and converter. The input parameters and cost assessment of different components are taken from [4]. Different combinations have been obtained from the simulation as shown in Table 2. There are many results obtained from the Homer software; only nine combinations have been shown here because of close comparisons and observed that from the different combinations, case 8 has the minimum cost of energy, i.e., 26.65 ₹/kWh. Therefore, it can be said that case 8 would be the optimal solution for the electrification of rural communities. As has been found, the optimal solution consists of 238 kW of SPV, 9 units of pumped storage, and 106 kW of converter. There are no wind turbine and biomass gasifier generators in the optimal solution obtained for rural areas. Table 3 shows the cost summary of the different components in HES.

As we know, the electrical energy generated from renewable energy resources are weather-dependent. In the proposed hybrid energy system, there is only SPV, pumped storage, and converter; therefore, the more sensitive parameter is solar radiations. Table 3 shows the sensitivity analysis of HES. As seen from the table that as the value of solar radiation increases, the cost values decrease. Therefore, it can be concluded that for the higher solar radiations there will be very less cost of electricity, and vice versa (Table 4).

5 Conclusion

This paper examines the technoeconomic aspects of renewable energy hybrid systems in order to find an ideal solution for fulfilling a rural community's demand for electrical loads. After the evaluation, it was found that on the basis of available local resources, SPV and biomass gasifier-based generators along with pumped hydro storage system are selected for providing electricity. Along with sizing optimization, the sensitivity analysis has been done and observed that with the increase in the sensitive parameters, levelized COE, total NPC, OC, and ICC decrease, and vice

Table 2 Different combinations obtained for the hybrid energy system

S. no.	PV (kW)	Wind (kW)	Biomass generator (kW)	Pumped storage (Quantity)	Converter (kW)	COE (₹/kWh)	NPV (₹*10^6)	OC (₹/year)	ICV (₹*10^6)
1	187	1	500	5	114	48.99	174.21	765572.43	164.54
2	199	4	500	5	120	51.74	177.19	608333.40	169
3	189	8	500	5	121	52.19	178.68	734675.56	169
4	185	3	500	5	132	52.49	179.43	1070446.41	166.02
5	242	13	0	9	105	29.18	99.76	1684363.59	78.17
6	181	15	500	6	130	55.02	188.36	1198947.63	172.72
7	187	21	500	6	111	55.17	188.36	928022.99	176.45
8	238	0	0	9	106	26.65	91.57	1614678.11	69.98
9	239	3	0	9	99.1	27.17	93.06	1626962.41	72.22

Table 3 Cost summary of different components in the proposed hybrid energy system

S. no.	Component	Capital cost (₹)	Replacement cost (₹)	Operation and maintenance cost (₹)	Salvage value (₹)	Total (₹)
1	Generic 245 kWh pumped hydro	14,741,100	475927.581	17324165.09	34395.9	32506796.77
2	Generic flat plate PV	53227563.68	0	2293667.512	0	55521231.19
3	System converter	2365268.311	1003521.229	0	188872.9495	3179917.334
4	System	70333931.99	1479449.554	19617832.6	223268.8495	91208312.07

Table 4 Sensitivity analysis of hybrid energy system with the change in solar radiations

S. no	Sensitive variable solar radiation	PV (kW)	Pumped storage (Quantity)	Converter (kW)	COE (₹/kWh)	NPV ($₹*10^6$)	OC (₹/year)	ICV (₹)
1	5.36	238	9	106	26.6531	91.5735	1614671.6	70333957.3
2	7	211	8	103	23.74955	81.1505	1439937.45	62481194.65
3	8	187	8	118	22.1861	75.939	1,430,929	57421274.85
4	9	189	7	112	21.21825	72.2165	1276147.45	56096437.1
5	10	174	7	129	20.3993	69.2385	1274956.25	53134816.1
6	12	169	6	106	18.7614	64.027	1104912.45	49962650.5
7	15	166	5	104	17.49575	59.56	948641.9	47563871.5

versa. One major finding of this study is the usage of biomass and pumped hydro storage in electricity generation, which brings a pollution-free environment into the system. This current proposal can improve living standards and boost the economic activity of rural communities in the state of Karnataka, India.

References

1. Kumar S, Kaur T, Arora MK (2019) Resource estimation and sizing optimization of P.P.V./micro hydro-based hybrid energy system in rural area of Western Himalayan Himachal Pradesh in India. Energy Source Part A 41:2795–2807
2. Upadhyay S, Sharma MP (2014) A review on configurations, control and sizing methodologies of hybrid energy systems. Renew Sustain Energy Rev 38:4763. ISSN 1364-0321
3. Singh R, Bansal RC (2018) Review of HRESs based on storage options, system architecture and optimization criteria and methodologies. IET Renew Power Gener 12(7):747–760

4. Mishra S, Panigrahi CK, Kothari DP (2016) Design and simulation of a solar–wind–biogas hybrid system architecture using HOMER in India. Int J Ambient Energy 37(2):184–191
5. Bagul AD, Salameh ZM, Borowy B (1996) Sizing of a standalone hybrid wind- photovoltaic system using a three-event probability density approximation. Sol Energy 56(4):323–335
6. Sharma S, Sood YR (2020) Optimal planning and sensitivity analysis of green microgrid using various types of storage systems. Wind Eng
7. Sharma S, Sood YR, Maheshwari A (2021) Techno-economic assessments of green hybrid microgrid. In: Gupta OH, Sood VK (eds) Recent advances in power systems. Lecture notes in electrical engineering, vol 699. Springer, Singapore
8. Aggarwal S et al (2020) Meta heuristic and evolutionary computation: algorithms and applications. Springer Nature, Berlin, p 949. https://doi.org/10.1007/978-981-15-7571-6. ISBN 978-981-15-7571-6
9. Yadav AK et al (2020) Soft computing in condition monitoring and diagnostics of electrical and mechanical systems. Springer Nature, Berlin, p 496. https://doi.org/10.1007/978-981-15-1532-3. ISBN 978-981-15-1532-3
10. Gopal et al (2021) Digital transformation through advances in artificial intelligence and machine learning. J Intell Fuzzy Syst Pre-press 1–8. https://doi.org/10.3233/JIFS-189787
11. Fatema N et al (2021) Intelligent data-analytics for condition monitoring: smart grid applications. Elsevier, p 268. ISBN 9780323855112
12. Smriti S et al (2018) Special issue on intelligent tools and techniques for signals, machines and automation. J Intell Fuzzy Syst 35(5):4895–4899. https://doi.org/10.3233/JIFS-169773
13. Jafar A et al (2021) AI and machine learning paradigms for health monitoring system: intelligent data analytics. Springer Nature, Berlin, p 496. https://doi.org/10.1007/978-981-33-4412-9. ISBN 978-981-33-4412-9
14. Sood YR et al (2019) Applications of artificial intelligence techniques in engineering, vol 1. Springer Nature, p 643. https://doi.org/10.1007/978-981-13-1819-1. ISBN 978-981-13-1819-1
15. Sood YR et al (2019) Applications of artificial intelligence techniques in engineering, vol 2. Springer Nature, p 647. https://doi.org/10.1007/978-981-13-1822-1. ISBN 978-981-13-1822-1
16. Vendoti S, Muralidhar M, Kiranmayi R (2020) Techno-economic analysis of off-grid solar/wind/biogas/biomass/fuel cell/battery system for electrification in a cluster of villages by HOMER software. Environ Dev Sustain
17. Krishan O, Sathans (2018) Design and techno-economic analysis of a HRES in a rural village. Procedia Comput Source 125:321–328
18. Rajanna S (2016) Integrated renewable energy system for a remote rural area. PhD thesis. Alternate Hydro Energy Center, IIT Roorkee
19. Aziz AS (2017) Techno-economic analysis using different types of hybrid energy generation for desert safari camps in U.A.E. Turk J Electr Eng Comput Sci 25:2122–2135

Analysis of the Impact of AC Faults and DC Faults on the HVDC Transmission Line

Deepak Singh, D. Saxena, and Rajeev Kumar Chauhan

Abstract Over the course of the year, the demand for electricity for industrial, residential, commercial, agricultural, and other purposes has been much greater. There is a need to transfer tremendous power over distances that are much too long. These days, the HVDC system is being proposed. For overhead transmission lines over 700 km and underground cables over 40 km, HVDC is economical. Yet, HVDC lags behind HVAC from a security point of view. Faults may occur in the HVDC system, and these seriously affect the converter and station inverter. To enhance system reliability, it is essential to isolate these faults. To isolate these faulty segments, there is a need to understand fault characteristics. Gird-side (AC-side) faults were analyzed in this paper as SLG fault, LLG fault, and LLLG fault. And HVDC-side faults were also analyzed in this paper as line to line and line to ground faults. In this paper, the PSCAD/EMTDC-Simulink model has been used to demonstrate the different fault characteristics.

Keywords VSC HVDC · Fault analysis · AC fault · DC fault · PSCAD · Reliability

1 Introduction

Where the gap is greater than 40 km for underground cables and 700 km for overhead lines, the HVDC system is economical to transmit bulk power [1]. For bulk power transfer, HVDC will be used in the future [2]. The capital investment of the HVDC system is more than the HVAC system [3]. In the HVAC system, there is a synchronization problem when it is connected to two areas. But in the HVDC system synchronization problem is not there.

D. Singh · D. Saxena (✉)
Department of Electrical Engineering, MNIT, Jaipur, India

R. K. Chauhan
IMS Engineering College, Ghaziabad, India

A. Tomar et al. (eds.), *Machine Learning, Advances in Computing, Renewable Energy and Communication*, Lecture Notes in Electrical Engineering 768,
https://doi.org/10.1007/978-981-16-2354-7_12

It is simpler to incorporate renewable energy into the HVDC system than the HVAC system [4]. The reliability of the HVDC system is one of the very interesting research areas. The HVDC system can be affected by several types of faults, such as direct current faults and alternating current faults [3]. In this case, there is a requirement to disconnect this faulty area from the system to improve system reliability. DC faults are more severe faults than AC faults. In the HVDC system, DC faults directly attack converter stations.

It is easier to design a circuit breaker in the HVAC system because there is an alternating current that changes polarity in every cycle; then when the fault occurs the circuit breaker is open, the arc is produced, and the current goes to zero, thus the arc automatically extinguishes. But in the HVDC system, there is no concept of alternating current, so in the HVDC, the system design of a circuit breaker is a very challenging task [5]. Moreover, advanced-level analysis of the impact of transmission line faults is represented in the recently published work of the digital domain [6–23].

This paper analyzes AC faults (LG, LLG, and LLLG) and DC faults (pole to pole). The VSC is used in this paper for simplicity. It introduces the VSC-based HVDC transmission line. PWM control circuit is used for VSC-based HVDC system. To simulate the VSC-based HVDC system, PSCAD/EMTDC software is used in this paper.

2 Basic Component of HVDC System

Figure 1 shows the two-terminal VSC-HVDC model.

(i) AC side: AC side shows AC power source and impedance. This is on both sides, sending end and receiving end.

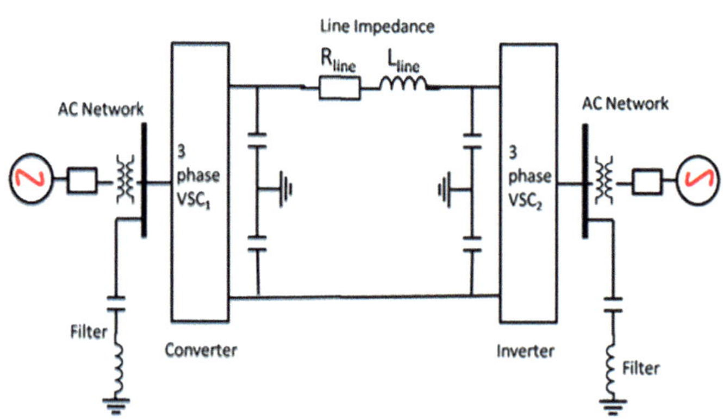

Fig. 1 Basic HVDC diagram

(ii) Passive filter: This is used to reduce harmonics on the AC side. In PWM control, harmonics are of very higher order. So, using a passive filter this can be eliminated.
(iii) Transformer: On the sending end, a step-up transformer is used which is of star-delta type with neutral ground.
(iv) Inverter/converter: Both sides of the converter station (sending end and receiving end) are working which converts AC to DC, or vice versa.
(v) DC side filter: On the DC side, a capacitor filter is used to reduce ripple.

3 Modeling of HVDC

HVDC transmission system requires a converter station. In this paper, VSC is used. This converter uses self-turn-off devices such as IGBT, GTO, etc. This is used because there is no requirement for the commutation circuit. In the voltage source converter thyristor is used, because in this case, it is self-turn off as reverse voltage is applied on the thyristor automatically [4]. In Fig. 1, a typical VSC-based HVDC transmission system is shown. Table 1 shows the system parameters list.

In VSC-HVDC if there is a requirement to improve the voltage rating of the converter, then connect switches in series according to the rating of the converter [4]. This is shown in Fig. 2.

Table 1 HVDC parameters and values

Parameter	Value
Rated AC voltage (line to line)	220 kV
Steady-state frequency	50 Hz
Rated DC line voltage	800 kV
Sending end transformer ratio	220 kV/540 kV
Receiving end transformer ratio	540 kV/220 kV
Switching frequency	5 kHz
Power rating	10 KVA

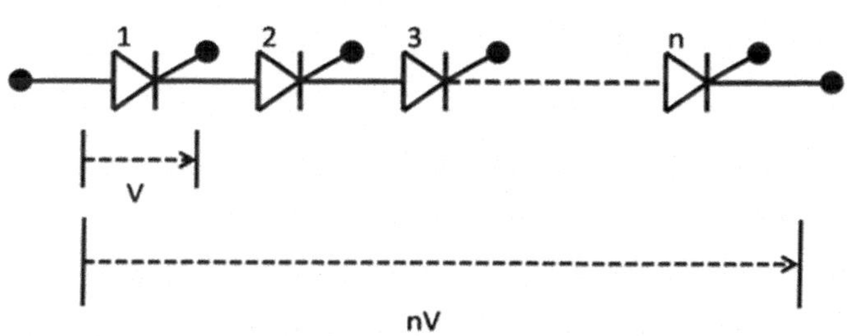

Fig. 2 Using string in VSC

4 DC Voltage Control

By changing the firing angle of the converter, controlling of DC-side voltage can be done. In VSC there is a relation between the output voltage and the firing angle, as shown in Eq. (1).

$$V_0 = \frac{3V_{ML}}{\pi} \cos \alpha \tag{1}$$

where V_0 and V_{ML} are output of DC voltage and maximum line voltage on AC side, respectively. α is the firing angle.

5 Faults in the HVDC System

In the HVDC system, there are various faults that exist, such as direct current faults, converter faults, and alternating current faults [24]. It is necessary to analyze DC faults because DC faults are the most dangerous faults than AC faults. DC line faults can be of two types. The line to line fault is the first one and the second one is the line to ground fault.

In DC line fault, line failure has a high probability. These faults are more powerful and are recurrent faults.

The AC fault can also occur in the HVDC system. On the AC side, these faults can occur and they can affect the DC line.

5.1 Direct Current Line Fault

HVDC transmission line faults cause external mechanical stress, lightning strikes, overloading of lines, and pollution. In the DC transmission system, two types of common fault can exist.

5.1.1 Line–Line Fault

When both wires of the HVDC line are connected together then the line–line fault is created. DC line to ground fault occurrence probability is very high than the line to line fault. Figure 3 shows an equivalent circuit of line to line fault.

DC short circuit current flow in three stages:

(a) Discharging of capacitor: In this stage, when the DC pole to pole fault occurs then the DC filter capacitor discharge through transmission line impedance. This circuit is equivalent to a mess loop without any source. When the fault

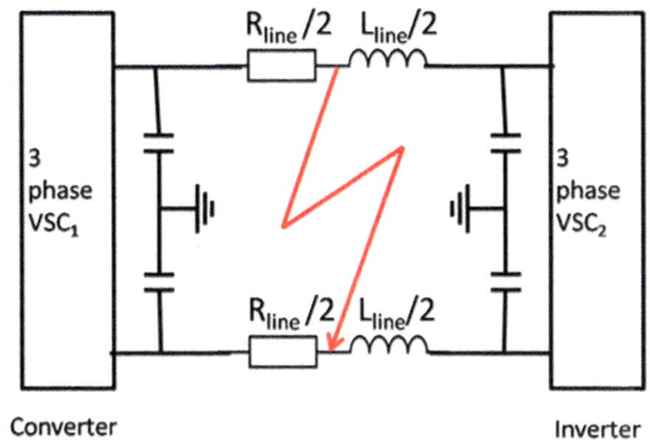

Fig. 3 HVDC line–line fault

occurs then capacitors are in the discharging stage and the capacitor voltage decreases. The capacitor discharging current is shown in Eq. (2).

$$I_f = I_0 e^{-(\frac{R_{\text{line}}}{L_{\text{line}}})t} \tag{2}$$

where I_f, R_{line}, and L_{line} are fault current and transmission line impedance parameters.

(b) Diode freewheeling stage: When the capacitor voltage decreases and reaches less than the grid voltage then the diode freewheeling stage starts, and in this stage, IGBT should be in a turn-off state for protection purpose. When the system goes in the freewheeling stage then the rectifier works as an uncontrolled rectifier.

(c) Grid feeding stage: In this stage, the converter works as a current source providing fault current to the line shown in Fig. 4. During the grid feeding stage the equations of sending end voltage and current are:

$$\begin{aligned} e_{SE} &= E_{SE}\sin(\omega_s t + \theta) \\ i_{SE} &= I_{SE}\sin(\omega_s t + \theta - \phi) + I_0 e^{-(\frac{R_{\text{line}}}{L_{\text{line}}})t} \end{aligned} \tag{3}$$

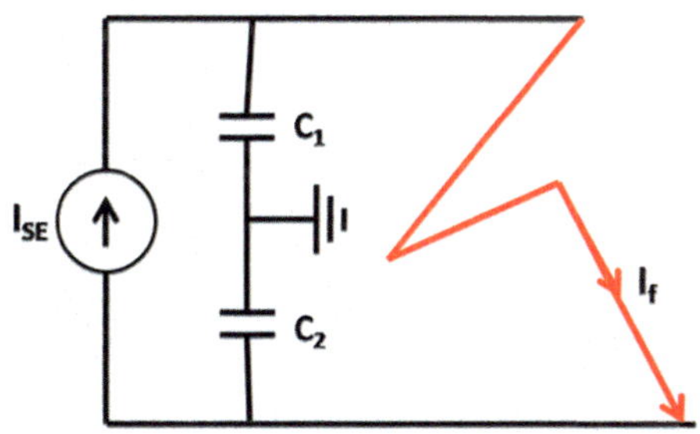

Fig. 4 Fault current feeding by grid

where E_{SE} is the magnitude of grid sending end voltage, I_{SE} is the magnitude of grid sending end current, ϕ is the phase angle of the grid, θ is the grid angle, and ω is the angular frequency.

If made to travel for long distances the voltage profile wherever it flows is affected. Thus, it is always recommended to provide support near to the point where needed.

When the initial condition is considered then Eq. (3) can be written as:

$$i_{SE} = I_{SE}\sin(\omega_s t + \theta - \phi) + I_{SE0}\sin(\theta - \phi_0) - I_0\sin(\theta - \phi)e^{-(\frac{R_{\text{line}}}{L_{\text{line}}})t} \quad (4)$$

where I_{SE0} and ϕ_0 are initial current of sending end and initial phase angle of sending end, respectively.

5.1.2 DC Line–Ground Fault

This is a more frequent fault. When one line is connected to the ground then the DC line to ground fault is created. In the case of the underground cable when one line insulation is failing then this line is connected to the sheath and then the line to ground fault is created. In Fig. 5, this equivalent circuit is shown.

5.2 *AC Faults*

When AC faults occur either on the inverter side or converter side then the DC transmission line is also affected by this fault. When the inverter side faces AC fault the commutation process can fail and then the flow of power may be interrupted. In AC faults there are three more frequent faults: AC line–ground fault, double line–ground fault, and triple line-ground fault.

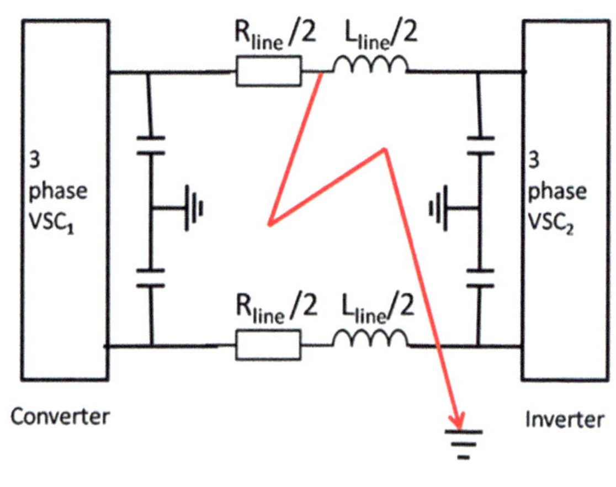

Fig. 5 HVDC system line to ground fault

5.2.1 AC Line–Ground Fault

The line–ground fault is the more repeating fault in the AC system. When a single-phase line out of the three phases falls on the ground then the single line to ground fault is created.

5.2.2 Double Line–Ground Fault

This fault is the second more frequent fault in the AC system. When a two-phase line falls on the ground then this fault is created.

5.2.3 Triple Line–Ground Fault

When all phase line falls on the ground then this type of fault is created. This is one of the most dangerous faults compared with the other two faults in the HVAC system.

6 Simulation Result

In this section, all simulation results are shown. In this, all simulation result considers that fault is occurring on 0.4 s and fault is clear after 0.05 s by the circuit breaker.

6.1 DC Faults

Two types of DC fault are described here.

6.1.1 DC Line–Line Fault

When both direct current line is short circuit then line–line fault is created. In this case, the fault simulation result is shown in Fig. 6. This result shows that when DC line to line fault is created at midpoint then the voltage drops up to 25.21 V on the first undershoot and then oscillates at 170–190 V, and in the current graph, we can see that current increases up to 42.41 A on the first overshoot and then oscillates at 17.21–18.25 A.

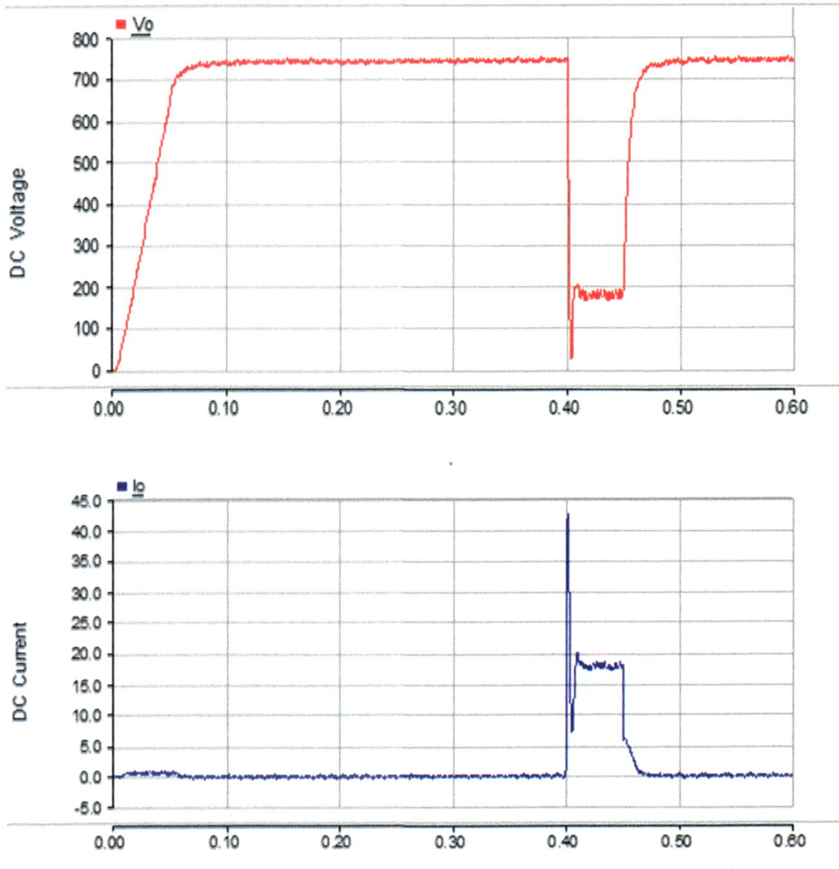

Fig. 6 DC line to line fault simulation result

6.1.2 DC Line–Ground Fault

When one wire short to ground in the DC system or underground cable wire insulation fails and connects to sheath then line to ground fault occurs and this fault is also known as a line–ground fault. DC line–ground fault simulation result is shown in Fig. 7. In this result, we can see that voltage is down up to 460 V and current increases up to 24.92 A. As a result, it is clearly shown that DC line–ground fault is less dangerous than DC line–line fault.

6.2 AC Faults

Three types of AC fault are described here.

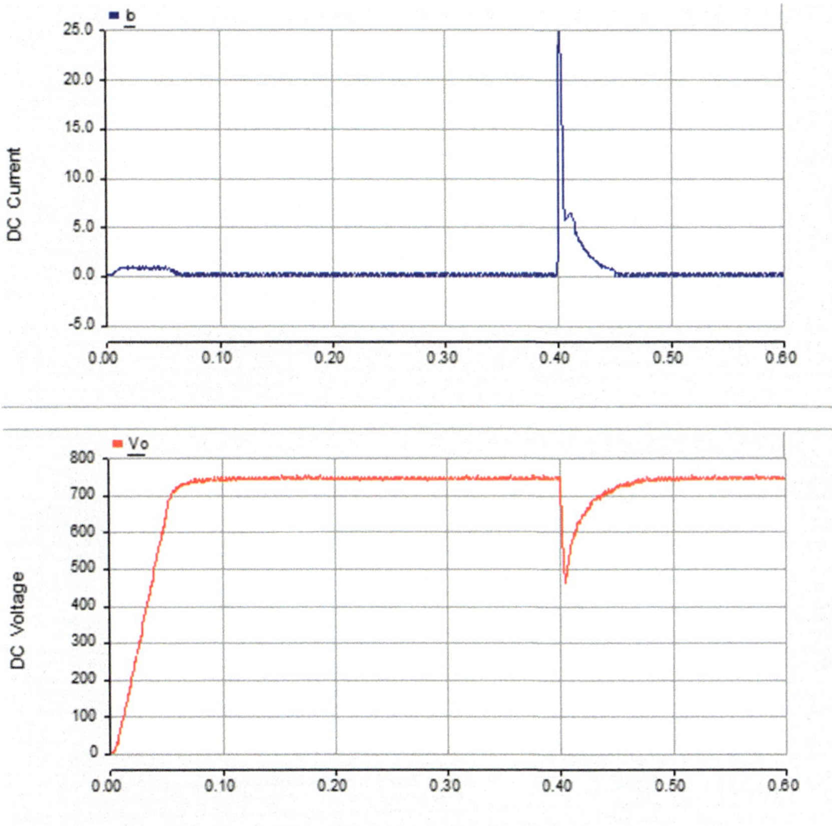

Fig. 7 DC line–ground fault simulation result

6.2.1 Line–Ground Fault

When grid-side one phase short to ground then this is called line–ground fault and this is a more repeating fault. When this fault occurs the customer experiences voltage sag. This fault also affects the DC transmission system. This fault pattern is shown in Fig. 8. When the AC fault occurs then this fault also affects our DC transmission line. In simulation result, when AC line to ground fault occurs then voltage decreases on DC side up to 537.80 V and current increases up to 7.73 A.

6.2.2 Double Line–Ground Fault

When two lines on the AC side join together and shorted with the ground then the double line to ground fault exists. This fault also affects the HVDC line. This fault pattern is shown in Fig. 9. In the simulation result, we can see that when the double

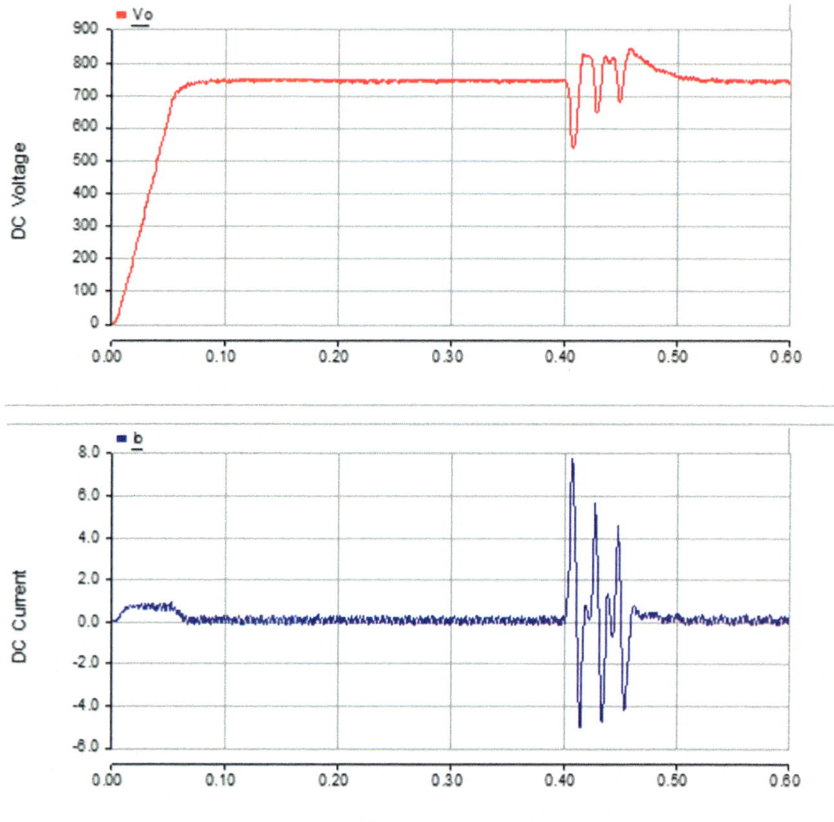

Fig. 8 AC line–ground fault simulation result

line to ground fault exists then the voltage drops up to 502.5 V and the current increases up to 8.58 A.

6.2.3 Triple Line to Ground Fault

When all three wires are shorted and touch through the ground then a three-phase–ground fault exists. This characteristic is displayed in Fig. 10.

In the simulation result, we can see that the voltage wave drops up to 443.44 V when the triple line to ground fault occurs, and the current maximum increases up to 10.50 A. As a result, we see that it is a second severe fault for the DC system.

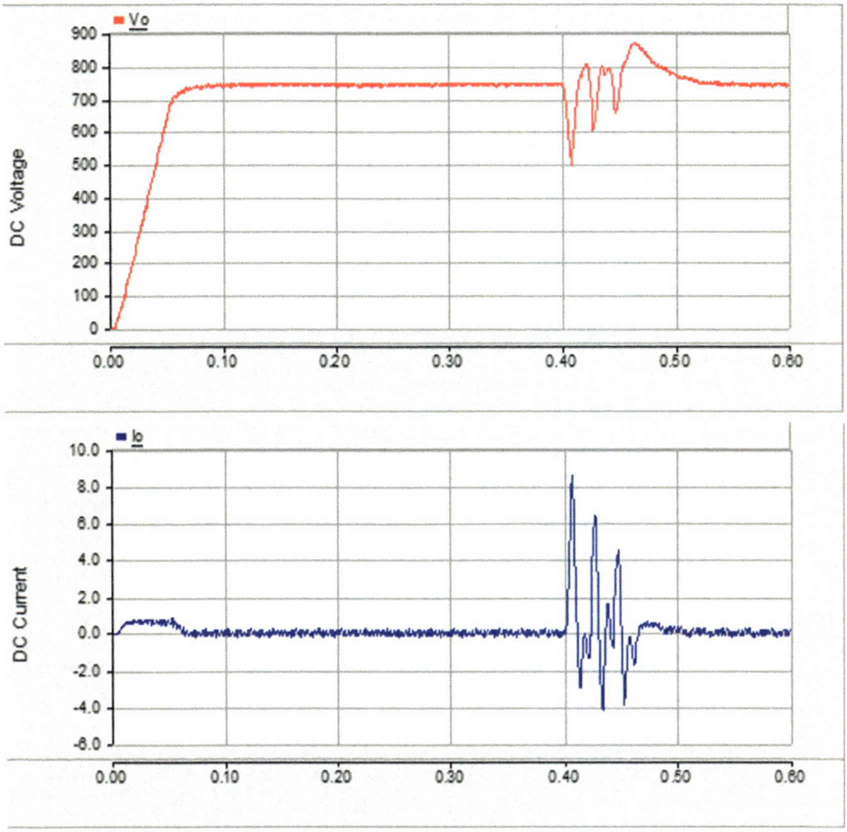

Fig. 9 AC double line–ground fault simulation result

6.3 *Results and Discussion*

Table 2 shows all per unit fault voltage and current. We can understand quite well from this discussion that DC line to line fault is more dangerous than any fault.

The most acute faults are: DC line to line fault, DC line to ground fault, AC triple line to ground fault, AC double line to ground fault, and AC single line to ground fault, as seen in the result. As an outcome, we have seen DC line to line fault and then DC line to ground fault for our HVDC line are the most severe faults, so there is a high-speed DC circuit breaker requirement for safety. We can use an IGBT-based circuit breaker in DC since its speed is very fast to break the circuit when the fault occurs, but because of no physical circuit disconnection in the normal state, it generates continuous power loss. Also, we can use hybrid circuit breakers in which both mechanical and switch operations are performed simultaneously.

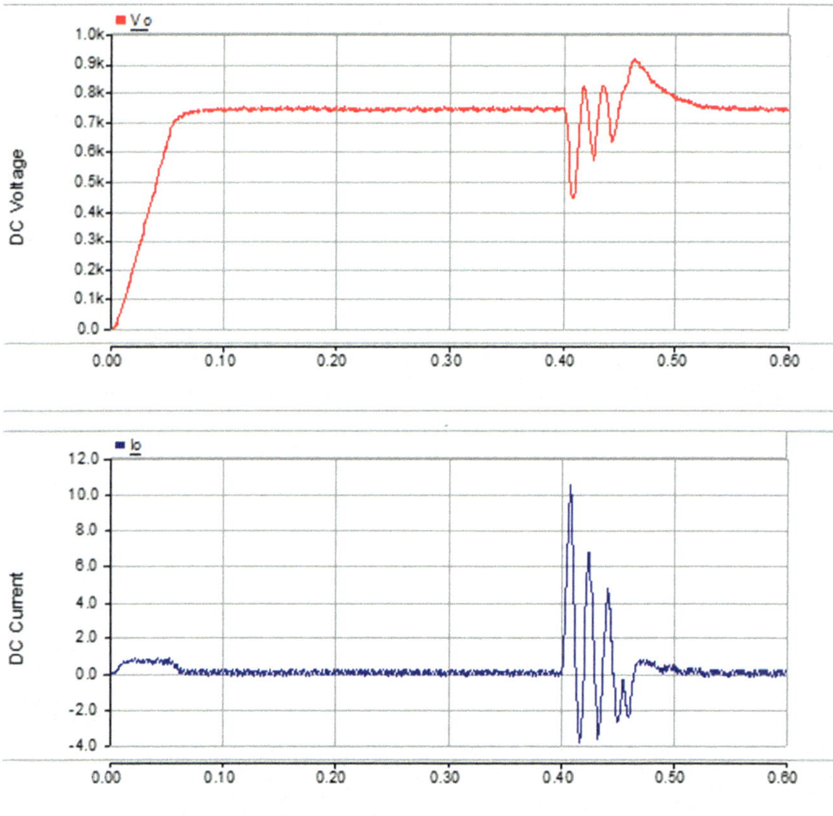

Fig. 10 AC triple line–ground fault simulation result

Table 2 Comparison of various faults

Fault type	Voltage drop up to (kV)	Current increase up to (kA)
DC line to line	25.21	42.41
DC line to ground	460	24.92
AC single line to ground	537.80	7.73
AC double line to ground	502.5	8.58
AC triple line to ground	443.44	10.50

7 Conclusion

HVDC system is economical than AC system, so in this paper AC and DC faults for HVDC system have been discussed. Because faults may occur in the HVDC system, and these faults seriously affect the converter and station inverter. To isolate these faulty segments, there is a need to understand fault characteristics. In this paper, AC and DC faults for the HVDC system have been analyzed using simulation results obtained on PSCAD/EMTDC. From the simulation result, we can see that DC faults have the highest current and therefore are the most dangerous faults. In the future, an IGBT-based high-speed hybrid circuit breaker can be designed to isolate such dangerous faults.

References

1. Wang P, Goel L, Liu X, Choo FH (2013) Harmonizing AC and DC: a hybrid AC/DC future grid solution. IEEE Power Energy Mag 11(3):76–83
2. Karthikeyan M, Yeap YM, Ukil A (2014) Simulation and analysis of faults in high voltage DC (HVDC) power transmission. In: 40th annual conference of the IEEE industrial electronics society, pp 1786–1791
3. El-Saady G, Ibrahim ENA, Okilly (2016) Analysis and control of HVDC transmission power system. In: Eighteenth international Middle East power systems conference (MEPCON), pp 190–198
4. Dessouky SS, Fawzi M, Ibrahim HA, Ibrahim NF (2018) DC pole to pole short circuit fault analysis in VSC-HVDC transmission system. In: Twentieth international Middle East power systems conference (MEPCON), pp 900–904
5. Belda NA, Plet CA, Smeets RPP (2017) Analysis of faults in multiterminal HVDC grid for definition of test requirements of HVDC circuit breakers. IEEE Trans Power Delivery 1(1):99
6. Liu D, Wei T, Huo Q, Wu L (2015) DC side line-to-line fault analysis of VSC-HVDC and DC-fault-clearing methods. In: International conference on electric utility deregulation and restructuring and power technologies (DRPT), pp 2395–2399
7. Göksu Ö, Cutululis NA, Zeni L (2017) Asymmetrical fault analysis at the offshore network of HVDC connected wind power plants. IEEE Manches Power Tech, pp 1–5, June
8. Arrillaga J, Liu YH, Watson NR (2007) Flexible power transmission: the HVDC options. Wiley
9. Tang L, Ooi BT (2007) Locating and isolating DC faults in multi-terminal DC systems. IEEE Trans Power Delivery 22(3):1877–1884
10. Wang Y, Zhang Z, Fu Y, Hei Y, Zhang X (2016) Pole-to-ground fault analysis in transmission line of DC grids based on VSC. In: IEEE 8th international power electronics and motion control conference (IPEMC-ECCE Asia), pp 2028–2032
11. Batra H, Khanna R (2013) Study of various types of converters station faults. Int J Eng Res Technol (IJERT) 2(6):3288–3293
12. Dulau Ll, Abrudean M, Bica D (2014) Effects of distributed generation on electric power systems. In: The 7th international conference interdisciplinarity in engineering (INTER-ENG 2013), vol 12, no 1, pp 681–686
13. Sharma R, Malik H (2017) Transmission line fault classification using modified fuzzy Q learning. IET Gener Transm Distrib 11(16):4041–4050. https://doi.org/10.1049/iet-gtd.2017.0331
14. Sharma R, Malik H (2017) EMD and ANN based intelligent fault diagnosis model for transmission line. J Intell Fuzzy Syst 32(4):3043–3050. https://doi.org/10.3233/JIFS-169247

15. Aggarwal A et al (2016) Feature extraction using EMD and classification through probabilistic neural network for fault diagnosis of transmission line. In: Proceedings of IEEE ICPEICES-2016, pp 1–6. https://doi.org/10.1109/ICPEICES.2016.7853709
16. Deepti C, Malik H (2018) A novel intelligent transmission line fault diagnosis model based on EEMD and multiclass PSVM. In: Applications of artificial intelligence techniques in engineering, advances in intelligent systems and computing, vol. 698, pp 85–92. https://doi.org/10.1007/978-981-13-1819-1_9
17. Aggarwal S et al (2020) Meta heuristic and evolutionary computation: algorithms and applications. Springer Nature, Berlin, 949 pp. https://doi.org/10.1007/978-981-15-7571-6. ISBN 978-981-15-7571-6
18. Yadav AK et al (2020) Soft computing in condition monitoring and diagnostics of electrical and mechanical systems. Springer Nature, Berlin, 496 pp. https://doi.org/10.1007/978-981-15-1532-3. ISBN 978-981-15-1532-3
19. Gopal et al (2021) Digital transformation through advances in artificial intelligence and machine learning. J Intell Fuzzy Syst 1–8 (Pre-press). https://doi.org/10.3233/JIFS-189787
20. Fatema N et al (2021) Intelligent data-analytics for condition monitoring: smart grid applications. Elsevier, 268 pp. ISBN: 9780323855112
21. Smriti S et al (2018) Special issue on intelligent tools and techniques for signals, machines and automation. J Intell Fuzzy Syst 35(5):4895–4899 (2018). https://doi.org/10.3233/JIFS-169773
22. Jafar A et al (2021) AI and machine learning paradigms for health monitoring system: intelligent data analytics. Springer Nature, Berlin, 496 pp. https://doi.org/10.1007/978-981-33-4412-9. ISBN 978-981-33-4412-9
23. Sood YR et al (2019) Applications of artificial intelligence techniques in engineering, vol 1. Springer Nature, 643 pp. https://doi.org/10.1007/978-981-13-1819-1. ISBN 978-981-13-1819-1
24. Khaimar AK, Shah PJ (2016) Study of various types of faults in HVDC transmission system. In: International conference on information computing and communication (ICGTSPICC), pp 480–484

Improving QoS of Cloudlet Scheduling via Effective Particle Swarm Model

Ankit Tomar, Bhaskar Pant, Vikas Tripathi, Kamal Kant Verma, and Saurabh Mishra

Abstract Cloud computing (CC) is an emerging area that includes the provisioning of dynamic and virtualized resources in a pay-as-you-go manner. Much exploration is required to enhance the scalability, effectiveness, and equilibrium in load balancing for better scheduling. Various scheduling algorithms are proposed to meet the user's requirements, but most of them failed to balance the load in critical resource demand hours. In this study, an effective particle swarm algorithm (EPSO) is addressed to solve the scheduling problem. For a faster discovery of resources, a reverse variation technique is employed in the proposed approach. The EPSO model is applied after readjustment and fine-tuning of its hyper-parameters to get precise and optimized results. This study provides a better quality of service (QoS) by optimizing resource utilization, service availability, and service-level agreement (SLA). Standard deviation (SD) is one of the critical statistical load distribution parameters computed to confirm the correct results. Results demonstrated in the form of graphs, tables, FCFS, RR, and SJF scheduling models confirm that EPSO provides a better outcome than the other state-of-the-art methods.

Keywords Load balancing · Particle swarm · Cloud computing · Meta-heuristic · Task scheduling · CloudSim

1 Introduction

For a decade, cloud computing achieved a significant potential of dynamic, omnipresent virtualized resources; it practiced ritualistic evolution in industry and academics. A massive amount of computing power is required for full utilization of VMs, hosts data centers, etc. Simultaneously, numerous users are engaged, so

Supported by Graphic Era Deemed University, Dehradun.

A. Tomar (✉) · B. Pant · V. Tripathi · S. Mishra
Graphic Era University, Dehradun, India

K. K. Verma
College of Engineering Roorkee, Roorkee, India

A. Tomar et al. (eds.), *Machine Learning, Advances in Computing, Renewable Energy and Communication*, Lecture Notes in Electrical Engineering 768,
https://doi.org/10.1007/978-981-16-2354-7_13

it is not easy to map the cloudlets (cloudlets) to allocated VMs manually under the cloud computing environment. In cloud computing, the mapping of cloudlets falls into well-defined NP class problems used to assign the cloudlets to available resources according to the precedence. A huge amount of cloud resources are rented out to exploit the profits in datacenters (DCs). As these resources are on-demand and dynamic, it is always a big deal in the cloud and entirely depends on allocating in an organized way. To run cloud applications at lower costs in dynamic load conditions, cloud users lease the resources to utilize them from cloud providers. For a particular cloudlet, every user wants many resources with full advantage with higher performance and has to be finished within a stipulated time. To enhance more elasticity of the cloud environment, no doubt, we require an efficient scheduling mechanism to map the cloudlets to resources [1]. This article's main aim is to enhance scheduling performance and good quality of service so that the load could be spread equally across the physical machines. A good scheduler always adapts to changes, and PSO is an appropriate nature-inspired approach to map the cloudlets in a dynamic environment [2]. We proposed an effective particle swarm optimization algorithm to schedule the upcoming cloudlets onto the VMs; the proposed model makes the load balancing fair enough to distribute the jobs across the processors with maximum throughput. The EPSO mode incorporates the simulated annealing concept and the reverse variation sampling technique to get faster discovery of resources.

1.1 Motivation

Conventional dynamic programming, branch and bound, divide and conquer, and other similar approaches produce global optimum solutions but are long-standing; therefore, these algorithms are not suitable to solve typical real-world problems of task scheduling in an optimized way. Researchers stated using dynamic and optimal schemes like an ant colony, genetic algorithms, tabu search optimization, honey bee optimization, and Lion optimization solve the scheduling problems within minimal time and maximal throughput [3, 22]. After analyzing the PSO-based articles of prestigious conferences and journals, we concluded that the particle swarm algorithm offers a better solution for scheduling issues than other meta-heuristic policies [2, 4–6]. One unique nature-inspired swarm optimization model is deployed here to strengthen scheduling and enhance the load balancing during cloud service provided to users. In earlier research work, PSO runs on different simulators in the same environment, but that does not produce satisfactory results; so, we deployed the CloudSim toolkit to investigate the impact of PSO results [7]. Simulated results confirm a lower failure rate and 100% performance to deal with cloud entities.

Cloud computing deploys its services with the highest degree of quality of service (QoS). To achieve this goal, the outcome of critical scheduling needs to be improved using various metrics like computation cost, throughput, workload, makespan, reliability, service availability, and utilization. After investigating the various state-of-the-art techniques, research articles, book chapters, and thesis, we found negligible

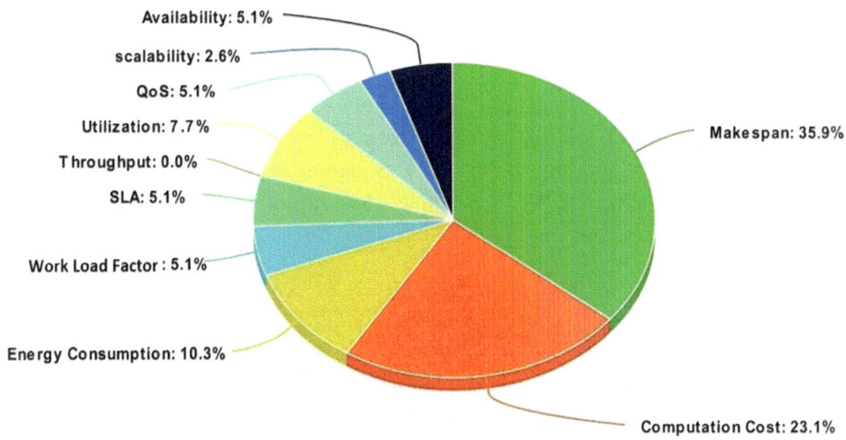

Fig. 1 Percentage of scheduling metrics implemented in earlier literature work

studies that emphasized these parameters. The computation of these metrics is very tough in a cloud atmosphere and might be researchers who have no desire to improve these criteria in this direction [8]. Hence this is a significant research gap in cloud load balancing, which we also tried to fill in this article. This study implemented resource availability, reliability, and resource utilization factors to gain the equilibrium in load among the cloud resources. The ultimate goal of this study is to provide better QoS. Figure 1 shows the previous work on scheduling parameters; we have some lower fraction metrics, which means almost little research was done on PSO algorithms.

1.2 Problem Formulation

Deployment of a resource-oriented scheduling policy at cloud sites is best suited to exaggerate utilization and revenue, and for cost and execution time, an application-centric policy is best suited. In a cloud computing environment, task scheduling plays an essential role in improving cloud services' reliability, efficiency, and flexibility. Any ideas of resource mapping policy assure competent and fair allocation of jobs, which are based on essential metrics as follows:

- Resource utilization: It defines the live execution of operations in processing elements (PEs); its measurement unit is the percentage of the time. It should be lower for a good scheduling algorithm [9–13]. In Eq. 1, a mathematical representation of resource utilization is given, where the execution time of each cloudlet is computed for *i* number of cloudlets.

$$\text{Utilization} = \frac{\text{Execution Time}_{\text{Task}_i}}{\text{Makespan}} \tag{1}$$

- Service availability: The availability of cloud resources depends upon free VMs at job submission time. Higher resource availability results in fewer active VMs, desired for ideal scheduling policy [14]. Equation 2 shows the total number of VMs, dented by VM_j and AVM_j is active VMs during simulation.

$$\text{Availability} = \text{VM}_j - \sum_{j=0}^{k} \text{AVM}_j \tag{2}$$

- Service-level agreement (SLA): It maintains the service provided by an internal or external provider to end-users; it is a contract between them. Its basic purpose is to define what customer measures [3, 6, 15]. Through Eq. 3 we showed its computation way.

$$\text{SLA} = \frac{\text{Total Number of Executed Cloudlets}}{\text{Total Number of Offered Resources}} \tag{3}$$

- Standard deviation (SD): It is an essential statistical metric of probability theory that shows the dispersion and variation from the mean value. High SD means data points diverge over a large range of solution values, and low SD indicates we tend the data points to be very close to the mean, shown in Eq. 4, where n are the number of samples, x is sample value and y is mean of all n values. If SD never tends to zero for cloudlet scheduling, we assured that the algorithm dynamically searches results [16, 17]; it enhances the novelty of this research article.

$$\text{Standard Deviation} = \sum \frac{(x-y)^2}{n-1} \tag{4}$$

- (QoS): Each user wants a full advantage of cloud services every time and everywhere since it is a pay-per-use service. In the CC domain, a good QoS is dependent upon resource flexibility, scalability, SLA, lower access time, higher resource availability, and maximum utilization. Aggregating these results of every scheduling metric compute QoS level. It can be measured by aggregating the results through Eqs. 1, 2, 3, and 4.

This paper's flow is as follows: In Sect. 2, the CC environment with CloudSim toolkit is discussed, and old literature is highlighted in the same section. Section 3 covers the proposed methodology of EPSO on cloudlet mapping. Results and discussion are conferred in Sect. 4, comparing the efficiency and implementation of EPSO with other algorithms presented in the same section. Finally, the conclusion with the future scope is shown in Sect. 5.

2 Background Study and Related Works

CC grows in a real-time environment for its deployment models because it follows the concept of on-demand service. It automatically accumulates and manages resources, providing a collective group of infrastructure entities, including data storage space, networks, and processing power. Nowadays world depends on cloud services to store public and personal information. Three essential layers are constituted to fulfill the user's requirements with enhanced scalability, availability, and fault tolerance [18].

SaaS (Software as a Service): Service providers configure this layer. At this level, people utilize up software-related utility without worrying about underlying technologies, platforms, and hardware.

IaaS (Infrastructure as a Service): This layer is responsible for the on-demand allocation of resources using pre-organized scheduling policies. Cloud users avail flexibility and top level of control by this layer.

PaaS (Platform as a Service): Without worrying about platform-related requirements and their availability, users use the organized and preserved on-demand application. At this level, the cloud provider's role is essential for allowing users to use cloud services independently.

2.1 CloudSim Toolkit

Although various simulators are available to compute the results in the cloud environment, particularly after 2010, most research studies adopted the CloudSim toolkit to simulate the cloud-related utility. CloudSim provides a real-like cloud atmosphere and bests too to simulate the CC-based environment [19]. This tool is the most favorable simulator implemented in java and can run on Eclipse or Net-Beans, and is an essential component of CloudSim shown in Table 1.

Though there are a lot of CloudSim components in terms of packages class, objects, and interfaces, we mentioned only essential elements which are as follows:

(i) Cloudlet: Any independent or dependent cloudlet submitted to the host in datacenter (DC) known as a cloudlet. Every cloudlet is created with some essential properties like processing elements, length, output file size, etc.
(ii) Cloud information service (CIS): CIS is a class that used to register an entity on DCs for searching of cloud resources
(iii) Cloudlet scheduler (CS): CS is used to implement time–space-shared cloudlet allocation policies to assess the processing power of VMs.
(iv) Datacenter (DC): To hold the VMs onto the hosts DC is used as an abstract class in CloudSim. Its primary aim is to allocate hardware infrastructures, like bandwidth requirements, RAM and storage, etc.
(v) Datacenter broker (DCB): It is a class used to submit, create, and delete the VMs class along with cloudlets; it functions as a mediator between cloud providers and software services to negotiate the issues.

Table 1 List of CloudSim entities, their attributes along with its numerical values

Entity	Attribute	Value/range
Cloudlet	Length	[500–1000] (MI)
Cloudlet	Count	[10–1000]
Cloudlet	MIPS	[500–10000]
VM	RAM	[512–2048] MB
VM	Allocation policy	Time or space shared
VM	Count	Less than 25
VM	MIPS	[100–1000]
VM	Bandwidth	1000
VM	Memory	[256–2048] MB
OS	Platform	XEN, Linux
Host	Count	Less than 3
Host	RAM	2048 MB
Host	Memory	1,000,000
Host	Allocation policy	Time or space shared
Host	Bandwidth	10,000

(vi) Host: It is used to model the physical servers (datacenters).
(vii) Virtual machine (VM): It is used to share the hosts with other VMs. Job scheduler jobs mapped to VMs with essential properties, like processing elements, storage, bandwidth, RAM, etc.
(viii) VM scheduler: It is used to divide the allocated hosts and VMs into essential time shared or space shared policies.

The entities of CloudSim with its attributes and values are shown in Table 1; during simulation, the same values are used for results computation.

2.2 *Related Work*

It is challenging to discuss all literature and applications of particle swarm methods for cloudlet scheduling; however, we include some classic PSO-based frameworks in the background study of this study to show the workflow from earlier to till now. With various versions, swarm optimization is a nature-inspired algorithm that was introduced by Kennedy and Eberhart in 1995 [20]. In a population-based meta-heuristic environment, PSO gained a lot of popularity attributable to its effectiveness, robustness, and ease. Regardless of the parameters and algorithm structure's decisions, and even though reasonable convergence properties, particle swarm optimization is still an iterative stochastic search process that works on problem hardness. It may need a large number of particles to revise and fitness estimation. It looks for the optimum of a cloudlet, termed fitness cloudlet, and the following rules are impressed by the behavior of flocks of birds searching for food.

Zhan et al. [21] addressed comprehensive evolutionary scheduling in the CC environment, which is a classification of various approaches. A classification of multiple scheduling policies using ACO, Lion optimization, GA at the SaaS layer is presented in this literature. Our last work is related to the useful PSO model worked with other scheduling metrics like computation cost, degree of imbalance, throughput, and makespan [22]. Feng et al. [23] used the PSO algorithm incorporating the Pareto domination principle (PDP) to get optimized results and to address the resource allocation in CC. Here the author considered QoS, resource reservation policy, and total execution time for each cloudlet (cloudlet). Furthermore, the cloudlet assignment problem is considered by Guo et al. [24] by using the PSO algorithm considering the cost and makespan of submitted cloudlets. Alkayal et al. [25] introduced a ranking-based PSO framework to reduce cloudlet waiting time and maximize throughput. Leena et al. [26] proposed a bi-objective PSO model to optimize makespan and computation cost to heighten CC cloudlet scheduling. Elhady et al. [27] compared a single-objective swarm optimization model for cloudlet scheduling in the CC background.

A hybrid GA-PSO algorithm is addressed by Beegom et al. [28] for dependent (homogeneous) cloudlet scheduling in the CC domain. In the MapReduce framework, Wang et al. [29] introduced a MOEA/D technique for multi-level objective cloudlet scheduling. For another solution of cloudlet scheduling in CC, Agrawal et al. [30] implemented a PSO scheme based on adaptive inertia weight technique incorporating cumulative binomial probability. However, there are some research gaps for challenging the CC framework. Shahid et al. [31] proposed a set of multi-objective optimization algorithms, including NSGA-II and PSO, to find optimal solutions in the field of a multi-processor system. Mostly cloudlet scheduling problems are focused on either single-objective or multi-objective. The single-objective functions of PSO-based schemes include computation cost and makespan; other scheduling metrics are covered under multi-objective PSO schemes. To overcome this completion, we implemented here more powerful PSO model to address the scheduling problem. Moreover, the reader may refer to [32–38] for more examples of this domain.

3 Proposed System

The major challenge in the CC domain is to use cloud resources proficiently with optimum mapping structure generating maximal revenue. A better scheduler always maps the cloudlets to the appropriate resource. The independent scheduler first finds allocated VMs, then VM maps the appropriate resource (host) in DCs [20]. PSO is a nature-inspired metaheuristic optimization policy that works on the concept of searching for adaptation or food. Here variables adjust their values closer to the closest member to any target at any particular moment. One will chip into food most relative to it, and other flocks swing nearby in search of a hidden food source, circling a herd of the swarm. In case any swarm comes near to target, then it tweets louder to others over toward him. This arrangement continues until all swarm gets the food that its algorithm states. If we relate this scenario to the algorithm, three global variables would be used. Decision (target) value, the best global value (gbest), indicates a particular particle's closest value to the target and if the target is not found, the algorithm will stop by a stopping value. A particle must have:

(1) pbest is particle's personal best value used to compare with gbest for indicating the nearest particle to the target.
(2) Velocity factor indicates how the data value is improved for a possible solution.

Till now, various nature-inspired algorithms presented by researchers to solve computation problems. PSO is one of the finest models suitable for dealing efficiently with scheduling problems in cloud scenarios. Due to its powerful randomicity, it gives highly improved results. This paper includes a simulated annealing technique incorporated in each iteration of PSO to increase its searching ability, enhancing the convergence ability and assuring the original solution. Novel reverse variation technique increases population diversity to avoid sinking of local optima. This new version of PSO discovers faster resource mapping, scheduling, and execution time. The flowchart of the proposed model is shown in Fig. 2. Initially, some random values of weight, inertia, and acceleration coefficients were taken. A sampled minimum and maximum length is taken to make a unidirectional flow of information; gbest is also used to pass information to nearby particles. This gbest is a critical parameter that strongly affects the performance of PSO. Poor scheduling ability for gbest sometimes results in the pre-maturity of an algorithm, so to overcome this situation gbest is sampled at every iteration. This solves the purpose of jumping out of local minima.

4 Results and Discussion

Scheduling is an essential step in parallel and cloud computing while balancing the load. Most articles simulated either a single or no more than two parameters for scheduling using PSO; however, these metrics are not enough to compute QoS provided by the scheduling mechanism. Dependent and pre-emptive cloudlets need to be scheduled but here we discussed only an independent assignment problem in the non-pre-emptive environment. We assumed that cloudlets are submitted mutually independent (heterogeneous) in nature, i.e., there is no precedence among them; therefore, no interference with processors' dependency is required during execution. Cloudlets are pre-emptive, so an interruption by any other processor is not possible during execution. Cloudlets and VMs have an essential characteristic that plays a core role in PSO performance.

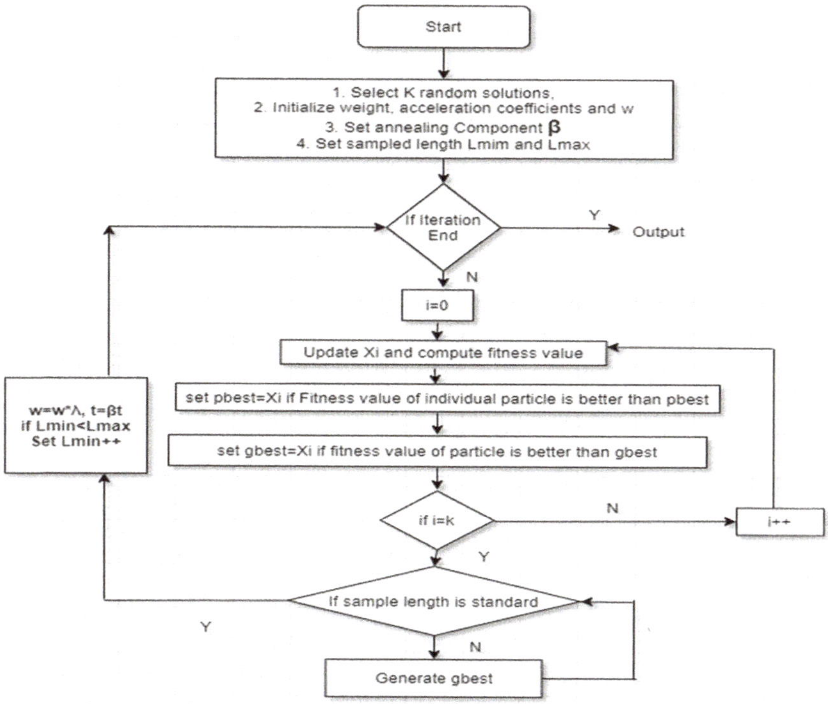

Fig. 2 Flowchart of effective particle swarm optimization along with a conceptual workflow for every entity represented in this diagram

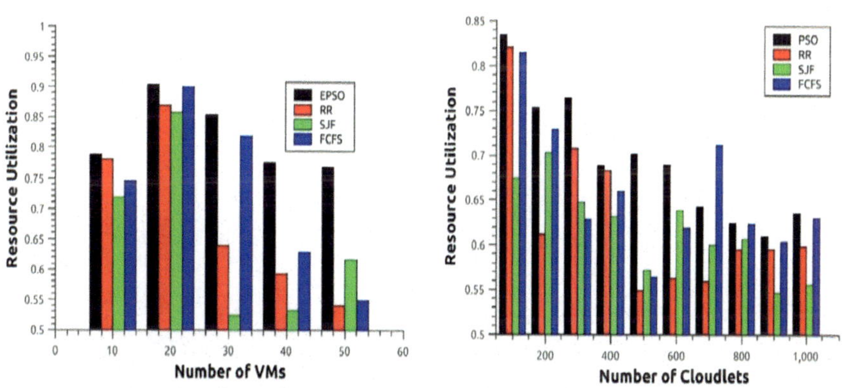

Fig. 3 Illustration of resource utilization against a varying number of cloudlets

4.1 Resource Utilization Comparison

In Fig. 3a, b, the utilization percentage is recorded against varying VMs and cloudlets, respectively. The utilization percentage is found higher in both situations in EPSO when compared with other algorithms. During cloudlet mapping to VMs, a higher rate of utilization is required for better scheduling performance.

4.2 Comparison of Resource Availability

We recorded the list of inactive VMs (Table 2) for different MIPS to investigate the performance of the proposed EPSO scheme for availability. Figure 4a shows that the

Table 2 Data obtained of inactive VMs for varying tasks and VMs

Cloudlets,VMs	FCFS	RR	SJF	EPSO
400, 400	146	147	139	154
500, 400	111	110	112	116
1000, 400	30	28	32	34
1500, 1000	206	222	230	242
2000, 1000	124	135	137	138
2500, 1000	80	76	89	84
3000, 2000	441	433	442	461
3500, 2000	341	329	358	340
4000, 2000	269	251	279	279
4500, 3000	649	649	652	683
5000, 3000	550	565	549	567

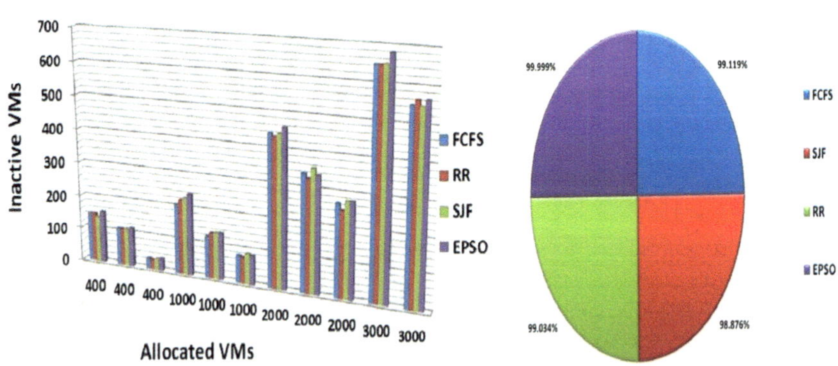

Fig. 4 Service-level agreement chart for various algorithms

proposed EPSO method uses less number of VMs to consolidate the execution of cloudlets to make off others and ensures the high availability of resources.

4.3 Comparison of Service-Level Agreement

SLA is the bond or agreement between cloud service users and cloud service providers that ensures maintenance of a minimum level of service. SLA is an important component retained between the provider and the users for providing unbreakable and on-demand service. It is computed by counting successfully executed cloud requests. Figure 4b shows the SLA percentage maintained by various scheduling policies and it confirmed that EPSO maintained a better level of SLA.

4.4 Comparison of Standard Deviation

SD of minimum execution time is computed using Eq. 4 for all algorithms, and the results are depicted in Fig. 5 for varying cloudlets ranging from 100 to 1000. It was observed that SD is lower in EPSO, which results in a lower fluctuation in terms of execution time than FCFS, SJF, and RR. The SD results confirm that EPSO has a higher possibility to find a better solution in a reasonable time.

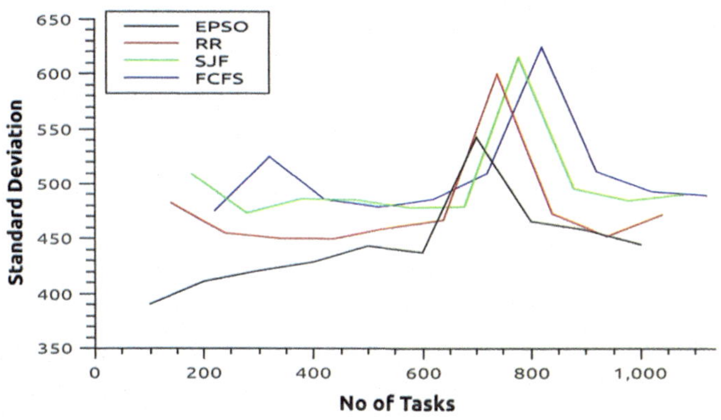

Fig. 5 Standard deviation of minimum execution time for various models

5 Conclusion

In the cloud computing domain, job scheduling is quite challenging for researchers as it considers different requirements compared to the traditional CC framework. The traffic pattern is highly unpredictable in the CC environment; therefore, a reverse variation technique based on EPSO is implemented to address the scheduling problem for such a complex domain. EPSO is applied to absorbing local optima policy to get results. Service availability, utilization, makespan, and SLA are CC metrics that optimized to provide better QoS. Experimental results confirm that EPSO discovers faster, efficient, and near-optimal results in all the test cases than other commonly used scheduling methods.

The EPSO framework can be explored in future work to reduce cloudlet's migration time for sustaining equilibrium in load balancing. Another important future scope of the proposed model is to deploy it for reducing the overall workload incorporating the degree of imbalance factor in real time.

References

1. Huang Q, Huang T (2010) An optimistic job scheduling strategy based on QoS for cloud computing. In: 2010 international conference on intelligent computing and integrated systems. IEEE
2. Zhan S, Huo H (2012) Improved PSO-based task scheduling algorithm in cloud computing. J Inf Comput Sci 9(13):3821–3829
3. Madni SHH et al (2017) Recent advancements in resource allocation techniques for cloud computing environment: a systematic review. Clust Comput 20 (3):2489–2533
4. Hassan R et al (2005) A comparison of particle swarm optimization and the genetic algorithm. In: 46th AIAA/ASME/ASCE/AHS/ASC structures, structural dynamics and materials conference
5. Madni SHH et al (2016) An appraisal of meta-heuristic resource allocation techniques for IaaS cloud. Indian J Sci Technol 9(4):1–14
6. Ramezani F et al (2013) Task scheduling optimization in cloud computing applying multiobjective particle swarm optimization. In: International conference on service-oriented computing. Springer, Berlin, Heidelberg
7. Bhatt K, Bundele M (2013) Study and impact of CloudSim on the run of PSO in cloud environment. Int J Innov Eng Technol 2(4):254–262
8. Milani AS, Navimipour NJ (2016) Load balancing mechanisms and techniques in the cloud environments: systematic literature review and future trends. J Netw Comput Appl 71:86–98
9. Zuo X et al (2014) Self-adaptive learning PSO-based deadline constrained task scheduling for hybrid IaaS cloud. IEEE Trans Autom Sci Eng 11(2):564–573
10. Abdi S et al (2014) Task scheduling using modified PSO algorithm in cloud computing environment. In: International conference on machine learning, electrical and mechanical engineering
11. Almezeini N, Hafez A (2017) Task scheduling in cloud computing using lion optimization algorithm. Algorithms 5:7
12. Xu L et al (2014) An improved binary PSO-based task scheduling algorithm in green cloud computing. In: 9th international conference on communications and networking in China. IEEE
13. Madni SHH et al (2017) Performance comparison of heuristic algorithms for task scheduling in IaaS cloud computing environment. PloS one 12(5):e0176321

14. Shelokar PS et al (2007) Particle swarm and ant colony algorithms hybridized for improved continuous optimization. Appl Math Comput 188(1):129–142
15. Tareghian S, Bornaee Z (2015) A new approach for scheduling jobs in cloud computing environment. Fen Bilimleri Dergisi (CFD) 36(3)
16. Tawfeek MA (2013) Cloud task scheduling based on ant colony optimization. In: 2013 8th international conference on computer engineering and systems (ICCES). IEEE
17. Zhang L et al (2008) A task scheduling algorithm based on PSO for grid computing. Int J Comput Intell Res 4(1):37–43
18. Wang L et al (2008) Scientific cloud computing: early definition and experience. In: 2008 10th IEEE international conference on high performance computing and communications. IEEE
19. Calheiros RN et al (2011) CloudSim: a toolkit for modeling and simulation of cloud computing environments and evaluation of resource provisioning algorithms. Softw Pract Exp 41(1):23–50
20. Eberhart R, Kennedy J (1995) A new optimizer using particle swarm theory. In: MHS'95. Proceedings of the sixth international symposium on micro machine and human science. IEEE
21. Zhan Z, Liu X, Gong Y, Zhang J, Chung HS, Li Y (2015) Cloud computing resource scheduling and a survey of its evolutionary approaches. ACM Comput Surveys 47(4):33 (Article 63)
22. Kalra M, Singh S (2015) A review of metaheuristic scheduling techniques in cloud computing. Egypt Informatics J 16(3):275–295
23. Feng M, Wang X, Zhang Y, Li J (2012) Multi-objective particle swarm optimization for reseource allocation in cloud computing. In: Proceedings of 2nd international conference on cloud computing and intelligent systems (CCIS), vol 3, pp 1161–1165
24. Guo L, Shao G, Zhao S (2012) Multi-objective task assignment in cloud computing by particle swarm optimization. In: Proceedings of 8th international conference on wireless communications, networking and mobile computing (WiCOM), pp 1–4
25. Alkayal ES, Jennings NR, Abulkhair MF (2016) Efficient task scheduling multi-objective particle swarm optimization in cloud computing. In: Proceedings of 41st IEEE conference on local computer networks workshops, pp 17–24
26. Leena VA, Beegom ASA, Rajasree MS (2016) Genetic algorithm based bi-objective task scheduling in hybrid cloud platform. Int J Comput Theory Eng 8(1):7–13
27. Elhady GF, Tawfeek MA (2015) A comparative study into swarm intelligence algorithms for dynamic task scheduling in cloud computing. In: Proceedings of 7th IEEE International Conference on Intelligent Computing and Information Systems, pp 362–369
28. Beegom ASA, Rajasree MS (2019) Integer-pso: a discrete pso algorithm for task scheduling in cloud computing systems. Evol Intell 12(2):227–239
29. Wang X, Wang Y (2012) An energy and data locality aware bilevel multiobjective task scheduling model based on map reduce for cloud computing. In: Proceedings of IEEE/WIC/ACM international conference on web intelligence and intelligent agent technology, pp 648–655
30. Agrawal A, Tripathi S (2018) Particle swarm optimization with adaptive inertia weight based on cumulative binomial probability. Evol Intell 1–9
31. Shahid A, Qadri MY, Fleury M, Waris H, Ahmad A, Qadri NN (2018) Ac-dse: approximate computing for the design space exploration of reconfigurable mpsocs. J Circuits Syst Comput 27(9):25
32. Aggarwal S et al (2020) Meta heuristic and evolutionary computation: algorithms and applications. Springer Nature, Berlin, p 949. https://doi.org/10.1007/978-981-15-7571-6. ISBN: 978-981-15-7571-6
33. Yadav AK et al (2020) Soft computing in condition monitoring and diagnostics of electrical and mechanical systems. Springer Nature, Berlin, p 496. https://doi.org/10.1007/978-981-15-1532-3. ISBN: 978-981-15-1532-3
34. Gopal et al (2021) Digital transformation through advances in artificial intelligence and machine learning. J Intell Fuzzy Syst (pre-press) 1–8. https://doi.org/10.3233/JIFS-189787
35. Fatema N et al (2021) Intelligent data-analytics for condition monitoring: smart grid applications. Elsevier, p 268. ISBN: 9780323855112

36. Smriti S et al (2018) Special issue on intelligent tools and techniques for signals, machines and automation. J Intell Fuzzy Syst 35(5):4895–4899. https://doi.org/10.3233/JIFS-169773
37. Jafar A et al (2021) AI and machine learning paradigms for health monitoring system: intelligent data analytics. Springer Nature, Berlin, p 49. https://doi.org/10.1007/978-981-33-4412-9. ISBN: 978-981-33-4412-9
38. Sood YR et al (2019) Applications of artificial intelligence techniques in engineering, vol 1. Springer Nature, p 643. https://doi.org/10.1007/978-981-13-1819-1. ISBN: 978-981-13-1819-1

Performance Evaluation of HHO Optimized Model Predictive Controller for AVR System and Its Comparison with Conventional Controllers

Vineet Kumar, Veena Sharma, and Vineet Kumar

Abstract An automatic voltage regulator (AVR) is treated as an essential part of a power system network and its proper functioning is required for the smooth and safe operation of the network. Work in this paper focuses on designing an efficient and robust control methodology for the AVR systems. Therefore, the Harris Hawks optimization (HHO)-based model predictive control (MPC) strategy has been discussed and implemented on MATLAB/SIMULINK and the transient response specifications have been compared with conventional PID and two degrees of freedom (2-DOF) PID controllers. Moreover, the robustness of discussed methodology has been discussed under parameter variation and external disturbance case.

Keywords Automatic voltage regulator · Model predictive controller · PID controller · Optimization · Harris Hawks optimization

1 Introduction

In any power generation network around the globe, the AVR is required by the power generating systems to safeguard terminal voltage stability throughout their operating region. The voltage stability disrupts because of a number of reasons, like changing load demand, disturbance at terminals, mismatches between reactive power demand and supply, etc. For an uninterrupted power supply and secure and safe operation of any power system network, it is vital that the network operates at a fixed value of terminal voltage. The voltage deviations beyond certain tolerable limits can cause significant damage to the equipment on the load and supply sides, thereby causing a great deal of damage to the reliability and security of the power system network [1].

Conventionally, the AVR systems included an amplifier to calibrate the field excitation of the rotor of synchronous generators, but in the quest to obtain a more precise and quick response, it is vital to have a dedicated control tool. In the past

V. Kumar (✉) · V. Sharma · V. Kumar
Electrical Engineering Department, NIT Hamirpur, Hamirpur, HP, India

A. Tomar et al. (eds.), *Machine Learning, Advances in Computing, Renewable Energy and Communication*, Lecture Notes in Electrical Engineering 768,
https://doi.org/10.1007/978-981-16-2354-7_14

few decades, major research on AVR control has elaborated the applications of classical controllers like PI and PID controllers [2]. Later, the researchers have also been using the distinctive variations of PID controllers (FOPID, PID-N) [3]. From the literature review, it has been established that the analytical parameter tuning approaches like Ziegler-Nichols (ZN) method and Cohen Coon method, etc. undergoes multiple bottlenecks, like inferior transient performance, failure to handle the uncertainties in the system dynamics, and external disruptions, hence model configuration complexities cannot be treated. With the stimulus of particle swarm optimization, an innovative approach of tuning PID variables had been implemented by Gaing [4]. PSO approach is the most researched meta-heuristic optimization procedure in the field of voltage control problems. Also, the researchers have tried to enhance the quality of time-domain response of the AVR system with the application of a PID controller tuned by numerous meta-heuristic procedures [5]. From the literature review, it can be concluded that the operational efficiency of AVR can be further upgraded with advanced control methods like MPC and sliding mode control (SMC). Therefore, further investigation is required in this direction. Moreover, the reader may refer to [8–14] for more examples and advanced applications.

In this study, an AVR system has been considered and it is controlled with the help of MPC, and its gains have been tuned with the assistance of Harris Hawks optimization approach. The efficacy of the MPC-HHO method has been evaluated in terms of transient response specification and it has been compared with the meta-heuristically tuned conventional PID controllers and 2-DOF PID controllers.

2 System Under Consideration

In this work, a simple AVR system has been considered with controller arrangement along with four basic components, amplifier, exciter, generator, and sensor. These components have been modeled with the use of first-order transfer function (T. F.).

2.1 *Amplifier*

The amplifier takes the mismatch between terminal voltage and a reference voltage; this error voltage is manipulated by the controller, and then it is fed to the amplifier. The gain of the amplifier is generally taken between 10 and 40 and its time constant is taken between 0.02 and 0.1 s.

$$\frac{V_r(s)}{V_e(s)} = \frac{K_a}{1 + T_a s} \tag{1}$$

Here, $V_r(s)$ and $V_e(s)$ are the values of desired terminal voltage and error signal voltage, respectively.

2.2 Exciter

The exciter takes the DC voltage as its input and it provides stationary RMF in the rotor field of the synchronous generator. Its modeling components are in the range of 0.8–1 and 0.1–1 s for gain and time constant values, respectively.

$$\frac{V_{exciter}(s)}{V_r(s)} = \frac{K_e}{1 + T_e s} \tag{2}$$

2.3 Generator

The generator model in this study is taken as a simple first-order transfer function without considering any nonlinearity or complex dynamics. The K_g and T_g values are taken in the range of 1–2 and 0.7–1 s.

$$\frac{V_{Gen}(s)}{V_{exciter}(s)} = \frac{K_g}{1 + T_g s} \tag{3}$$

2.4 Sensor

The first-order T. F. of sensor component is provided as:

$$\frac{V_s(s)}{V_t(s)} = \frac{K_r}{1 + T_r s} \tag{4}$$

Here, $V_s(s)$ is the sensor output and $V_t(s)$ is the terminal voltage output from the generator (Fig. 1).

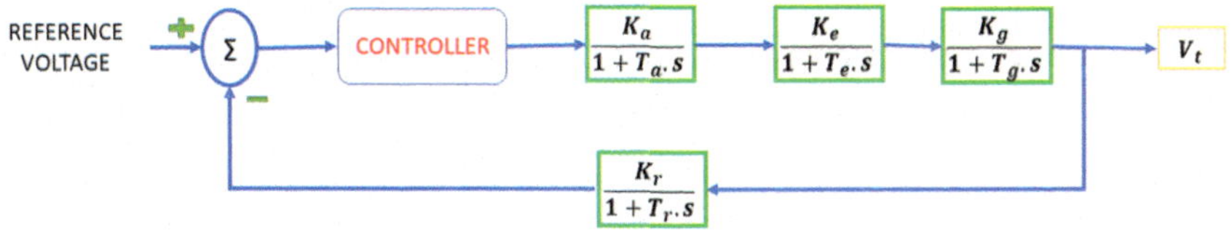

Fig. 1 AVR block diagram

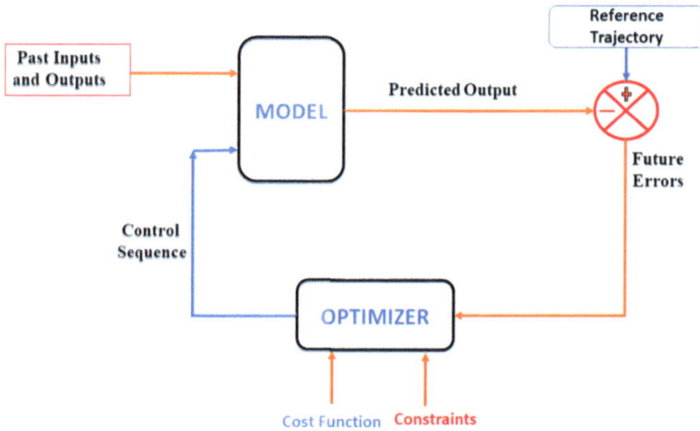

Fig. 2 Basic block diagram representation of MPC

3 Model Predictive Controller

MPC is a very advanced controller where a process model is inducted to foresee the future values of the output, based on the past and current measurements of input and output values. These future outputs are compared with nominal outputs to obtain a future error and are then fed to the optimizer, where an online optimization problem is simplified at every sample of time (k) to calculate a control sequence. The performance index is given in Eq. (5).

$$J = \sum_{m=1}^{P} y1[\hat{y}(k+m/k) - y_r(k+m)]^2 + \sum_{m=1}^{C} u1[\Delta u(k+m-1)]^2 \quad (5)$$

Here, P is the length of the predicted output sequence and C is the length of the control sequence [18]. The cost function weights y1 and u1 can be tuned to get the desired performance from the model predictive controller [6] (Fig. 2).

4 Harris Hawks Optimization Algorithm

It is a nature-inspired meta-heuristic approach, which offers a great balance between exploration and exploitation stages. And by doing so, it reduces the chances of the solution getting stuck into the local optimum. This enables the user to get an optimal solution in a minimum number of iterations and guarantees a good solution in a single run. The HHO algorithm has very few parameters to specify, hence it has reduced complexity.

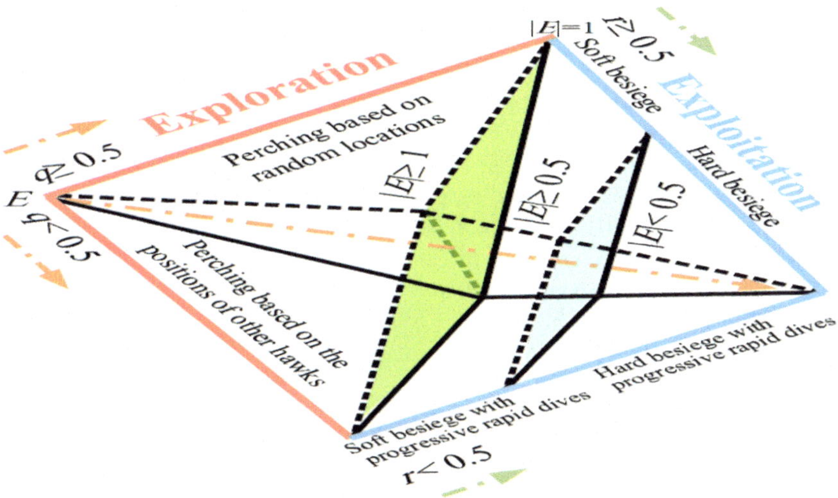

Fig. 3 Various stages and searching techniques followed by HHO [20] [7]

The algorithm follows a number of steps based on the chances of the rabbit escaping the pray (r) and the rabbit's own energy (E). Figure 3 briefly explains the algorithmic steps of HHO, and it can be seen that an exploration stage has both randomness and organized search, and in the exploitation stage, the Hawks will follow numerous attacking patterns, soft besiege, hard besiege, etc. The parameters q and r are arbitrarily selected numbers ranging from 0 to 1.

5 Results and Discussions

After considering the below-given parameter values for the AVR system, the simulation has been performed using MATLAB/Simulink. Afterward, various cases have been taken into the account for performance evaluation of the proposed methodology (Table 1).

Table 1 Parameter values of the AVR system

Parameters	K_a	T_a	K_e	T_e	K_r	T_r	K_g	T_g
Values	10	0.1	1	0.4	1	0.01	1	1

5.1 Basic Performance Analysis

Under this test, the HHO optimized MPC has been used to produce optimal performance for the AVR system. Also, the transient performance of the proposed methodology has been compared with that of HHO tuned PID and 2-DOF PID controllers. Table 2 shows the HHO optimized parameter values of MPC (y1, u1, du1), PID (Kp, Ki, Kd), and 2-DOF PID (Kp, Ki, Kd, b, c) controllers.

In Fig. 4, it can be recognized that the recommended MPC controller delivers improved transient performance for the AVR system discussed in this paper. Table 3 shows that the MPC-HHO controller gives better settling time (ST), rise time (RT), and peak overshoot (PO) when compared with the PID and 2-DOF PID controllers.

Table 2 Parameter values of PID, 2-DOF PID, and MPC controllers

Controller	Parameter values tuned by HHO algorithm	Fitness value (ISE)
PID	Kp = 1; Ki = 0.2181; Kd = 0.2137	0.3507
2 DOF PID	Kp = 1; Ki = 0.4042; Kd = 0.206; b = 0.928; c = 0.0813	0.3410
MPC	y1 = 0.3975; u1 = 0.0100; du1 = 0.0	0.1305

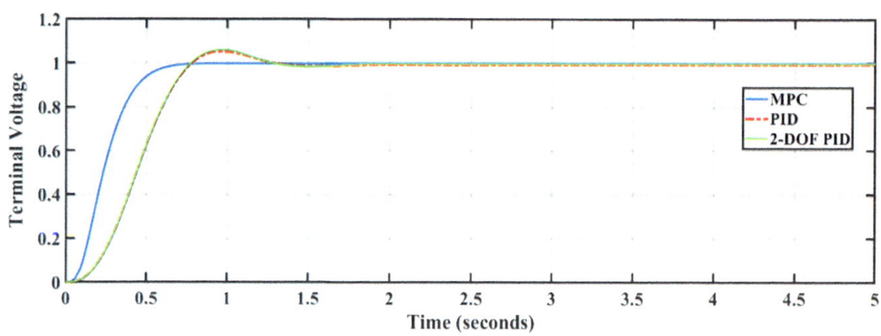

Fig. 4 Transient response curve of terminal voltage obtained from AVR system controlled by MPC, PID, and 2-DOF PID controllers

Table 3 Transient response specifications obtained from the controlled AVR system

Controller	RT	ST	PO
PID	0.458 s	1.222 s	5.6%
2-DOF PID	0.451 s	1.196 s	6.1%
MPC	0.336 s	0.602 s	0%

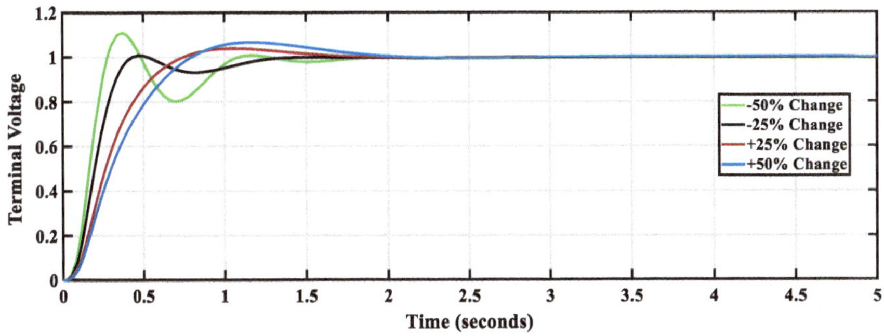

Fig. 5 Time response plots of terminal voltage obtained after varying the exciter parameters

Table 4 Transient response after adding parameter uncertainty

Change in T_e (in %)	Rise time (sec)	Settling time (sec)	Overshoot (%)
−50	0.163	1.019	10.70
−25	0.236	1.173	−8.17
+ 25	0.430	1.452	4.1
+ 50	0.505	1.773	6.9

5.2 Robustness Analysis Under Parameter Variation Case

In this case, the parameter values of the AVR model have been perturbed and its response has been noted with MPC controller with the same settings and parameters.

Table 4 shows that the MPC-HHO controller gives excellent robustness against parameter variation in the AVR model. Fig. 5 shows the obtained characteristic plot of the system after varying exciter parameter values (Table 4).

5.3 Robustness Analysis Under External Disturbance Case

In this test case, the strength of the suggested algorithm has been re-verified with respect to external disturbance. Here, the step disturbance has been added on the output side, and the amplitude of the step function is taken as ±10% of the nominal value (0.1 pu) (Fig. 6).

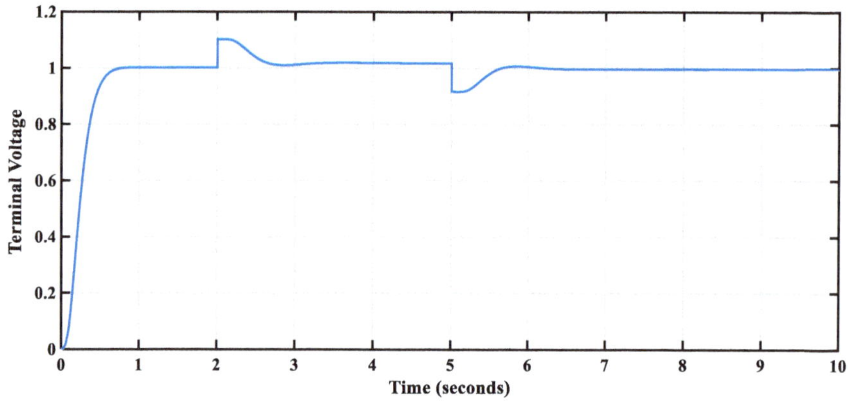

Fig. 6 Obtained output plot after adding disturbance into the AVR model

6 Conclusion

The HHO-optimized MPC has been successfully carried out for the performance improvement of the AVR system using MATLAB/SIMULINK 2018a software. The obtained results have shown remarkable performance when compared with conventional PID and 2-DOF PID controllers. Later on, the robustness analysis has been executed after varying the specifications of the AVR system. Also, in a similar test, the robustness has been re-investigated after adding external disturbance into the system. Under both cases, the proposed MPC-HHO controller (with the same controller settings) was able to produce good transient characteristics.

Acknowledgments The research in this paper has been funded by H. P. Council for Science, Technology & Environment (HIMCOSTE), SCST & E (R&D)/2019-20, under project grant no. STC/F(8)-6/2019(R&D 2019-20)-408, H. P., India, investigated by the second author.

References

1. Saadat H (2004) Power system analysis. McGraw-Hill
2. Faiz J, Shahgholian G, Arezoomand M (2007) Analysis and simulation of the AVR system and parameters variation effects. In: 2007 international conference on power engineering, energy and electrical drives. Setubal, Portugal, pp 450–453
3. Sikander A, Thakur P, Bansal RC, Rajasekar S (2018) A novel technique to design cuckoo search based FOPID controller for AVR in power systems. Comput Electr Eng 70, 261–274
4. Gaing ZL (2004) A particle swarm optimization approach for optimum design of PID controller in AVR system. IEEE Trans Energy Convers 19(2):384–391
5. Bingul Z, Karahan O (2018) A novel performance criterion approach to optimum design of PID controller using cuckoo search algorithm for AVR system. J Franklin Inst 355(13):5534–5559

6. Kumar V, Sharma V, Rahi OP, Kumar U (2019) Optimal position tracking for an AC servomotor using linear quadratic and model predictive control. In: 2019 4th international conference on information systems and computer networks (ISCON). Mathura, India
7. Heidari A, Mirjalili S et al (2019) Harris hawks optimization: algorithm and applications. Future Gener Comput Syst 97:601–610
8. Aggarwal S et al (2020) Meta heuristic and evolutionary computation: algorithms and applications. Springer Nature, Berlin, 949 p https://doi.org/10.1007/978-981-15-7571-6. (ISBN 978-981-15-7571-6)
9. Yadav AK et al (2020) Soft computing in condition monitoring and diagnostics of electrical and mechanical systems. Springer Nature, Berlin, 496 p https://doi.org/10.1007/978-981-15-1532-3. ISBN 978-981-15-1532-3
10. Gopal et al (2021) Digital transformation through advances in artificial intelligence and machine learning. J Intell Fuzzy Syst, Pre-press, 1–8. https://doi.org/10.3233/JIFS-189787
11. Fatema N et al (2021) Intelligent data-analytics for condition monitoring: smart grid applications. Elsevier, 268 p ISBN: 978-0-323-85511-2. https://www.sciencedirect.com/book/978 0323855105/intelligent-data-analytics-for-condition-monitoring
12. Smriti S et al (2018) Special issue on intelligent tools and techniques for signals, machines and automation. J Intell Fuzzy Syst 35(5):4895–4899. https://doi.org/10.3233/JIFS-169773
13. Jafar A et al (2021) AI and machine learning paradigms for health monitoring system: intelligent data analytics. Springer Nature, Berlin, 496 p https://doi.org/10.1007/978-981-33-4412-9. ISBN 978-981-33-4412-9
14. Sood YR et al (2019) Applications of artificial intelligence techniques in engineering, vol 1. Springer Nature, 643 p https://doi.org/10.1007/978-981-13-1819-1. ISBN 978-981-13-1819-1

Squared Error Autocorrelation-Based VSS-LMS Control Algorithm for Grid-Integrated Solar Photovoltaic System

Shreya Chaudhary, Rachana Garg, and M. Rizwan

Abstract The electrical power generated from the solar photovoltaic system is not constant due to the changing atmospheric parameters which are not desirable for reliable power supply and grid management. In this paper, a variable step size least mean square using squared error autocorrelation (VSS-SEA)-based control algorithm for a two-stage grid-integrated photovoltaic system is proposed. The value of weights is small during steady-state for low misadjustment level and increased during the transient state for fast convergence incremental conductance control for MPPT. Results are obtained for a 25-kW solar photovoltaic system using VSS-SEA under load unbalancing and varying solar irradiance for different loads and compared with the least mean square algorithm. Performance of the system using VSS-SEA is faster, smoother and more stable as compared to LMS under dynamically changing conditions. Moreover, the total harmonic distortion is found to be 1.76%, which is under IEEE-519 limits.

Keywords Solar photovoltaic system · Sustainable power generation · Voltage source converter · Adaptive control algorithm · Power quality

1 Introduction

The world is metamorphosing, and electricity is an important requirement for this change. A major share of electricity is produced using conventional fuels which leads to the emission of gases, like CO_2, SO_2, NO_2 and suspended particulate matter (SPM), which are harmful to the environment and human health as well. Renewable energy resources such as solar energy and wind energy have stepped in as a good replacement for conventional energy sources. Solar photovoltaic (SPV) technology is gaining popularity due to its abundance, economic reliability and increasing efficiency [1]. The peak power produced by SPV can vary depending on aerial factors, like solar irradiance, cloud cover, wind speed, humidity, dew point and temperature. Its value

S. Chaudhary (✉) · R. Garg · M. Rizwan
Department of Electrical Engineering, Delhi Technological University, Delhi 110042, India

A. Tomar et al. (eds.), *Machine Learning, Advances in Computing, Renewable Energy and Communication*, Lecture Notes in Electrical Engineering 768,
https://doi.org/10.1007/978-981-16-2354-7_15

follows a point on the I–V or P–V curve, and the maximum value is called the maximum power point (MPP) [2]. Various MPPT algorithms have been developed by researchers [3]. The incremental conductance (IC) algorithm has several advantages over the other types of MPPT techniques. The IC-based MPPT is used as it can reach MPP under fast varying atmospheric conditions and stop perturbing after reaching the operating point unlike P&O and reduce power loss [4]. Once the MPP is obtained the electrical power generated from SPV requires conversion from DC to AC for grid integration. Moreover, for the proper synchronization, inverter current control algorithms are used which can be classified into conventional, adaptive and intelligent techniques. Adaptive algorithms use present value to determine the future value and offer a faster response than conventional control.

Adaptive control, like least mean square (LMS), though has fast current convergence than conventional algorithms and low computational complexity, due to fixed step size, it can either have high convergence speed or low misadjustment error [5]. To overcome the shortcomings of fixed step size in LMS, variable step size is used to calculate the weights based on the error observation criterion. In [6], a VSS algorithm using squared error and error autocorrelation (VSS-SEAE) is presented. However, in an abruptly changing environment, VSS-SEAE depicts weak performance. A gradient-based VSS (GVSS-LMS) was presented recently in [7], with high convergent speed and low misadjustment level but high computational complexity. Moreover, the reader may refer [9–18] for more examples with advanced applications.

In this paper, the VSS-SEA algorithm is developed and utilized using the step size which is smoothly adjusted such that in the transient state it has a large value for fast convergent speed and in steady-state it has a small value to maintain low misadjustment levels. It provides a smoother convergence of step size under both transient and steady-state conditions. Using squared error autocorrelation-based cost function also reduces uncorrelated noise.

The developed algorithm is applied to a 25-kW grid-connected solar PV system as presented in Sect. 2. The developed control algorithm is presented in Sect. 3 The performance of the developed algorithm for different loads under changing solar irradiance is presented and compared with the established algorithm like LMS. Comparison is drawn based on their current convergence rate and current THD. Results and observations for both VSS-SEA and LMS are given in Sect. 4 followed by the conclusion. Component ratings are given in the Appendix.

2 System Design

Figure 1 shows the proposed grid-connected solar PV generating system. PV system is designed to produce 25 kW power at standard conditions. The peak power is tracked via MPPT utilizing the IC algorithm. MPPT is implemented through the medium of a DC-DC boost converter. It boosts PV voltage to 800 V. PI controller is used to maintain V_{dc} during fluctuation. DC to AC conversion for grid integration of the PV array is done with the help of a VSC.

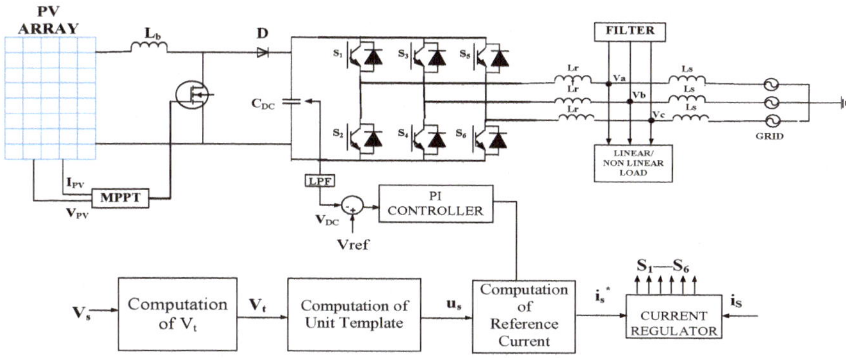

Fig. 1 Proposed grid-connected SPV generating system

3 Control Algorithm

In this section, the basic LMS control algorithm is presented. Further, the LMS-based novel VSS-SEA control algorithm is developed for a grid-connected SPV system.

3.1 *Least Mean Square*

The peak of the amplitude of PCC voltage V_t is calculated by

$$V_t = \sqrt{\left[\frac{2}{3}\left(v_a^2 + v_b^2 + v_c^2\right)\right]} \tag{1}$$

The unit vectors are derived using

$$u_a = \frac{v_a}{V_t}, u_b = \frac{v_b}{V_t}, u_c = \frac{v_c}{V_t} \tag{2}$$

Active weights related to each phase are calculated using the load current ($i_L(t)$) component as input and the gap between input and output ($y(t)$) is known as error signal ($e(t)$) [6]. For phase a,

$$e_a(t) = i_{La}(t) - y_a(t) \tag{3}$$

$$y_a(t) = \lambda_a(t)u_a(t) \tag{4}$$

where $\lambda_a(t) = \int_0^t e_a(t)u_a(t)dt$ is the active weight component for phase a. Similarly, λ_b and λ_c are calculated.

$$\lambda_{avg} = \frac{1}{3}(\lambda_a + \lambda_b + \lambda_c) \tag{5}$$

DC loss weight component (λ_{cp}) is given by

$$\lambda_{cp}(m+1) = \lambda_{DC}(m) + k_P[V_e(m+1) - V_e(m)] + k_I V_e(m+1) \tag{6}$$

where $V_e(m+1) = V_{DC}^*(m+1) - V_{DC}(m+1)$. The resultant active weight component is obtained from

$$\lambda_r = \lambda_{avg} + \lambda_{cp} \tag{7}$$

I_S^* is calculated from

$$I_{sa}^* = \lambda_r u_a,\ I_{sb}^* = \lambda_r u_b,\ I_{sc}^* = \lambda_r u_c \tag{8}$$

3.2 Variable Step Size LMS Using Squared Error Autocorrelation

V_t and unit template are evaluated from Eqs. (1) and (2), respectively. The DC loss element, λ_{cp} is obtained from (6).

To ensure the stability of this adaptive control algorithm, λ_{pv} is defined by

$$\lambda_{pv} = \frac{2P_{pv}}{3V_t} \tag{9}$$

In VSS-SEA, the evaluation of active component for phase 'a' I_L at mth instant is updated

$\lambda_{pa}(m+1) = \lambda_{pa}(m) + \mu_{pa}(m)u_a(m)e_{spa}(m)$(10) where $\lambda_{pa}(m)$ is the estimated active component at mth instant;$\mu_{pa}(m)$ is used to calculate step size using [8].

$$\mu_{pa}(m+1) = \alpha\mu_{pa}(m) + \gamma P_{pa}(m) \tag{11}$$

where the convergence time of this algorithm is dependent on $0 < \alpha < 1$ and $\gamma > 0$. They are constant parameters for controlling exponential regress and fluctuations in step size parameters, respectively. $P_{pa}(k)$ is the squared error autocorrelation function.

$$P_{pa}(m+1) = \beta P_{pa}(m-1) + (\beta - 1)[e(m)e(m-1)]^2 \tag{12}$$

β is a constant value that denotes an exponential weighting function responsible for smoothing the function. Its value lies between 0 and 1.

$E_{spa}(m)$ is the prediction error estimated using,

$$e_{spa}(m) = i_{La}(m) - u_{pa}(m)\lambda_{pa}(m) \tag{13}$$

Similarly, weight vector, μ and squared error autocorrelation for phases 'b and c' are calculated (11)–(13). The average is calculated from Eq. (5). The resultant weight vector is calculated by

$$\lambda_{rp} = \lambda_{cp} + \lambda_{Pavg} - \lambda_{pv} \tag{14}$$

The real reference grid current I_s^* is calculated and is given by Eq. (8).

3.3 PWM Current Controller

Reference current I^*_s obtained from the above techniques are matched with the sensed currents I_S via PWM current controller. The resultant error on amplification is compared to a triangular wave of 10 kHz frequency to produce a gate signal for the IGBTs (or MOSFETs) of VSC.

4 Results and Discussions

MATLAB model is developed for the proposed system. The voltage at PCC (V_s), DC link voltage (V_{dc}), source current (I_s), load current (I_L), inverter current (I_{inv}), PV voltage (V_p), current (I_p), PV power (P_{pv}), real (P) and reactive power(Q) of the grid are plotted for different loads. Unbalance is created by disconnecting phase b from 0.3 to 0.4 s. The solar irradiance is maintained constant at 1000 W/m^2 till 0.5 s after which it is reduced to 500 W/m^2. However, the temperature is maintained as constant at 25 °C.

Figure 2a and b exhibits the results for VSS-SEA control with linear and nonlinear loads, respectively. Similarly, Fig. 3a and b exhibits the results for LMS control algorithms. The rated capacity of the PV system is 25 kW but the power generated at MPP is found to be 24.910 kW. The voltage of PV at MPP is 505.6 V and the current is 49 A. The system operates in unity power factor (UPF) mode (Fig. 4).

As phase b is disconnected and unbalance occurs there is distortion in I_s and its amplitude increases. It is observed that even though I_s is sinusoidal in all both controls, in VSS-SEA it is smoother and more balanced. Also, during this condition load requirement reduces and hence more power is injected into the grid. When the

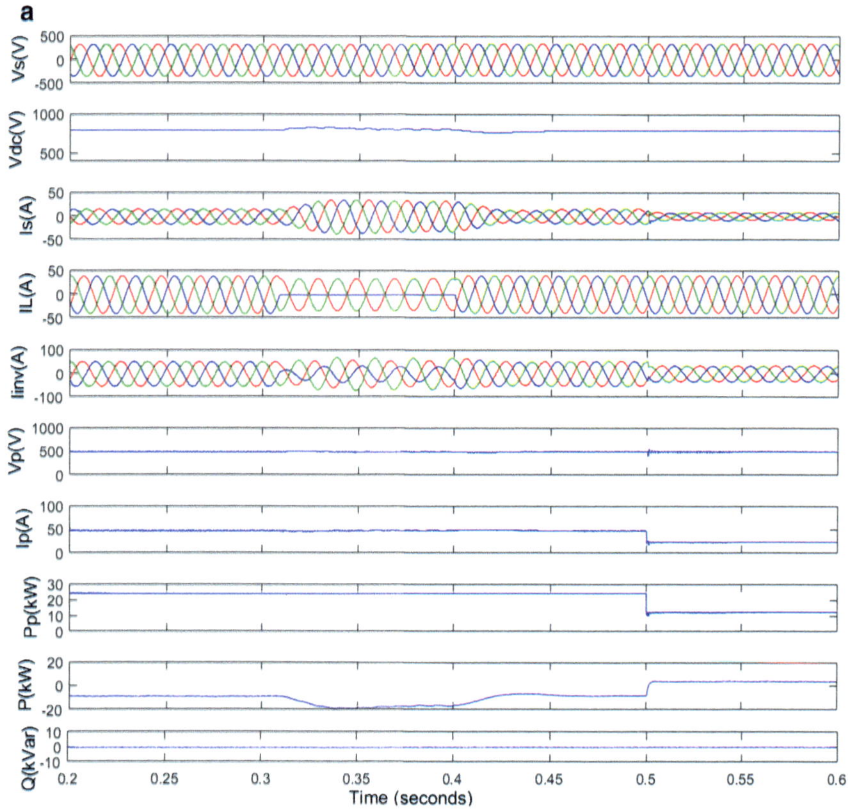

Fig. 2 **a** System performance at unbalanced linear load and varying solar irradiance using VSS-SEA. **b** System performance at unbalanced linear load and varying solar irradiance using VSS-SEA

solar irradiance is reduced to 500 W/m^2 at 0.5 s I_{pv} and P_{pv} are reduced to half and can no longer supply the entire linear load, thus the grid feeds the remaining power to the load. VSS-SEA offers a much faster response than LMS during such dynamic conditions, as evident from Fig. 2a and b. A linear load of 20 kVA at 0.8 lagging pf is connected.

V_{dc} is more stable in VSS-SEA than LMS. Disturbance in P is also reduced in VSS-SEA. Q remains constant at zero which indicates that it is entirely supplied by PV. THD of I_s using LMS is 2.06% and VSS-SEA is 1.76% which are within the IEEE-519 [9] prescribed limit. The current THD for I_L is 30.55%.

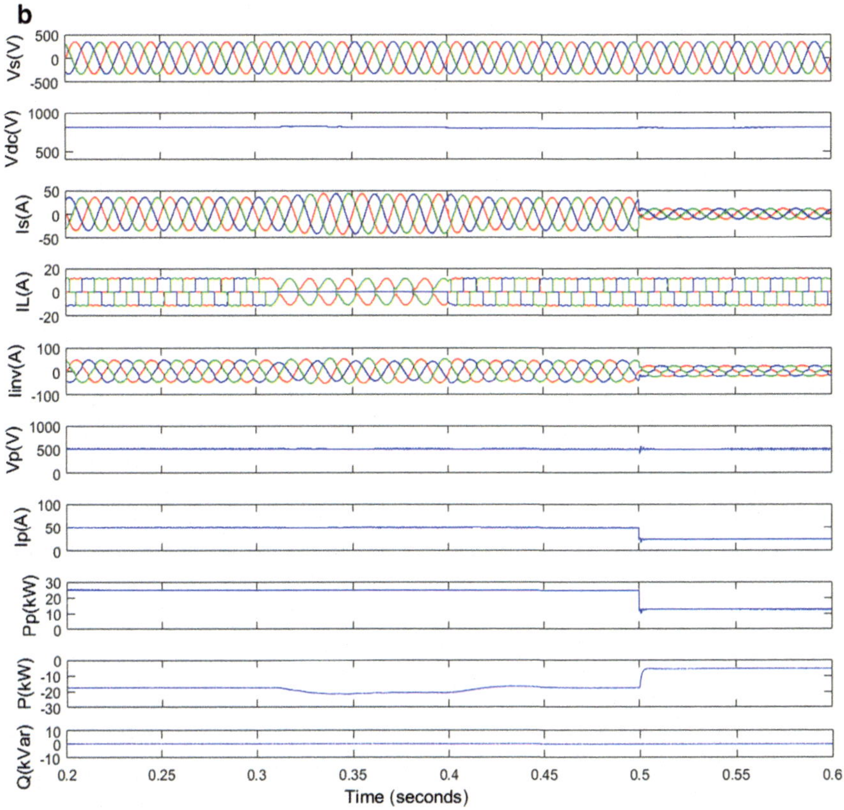

Fig. 2 (continued)

5 Conclusion

The performance of the grid-connected SPV system controlled by the VSS-SEA approach is found to be satisfactory for load balancing, reactive power compensation, harmonics elimination in UPF mode for different load under unbalancing and varying solar irradiance. It offers a faster response under dynamically changing conditions and a smooth and balanced response during unbalance. The grid current THD is 1.76% which is less than the limit prescribed by IEEE-519. I_s in VSS-SEA settles immediately at 0.5 s, indicating the high convergence speed of the algorithm in dynamically changing circumstances. During unbalance also, a more balanced grid current is seen in VSS-SEA as compared to LMS. Therefore, it can be concluded that even though VSS-SEA has more computational complexity than LMS, it offers a faster, smoother and more balanced response than LMS.

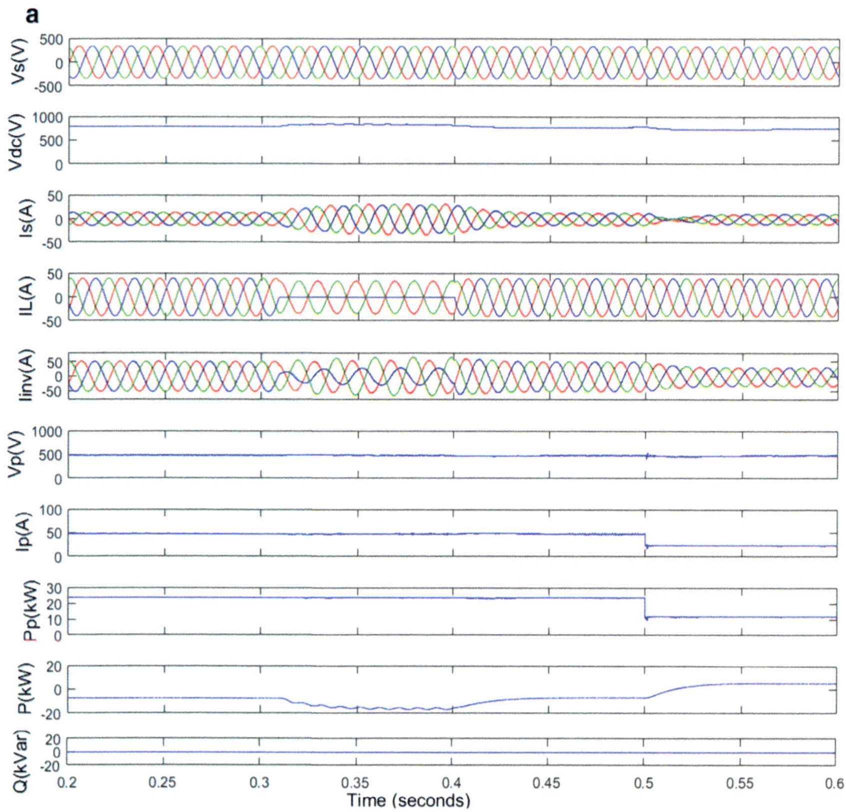

Fig. 3 **a** System performance at unbalanced linear load and varying solar irradiance using LMS. **b** System performance at unbalanced nonlinear load and varying solar irradiance using LMS

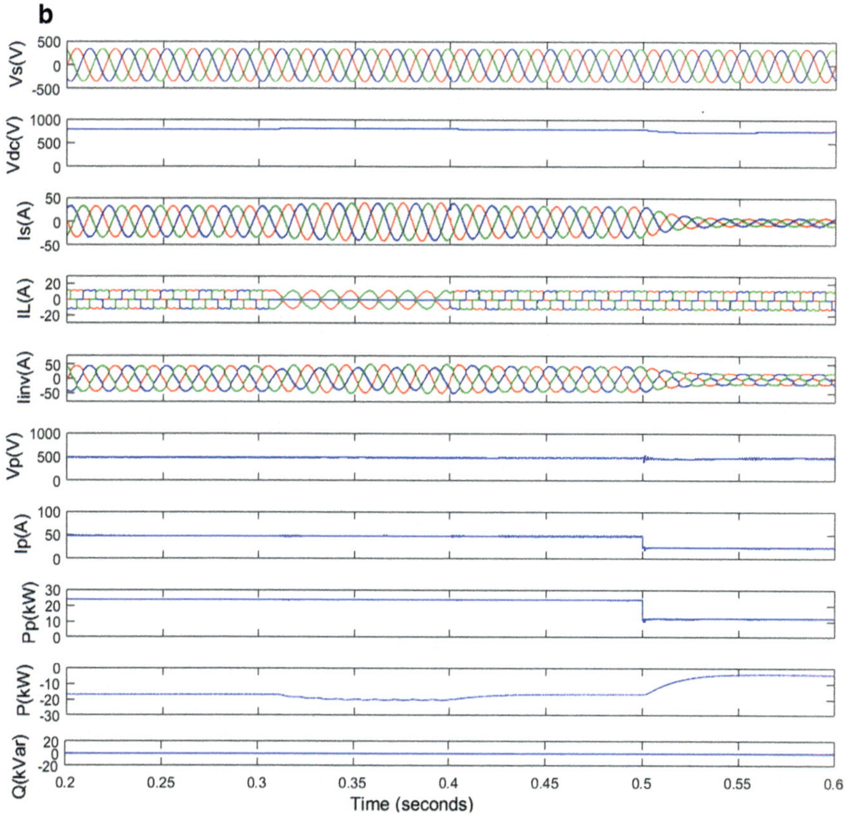

Fig. 3 (continued)

Appendix

$P_{mp} = 25$ kW, $V_{mp} = 505.6$ V, $I_{mp} = 49$A, Boost converter inductor (L_b) = 0.45 mH, $f_{sw} = 10$ kHz, $V_{dc} = 800$ V, $C_{dc} = 3000$ μF, $L_r = 4.5$ mH

Grid parameters = 415 V(rms), 50 Hz, 0.1 Ω, 1 mH

Linear load = 20 kVA, 0.8 lagging pf and nonlinear load = three-phase diode rectifier $R_{dc} = 50$ Ω and L = 50 mH.

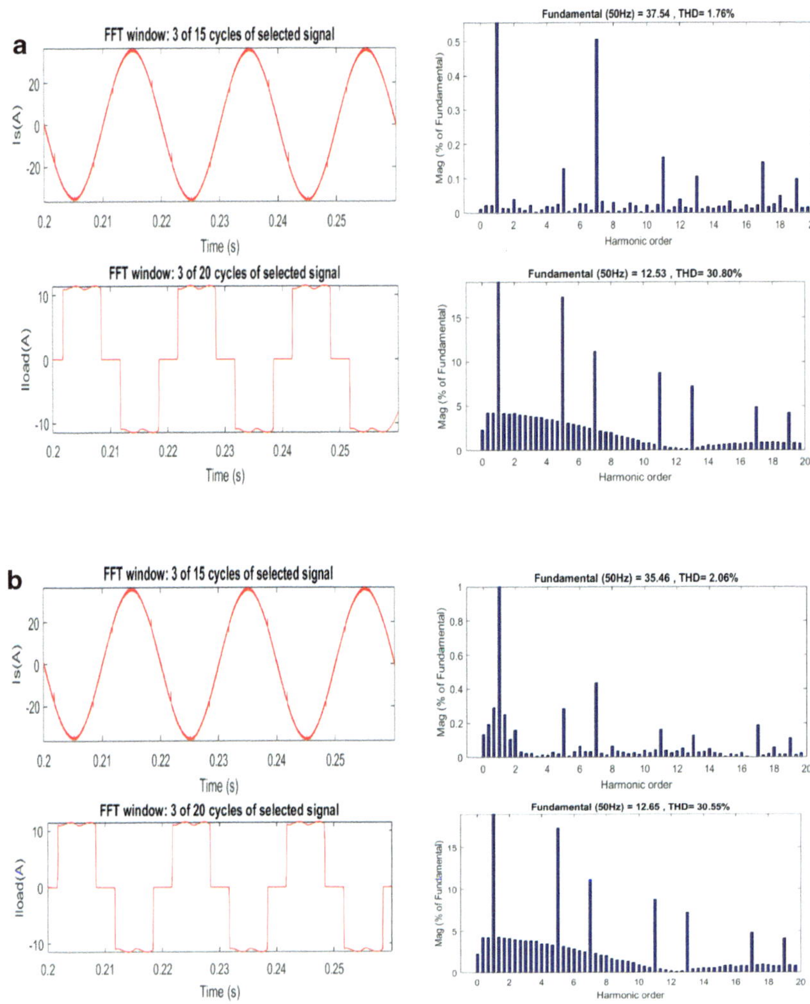

Fig. 4 **a** FFT study of grid and load currents for VSS-SEA, respectively. **b** FFT study of grid and load currents for LMS, respectively

References

1. Bo Y, Wuhua L, Yi Z, Xiangning H (2010) Design and analysis of a grid-connected photovoltaic power system. IEEE Trans Power Electron 25(4):992–1000
2. Elnagi MEM, Zaki AAD, Kotin DA (2018) Simulation and experimental validation of two-diode model of photovoltaic (PV) modules. In: XIV international scientific-technical conference on actual problems of electronics instrument engineering (APEIE). Novosibirsk, pp 244–251
3. Singh O, Gupta, SK (2018) A review on recent Mppt techniques for photovoltaic system. In: IEEMA engineer infinite conference (eTechNxT). New Delhi

4. Moreira HS, Gomes dos Reis MV, de Araujo LS, Perpetuo e Oliveira T, Villalva MG (2017) An experimental comparative study of perturb and observe and incremental conductance MPPT techniques for two-stage photovoltaic inverter. In: Brazilian power electronics conference (COBEP). Juiz de For a, pp 1–6
5. Ahmad M, Kirmani S (2019) Performance analysis of LMS based control algorithm for power quality improvement in three phase grid connected system for linear/non-linear load. Int J Power Electron Drive Syst (IJPEDS) 10, 1944–1950
6. Woo HC (2009) Variable step size LMS algorithm using squared error and autocorrelation of error. In: IEEE Conference on industrial electronics and applications, pp 2699–2703
7. Zhang Y, Li N, Chambers JA, Hao Y (2008) New Gradient-based variable step size LMS algorithm. EURASIP J Adv Signal Process 2008(105):1–9
8. Loedwassana W (2019) A variable step size algorithm of lms algorithm based on squared autocorrelation criterion. In: 7th international electrical engineering congress (iEECON). Hua Hin, Thailand, pp 1–4
9. IEEE recommended practice and requirements for harmonic control in electric power systems. In: IEEE Std 519–2014 (Revision of IEEE Std 519–1992), pp 1–29 (2014)
10. Aggarwal S et al (2020) Meta heuristic and evolutionary computation: algorithms and applications, Springer Nature, Berlin, 949 p https://doi.org/10.1007/978-981-15-7571-6. ISBN 978-981-15-7571-6
11. Ahmad MW et al (2020) A fault diagnostic and post-fault reconfiguration scheme for interleaved boost converter in PV-based system. IEEE Trans Power Electron, Early Access https://ieeexplore.ieee.org/document/9173793
12. Iqbal A et al (2021) Chapter 2—intelligent data analytics for PV fault diagnosis using deep convolutional neural network (ConvNet/CNN), intelligent data-analytics for condition monitoring. Academic Press, pp 31–43, ISBN 9780323855105, https://doi.org/10.1016/B978-0-323-85510-5.00002-8
13. Yadav AK et al (2020) Soft computing in condition monitoring and diagnostics of electrical and mechanical systems. Springer Nature, Berlin, 496 p https://doi.org/10.1007/978-981-15-1532-3. ISBN 978-981-15-1532-3
14. Gopal et al (2021) Digital transformation through advances in artificial intelligence and machine learning. J Intell Fuzzy Syst, Pre-press, pp 1–8. https://doi.org/10.3233/JIFS-189787
15. Fatema N et al (2021) Intelligent data-analytics for condition monitoring: smart grid applications, Elsevier, 268 p ISBN: 978-0-323-85511-2. https://www.sciencedirect.com/book/9780323855105/intelligent-data-analytics-for-condition-monitoring
16. Smriti S et al (2018) Special issue on intelligent tools and techniques for signals, machines and automation. J Intell Fuzzy Syst 35(5):4895–4899. https://doi.org/10.3233/JIFS-169773
17. Jafar A et al (2021) AI and machine learning paradigms for health monitoring system: intelligent data analytics. Springer Nature, Berlin, 496 p https://doi.org/10.1007/978-981-33-4412-9. ISBN 978-981-33-4412-9
18. Sood YE et al (2019) Applications of artificial intelligence techniques in engineering, vol 1, Springer Nature, 643 p https://doi.org/10.1007/978-981-13-1819-1. ISBN 978-981-13-1819-1

Optimal Power Flow and Its Vindication in Deregulated Power Sector

Ankur Maheshwari, Yog Raj Sood, Sumit Sharma, and Naveen Kumar Sharma

Abstract Optimal power flow (OPF) is one of the most important research fields in the deregulated power sector. OPF problems aim to optimize particular power system goals by modifying certain power system variables, while all the system equality and inequality constraints are satisfied. Traditional methods for solving the OPF problems are highly susceptible to initial points and frequently converge locally to optimized solutions. Addressing these issues, this paper aims to solve the OPF problems using a metaheuristic technique called particle swarm optimization (PSO) considering two objectives that minimize fuel cost with and without thermal generating unit's emission cost. The proposed approach has been evaluated on the standard IEEE-30 bus system to manifest its effectiveness.

Keywords Deregulated power sector · Power system optimization · Metaheuristic

1 Introduction

The deregulated power sector has been embraced by many developed nations such as the USA, Britain, Canada, and Australia, and India as an emerging economy is also moving toward deregulation to escalate power production and consumption efficiency. Although the system is escalating toward deregulation, electricity companies have become more aware of the generation cost, power quality issues, and system reliability. In order to decide rates for energy and congestion control, OPF is therefore gaining significance in the deregulated world.

The operators must decide on various priorities in the operation and planning of power systems. Therefore, to assist the operators, several tools have been created. One of them is OPF, which allows operators to operate the system optimally under specific

A. Maheshwari (✉) · Y. R. Sood · S. Sharma
NIT, Hamirpur, Hamirpur, HP 177005, India
e-mail: maheshwari.ankur@nith.ac.in

N. K. Sharma
I.K Gujral Punjab Technical University, Jalandhar, Punjab 144603, India

A. Tomar et al. (eds.), *Machine Learning, Advances in Computing, Renewable Energy and Communication*, Lecture Notes in Electrical Engineering 768,
https://doi.org/10.1007/978-981-16-2354-7_16

restrictions or constraints. A lot of research has been done in this area from the early 1960s onwards to minimize the cost of generation. Following the 1990 amendments to the Clean Air Act (Kyoto Protocol), working at minimal costs while maintaining protection is no longer an appropriate condition for the dispatch of electricity. In many nations, the reduction of polluted gases is now becoming mandatory for the power generating utility. Therefore, the OPF problem becomes a problem of multi-objective optimization.

The OPF optimizes the goals of the power system and complies with all the imposed system constraints. It is highly nonlinear due to nonlinearity present in the power flow equations and highly non-convex because of the thermal generator's cost function, and non-convexity increased while considering the valve-point loading (VPL) effect of the thermal generators [1]. It is highly constrained because of restrictions imposed on operating zones of the power system components. Many researchers have suggested distinct deterministic methods such as nonlinear programming, interior point method, quadratic programming, but these techniques often converge locally to optimized solutions during OPF problems evaluation. In deterministic methods, if the initial guess was close to the optimal global value, the solution converges optimally globally; otherwise, the solution will likely converge optimally at the local level [2].

The researchers were inspired by these disadvantages and shift from deterministic techniques to evolutionary techniques. A wide range of optimization methods is employed to solve OPF problems using evolutionary techniques such as PSO [3], elephant herding optimization [4], salp swarm algorithm [5], firefly algorithm [6], black hole optimization [7], Harris' Hawk optimization [8], water evaporation algorithm [9], Coyote optimization [10], Hamiltonian technique [11], whale optimization [12], and many more. Researchers worldwide have comprehensively studied the OPF problem considering minimizing the operating cost of thermal power generators. Addressing these issues, a metaheuristic technique called PSO is used to solve the OPF problem in this paper. The problem formulation is done as a mild constraint OPF problem. Two distinct objectives are considered in this study: minimizing fuel cost without considering the costs of emissions and minimizing fuel cost along with thermal power plant emission costs. The solution proposed was investigated and tested using the IEEE 30-bus standard framework. Moreover, the reader may refer [13–20] for other examples with advanced AI and machine learning applications.

The remainder of the paper is arranged as follows. For the OPF problem, the mathematical formulation is carried out in Sect. 2. In Sect. 3, the fundamental concept of the PSO is clarified. The outcome of this is addressed in Sect. 4. Final remarks shall be concluded in Sect. 5.

2 Mathematical Model

Fewer details of the IEEE-30 bus network considered in this work are described in reference [21]. The overall generating cost is equal to the summation of the fuel cost of all thermal generating units (TGU) in operation, including the penalty cost.

2.1 Thermal Generating Units Cost Model

Generating units required fossil fuel for their service. The mathematical relation between fuel costs in (\$/h) and power produced in (MW) is given in Eq. 1.

$$C(P_T) = \sum_{t=1}^{NGU} a_t + b_t P_{Tt} + c_t P_{Tt}^2 \tag{1}$$

where a_t, b_t, and c_t are the cost coefficient of generating units, P_T is the power generation of and NGU denotes the overall number of TGU.

The VPL effect should be considered for a more feasible and reliable cost function estimation. TGU with valve point effect provides a broader range of fuel-cost functions. The overall cost of generating electricity in (\$/h) will become as

$$C(P_T) = \sum_{t=1}^{\mathrm{NGU}} \left(a_t + b_t P_{Tt} + c_t P_{Tt}^2\right) + |d_t * \sin(e_t.(\min(P_{Tt}) - P_{Tt}))| \tag{2}$$

The VPL effect is indicated by the coefficients d_t and e_t. The minimal power generation of the thermal generating unit during operation is denoted by $\min(P_{Tt})$.

2.2 Emission and Carbon Tax

It is worthless to mention that electricity generation from traditional energy sources, such as thermal generating units, releases poisonous gases into the atmosphere. Thermal power plant emissions are calculated in tons/hour (t/h) and are given by

$$E = \sum_{t=1}^{\mathrm{NGU}} \left[\alpha_t + \beta_t P_{Tt} + \chi_t P_{Tt}^2\right] \times 0.01 + \left(\mu_t e^{\omega_t . P_t}\right) \tag{3}$$

where α, β, χ, μ, and ω are emission coefficients, and these values are taken from Ref. [22].

Many developed nations already imposed colossal pressure on the power sector by the levied carbon emission tax (C_{tax}) per unit volume of emitted greenhouse gases (GHG) to reduce carbon emissions and to encourage the incorporation of renewable energy sources (RES). The integration of RES into power supply networks has a range of advantages, such as reducing GHG emissions, mainly CO_2, thereby helping to address the global warming crisis. However, renewable power plant's capital costs are very high compared to the traditional power plants, while operational and maintenance costs are less costly and will continue to decrease with recent technological advances [2]. The cost of emissions is defined as:

$$\text{Emission Cost, } C_{\text{emiss}} = C_{\text{tax}} \times E \tag{4}$$

2.3 Optimization Objectives

The objectives are constructed considering all costs discussed above.

Primary objective

The primary objective function is contemplated without considering emission cost and is given as

$$C_1 = \text{minmize}\big(C(P_T) + C_{\text{penalty}}\big) \tag{5}$$

where the penalty cost C_{Penalty} is imposed for defying the constraints that are considered. A penalty is formulated as:

$$C_{\text{penalty}} = \begin{bmatrix} \psi_j\big(q_{p\max} - q_p\big)^2 & \text{if} \quad q_p > q_{p\max} \\ \psi_j\big(q_p - q_{p\min}\big)^2 & \text{elseif } q_p < q_{p\min} \\ 0 & \text{else} \quad \text{instead} \end{bmatrix} \tag{6}$$

where φ_j is the penalty factor constant to disobey the inequality constraint p.

Second objective

Emission cost is incorporated with the primary objective function for examining the transition in the power generation of TGU and to perceive the variation in their operating cost.

$$C_1 = \text{minimize}[C_1 + C_{\text{emiss}}] \tag{7}$$

Constraints

The constraints are watt and wattless power limits on generator buses, generator and load bus voltage limit, transformer tap setting limit, var compensation limit, and line flow limits, and these are referred to as inequality constraints. Equations of power balance are commonly referred to as equality constraints. In [23, 24], both constraints are mathematically explained.

3 PSO Algorithm

PSO is a population-based, smart optimization approach that is simple and efficient. In PSO, possible solutions are meant to refer to as particles, and the swarm is called the particle population. Based on their own experience and neighboring particles' experiences, each particle in a swarm moves toward the optimum or almost the optimum solution within the search space area. Let us describe the search space S in the n dimension, and there are N particles in the swarm. Each particle i constitutes position xp, velocity vl, its exclusive best position p_{best}, and global best position g_{best}. The particle's location demonstrates its fitness, and the greater the fitness, the closer the solution is to the optimum value [25]. The velocity and position of all particles were modified for each iteration to achieve better fitness. In any iteration, the particle velocity and location can be determined by Eq. 8 and Eq. 9.

$$vl_{kn}^{itr+1} = w \cdot \left(vl_{kn}^{itr}\right) + c_1 \cdot rnd_{1n}^{itr} \cdot \left(p_{best\ kn}^{itr} - xp_{kn}^{t}\right) + c_2 \cdot rnd_{2n}^{itr} \cdot \left(g_{best\ n}^{itr} - xp_{kn}^{itr}\right) \tag{8}$$

$$xp_{kn}^{itr+1} = xp_{kn}^{itr} + vl_{kn}^{itr+1} \tag{9}$$

$k = 1 \ldots N, itr$ denotes the number of iteration and w denotes the inertial weight. The rnd_1 and rnd_2 are the arbitrary numbers ranging from [0 1] and are uniformly distributed. c_1 is called the cognitive parameter, whereas constant c_2 is called the social parameter. The parameter c_1 pushes the particle toward the local best position and parameter c_2 toward the global best position. Typically, the range of c_1 and c_2 values lies between 0 and 4. The PSO parameters considered in this work are as follows: Population size is 40, and the values of w, c_1, and c_2 are 0.1617, 1.7901, and 3.1125, respectively. The velocity vl, as well as the location xp of a particle, is usually restricted by the correct limits shown as:

$$\begin{aligned} vl_{\min} &< vl_{kd} < vl_{\max} \\ xp_{\min} &< xp_{kd} < xp_{\max} \end{aligned} \tag{10}$$

The particles utilized the following equation to update their position.

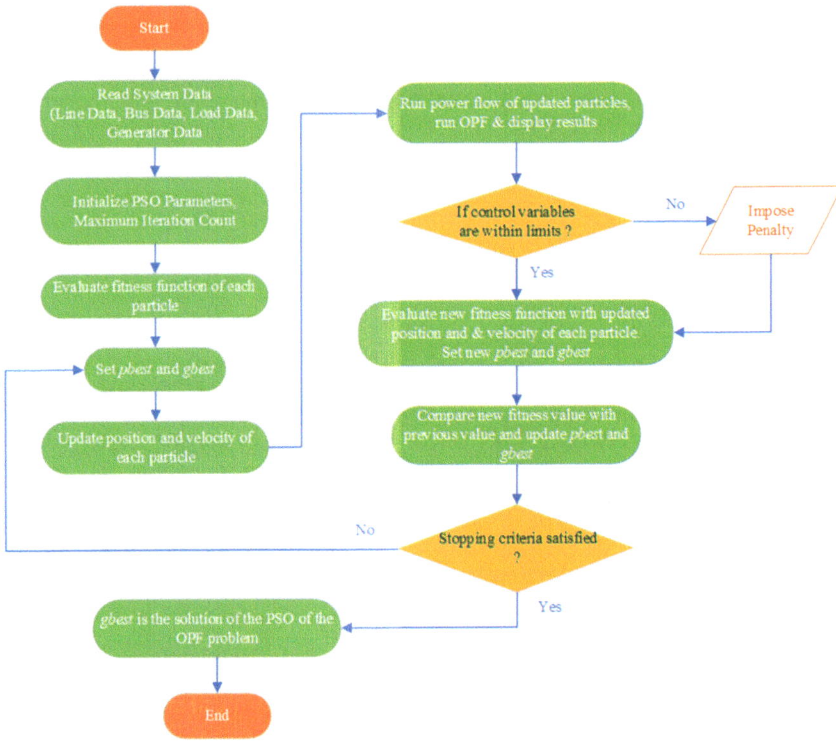

Fig. 1 The workflow of PSO-based OPF

$$p_{best\ k}^{t+1} = \begin{cases} p_{best\ k}^{t} \text{ if } C\left(p_{best\ k}^{t}\right) \le C\left(xp_{k}^{t+1}\right) \\ xp_{k}^{t+1} \text{ if } C\left(p_{best\ k}^{t}\right) > C\left(xp_{k}^{t+1}\right) \end{cases} \tag{11}$$

The global swarm is updated as:

$$g_{best\ k}^{t+1} = \left|C\left(p_{best\ k}^{t+1}\right)\right| \tag{12}$$

where C assesses the fitness value of a position. Figure 1 shows the workflow PSO-based OPF.

4 Results and Discussion

Two cases are being carried out. Table 1 tabulates the optimum setting of the control variable (CV) of all the cases with the PSO application. Each case is executed 20 times for 500 iterations to get the optimum value of considered objectives and to find the best value of CVs.

Table 1 Control variables

CV	Min	Max	Case 1	Case 2	CV	Min	Max	Case 1	Case 2
P_{T1}	50	200	177.9848	153.2761	T_{12}	0.9	1.1	1.0994	1.0641
P_{T2}	20	80	48.5939	52.8855	T_{15}	0.9	1.1	1.0197	1.0228
P_{T5}	15	50	21.2553	23.1541	T_{36}	0.9	1.1	0.9763	1.0127
P_{T8}	10	35	21.7813	32.2575	Q_{10}	0	5	4.9741	4.7107
P_{T11}	10	30	10.5429	14.2257	Q_{12}	0	5	4.6784	4.9978
P_{T13}	12	40	12.1001	15.0212	Q_{15}	0	5	2.6369	0.0433
V_{T1}	0.95	1.10	1.0999	1.1	Q_{17}	0	5	3.6102	4.1123
V_{T2}	0.95	1.10	1.0856	1.0875	Q_{20}	0	5	0.0012	4.9103
V_{T5}	0.95	1.10	1.0623	1.0627	Q_{21}	0	5	1.0654	4.9731
V_{T8}	0.95	1.10	1.0686	1.066	Q_{23}	0	5	1.9007	1.3537
V_{T11}	0.95	1.10	1.0842	1.0388	Q_{24}	0	5	3.0508	4.9939
V_{T13}	0.95	1.10	1.099	0.9826	Q_{29}	0	5	2.65	0.9533
T_{11}	0.9	1.1	0.9003	1.0727	Cost ($/hr.)			799.7335	841.2381

4.1 *Case 1*

In this case, the primary objective is evaluated, including only the fuel cost of the thermal generating units described in Eq. 5. The overall cost of generation obtained is **$ 799.7335**/hour.

4.2 *Case 2*

In this case, the objective involves the fuel and emissions cost of the thermal generating units described in Eq. 7. The aggregate cost of generation obtained is **$ 841.2381**/hour. The cost obtained in this case is **5.189%** higher than in the former case. The carbon tax imposed is considered to be **$ 30/ton** [22].

Table 1 mentioned the optimum setting of CV obtained for Cases 1 and 2, and it is observed that all CVs are within their limits. Figure 2 indicates the cost convergence characteristic for both the cases and it is noticeable that in both cases the overall fuel cost of thermal generating units becomes almost constant when the PSO algorithm converges after a significant number of iterations. The load bus voltage profile is also shown in Fig. 3. It can be concluded from Fig. 3 that the load bus voltage is also within their prescribed limits, i.e., (0.95–1.1) per unit. It is also noticeable from Table 1 that the generation changes marginally in case 2 from case 1 with the reduction in the environmental contamination content.

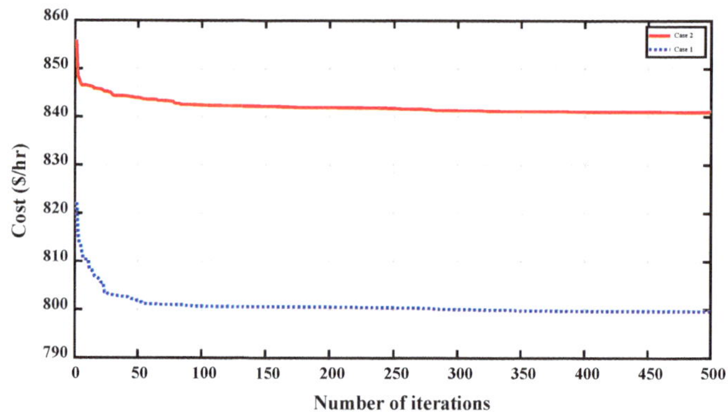

Fig. 2 Cost convergence curve

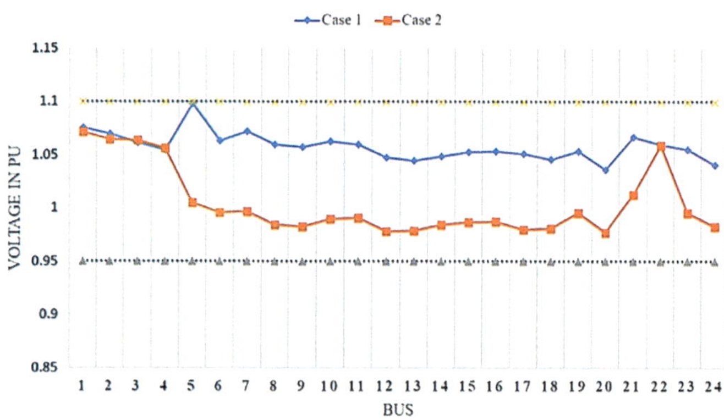

Fig. 3 Load bus voltage profile

5 Conclusion

This paper executed and effectively utilized a method of optimization inspired by nature to solve the OPF problem with two competitive objectives, generation costs without considering emission costs and generation costs considering thermal power plant emission cost. To solve these optimization problems, the PSO technique is used. The results of the simulation demonstrate that PSO guarantees strong convergence and seeks better solutions. This OPF approach can optimize any number of objectives and help system operators choose judicious decisions to operate the system effectively and economically. For further research work in the field, this paper's outcome can be used as a benchmark.

References

1. Sood YR (2007) Evolutionary programming based optimal power flow and its validation for deregulated power system analysis. Int J Electr Power Energy Syst 65–75
2. Khaled U, Eltamaly AM, Beroual A (2017) Optimal power flow using particle swarm optimization of renewable hybrid distributed generation. Energies 10(7):1013
3. Kumar P, Pukale R (2015) Optimal power flow using PSO. Computational intelligence in data mining, vol. 1, conference 2015, New Delhi. Springer, India, pp 109–121
4. Li W, Wang G-G, Alavi AH (2020) Learning-based elephant herding optimization algorithm for solving numerical optimization problems. Knowl Based Syst 195:105675
5. El-Fergany AA, Hasanien HM (2020) Salp swarm optimizer to solve optimal power flow comprising voltage stability analysis. Neural Comput Appl 32(9):5267–5283
6. Wahid F, Ghazali R (2021) A hybrid approach of firefly and genetic algorithm for solving optimisation problems. Int J Comput Aided Eng Technol 14(1):62
7. Velasquez OS, Montoya Giraldo OD, Garrido Arevalo VM, Grisales Norena LF (2019) Optimal power flow in direct-current power grids via black hole optimization. Adv Electr Electron Eng 17(1)
8. Hussain K, Zhu W, Mohd Salleh MN (2019) Long-term memory Harris' hawk optimization for high dimensional and optimal power flow problems. IEEE Access 7:147596–147616
9. Saha A, Das P, Chakraborty AK (2017) Water evaporation algorithm: a new metaheuristic algorithm towards the solution of optimal power flow. Eng Sci Technol Int J 20(6):1540–1552
10. Li Z, Cao Y, Van Dai L, Yang X, Nguyen TT (2019) Optimal power flow for transmission power networks using a novel metaheuristic algorithm. Energies 12(22):4310
11. Tehzeeb-Ul-Hassan H, Tahir MF, Mehmood K, Cheema KM, Milyani AH, Rasool Q (2020) Optimization of power flow by using Hamiltonian technique. Energy 6:2267–2275
12. Medani KBO, Sayah S, Bekrar A (2017) Whale optimization algorithm based optimal reactive power dispatch: a case study of the Algerian power system. Electric Power Syst Res
13. Aggarwal S et al. (2020) Meta heuristic and evolutionary computation: algorithms and applications. Springer Nature, Berlin, p 949.https://doi.org/10.1007/978-981-15-7571-6. (ISBN 978-981-15-7571-6)
14. Yadav AK et al. (2020) Soft computing in condition monitoring and diagnostics of electrical and mechanical systems. Springer Nature, Berlin, p 496. https://doi.org/10.1007/978-981-15-1532-3. ISBN 978-981-15-1532-3
15. Gopal et al. (2021) Digital transformation through advances in artificial intelligence and machine learning. J Intell Fuzzy Syst 1–8. https://doi.org/10.3233/JIFS-189787
16. Fatema N et al. (2021) Intelligent data-analytics for condition monitoring: smart grid applications. Elsevier, p 268. ISBN: 978-0-323-85511-2. https://www.sciencedirect.com/book/9780323855105/intelligent-data-analytics-for-condition-monitoring
17. Smriti S et al (2018) Special issue on intelligent tools and techniques for signals, machines and automation. J Intell Fuzzy Syst 35(5):4895–4899. https://doi.org/10.3233/JIFS-169773
18. Jafar A et al. (2021) AI and machine learning paradigms for health monitoring system: intelligent data analytics. Springer Nature, Berlin, p 496. https://doi.org/10.1007/978-981-33-4412-9. ISBN 978-981-33-4412-9
19. Sood YR et al. (2019) Applications of artificial intelligence techniques in engineering, vol 1. Springer Nature, p 643. https://doi.org/10.1007/978-981-13-1819-1. (ISBN 978-981-13-1819-1
20. Sharma NK et al. (2021) Optimal design of passive power filter using multi-objective Pareto-based firefly algorithm and analysis under background and load-side's nonlinearity. IEEE Access. https://doi.org/10.1109/ACCESS.2021.3055774
21. Biswas PP, Suganthan PN, Amaratunga GAJ (2017) Optimal power flow solutions incorporating stochastic wind and solar power. Energy Convers Manag 148:1194–1207
22. Bouchekara HREH, Chaib AE, Abido MA, El-Sehiemy RA (2016) Optimal power flow using an improved colliding bodies optimization algorithm. Appl Soft Comput 42:119–131

23. Umapathy P, Venkataseshaiah C, Arumugam MS (2010) Particle swarm optimization with various inertia weight variants for optimal power flow solution. Discrete Dyn Nat Soc 2010:1–15
24. Kumar V, Naresh R (2020) Application of BARON solver for solution of cost based unit commitment problem. Int J Electr Eng Inform 12(4):807–827
25. Kumar V, Sharma RN, Sikarwar SK (2018) Multi-area economic dispatch using dynamically controlled particle swarm optimization. In: Lecture notes in electrical engineering. Springer Singapore, Singapore, pp 151–163

Distribution Automation and Energy Management System in Smart Grid

Devki Nandan Gupta, Abhishek Sharma, and Deepak Verma

Abstract The conventional energy system dependent on exhaustible resources which use single-way interfacing between the source and the consumer via generation, transmission, distribution and consumption is inefficient, unsustainable and non-reliable due to the high energy needs of the digital society. A new system must be defined which overcomes all the above disadvantages and is self-reliable and manageable. The objective of this paper is to compare the conventional electricity grid system with the modern (smart grid) system, to define the smart grid, its components and their impact on the future of the electrical system. There is a need to use electrical energy in a more efficient and sustainable manner. The aim is to define a system that will be cyber secured and a two-way interface between the supplier and customers. This system will work on real-time data and also integrate it with technologies, tools and techniques to identify faults and reconfigure them automatically based on proper diagnostics. Also, several new components such as distribution generators and mini-grids will be introduced which will run the system uninterruptedly during the outages. This paper gives a brief about the distribution automation and energy management system post comparing conventional grid system with the smart grid system.

Keywords Smart grid · Distribution automation · Energy management system

1 Introduction

The idea of smart grid is the efficient, reliable and secure use of electric energy using renewable resources, viz., wind, thermal, solar, etc. since the carbon containing resources are exhausting, making the power supply cost-effective and less reliable [1].

Adding the new infrastructure (such as energy storage system), advanced communication technologies, security and intelligence to the conventional grid system along

D. N. Gupta (✉) · A. Sharma · D. Verma
Department of Electrical & Electronics Engineering, Birla Institute of Technology, Mesra, Jaipur Campus, Jaipur, Rajasthan, India

A. Tomar et al. (eds.), *Machine Learning, Advances in Computing, Renewable Energy and Communication*, Lecture Notes in Electrical Engineering 768,
https://doi.org/10.1007/978-981-16-2354-7_17

with relying on renewable resources like wind, PV cells, thermal energy etc. rather than the traditional ones will lead to the smart grid system [2].

Smart grid system not only upgrades the system but also empowers customer services and creates new products. Installing the smart meters to the customer's side is a two-way communication source between the demand and supply side [3]. The meter also provides complete knowledge about how much energy used when it is used and how much they are paying [2]. Also, it alerts about not using electricity during peak load time and hence maintains the demand/load curve smooth. The smart grid system collects the real-time data and analyze it using the technology and informs both the side which helps in alerting about faults/overloads or blackouts hence increasing reliability. Apart from this, the reader may refer to book and/or journal special issues [4–10] for more examples.

2 Smart Grid

2.1 Definition

Smart grid is an electrical system that uses two-way cyber secure communication technologies modern intelligence in an integrated way across major components of power system, viz., generation, transmission, distribution and consumer, making the system efficient, secure, clean, dependent and sustainable [1, 2].

2.2 Comparing Conventional Grid System and Smart Grid System [2, 11]

The conventional grid system and smart grid system have various differences on the basis of various factors which are shown in the following table (Table 1).

2.3 Challenges

- **Use of Renewable Resources**—Accurate wind and solar forecast are not possible with the present technology.
- **Demands**—Smart grid uses HVDC for demand response application. Building it will be a huge challenge to the smart grid system.
- **Cost**—This system contains various components such as advanced metering system, energy storage and metering system, advanced communication technology, etc. which will require a huge investment and security too.

Table 1 Difference between conventional grid system and smart grid system

Characteristics	Conventional grid	Smart grid
1. Generation	Generation of electricity in Central Power Station	Generation is a mix of centralized system and the distributed generators
2. Source of energy	Diesel engine, hydrocarbons like coal, petroleum	Traditional sources integrating with renewable energy sources like wind, solar, thermal, etc.
3. Power flow and communication	Unidirectional power flow/communication from supplier to customer	Two-way cyber secured power flow using smart meter and advanced telecommunication technologies (Digital Communication)
4. Equipments	Electromechanical sensors and relays	Digital relays, sensors, etc.
5. Monitoring and repairing	Manual monitoring of power and fault issues, hence slow reaction time Manual repairing from control center	Self-monitoring is done using real time data. These are fast self-healing grids
6. Impact of blackouts/emergencies	Damage and loss suffering (no alternative)	Network islanding is an option or using the mini grids or distribution generators when main generators suffer outages
7. Efficiency	Less energy efficiency and high loss rate	Increase in efficiency and decrease in loss rate
8. Power quality and control	Low power quality and complex control	High power quality and easy control

- **Strength**—Since, the electrical system is highly complex, vast and involves humans, strength of grid will be an important challenge otherwise it could be the risk of life.
- **Communication**—The two-way traversal and security of real-time data is not an easy task. It includes testing on various levels and in normal to the extreme weather conditions of India; hence it is to be taken care of.
- **Preparing of Plug-in hybrid vehicles**—Integration of a field that is in a developing stage will increase its customers; along with its versatile nature it will be hard to use vehicle batteries as a support to the grid system [2, 12].

3 Distribution Automation

Distribution automation is a technology that includes digital switches, sensors, processors, information technology with advanced control and communication to collect, analyze and optimize the data to automate feeder switching, voltage regulation, equipment health monitoring and outage management. It improves speed, cost and accuracy, thus improving efficiency and reliability of the distribution system and customer satisfaction [13].

3.1 Benefits of Distribution Automation

3.1.1 Reliability

Distribution automation provides fault location, isolation and service restoration (FLISR) which automatically locates and isolates the fault; hence restores power, reducing the outage duration. It also conveys real-time information to the consumer about the outage status; hence the customer does not have to report since the automation system notifies and using the GPRS, repair crews can dispatch at the location reducing the vehicle miles [13, pp. 4–7].

3.1.2 Voltage and Reactive Power Management

The automation reduces the peak demand and improve power factor, hence the power quality increases the efficiency of the distribution system and reduces the customer bills. It enables the voltage and reactive power to improve the power factor which manages the reactive power flow and reduces losses [13, pp. 4–7].

All the components are attached with digital sensors which give real-time data, real-time alerts during abnormal conditions (such as fault/overloads) which helps in diagnosing the data and preparing proper repair techniques. This helps in improving operational efficiency since equipment gets extra life [13, pp. 4–7].

3.1.3 Distribution Energy Resources Integration

It uses various technologies such as diesel engines, PV cells, wind turbines, energy storage systems, load control systems etc. for onsite power generation and storage. Here, the consumers can also act as producers, hence reduction in bill amount; also the use of renewable resources reduces carbon emission and decreases the dependence on the central generation system [13, pp. 4–7].

3.2 *Distribution Automation Field Devices*

3.2.1 Remote Fault Indicators

These are the sensors that distinguish and tell about the overloading of voltage and current or fault conditions. They are connected with visual displays and communication technology like SCADA, increasing accuracy. This makes the system reliable and efficient reducing the time and cost of the repair crew [13].

3.2.2 Smart Relays

They are software-based technology that accurately detects a fault, isolates it and analyzes it. They can be used for switching according to the algorithm as well as for protection. These devices also keep the track of data and send it to operators for analysis [13, pp. 14–15].

3.2.3 Automated Feeder Switches and Monitors

The feeder switches open and close for isolation and automatically reconfigure the fault part to restore power, increasing customer satisfaction and efficiency. The switches are operated with smart relays by distributional management systems. When some object comes in touch with a power line during high winds, these switches automatically open and close reducing the damages. Automatic monitors measure the load on distribution lines and notify if the damaging level is nearby. They collect real-time data and pass it to departments and hence control the damage as operators can take actions accordingly [13, pp. 14–15].

3.2.4 Automated Capacitor and Voltage Regulator

Capacitors compensate for the reactive power caused by the inductive loads reducing energy wastage due to reactive power in the feeder. The distribution capacitor bank is the group of several capacitors connected and the size depends on the amount of kVAR. The voltage regulator raises or lowers the voltage level according to the change in load [13, pp. 14–15].

3.2.5 Transformer Monitoring

These give information about the transformers or any abnormal condition which can lead to damage [13, pp. 14–15].

3.2.6 Communication Network

Various advanced communication technologies have been adopted for collecting, processing and traversal of data. These are further connected with automation systems such as SCADA for analyzing or operations. Various wireless and wired communication systems have been selected according to their features such as 4G/LTE, coaxial cable, Ethernet, 2G, Bluetooth etc. The demand is for two-way cyber secure, fast and highly reliable communication [13, pp. 14–15].

3.3 Case Study of Jinan Grid

Jinan power grid with a power supply area of 8177 km^2 serves 1.93 million customers in 11 districts and countries in Jinan. The 10 kV distribution grid of length 13,525.291 km consists of 1429 lines all of which are intelligent (distribution automation system) upgraded [14].

Before this upgrade, 10 kV Luodong line of the Jinan grid was connected to the 10 kV #1 bus of 110 kV Luokou station by a circuit breaker consisting of three sectionalizing switches. The single line diagram of the line after the upgrade is shown in Fig. 1.

On 25 July 2011, a permanent fault occurred in the Luodong line due to the thunderstorm weather, the fault point was found 3 h and 31 min after the fault has occurred and the power supply for the non-fault section was restored after 224 min of occurrence of a fault. Also, the range of blackout expanded due to n interconnection lines [14].

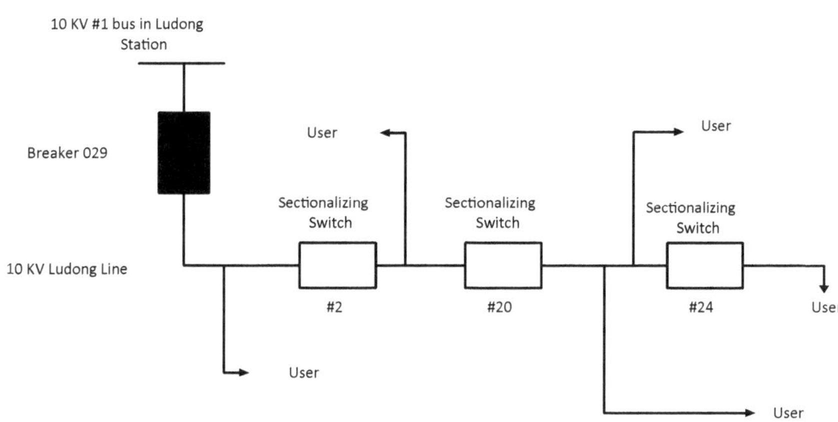

Fig. 1 The single line diagram of 10 kV Ludong line before the intelligent upgrade

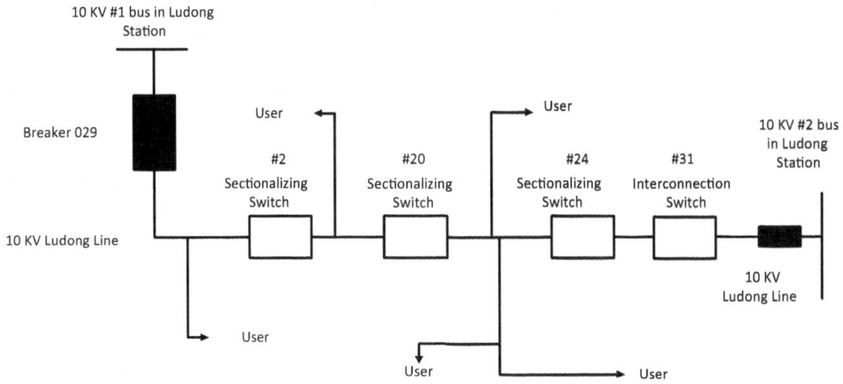

Fig. 2 The single line diagram of 10 kV Ludong line after the intelligent upgrade

In September 2013, the 10 kV Luodong line was upgraded with the centralized feeder automation DTS and self-healing ability [14]. The single line diagram of the line after the upgrade is shown in Fig. 2.

On 6 May 2014, around 11:06:09 p.m., a permanent fault occurred in the Luodong line. Circuit breaker 029 tripped and the reclosure failed. The DAS station received the fault information by intelligence in the positions of #20 and #24 poles. Within 36 s, the station located the fault section between #20 and #24 poles, isolated it and also transferred the load [14]. In 5 s, the switches of #20 and #24 poles were turned off and locked. At 11:07:06 p.m., the 029 breaker was turned on to restore the power supply of the user line in front of the #20 pole. In 4 s, the interconnection switch of #31 poles was turned on by remote control which restored the power supply behind #24 pole sections [14].

Comparing the system in 2011 and 2014, we find that the power supply recovery time reduced from 224 min to 61 s in the non-fault section improving the efficiency by 99.55% and the fault detection time also reduced to 29 min from 211 min improving the efficiency by 86.26% [14].

4 Energy Management System

Energy management in smart grid energy management is becoming more and more important as the energy demand is continuously rising and the power system is expanding. To fulfill the demand, the energy resources (renewable and non-renewable) are integrated together and this makes the power system bigger, more complex than ever before [3]. Hence there is a very high requirement of some methodologies and smart techniques so that the complexity in the system can be efficiently dealt with in cases of malfunctions and faulty operations in the power system. Though the energy can be reproduced by the renewable energy resources but the transmission and distribution are also the major part of the power system, which includes several losses, and to minimize such power losses there must be an algorithm (software or hardware) that can help to achieve desirable efficiency according to supply and demand [15]. This smart algorithm must use the data and maintains a balance between supply and demand and the study and implementation of these techniques and algorithms are the building blocks of the energy management concept in smart grid [16].

4.1 *Definition*

Energy management in the smart grid refers to the techniques (hardware and software) such as intelligent system drivers like microprocessors and microcontrollers by which a balance is maintained between supply and demand (production and consumption) by acquiring the data and combining it with advanced communication and networking techniques and then taking a suitable action for carrying out the desired performance the power system [3, 15]. In simple words, energy management refers to the efficient production and consumption of energy and maintaining a balance between the different operations of the power system using intelligent systems to meet the efficiency and energy requirements [3, 17].

4.2 *Need of Energy Management System*

- Due to the integration of the renewable energy resources (e.g., wind, solar, etc.) in the smart grid the energy obtained is fluctuating in nature, at the same time, the loads such as electric vehicles are also fluctuating in nature and this creates a mismatch between supply and demand (load side and source side). This decreases the overall performance and the stability of the power system [3, 17].
- To deal with the issue of power demand/supply mismatch, the electric utility has the following two options:

    1. Increase the size of the power system by adding new generation plants and expanding the grid by integrating more and more renewable sources of energy [3, 17].
    2. The other method deals with the efficient and smart transmission, distribution, consumption and storage of the energy produced [3, 17].

- Now, among the two options, the second option is more reasonable, economical and efficient. Due to the increasing demand for energy, it is not only important to increase energy production, but at the same time, it is also required to efficiently manage the energy which is being produced [17].

- If we would not manage the energy produced in a smart way then the losses will continue to rise and lowering the performance of the power system, thus the energy management system of the smart grid enables the ability of the smart grid to be more efficient. That's why energy management is very much important in a smart grid [3].

4.3 Types of Energy Management

(a) Energy management on the supply side

The generation can be balanced according to the demand by turning on and off the generators to meet the demand. For example, if the need of the consumer can be fulfilled by one generator, then the other generators must be off, and if demand increases the additional generators are turned on [17].

(b) Energy management on consumer side

This is also called energy management on the demand side and it includes the smart consumption and storage of the energy by the end-users to make the demand/supply balanced [17].

4.4 Tools and Techniques Used for Energy Management in Smart Grid

Now the main task to make the smart grid smarter and efficient is to implement the tools for energy management which includes modern hardware and software techniques such as advanced communication and intelligent computer-based systems to acquire the data and process of that data to take a suitable action [3, 16]. Here are some of the most used techniques which are used for energy management in smart grids.

1. Programmable Logic Controller (PLC)

A programmable logic controller or PLC is an industrial digital computer that controls and automates a system and provides the desired functions [3]. A PLC automates a specific process, functions or even an entire production line. The working of a PLC is based on processing the data given to it and implementing that data to carry out the desired result [15].

2. Supervisory Control and Data Acquisition (SCADA)

Supervisory control and data acquisition or SCADA is an industrial computer system that monitors a process and acquires the data at different instants. SCADA monitors the transmission, distribution and consumption part of the electrical power system. This tool also provides a cyber-secure acquisition of the data [3, 15].

3. Battery Management System (BMS)

A battery management system or BMS is an electronic system used to manage and protect the storage of the energy in batteries and their safe operation. The nature of the electric energy obtained from the renewable energy sources is dc that's a requirement arise to store and then use the energy [18]. For the relays to operate dc supply is required and for that battery is used. So battery management system becomes important in smart grid important [3].

4. Building Management System

A building management system or building automation system is a computer-based intelligent control system that controls and monitors the building's mechanical and electrical equipment such as ventilation, lightning and energy usage. Building management system is important for consumer-side energy management [3, 17].

5. Home automation systems

Home automation system includes latest technologies such as internet of things (IOT) and artificial intelligence which makes the energy consumption more efficient by reducing the energy wastage and power consumption. Home automation systems are the major part of the smart grid at the end-user [3, 17].

4.5 Benefits of Energy Management in Smart Grid [16, 19]

First, the energy management in the smart grid increases the overall power system efficiency by decreasing the transmission and distribution losses, making the energy consumption more intelligent using modern tools and techniques. It has the following advantages:

- It automates the system and reduces human efforts.
- The results obtained are accurate and precise so the action to be taken by the data available will also be accurate.
- Better utilization of the energy resources and generation units and decreases the cost of production of energy.
- The energy losses are less during transmission so the cost of transmission reduces.
- The power system is stabilized from both the ends i.e., supply end and demand end.
- The end-user also helps in maintaining a balance between supply and demand.
- The data is cyber-secure.
- It gives accurate predictions which are very important in load forecasting.
- It stabilizes the power system economically and environmentally.

4.6 *Electricity Tariff System*

The rate at which the electricity is sold to the consumers by the electric utilities is called electricity tariff. To minimize the overall cost of energy the traditional energy tariff i.e., fixed tariff is not sufficient [16]. So to overcome the requirements many new tariff concepts are being created such as demand response programs [3]. The electric tariff varies with time and if the consumers exceed a certain limit they will have to pay a penalty [16].

5 Conclusion

Since the conventional sources of energy are less reliable, less efficient and are for a limited time, the use of renewable energy as a source of energy has started and is increasing in the past few years. Due to this increase in the demand for renewable sources of energy, an intelligent system is required that can reduce the complexity of the system, at the same time, increase the overall efficiency and show real-time data. Implementation of this system requires modern tools and techniques.

Distribution automation and energy management in the smart grid system are techniques that play a vital role in making the future grids more intelligent, reliable and efficient. DAS reduces the blackout time due to fault in the distribution grid and also improves the power supply reliability [14]. During January–April 2014, the DAS self-healing function started 47 times, saving economic loss of 13.22 million and electricity of 0.28 million kWh [14, 20].

There are some challenges in implementing these concepts due to lack of research and knowledge but continuous efforts and research in exploring smart grid systems will definitely lead to a prominent future in the renewable energy revolution.

References

1. Wikipedia contributors. Smart grid. Wikipedia, The Free Encyclopedia. Wikipedia, The Free Encyclopedia, 20 Jun 2020. Web. 14 Jul 2020
2. Anderson R, Ghafurian R, Gharavi H (2018) Smart grid the future of the electric energy system
3. Nanda P, Panigrahi CK, Dasgupta A energy management system in smart grid: an overview
4. Aggarwal S et al (2020) Meta heuristic and evolutionary computation: algorithms and applications. Springer Nature, Berlin, p 949. https://doi.org/10.1007/978-981-15-7571-6. ISBN 978-981-15-7571-6
5. Yadav AK et al (2020) Soft computing in condition monitoring and diagnostics of electrical and mechanical systems. Springer Nature, Berlin, p 496. https://doi.org/10.1007/978-981-15-1532-3. ISBN 978-981-15-1532-3
6. Gopal et al (2021) Digital transformation through advances in artificial intelligence and machine learning. J Intell Fuzzy Syst, Pre-press 1–8. https://doi.org/10.3233/JIFS-189787

7. Fatema N et al (2021) Intelligent data-analytics for condition monitoring: smart grid applications, Elsevier, p 268. ISBN: 978-0-323-85511-2. https://www.sciencedirect.com/book/9780323855105/intelligent-data-analytics-for-condition-monitoring
8. Smriti S et al (2018) Special issue on intelligent tools and techniques for signals, machines and automation. J Intell Fuzzy Syst 35(5):4895–4899. https://doi.org/10.3233/JIFS-169773
9. Jafar A et al (2021) AI and machine learning paradigms for health monitoring system: intelligent data analytics, Springer Nature, Berlin, p 496. https://doi.org/10.1007/978-981-33-4412-9. ISBN 978-981-33-4412-9
10. Sood YR et al (2019) Applications of artificial intelligence techniques in engineering, vol 1, Springer Nature, p 643. https://doi.org/10.1007/978-981-13-1819-1.ISBN 978-981-13-1819-1
11. Zahedi A (2011) Developing a system model for future smart grid. In: Proceedings of the 2011 IEEE PES innovative smart grid technologies. Perth, WA, pp 1–5. https://doi.org/10.1109/ISGT-Asia.2011.6167326
12. Gözde H, Taplamacıoğlu MC, Arı M, Shalaf H (2015) 4G, LTE technology for smart grid communication infrastructure. In: 3rd international Istanbul Smart Grid Congress and Fair (ICSG), Istanbu, pp 14. https://doi.org/10.1109/SGCF.2015.7354914
13. Distribution Automation by U.S. Department of Energy, SMARTGRID.GOV, September 2016
14. Xu B, Li T (2010) Investigation to some distribution automation issues. Autom Electr Power Syst 34(9):81–86
15. Rathor SK, Saxena D Energy management system for smart grid: an overview and key issues
16. El-Bayeh CZ, Alzaareer K Energy management in smart grid
17. Nozaki Y, Tominaga T, Iwasaki N, Takeuchi A (2011) A technical approach to achieve smart grid advantages using energy management systems. In: Proceedings of the international conference on wireless communications and signal processing (WCSP), Nanjing, China, 9–11 November 2011, pp 1–5
18. Wikipedia contributors. Battery management system. Wikipedia, The Free Encyclopedia. Wikipedia, The Free Encyclopedia, 3 Jun 2020. Web. 15 Jul 2020
19. Amaral J, Reis C, Brandão RFM, Energy management systems. In Proceedings of the 2013 48th international universities' power engineering conference (UPEC) Dublin, pp 1–6 doi:https://doi.org/10.1109/UPEC.2013.6715015
20. Fu Y, Xuan Z, Jinguo D, Chao Q, Rongfei Q (2014) Research and application of distribution grid self-healing based on distribution automation. In: 2014 China international conference on electricity distribution (CICED), Shenzhen, pp 1211–1214. https://doi.org/10.1109/CICED.2014.6991899

Optimal Planning of Green Hybrid Microgrid in Power Industry

Naveen Kumar Sharma, Sumit Sharma, Yog Raj Sood, Ankur Maheshwari, and Anuj Banshwar

Abstract In supply-side planning for microgrids, renewable energy sources will be recognized gradually as major options. This research paper proposed a green microgrid system consisting of a solar photovoltaic, hydro turbine, battery, diesel generator (DG) and converter. Four different cases are studied in a simulation environment, to compare and evaluate the most feasible solution based on the cost parameters of the system. The analysis was also carried out to find out the electrical power production and environmental pollutants of different components for the typical Iraqi rural village, i.e., Sakran in district Choman.

Keywords Microgrid · Hybrid energy system (HES) · Solar photovoltaic (SPV)

1 Introduction

The uses of renewable sources of energy have received increasing attention in the production of electricity as an alternative to conventional energy sources. In order to minimize greenhouse gas emissions, renewable energy sources incorporated into the fossil fuel systems could have a vital role. Because of the high cost of transmission in remote communities, there is a necessity for the use of green HES. In comparison to conventional sources, a green HES would be a more stable energy supply. The previous analysis shows that the hybrid system has a very less cost solution in comparison with the isolated microgrid system [1]. Also, the hybrid system consisting of more renewable energy sources has less environmental pollution. Many researchers

N. K. Sharma
Electrical Engineering Department, I.K Gujral Punjab Technical University, Jalandhar, Punjab, India

S. Sharma (✉) · Y. R. Sood · A. Maheshwari
Electrical Engineering Department, National Institute of Technology, Hamirpur, H.P., India
e-mail: sumitsharma8882@nith.ac.in

A. Banshwar
Government Polytechnic, Puranpur (Pilibhit), U.P, India

A. Tomar et al. (eds.), *Machine Learning, Advances in Computing, Renewable Energy and Communication*, Lecture Notes in Electrical Engineering 768,
https://doi.org/10.1007/978-981-16-2354-7_18

found that renewable energy is the vital solution for electricity in rural locations [3, 4]. The remote rural communities in Iraq are not much developed and have an inappropriate amount of electricity. Most of the rural places are dependent on diesel generators for electricity [5]. As the natural gas and crude oil availability is abundant, besides these there is a great potential for renewable energy sources in Iraq [6]. Due to environmental uncertainties, asperity results in more usage of renewable energy sources. To mitigate the shortcomings of individual energy sources, HES has been developed. Such types of systems can be more efficient, viable and weather-friendly [7]. The author in reference [8] found that hybrid system has very less environmental emission compared to DG alone system. Mostly HES combinations are effectively and efficiently at meeting future energy demand [9]. The results obtained demonstrated that in both Enugusite and Maiduguri, HES was the most profitable structure to power rural health centres. A hybrid energy system is an authentic and viable option, particularly in locations where the environmental conditions are not constant.

Many factors would be considered in the optimal planning of green hybrid microgrid, like optimal sizing and capacity of energy sources, different cost factors etc [10]. In this research paper, optimal planning of green hybrid microgrid is done in order to meet load demand effectively and efficiently in the selected location. We consider green hybrid microgrids because of the more use of renewable energy sources. Apart from this, the reader may refer to book/journal special [11–18] for more examples based on AI and machine learning applications.

The green microgrid system will supply the load in a specified location by using solar photovoltaic (SPV) as well as hydropower as the main generation source together with battery bank and DG are the backup power source. Section 2 will present the load profile, different system resources, system components and generation modeling. Results and discussion of the study will be presented in Sect. 3. Section 4 will present the conclusion of this article.

2 Input Parameters

It is assumed that hybrid energy system supply electricity to many electrical appliances like fans, air conditioning, geyser, iron etc. Figure 1 shows the daily load profile

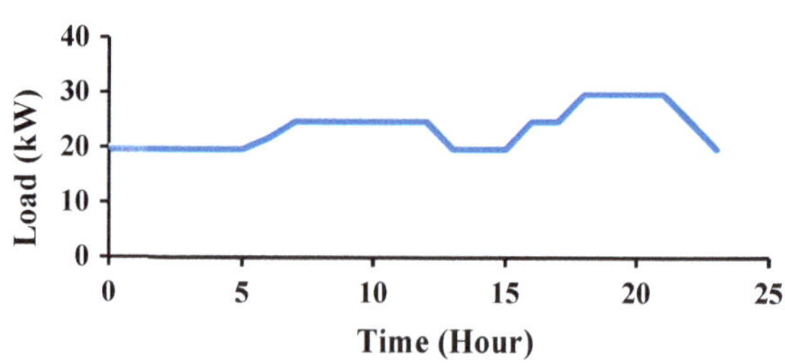

Fig. 1 Electrical load for 24 h

for selected location. To provide better reliability a random variability of 5% is taken into account.

2.1 Resources

The amount of power production from the solar photovoltaic (SPV) is mainly affected by the temperature and radiations. The monthly average temperature shown in the Fig. 2 with scaled annual average value of 1100 C. Figure 3 Shows the monthly daily radiations in kWh/m^2/day. The stream flow is shown in the Fig. 4 with scaled annual average value of 372 L/s.

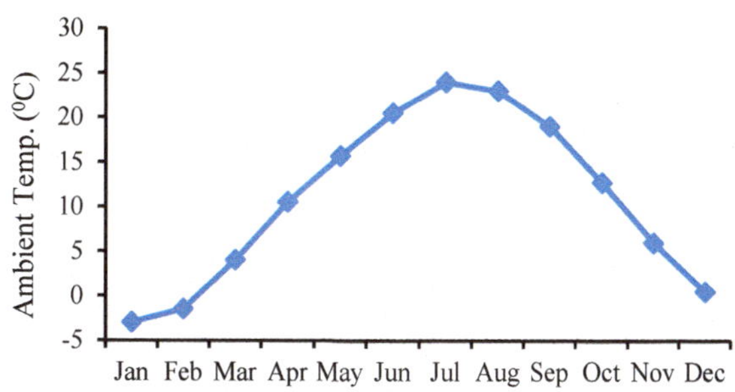

Fig. 2 Monthly temperature variations

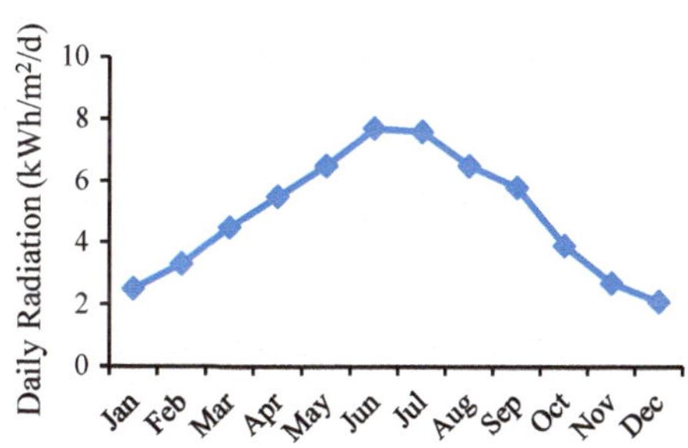

Fig. 3 Average solar irradiance for different months

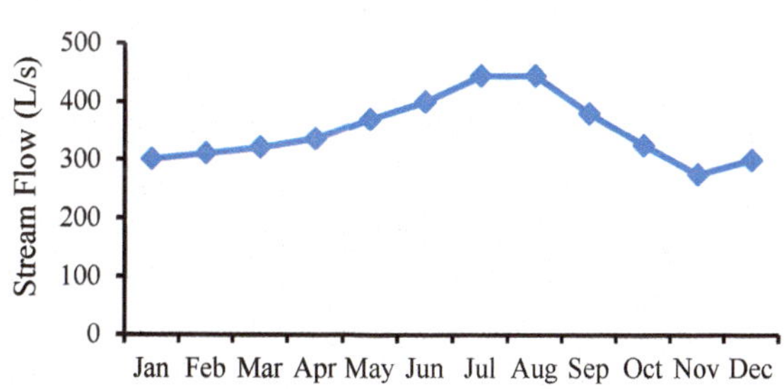

Fig. 4 Average stream flow for different months

2.2 System Components

Different system components are to be used in this study like SPV, hydropower, converter, DG, battery bank. Table 1 shows the input parameters and different cost values taken into consideration in this study.

Table 1 Different cost parameters of components

S. No	Description	Specification
1	*Solar photovoltaic (SPV)*	
	Capital cost	$1500/kW
	Lifetime	25 years
	Maintenance cost	$5/kW/year
	Replacement cost	$1000/kW
2	*Hydropower*	
	Initial cost	$1700/kW
	Efficiency	75%
	Cost of operation and maintenance	$51/kW-year
	Replacement value	$500/kW
3	*Diesel Generator (DG)*	
	Initial cost	$500/kW
	Operational cost	$0.02/kW/hour
	Replacement value	$450/kW
	Lifetime	15,000 h
4	*Battery*	
	Model	Surrette 6CS25P
	Nominal capacity	1156 Ah (6.94 kWh)
	Nominal voltage	6 V
	Capital cost	$1100
	Operational cost	$10/year
	Replacement value	$1000
	Life throughout	9645 kWh
5	*Converter*	
	Initial cost	$550/kW
	Operational cost	$5/kW/year
	Replacement value	$450/kW
	Life	$450/kW

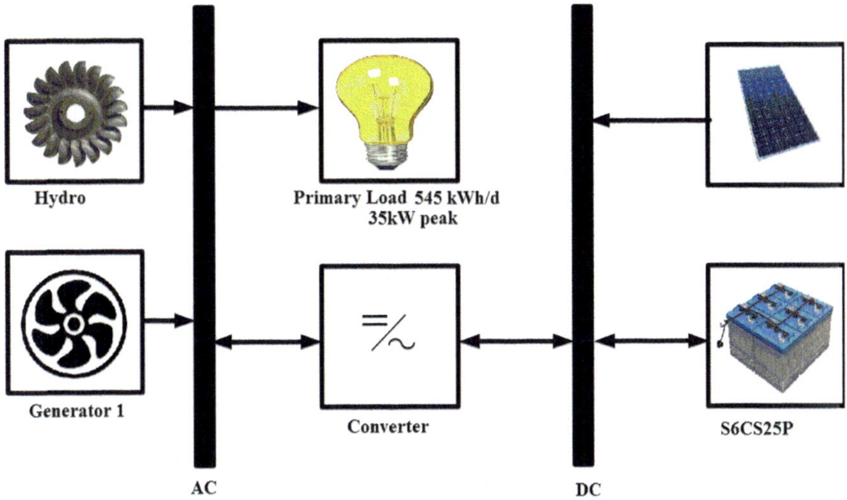

Fig. 5 Block diagram of SPV, hydro turbine, DG, converter and battery green hybrid microgrid

2.3 Generation Modelling

The software used in this research article is HOMER and the dispatch strategy is cycle charging. The suggested green HES is shown in Fig. 5. The SPV produces the direct current (DC), while hydro and DG producing the alternating current (AC). For the conversion of DC to AC a converter is used.

3 Result and Discussion

In this research paper, the software employed is HOMER to find out the sizing and different cost parameters for HES. For performing the simulation, a case study of Iraq has been taken for observation [2]. Four different cases are to be studied in simulation environments such as hybrid system except for SPV, battery and converter, hybrid system except for battery and a hybrid system consisting of all components. Some assumptions are also taken, such as the maximum value of capacity shortage is taken to be 2.5%, the annual real interest rate to be 7.85% and the project lifetime is 20 years. HES is consisting of green energy sources with diesel as a backup power source. Because of DG used in this paper, environmental emissions would be occurring. The Table 2 represents the cost and sizing values of different components. The cost summary of the most feasible solution is represents in the Table 3. The electrical power production through different components in feasible solution is shown in Table 4. Different atmospheric emissions values are presented in Table 5. The main source of emission is carbon dioxide (CO_2), much larger value in comparison with

Table 2 Cost and sizing values of different combinations

S. No	SPV (kW)	Hydro (kW)	DG (kW)	Battery (Qty.)	Converter (kW)	Cost of operation ($/year)	Total net present value ($)	Energy cost ($/kWh)	Diesel (L)	Renewable fraction (%)
1	–	14.7	7	7	6	15,549	170,580	0.084	15,436	79
2	–	14.7	11	–	– –	20,666	212,384	0.104	19,982	77
3	11	14.7	11	–	5	18,697	212,086	0.104	17,762	81
4	12	14.7	5	11	9	11,468	153,115	0.076	11,043	85

Table 3 Components cost summary in HES

S. No	Component	Capital cost ($)	Cost of replacement ($)	Cost of operation and maintainence ($)	Cost of fuel ($)	Salvage value ($)	Total ($)
1	SPV	18,000	0	596	0	−529	18,066
2	Hydro	1700	0	506	0	−22	2184
3	DG	2500	10,179	7360	87,715	−58	107,696
4	Battery	12,100	8018	1092	0	−2147	19,064
5	Converter	4950	1304	447	0	−596	6105
6	System	39,250	19,501	10,001	87,715	−3352	153,115

Table 4 Electrical power production of different components

S. No	Components	Power production (kWh/year)	Amount (%)
1	SPV	16,181	8
2	Hydro turbine	163,963	77
3	DG	32,310	15
4	Total	212,455	100

Table 5 Environmental emission pattern for HES

Serial	Atmospheric emission	Emission (kilogram/year)
A	CO_2	29,080
B	CO	71.8
C	UH	7.95
D	PM	5.41
E	SO_2	58.4
F	NO	640

other emission factors, i.e., 29,080 kg/year and followed by carbon monoxide (CO), unburned hydrocarbon (UH), Particulate matter (PM), Sulphur dioxide (SO_2) and nitrogen oxides (NO).

4 Conclusion

This paper provides a comprehensive study of the sizing, techno-economic and environmental viability of standalone hybrid energy systems for a rural village. Four design cases are suggested and tested on the basis of different combinations of SPV, hydro, diesel and battery. As has been found, green hybrid microgrids consisting of all the components have the most feasible solution with an operating cost of $11,468.

Since the renewable fraction of the system is 85%, the HES consumes 11,043 L of diesel fuel and the emissions results show that carbon dioxide is mainly responsible for environmental emission. The proposed scheme in this research article may be considered for the sizing of a microgrid in future.

References

1. Hafez O, Bhattacharya K (2012) Optimal planning and design of a renewable energy based supply system for microgrids. Renew Energy 45:7–15. ISSN 0960-1481
2. Aziz AS, Tajuddin MFN, Adzman MR, Azmi A, Ramli MAM (2019) Optimization and sensitivity analysis of standalone hybrid energy systems for rural electrification: a case study of Iraq. Renew Energy 138:775–792. ISSN 0960-1481
3. Halabi LM, Mekhilef S, Olatomiwa L, Hazelton J (2017) Performance analysis of hybrid PV/diesel/battery system using HOMER: a case study Sabah, Malaysia. Energy Convers Manage 144:322–339
4. Rajbongshi R, Borgohain D, Mahapatra S (2017) Optimization of PV-biomass-diesel and grid base hybrid energy systems for rural electrification by using HOMER. Energy 126:461–474. ISSN 0360-5442
5. Aziz AS, Khudhier SA (2017) Optimal planning and design of an environmentally friendly hybrid energy system for rural electrification in Iraq. Am J Appl Sci 14:157–165
6. Alasady AMA (2016) Solar energy the suitable energy alternative for Iraq beyond oil. In: Proceedings of the international conference on petroleum and sustainable development (IPCBEE), vol 26, Singapore
7. Garni HZA, Awasthi A, Ramli MA (2018) Optimal design and analysis of gridconnected photovoltaic under different tracking systems using HOMER. Energy Convers Manag 155:42–57
8. Aziz AS (2017) Techno-economic analysis using different types of hybrid energy generation for desert safari camps in UAE. Turk J Electr Eng Comput Sci 25:2122–2135
9. Nag AK, Sarkar S (2018) Modeling of hybrid energy system for futuristic energy demand of an Indian rural area and their optimal and sensitivity analysis. Renew Energy 118:477–488
10. Olatomiwa L (2016) Optimal configuration assessments of hybrid renewable power supply for rural healthcare facilities. Energy Rep 2:141–146
11. Aggarwal S et al (2020) Meta heuristic and evolutionary computation: algorithms and applications. Springer Nature, Berlin, p 949. https://doi.org/10.1007/978-981-15-7571-6. ISBN 978-981-15-7571-6
12. Yadav AK et al (2020) Soft computing in condition monitoring and diagnostics of electrical and mechanical systems. Springer Nature, Berlin, p 496. https://doi.org/10.1007/978-981-15-1532-3. ISBN 978-981-15-1532-3
13. Gopal et al (2021) Digital transformation through advances in artificial intelligence and machine learning. J Intell Fuzzy Syst, Pre-press 1–8. https://doi.org/10.3233/JIFS-189787
14. Fatema N et al (2021) Intelligent data-analytics for condition monitoring: smart grid applications. Elsevier, p 268. ISBN 978-0-323-85511-2. https://www.sciencedirect.com/book/9780323855105/intelligent-data-analytics-for-condition-monitoring
15. Smriti S et al (2018) Special issue on intelligent tools and techniques for signals, machines and automation. J Intell Fuzzy Syst 35(5):4895–4899. https://doi.org/10.3233/JIFS-169773
16. Jafar A et al (2021) AI and machine learning paradigms for health monitoring system: intelligent data analytics. Springer Nature, Berlin, p 496. https://doi.org/10.1007/978-981-33-4412-9. ISBN 978-981-33-4412-9
17. Sood YR et al (2019) Applications of artificial intelligence techniques in engineering, vol 1, Springer Nature, p 643. https://doi.org/10.1007/978-981-13-1819-1. ISBN 978-981-13-1819-1

18. Sharma NK et al (2021) Optimal design of passive power filter using multi-objective pareto-based firefly algorithm and analysis under background and load-side's nonlinearity. IEEE Access. https://doi.org/10.1109/ACCESS.2021.3055774

Open Access Same-Time Information System (OASIS) of New York

Chandransh Singh and Yog Raj Sood

Abstract There is a web-based OASIS of New York. OASIS is a real-time non-discriminative specified in FERC request 889. Before OASIS there is a monopoly in generation and distribution. OASIS is a web-based administration and gives data about accessible transmission ability for point to point administrations. It empowers transmission suppliers and transmission clients to communicate requests and reactions to purchase and sell accessible transmission tariff offered. They direct showcases and keep up dependability straightforwardly, giving information, examinations, and data relating to New York's capacity framework to policymakers, partners, and the overall population.

Keywords Federal Energy Regulatory Commission (FERC) · Locational-based marginal pricing · Total transfer capability · Available transfer capability · Load bidding · Marginal cost · Congestion

1 Introduction

To advance discount rivalry through non-oppressive open access, Federal Energy Regulatory Commission (FERC) required every transmission possessing open utility or its administrator to make an ongoing data system to scatter data about the accessibility and cost of transmission administrations.

In Order 889, FERC diagrams key data necessities for OASIS. These prerequisites might be gathered into four classes. These classifications and the data substance are:

- Transmission framework data—ATC, framework unwavering quality, reaction to framework conditions, and the date and time stamp for all the data.
- Transmission administration data—complete levy, administration limits, subordinate administrations, and current working and financial conditions.

C. Singh (✉) · Y. R. Sood
Department of Electrical Engineering, NIT Hamirpur, Hamirpur, Himachal Pradesh, India

A. Tomar et al. (eds.), *Machine Learning, Advances in Computing, Renewable Energy and Communication*, Lecture Notes in Electrical Engineering 768,
https://doi.org/10.1007/978-981-16-2354-7_19

- Transmission administration solicitation and reaction information—booking of intensity moves, administration interferences and diminishing, administration gatherings' personalities, and review the log for optional activities.
- General data—declarations and worth included administrations.

Figure 1 presents the applied structure of the internet-based OASIS network. Moreover, Fig. 2 shows the engineering of an OASIS node. Power marketers who become signatories to a transmission supplier's OATT acquire total access so that they can see existing transmission and administration accessibility and help demands made by different gatherings. There are additionally advertised eyewitnesses who have perused just access and may see action yet not demand administrations. Apart from this, more advanced examples are represented in [1–7].

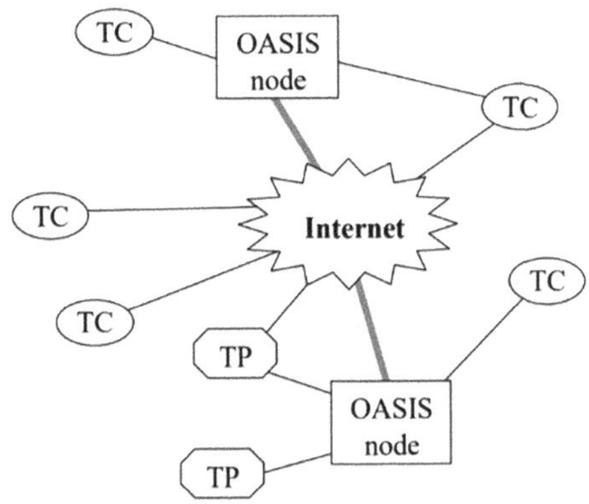

Fig. 1 The OASIS network

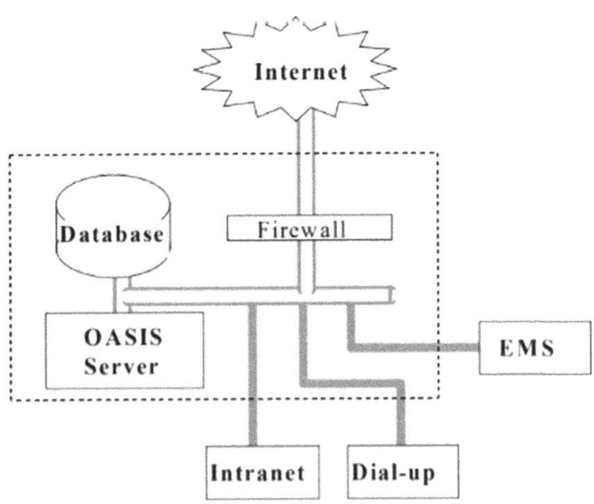

Fig. 2 OASIS node

2 New York Independent System Operator (NYISO)

The NYISO is the association answerable for overseeing New York's electric framework and its serious discount electric commercial centre. They don't create power or own transmission lines, yet they work with power makers, service organizations, and partners to give the capacity to meet New Yorkers' power needs on a day by day, hourly, and minute-to-minute premise. We are focused on a reasonable, straightforward market system since it encourages us to convey the most reduced cost discount power arrangements.

The NYISO is accused of dependably working on New York's capacity matrix, fulfilling the toughest guidelines in the country, under severe administrative oversight. The NYISO plans the force system for the future, more than one, five and ten years, considering to keep up long haul unwavering quality, decrease blockage on the transmission system, and meet open arrangement needs calling for a new transmission, for example, lines to carry renewable assets to clients. The NYISO manages advertises and keeps up unwavering quality straightforwardly, giving information, investigations, and data relating to New York's capacity system to policymakers, partners, and the overall population.

The NYISO is a not revenue-driven, independent organization unaffiliated with any state or government office or vitality organization. The NYISO's governmentally endorsed taxes contain exact prerequisites for our Board of Directors and all workers to have no monetary relationship with any of the organizations that take an interest in our discount vitality markets. The NYISO is committed to straightforwardness by the way we work, the data we give, and our job as a fair-minded system operator, organizer, and agent of New York's discount power markets (Fig. 3) [8–11].

2.1 *NYISO Work*

It manages how to stream power throughout New York guaranteeing power is delivered in adequate amounts and transmitted where it needs to go—precisely when it is required.

This includes:

- Balancing the accessible gracefully of intensity, like clockwork, from many force plants more than a large number of miles of transmission lines.
- Matching offers from vitality makers with buyer utility interest to flexibly control for the state as productively and cost-viably as could be expected under the circumstances.
- Overseeing the conveyance of intensity from generators to the service organizations that serve a great many New York power shoppers 24 h per day, 7 days per week.
- Evaluating system requirements for what's to come.

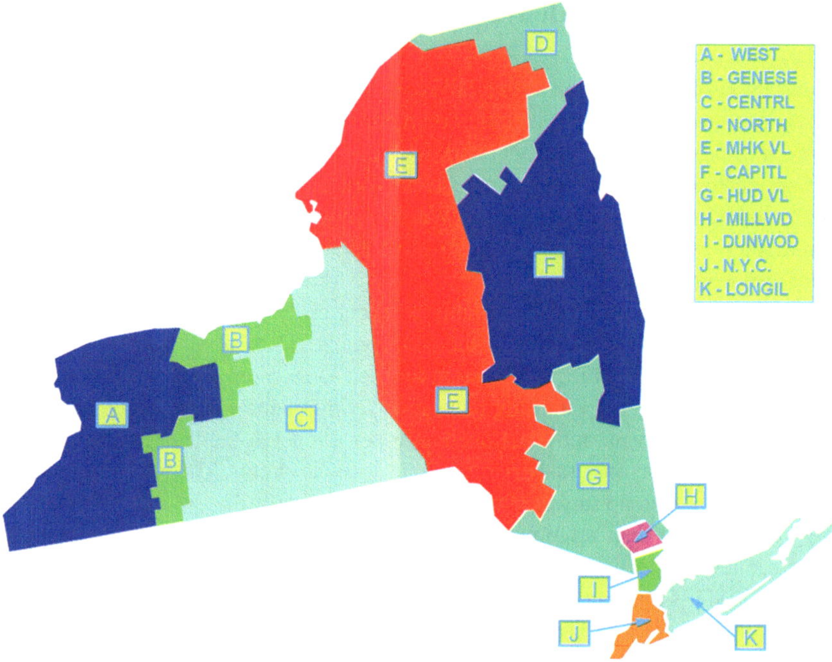

Fig. 3 New York load zone

The NYISO works nonstop to support and improve provincial unwavering quality, plan a progressively successful and productive force system for the future, and give target data and information to policymakers, partners, and financial specialists relating to New York's capacity system and electric framework.[11–14]

3 Auction and Load Bidding

3.1 Auction

An auction is an allotment system dependent on an exact assessment measure indicated by the auctioneer, and a pre-characterized openly accessible arrangement of rules intended to distribute or grant articles or items (for example contracts) based on a money-related offer. It is straightforward because of the way that it depends on a lot of rules dictated by the auctioneer and known by the bidders before the auction. An auction might be depicted by its three key guidelines, in particular (i) offering, (ii) clearing, and (iii) pricing.

3.2 Load Bidding

The screening to direct providers for physical and monetary retention is significant the conduct of purchaser's persuasions vitality costs. So load offering is reliable with functional rivalry. The load can be planned for one of the accompanying different ways:

(i) Physical bilateral contracts
(ii) Day-ahead fixed load
(iii) Price-capped load bids
(iv) Virtual load bids
(v) Virtual exports.

3.3 Interface Data

PJM

PJM is a truncation of Pennsylvania, New Jersey, and Maryland after the domains where the principal utilities combined. Today, the PJM incorporates all or parts of Pennsylvania, New Jersey, Maryland, Delaware, Ohio, Virginia, Kentucky, North Carolina, West Virginia, Indiana, Michigan, and Illinois. It oversees power conveyance for in excess of 60 million individuals and $42 billion worth of power. PJM is responsible for dealing with the transportation of power from power plants to the different utilities in its region.

Hydro-Quebec (HQ)

Hydro-Quebec is an open utility that deals with the age, transmission, and dissemination of power in the Canadian territory of Quebec, just as the fare of capacity to segment the Northeast United States.

4 Results

4.1 Day-Ahead Market Zone LBMP

See Fig. 4 and Table 1.

4.2 Real-Time Market Zonal LBMP

See Fig. 5 and Table 2.

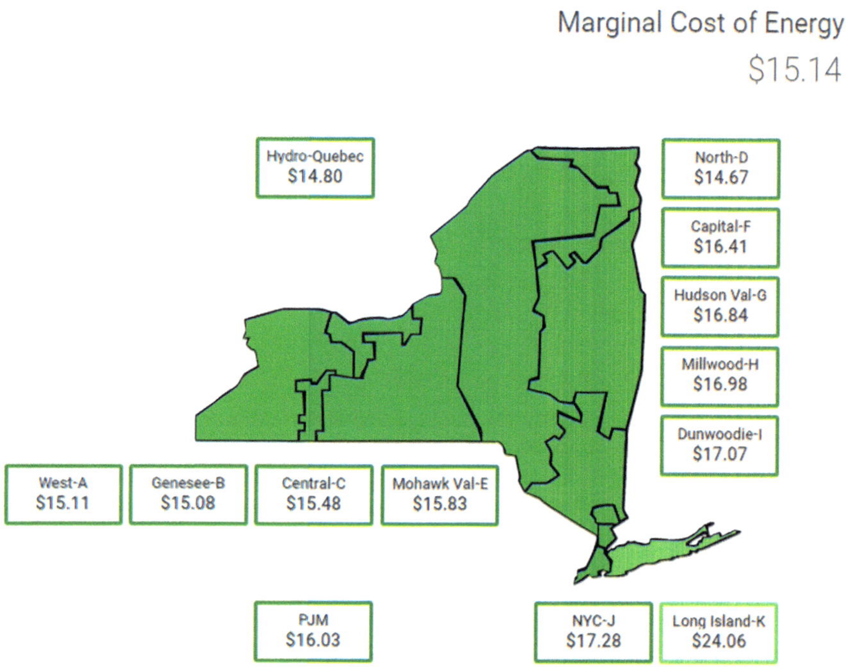

Fig. 4 Day-ahead market marginal cost of energy flow on July 30, 2020 at time 4:00 EDT

Table 1 Day-ahead market zones/interface various data LBMP, losses, and congestion on July 30, 2020 at 4:00 EDT

Zones/Interface	LBMP ($)	Losses ($)	Congestion ($)
West-A	15.11	0.21	0.00
Genesee-B	15.08	0..25	0.00
Central-C	15.48	0.15	0.00
North-D	14.67	0.66	0.00
Mohawak Val-E	15.83	0.50	0.00
Capital-F	16.41	1.09	0.00
Hudson Val-G	16.84	1.52	0.00
Milwood-H	16.98	1.66	0.00
Dunwoodie-I	17..07	1.75	0.00
NYC-J	17.28	1.95	0.00
Long Island-K	24.06	2.38	6.36
PJM	16.03	0.71	0.00
Hydro-Quebec	14.80	0.52	0.00

4.3 ATC and TTC

Interface: NYISO-PJM

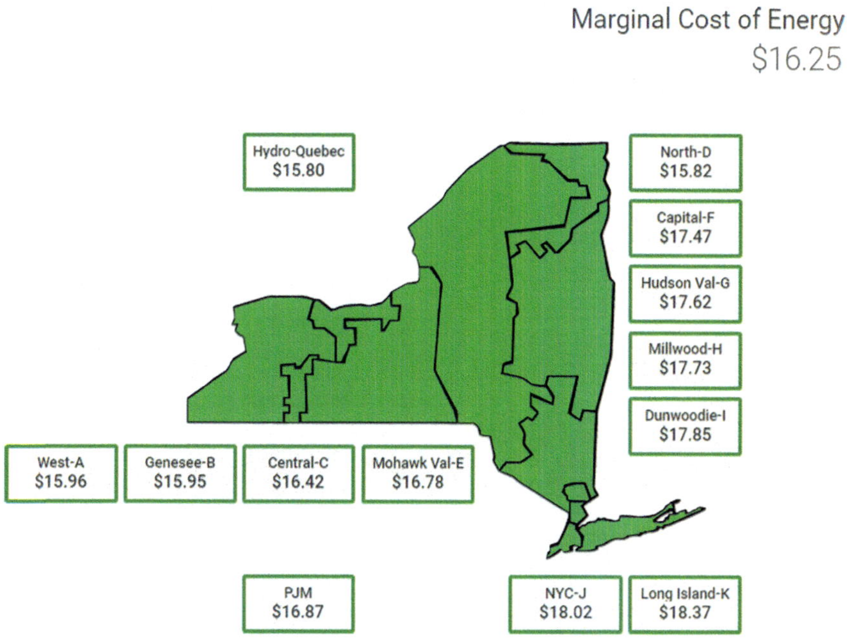

Fig. 5 Real-time market marginal cost of energy flow on July 30, 2020 at 4:00 EDT

Table 2 Real-time market zones/interface various data LBMP, losses and congestion on July 30, 2020 at 4:00 EDT

Zones/Interface	LBMP ($)	Losses ($)	Congestion ($)
West-A	15.96	0.29	0.00
Genesee-B	15.95	0.31	0.00
Central-C	16.42	0.16	0.00
North-D	16.78	0.44	0.00
Mohawak Val-E	15.82	0.52	0.00
Capital-F	17.47	1.22	0.00
Hudson Val-G	17.62	1.37	0.00
Milwood-H	17.73	1.48	0.00
Dunwoodie-I	17.85	1.59	0.00
NYC-J	18.02	1.76	0.00
Long Island-K	18.37	2.11	0.01
PJM	16.87	0.62	0.00
Hydro-Quebec	15.87	0.46	0.00

Table 3 Hourly data representation on ATC and TTC of the interface NYISO-PJM

	DAM									
Time	TTC	ATC	TTC :00	ATC :00	TTC :15	ATC :15	TTC :30	ATC :30	TTC :45	ATC :45
00:00 EDT	1000	700	1000	700	1000	700	1000	700	1000	700
01:00 EDT	1000	700	1000	700	1000	700	1000	700	1000	700
02:00 EDT	1000	700	1000	700	1000	700	1000	700	1000	700
03:00 EDT	1000	700	1000	700	1000	700	1000	700	1000	700
04:00 EDT	1000	700	1000	700	1000	700	1000	700	1000	700
05:00 EDT	1000	700	1000	700	1000	700	1000	700	1000	700
06:00 EDT	1000	700	1000	700	1000	700	1000	700	1000	700
07:00 EDT	1000	700	1000	700	1000	700	1000	700	1000	700
08:00 EDT	1000	700	1000	700	1000	700	1000	700	1000	700
09:00 EDT	1000	700	1000	700	1000	700	1000	700	1000	700
10:00 EDT	1000	700	1000	700	1000	700	1000	700	1000	700
11:00 EDT	1000	700	1000	700	1000	700	1000	700	1000	700
12:00 EDT	1000	700	1000	700	1000	700	1000	700	1000	700
13:00 EDT	1000	700	1000	700	1000	700	1000	700	1000	700
14:00 EDT	1000	647	1000	700	1000	700	1000	700	1000	700
15:00 EDT	1000	568	1000	700	1000	700	1000	700	1000	700
16:00 EDT	1000	398	1000	700	1000	700	1000	700	1000	629
17:00 EDT	1000	565	1000	700	1000	700	1000	700	1000	700
18:00 EDT	1000	615	1000	700	1000	700	1000	700	1000	700
19:00 EDT	1000	700	1000	700	1000	700	1000	700	1000	700
20:00 EDT	1000	700	1000	700	1000	700	1000	700	1000	700
21:00 EDT	1000	700	1000	700	1000	700	1000	700	1000	700
22:00 EDT	1000	700	1000	700	1000	700	1000	700	1000	700
23:00 EDT	1000	700	1000	700	1000	700	1000	700	1000	700

See Table 3.

Interface: PJM-NYISO

See Table 4.

Interface: NYISO HQ

See Table 5.

Interface: HQ-NYISO

See Table 6.

4.4 Load Bidding of Different Zones

See Table 7.

Table 4 Hourly data representation on ATC and TTC of the interface PJM-NYISO

	DAM									
Time	TTC	ATC	TTC :00	ATC :00	TTC :15	ATC :15	TTC :30	ATC :30	TTC :45	ATC :45
00:00 EDT	2000	660	2000	617	2000	610	2000	611	2000	627
01:00 EDT	2000	770	2000	612	2000	618	2000	608	2000	613
02:00 EDT	2000	787	2000	538	2000	541	2000	537	2000	543
03:00 EDT	2000	778	2000	502	2000	496	2000	503	2000	500
04:00 EDT	2000	729	2000	500	2000	502	2000	502	2000	500
05:00 EDT	2000	758	2000	501	2000	548	2000	552	2000	496
06:00 EDT	2000	744	2000	615	2000	615	2000	612	2000	614
07:00 EDT	2000	411	2000	471	2000	414	2000	495	2000	488
08:00 EDT	2000	466	2000	588	2000	588	2000	587	2000	580
09:00 EDT	2000	520	2000	366	2000	473	2000	534	2000	488
10:00 EDT	2000	466	2000	417	2000	549	2000	428	2000	527
11:00 EDT	2000	687	2000	669	2000	556	2000	325	2000	569
12:00 EDT	2000	872	2000	516	2000	537	2000	868	2000	872
13:00 EDT	2000	1347	2000	1011	2000	1013	2000	985	2000	1012
14:00 EDT	2000	1700	2000	1387	2000	1283	2000	1298	2000	1365
15:00 EDT	2000	1700	2000	1152	2000	1408	2000	1303	2000	1396
16:00 EDT	2000	1700	2000	1617	2000	1623	2000	1623	2000	1700
17:00 EDT	2000	1700	2000	1490	2000	1586	2000	1580	2000	1435
18:00 EDT	2000	1700	2000	1620	2000	1463	2000	1464	2000	1470
19:00 EDT	2000	1279	2000	1238	2000	1331	2000	1339	2000	1324
20:00 EDT	2000	818	2000	990	2000	1069	2000	927	2000	968
21:00 EDT	2000	881	2000	899	2000	1022	2000	919	2000	970
22:00 EDT	2000	1234	2000	1296	2000	1180	2000	1321	2000	1308
23:00 EDT	2000	1448	2000	1529	2000	1525	2000	1524	2000	1503

Table 5 Hourly data representation on ATC and TTC of the interface NYISO-HQ

	DAM									
Time	TTC	ATC	TTC :00	ATC :00	TTC :15	ATC :15	TTC :30	ATC :30	TTC :45	ATC :45
00:00 EDT	1000	1000	1000	1000	1000	1000	1000	1000	1000	1000
01:00 EDT	1000	1000	1000	1000	1000	1000	1000	1000	1000	1000
02:00 EDT	1000	1000	1000	1000	1000	1000	1000	1000	1000	1000
03:00 EDT	1000	1000	1000	1000	1000	1000	1000	1000	1000	1000
04:00 EDT	1000	1000	1000	1000	1000	1000	1000	1000	1000	1000
05:00 EDT	1000	1000	1000	1000	1000	1000	1000	1000	1000	1000
06:00 EDT	1000	1000	1000	1000	1000	1000	1000	1000	1000	1000
07:00 EDT	1000	1000	1000	1000	1000	1000	1000	1000	1000	1000
08:00 EDT	1000	1000	1000	1000	1000	1000	1000	1000	1000	1000
09:00 EDT	1000	1000	1000	1000	1000	1000	1000	1000	1000	1000
10:00 EDT	1000	1000	1000	1000	1000	1000	1000	1000	1000	1000
11:00 EDT	1000	1000	1000	1000	1000	1000	1000	1000	1000	1000
12:00 EDT	1000	1000	1000	1000	1000	1000	1000	1000	1000	1000
13:00 EDT	1000	1000	1000	1000	1000	1000	1000	1000	1000	1000
14:00 EDT	1000	1000	1000	1000	1000	1000	1000	1000	1000	1000
15:00 EDT	1000	1000	1000	1000	1000	1000	1000	1000	1000	1000
16:00 EDT	1000	1000	1000	1000	1000	1000	1000	1000	1000	1000
17:00 EDT	1000	1000	1000	1000	1000	1000	1000	1000	1000	1000
18:00 EDT	1000	1000	1000	1000	1000	1000	1000	1000	1000	1000
19:00 EDT	1000	1000	1000	1000	1000	1000	1000	1000	1000	1000
20:00 EDT	1000	1000	1000	1000	1000	1000	1000	1000	1000	1000
21:00 EDT	1000	1000	1000	1000	1000	1000	1000	1000	1000	1000
22:00 EDT	1000	1000	1000	1000	1000	1000	1000	1000	1000	1000
23:00 EDT	1000	1000	1000	1000	1000	1000	1000	1000	1000	1000

Table 6 Hourly data representation on ATC and TTC of the interface HQ-NYISO

	DAM									
Time	TTC	ATC	TTC :00	ATC 00	TTC :15	ATC :15	TTC :30	ATC :30	TTC :45	ATC :45
00:00 EDT	1500	33	1400	0	1400	0	1400	0	1400	0
01:00 EDT	1500	28	1400	34	1400	34	1400	34	1400	34
02:00 EDT	1500	63	1400	0	1400	0	1400	0	1400	0
03:00 EDT	1500	63	1400	25	1400	25	1400	25	1400	25
04:00 EDT	1500	97	1400	0	1400	0	1400	0	1400	0
05:00 EDT	1500	63	1400	0	1400	0	1400	0	1400	0
06:00 EDT	1500	16	1400	0	1400	0	1400	0	1400	0
07:00 EDT	1500	0	1400	0	1400	0	1400	0	1400	0
08:00 EDT	1500	0	1400	0	1400	0	1400	0	1400	0
09:00 EDT	1500	0	1400	0	1400	0	1400	0	1400	0
10:00 EDT	1500	6	1400	0	1400	0	1400	0	1400	0
11:00 EDT	1500	5	1400	0	1400	0	1400	0	1400	0
12:00 EDT	1500	5	1400	0	1400	0	1400	0	1400	0
13:00 EDT	1500	53	1400	0	1400	0	1400	0	1400	0
14:00 EDT	1500	41	1400	0	1400	0	1400	0	1400	0
15:00 EDT	1500	41	1400	0	1400	0	1400	0	1400	0
16:00 EDT	1500	42	1500	100	1500	100	1500	100	1500	100
17:00 EDT	1500	66	1500	0	1500	0	1500	0	1500	0
18:00 EDT	1500	66	1500	0	1500	0	1500	0	1500	0
19:00 EDT	1500	65	1500	0	1500	0	1500	0	1500	0
20:00 EDT	1500	40	1500	0	1500	0	1500	0	1500	0
21:00 EDT	1500	51	1500	81	1500	81	1500	81	1500	81
22:00 EDT	1500	26	1500	64	1500	64	1500	64	1500	64
23:00 EDT	1500	26	1500	150	1500	150	1500	150	1500	150

Table 7 Hourly representation of day-ahead bids on July 30, 2020

Time (EDT)	Zone-A (MW)	Zone-B (MW)	Zone-C (MW)	Zone-D (MW)	Zone-E (MW)	Zone-F (MW)	Zone-G (MW)	Zone-H (MW)	Zone-I (MW)	Zone-J (MW)	Zone-K (MW)
00:00	1262	1169	1654	131	730	1883	1215	356	547	3570	3425
01:00	1098	1156	1674	134	747	1865	1201	376	538	3514	3429
02:00	1156	1185	1680	141	761	1871	1185	388	533	3405	3477
03:00	1235	1182	1668	140	763	1841	1175	337	517	3204	3378
04:00	1212	1143	1609	136	734	1788	1132	315	493	3101	3162
05:00	1276	1128	1584	132	720	1767	1096	305	457	3016	3029
06:00	1290	1098	1542	122	687	1702	1021	299	404	2937	2869
07:00	1207	1034	1439	112	613	1567	937	217	320	2827	2568
08:00	1103	932	1292	101	535	1432	851	184	229	2642	2233
09:00	1021	857	1238	95	524	1345	789	181	192	2506	1834
10:00	962	798	1159	90	468	1219	739	187	181	2419	1636
11:00	902	751	1109	85	441	1134	701	170	171	2362	1494
12:00	876	727	1073	84	426	1060	678	174	164	2316	1402
13:00	872	719	1060	83	427	1069	661	113	168	2315	1373
14:00	894	746	1107	87	462	1130	672	110	174	2417	1430
15:00	939	803	1174	93	496	1237	705	155	225	2624	1600
16:00	1013	881	1250	101	546	1343	764	210	342	2912	1877
17:00	1067	915	1285	106	571	1420	822	270	429	3094	2129
18:00	1117	955	1351	108	604	1478	917	318	465	3304	2370
19:00	1173	994	1412	116	622	1536	1003	339	489	3422	2632
20:00	1113	1037	1467	121	651	1585	1079	317	512	3513	2862
21:00	1158	1073	1523	126	661	1611	1125	331	535	3553	3041
22:00	1206	1105	1553	127	669	1641	1187	352	543	3635	3190
23:00	1246	1123	1575	125	683	1664	1199	346	545	3632	3295

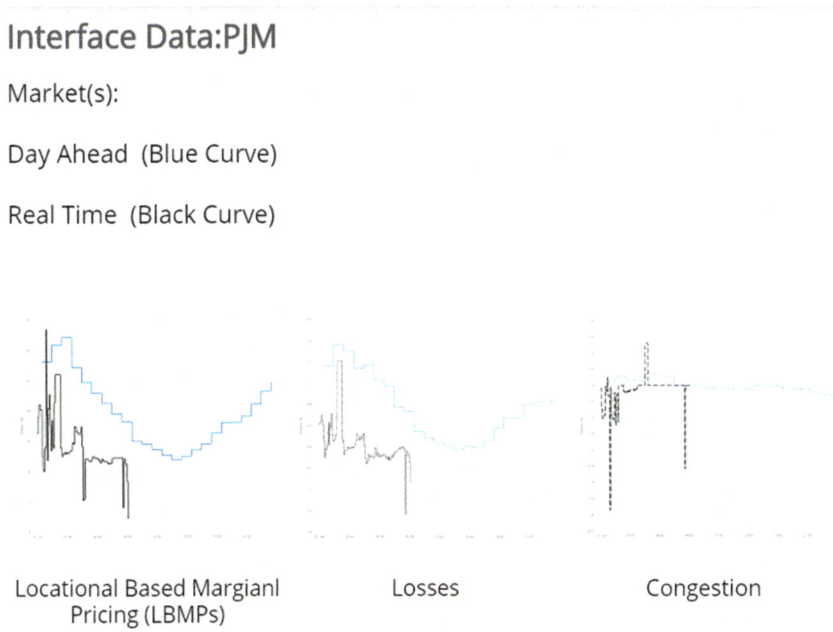

Fig. 6 Interface data window of PJM shows that LBMP graph, losses graph, and congestion graph on July 30, 2020, and the x-axis shows the hourly duration and the y-axis shows the dollars ($)

4.5 Interface Data Window of PJM and Hydro-Quebec

Interface data window of PJM

See Fig. 6.

Interface data window of Hydro-Quebec

See Fig. 7.

5 Conclusion

NYISO is a not revenue-driven enterprise liable for working to New York's mass power system, overseeing wholesale power markets and directing system arranging. It is subjected to the oversight of FERC and controlled in certain aspects by the New York State Public Service Commission. NYISO activities are likewise managed by electric system unwavering quality controllers, including the NERC, Northeast Power Coordinating Council, and the New York State Reliability Council. The NYISO's wholesale competitive power markets have conveyed financial and natural advantages for New York.

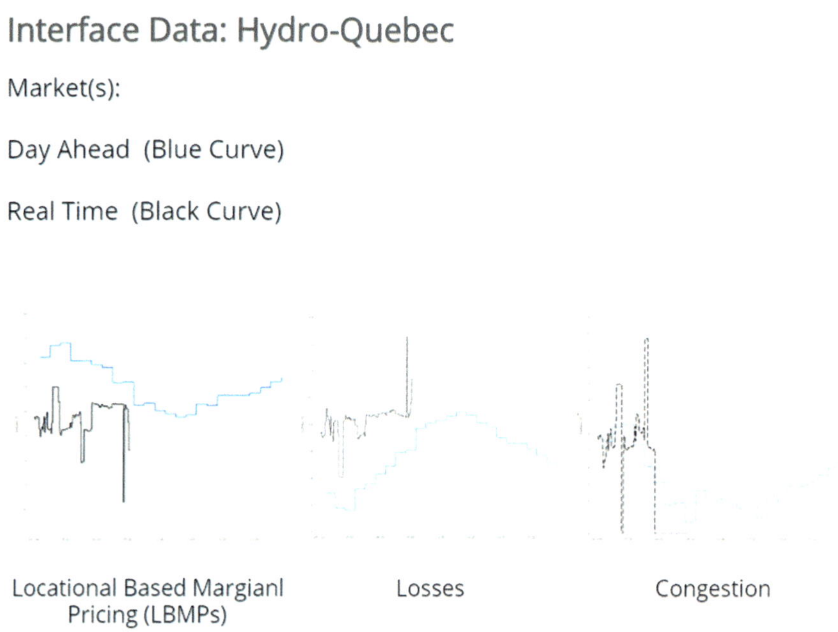

Fig. 7 Interface data window of Hydro-Quebec shows that LBMP graph, losses graph, and congestion graph on July 30, 2020, and the x-axis shows the hourly duration and the y-axis shows the dollars ($)

References

1. Aggarwal S et al (2020) Meta heuristic and evolutionary computation: algorithms and applications. Springer Nature, Berlin, 949 pp. https://doi.org/10.1007/978-981-15-7571-6. ISBN 978-981-15-7571-6
2. Yadav AK et al (2020) Soft computing in condition monitoring and diagnostics of electrical and mechanical systems. Springer Nature, Berlin, 496 pp. https://doi.org/10.1007/978-981-15-1532-3. ISBN 978-981-15-1532-3
3. Gopal et al (2021) Digital transformation through advances in artificial intelligence and machine learning. J Intell Fuzzy Syst (Pre-press) 1–8. https://doi.org/10.3233/JIFS-189787
4. Fatema N et al (2021) Intelligent data-analytics for condition monitoring: smart grid applications. Elsevier, 268 pp. ISBN 978-0-323-85511-2. https://www.sciencedirect.com/book/9780323855105/intelligent-data-analytics-for-condition-monitoring
5. Smriti S et al (2018) Special issue on intelligent tools and techniques for signals, machines and automation. J Intell Fuzzy Syst 35(5):4895–4899. https://doi.org/10.3233/JIFS-169773
6. Jafar A et al (2021) AI and machine learning paradigms for health monitoring system: intelligent data analytics. Springer Nature, Berlin, 496 pp. https://doi.org/10.1007/978-981-33-4412-9. ISBN 978-981-33-4412-9
7. Sood YR et al (2019) Applications of artificial intelligence techniques in engineering, vol 1. Springer Nature, 643 pp. https://doi.org/10.1007/978-981-13-1819-1. ISBN 978-981-13-1819-1
8. Min L, Abur A (2006) Total transfer capability computation for multi-area power systems. IEEE Trans Power Syst 21(3):1141–1147

9. Mohammed OO, Mustafa MW, Mohammed DSS, Otuoze AO (2019) Available transfer capability calculation methods: a comprehensive review. Int Trans Electr Energy Syst 29(6):e2846
10. Singh HARRY (2008) Transmission markets, congestion management, and investment. In: Competitive electricity markets. Elsevier, pp 141–178
11. Mughal SN, Sood YR (2019) A Proposal to regulate frequency in deregulated power system using unscheduled interchange tariff. In: International conference on innovation in modern science and technology. Springer, Cham, pp 269–277
12. Padhy NP, Sood YR (2004) Advancement in power system engineering education and research with power industry moving towards deregulation. In: IEEE power engineering society general meeting. IEEE, pp 71–76
13. Sood YR (2007) Evolutionary programming based optimal power flow and its validation for deregulated power system analysis. Int J Electr Power Energy Syst 29(1):65–75
14. Sood YR, Padhy NP, Gupta HO (2007) Deregulated model and locational marginal pricing. Electr Power Syst Res 77(5–6):574–582

Open-Access Same-Time Information System: Extended to Indian Power Market

Nivedita Singh and Yog Raj Sood

Abstract This research paper is an attempt to analyse and explore the huge potential of open-access same-time information system in the Indian power market. In the present scenario, India is swiftly moving towards complete deregulation in all aspects, i.e. generation, transmission and distribution. OASIS, in layman terms, is basically a system based on the internet for obtaining services that are related to electrical power transmission. This paper also contains a comprehensive case study attempting to explain the data contents, interface and working of an OASIS website. In deregulation, open access is given to all power-related entities in order to raise the competition and keep the monopoly in check. A few proposed strategies are discussed for the betterment of already existing OASIS websites for more economic, practical and reliable operation.

Keywords Deregulation · Open access · Power market · Comprehensive · Entities · Monopoly

1 Introduction

In recent times, the privatization of the power market is taking place in India. Earlier, all distributions, transmissions and distributions were owned by the government but now with India moving towards a developed economy, more and more private players are coming into the picture in all sectors.

This issues the requirement for an internet-based, transparent and non-discriminatory system. All the developed countries with deregulation are using OASIS since it is basically the means by which high-voltage transmission lines are used for moving wholesale quantities of electricity in transmission lines.

On an OASIS website, bidding auctions take place. All genco's and disco's bid in the form of some parameters. These bids are stored and processed by the OASIS

N. Singh (✉) · Y. R. Sood
Department of Electrical Engineering, NIT Hamirpur, Hamirpur, Himachal Pradesh, India

A. Tomar et al. (eds.), *Machine Learning, Advances in Computing, Renewable Energy and Communication*, Lecture Notes in Electrical Engineering 768,
https://doi.org/10.1007/978-981-16-2354-7_20

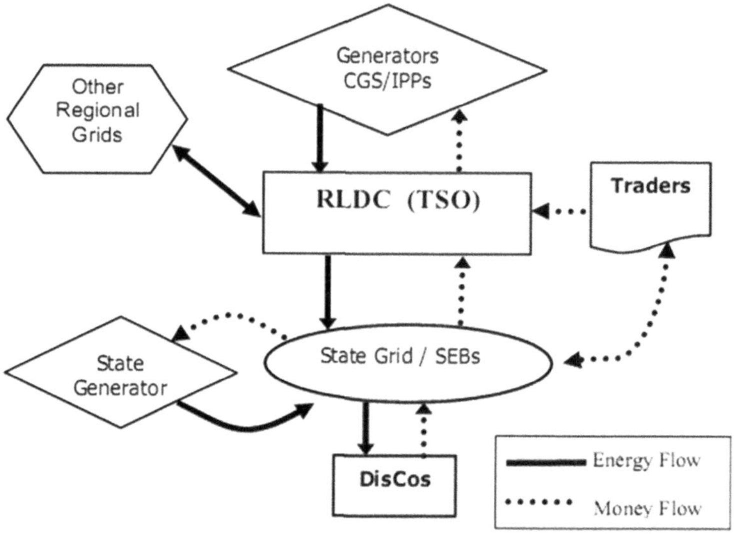

Fig. 1 Current Indian power market

and the result of the auction is declared which reflects the dispatch of power and its cost (Fig. 1).

As a case study for proposed strategies discussed in this paper, a prototype of OASIS website is made. This website is dedicated in an attempt to understand the basics of bidding auctions as per the proposed Indian power system market model (Fig. 2).

The website is based on the power sector of Columbia and gives the basic idea of how open-access same-time information system works when applied to the power sector. This case study and website can be further extended to the Indian power market. Moreover, some recent examples of power market are represented in the book and journal [1–7].

2 Historical Review

A brief history of market-related legislations in the Indian power market is given in Fig. 3.

3 OASIS: An Open Platform

OASIS serves differently for different aspects of the power market since the constraints are different for all generation, transmission and distribution.

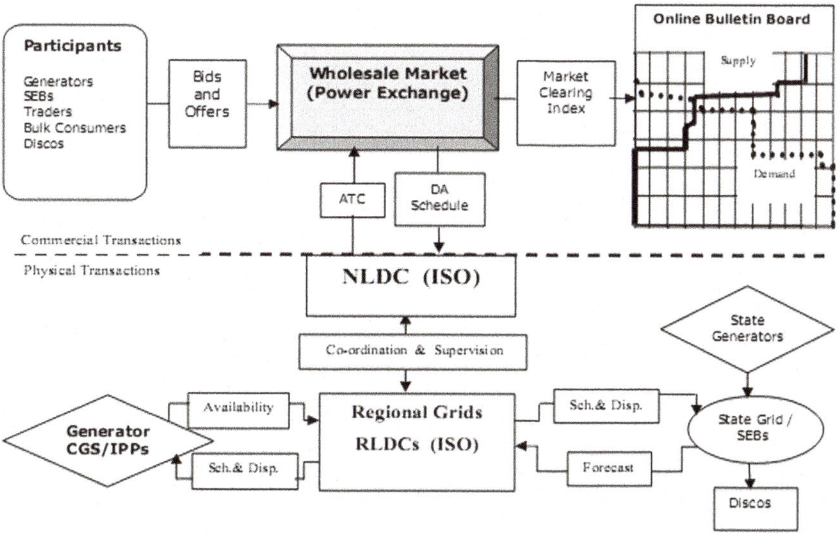

Fig. 2 Proposed Indian power market system

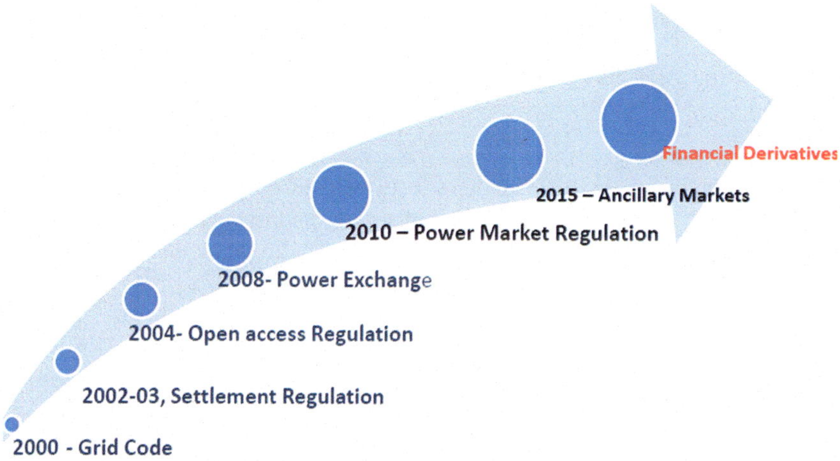

Fig. 3 Evolution of electricity market

For generation, there are constraints regarding the minimum and maximum loading of generators, so economic dispatch needs to be taken into consideration.

For transmission, the main consideration is the line loading limit. If transmission lines are loaded more than their capacity, it may result in congestion of line, which if not handled properly may result in the outage of line. It will result in burdening of the system, which can lead to cascaded outages, and if not taken care of will convert into blackouts [8–13].

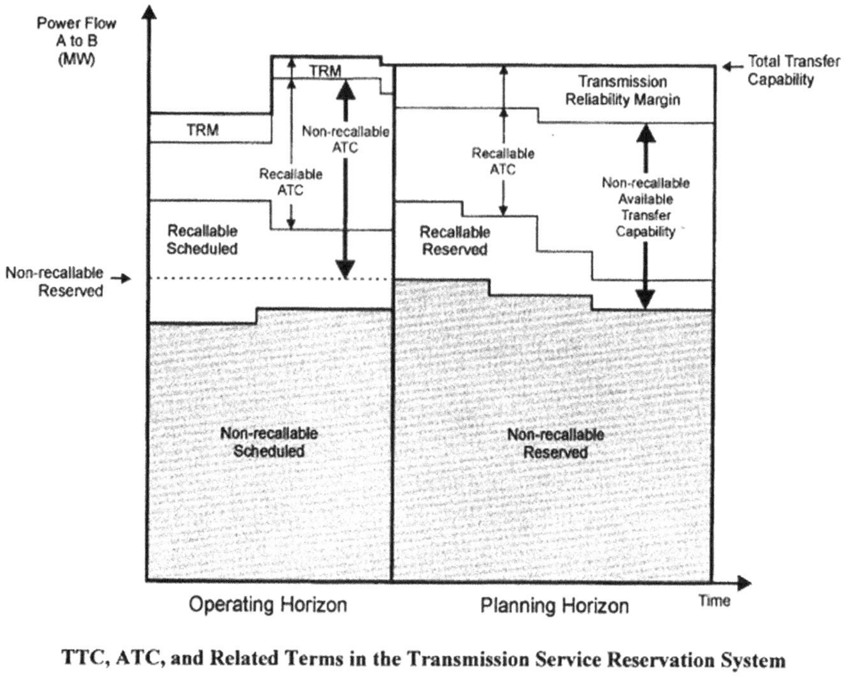

Fig. 4 Transfer capabilities (*Source* Available transfer capabilities definitions and determinations by NERC)

For distribution, the major challenges are to maintain power quality, voltage profile of the dead-end customer, droop control of voltage and frequency. On the transmission side, we generally talk in terms of transfer capabilities that includes as given in Fig. 4.

4 Prototype Website

For the case study, a website is developed which demonstrates the working of the OASIS website of Columbia, i.e. PJM (www.pjm.com).

Prototype Website Created—https://oasisprototype.simdif.com/

The website created is a prototype version of the original website as an attempt to replicate the interface and information of the original site.

The home page of the prototype website is shown in Fig. 5.

The contents of this website are as follows:

- About—This section includes the basic information about the working of this website.

Fig. 5 Home page of the prototype website

- Vision—This section includes the vision for which this company is working towards.
- Mission—The mission of this website is discussed at length in this section.
- History—The major developments since the website was established are discussed in this section.
- Territory served—The area in which this website is working is discussed under this section.
- Planning—This includes the future plans of this company.
- Innovation—The new developments that are going on in the field are discussed.
- Contact page—Contact information
- Data snapshots—This includes the total data values of generation and consumption.
- Transfer capability—It includes the value of ATC, TTC, CBM, TRM, curtailabilty, recallabilty, NATC and RATC etc.
- Bidding values (includes all real-time, historical and day-ahead bids)—This includes the bidding values of all Disco's and Genco's. The total MW generated by the Genco's and at what price they are willing to sell their units. It also includes the total demand by Disco's and at what rate they are willing to purchase it. All of this information is present in form of bids. The bidding values of real-time, historic and day-ahead are also displayed on the website page.
- Operation—All the bidding values are collected 24-h ahead. These values are processed as per the demand, supply, cost and price of the contemporary market. The Genco's and Disco's which have overbid and underbid are eliminated.
- Result—The result is then displayed on the website which contains what Genco will dispatch what amount of power at what price and what Disco will purchase what amount of units and at what cost.
- Market—The market ranging from all Genco's and Disco's is displayed.

- Load forecast—The day-ahead bid is collected on the basis of load forecasting. This process is based on probability.
- Locational marginal pricing—The price of the power dispatched is different for different locations. In some places, the cost per unit might be greater than the other places. So the model of locational marginal pricing is employed.
- Instantaneous load—The total amount of load that is present in the system is simultaneously displayed on the website page.
- Forecast load history—Day-ahead load is forecasted and displayed on the website page and the bids for this load are collected.
- Tie flows—The total amount of power flowing in tie-lines is displayed.
- Dispatch rates—The cost of the unit that is dispatched is given.
- Constraints—Some constraints are present in the bidding process in the form of minimum and maximum dispatches. Pricing also contains constraints in the form of minimum and maximum costs.
- Reactive power transfer interface—The amount of reactive power flowing in the lines is also displayed.
- Zonal aggregate LMPs—Locational marginal pricing as per the area zones is also given.
- Ancillary services—Apart from generation and consumption of electrical power, this website provides some other services which are listed under ancillary services.
- Reserve quantities—Some dispatch and load are kept in reserve in order to make this system more reliable towards dynamic changes.
- Market clearing price—This price at which all the generated units are consumed and all the loads are fulfilled.

5 Proposed Strategies

Some strategies are suggested for better working and optimal futuristic approach of OASIS system:

- Taking environmental factors into consideration, a lot of stress is being given to the use of RES in power market. After due consideration, 40% quota should be reserved for RES-based power in the OASIS policy.
- The Disco's which are causing congestion and overloading in transmission lines should be penalized.
- The households which are responsible for polluting the power supply by injecting harmonics in it should be penalized accordingly.
- Although the system is completely open, ISO should issue caps over all dispatches and prices [14–20].

6 Future Scope in the Indian Power Market

As India is moving towards complete deregulation, privatization of the power sector is happening. The era of new evolutions in the field of the electric market is rapidly growing. The older system of state boards is replaced by new systems in which the departments of generation, transmission and distribution are separately looked after. The distribution systems are further broken down into sub-categories and departments under which billing and accountability of work are separately handled. In a few years, our power market will be completely open. For accessing this open web-based platform, OASIS systems will be used where all the power and prices will be managed on a non-discriminatory basis. The OASIS systems will be used in future to harness the huge potential of the Indian power market. With more and more private players coming into the picture, there is a need for an open platform to collect the wholesale energy power market in one place which is accessible to all.

References

1. Aggarwal S et al (2020) Meta heuristic and evolutionary computation: algorithms and applications. Springer Nature, Berlin, 949 pp. https://doi.org/10.1007/978-981-15-7571-6. ISBN 978-981-15-7571-6
2. Yadav AK et al (2020) Soft computing in condition monitoring and diagnostics of electrical and mechanical systems. Springer Nature, Berlin, 496 pp. https://doi.org/10.1007/978-981-15-1532-3. ISBN 978-981-15-1532-3
3. Gopal et al (2021) Digital transformation through advances in artificial intelligence and machine learning. J Intell Fuzzy Syst (Pre-Press) 1–8. https://doi.org/10.3233/JIFS-189787
4. Fatema N et al (2021) Intelligent data-analytics for condition monitoring: smart grid applications. Elsevier, 268 pp. ISBN 978-0-323-85511-2. https://www.sciencedirect.com/book/9780323855105/intelligent-data-analytics-for-condition-monitoring
5. Smriti S et al (2018) Special issue on intelligent tools and techniques for signals, machines and automation. JIntell Fuzzy Syst 35(5):4895–4899. https://doi.org/10.3233/JIFS-169773
6. Jafar A et al (2021) AI and machine learning paradigms for health monitoring system: intelligent data analytics. Springer Nature, Berlin, 496 pp. https://doi.org/10.1007/978-981-33-4412-9. ISBN 978-981-33-4412-9
7. Sood YR et al (2019) Applications of artificial intelligence techniques in engineering, vol 1. Springer Nature, 643 pp. https://doi.org/10.1007/978-981-13-1819-1. ISBN 978-981-13-1819-1
8. Annual Report, 2002–03, Ministry of Power, India
9. Government of India "The Electricity Act, 2003", The Gazette of India, Extraordinary, 2003, Part I 1 Section 3 Sub-section (ii), New Delhi, Ministry of Power, June IO, 2003
10. Central Electricity Authority of India Website. http://www.cea.nic.in
11. Published in Gazette of India The Electricity Act, 2003. Universal Law Publication Company Pt. Ltd., India
12. "Available transfer capabilities definitions and determinations" by NERC
13. Singh R, Sood YR (2011) Evolutionary programming based optimal model of congestion management for deregulated environment of power sector. Eng Intell Syst
14. Parkash V, Sood YR (2013) Penalization for unscheduled interchange based on availability based tariff in competitive electricity market. Trends Electr Eng

15. Sood YR (2011) Availability based tariff: a mean of resource adequacy and reliability in Indian power system. In: Proceeding of the international conference on science and engineering
16. Sood YR, Padhy NP, Gupta HO (2002) Restructuring of power industry-a bibliographical survey. In: Proceedings IEEE power engineering society winter meeting conference
17. Mughal S, Sood YR (2011) Restructuring developments and issues in indian power system. Int J Recent Trends Eng Technol
18. Khaparde SA, Kulkarni SV, Karandikar RG, Agalgaonkar AP (2003) Role of distributed generation in Indian scenario. In: Proceedings of South Asia regional conference, New Delhi, India
19. Agalgaonkar AP, Kulkarni SV, Khaparde SA (2003) Impact of wind generation on losses and voltage profile in a distribution system. In: Proceedings of the IEEE TENCON conference, Bangalore, India
20. Vindal SS, Saxena NS, Srivastava SC (2002) Industry structure under deregulated wholesale power markets in India. In: Proceedings of international conference on present and future trends in transmission and convergence, New Delhi, India

Enhancing Security Using Quantum Computing (ESUQC)

Mritunjay Shall Peelam and Rahul Johari

Abstract In a classical computer, the computation is based on transistors which encode the information in bits and usually vulnerable to attack by professional attackers. In quantum computing, the computation is based on qubits. Using single-qubit the information is encoded in the combination of two bits. Quantum computing is a completely new technology that holds all the characteristic of quantum mechanics to solve a particular problem faster than a classical computer. For solving these problems, a scientist uses a "logical qubit" since classical computer bits consist of mechanical relays or vacuum tubes which is possible to flip unexpectedly. It is possible to create quantum algorithms using qubit that run faster than classical algorithms and these algorithms reduce the time complexity and also it is impossible for the attackers to attack. There are some most popular algorithms which are possible in the quantum world, and they are Grover's algorithm and Shor's algorithm.

1 Introduction

The first quantum computing ideas were introduced by Stephen Wiesner, which is named "Conjugate Coding " in 1960. Conjugate coding is a tool based on the concept of "transmitting the multiple messages in such a way that reading one destroys the others".

Quantum computing was first recognized in the 1980s by Paul Benioff. According to Paul Benioff, quantum computers are theoretically possible. After that in 1985, David Deutsch shows what is mathematically possible on a quantum computer. Quantum computers are difficult to program and build. They produce error in the form of noise and loss of quantum coherence. The loss of quantum coherence is also called the decoherence problem. This paper covers all the basic aspects of a

M. S. Peelam (✉) · R. Johari
SWINGER(Security, Wireless, IoT Network Group of Engineering and Research) Lab, USICT, Guru Gobind Singh Indraprastha University, Dwarka, Village and Post - Pandari, 16, Mirzapur, Uttar Pradesh, India

A. Tomar et al. (eds.), *Machine Learning, Advances in Computing, Renewable Energy and Communication*, Lecture Notes in Electrical Engineering 768,
https://doi.org/10.1007/978-981-16-2354-7_21

quantum algorithm in the real quantum world and shows the weakness of modern quantum-based algorithm.

2 AIMS and Objectives

In quantum computing, a quantum bit (qubit) has three possible states 0, 1 and 0 or 1. The last one is called "coherent state"; it can perform an operation at the same time on two different values, and it creates a "de-coherence" problem. It will become more difficult to perform the computation using quantum computers.

There are mainly four aims and objectives of quantum cryptography [2]:

1. **Information integrity**: In this activity, the receiver must be able to determine the messages if the messages are altered during the transmission.
2. **Authentication**: In this activity, the receiver must be able to identify the sender.
3. **Immutability**: In this activity, the sender should not be possible to deny the creation of messages.
4. **Confidentiality**: In this activity, only the authenticated receiver should be able to extract the cipher.

3 Literature Review

A classical computer performs any type of operation by using classical bits, that is, binary digits, either 0 or 1, whereas quantum nodes use quantum bits or qubits that can be both 0 or 1 at the same time. Due to this property quantum nodes have more computing processing capability than a classical computer. Quantum computing allows to make a superposition of each one of the four states and it can be written in the form of quantum mechanical state which is perfectly legitimate; that is in the classical computer two qubits contain four bits of information (Fig. 1).

In 1917, one time pad (OTP) encryption was introduced by Gilbert Verman. OTP demands a very long key just like plain text. In order to implement this, it is very difficult and the cipher text should not leak the information about the plain text. In 1940, Shannon changed the look of cryptography. In one time pad the length of messages being encrypted must be greater than or equal to the length of keys. In this

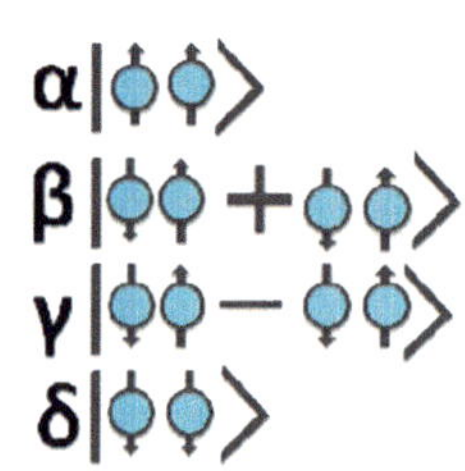

Fig. 1 Quantum mechanical state

way, the security of the stream cipher lies with the pseudo-random generator which holds the property of being unpredictable.

4 Methodology

This research paper primarily focuses on the understanding and presents an in-depth analysis of only two algorithms, viz., Grover's algorithm and Shor's algorithm from the host of many algorithms that include: (1) Josephson junction [3], (2) Grover's search algorithm [4], (3) Shor's algorithm [5], (4) elliptic curve cryptography (ECC), and (5) zero-knowledge protocol [6].

4.1 Grover's Algorithm

The "Grover's Algorithm" is the fastest quantum algorithm based on searching. It was proposed by Lov Grover in 1996. Classically, for single-element search within N elements, one needs to lookup $\frac{N}{2}$ elements on average before finding the non-zero output that is, it takes $\boldsymbol{O(N \log_2 N)}$. Grover's algorithm performs the searching by looking at an average of $\boldsymbol{N^{\frac{1}{2}}}$ times. It is useful for search in an unstructured database. Examples are searching a needle in a haystack, searching the name of a person living at a known address from a regular telephone directory, and salesman route plan for selling his item. In the classical case, for the above problems, it takes $\boldsymbol{O(N)}$ time but these searches can be speedup by a quadratic factor $\boldsymbol{O(\surd 2)}$ by choosing the quantum search algorithm (Grover's search algorithm).

4.1.1 Grover's Iterations

It uses a set of repeated iterations of quantum functions popularly known as "Grover iteration". It constitutes the following steps: (1) First, apply the Oracle function "O". (2) Secondly, apply the Hadamard transform. (3) Compute a conditional phase shift, usually with every computational basis state except for $|0\rangle$ receiving a phase shift of -1 that is

$$\mathbf{z} > \Big| \rightarrow -(-\mathbf{1})^{\delta_{z0}} \mathbf{Z} \Big\rangle$$

Again apply the Hadamard transform.

4.1.2 Working of Grover's Algorithm?

Grover's algorithm starts with a quantum register of size **n** qubits, and the size of searched space is 2^n and will be initialized to $|\mathbf{0}\rangle$, that is

$$(|\mathbf{0}\rangle)^{\otimes n} = |\mathbf{0}\rangle \tag{1}$$

Now by applying the Hadamard transform place the system into an equal superposition state that is $\boldsymbol{H}^{\otimes n}$.

For the list of $\boldsymbol{N}$ elements, the performance of Grover's algorithm will be (Fig. 2).

O($\sqrt{\text{N}}$) means that list of 10^4 elements will take the searching with order ***100***. Let the search space be

$\boldsymbol{S} = \{|\mathbf{0}\rangle, |\mathbf{1}\rangle, \ldots |\boldsymbol{N} - \mathbf{1}\rangle\}$, and $|\boldsymbol{x}_0\rangle \in \boldsymbol{S}$ is the only solution for search. Let the Hadamard gate for n application is:

$$|\psi\rangle = \mathbf{H}^{\otimes n}(|\mathbf{0}\rangle)^{\otimes n} = \frac{1}{\sqrt{N}} \sum_{x=0}^{N-1} |x\rangle \tag{2}$$

Let's take an example of n = 2 qubit (Fig. 3):

Since $\boldsymbol{N} = 2^n => \boldsymbol{N} = \mathbf{4}$ now from Eq. (2)

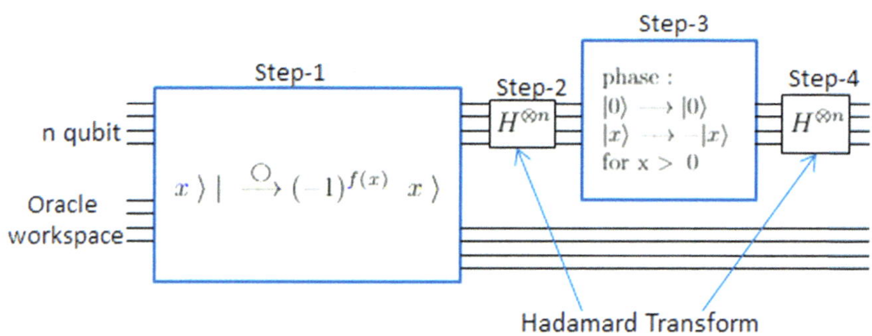

Fig. 2 Grover's algorithm iteration

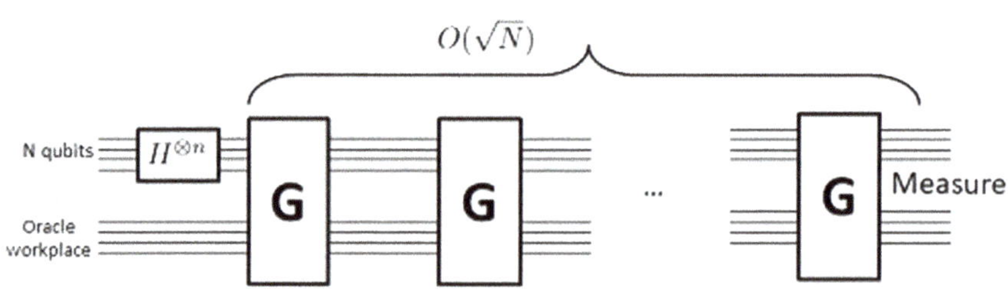

Fig. 3 Working of Grover's algorithm

$$|\psi\rangle=\frac{1}{2}|00\rangle+\frac{1}{2}|01\rangle+\frac{1}{2}|10\rangle+\frac{1}{2}|11\rangle$$

Hence

$$|\psi\rangle=\frac{1}{2}(|00\rangle+|01\rangle+|10\rangle+|11\rangle)$$

4.2 Shor's Algorithm

Shor's algorithm is one of the most popular "quantum factorization algorithm" for factoring of integer, and it is invented by "Peter Shor" in 1994. It takes $\boldsymbol{O}((\log N)^3)$ time complexity for factoring an integer $\boldsymbol{N}$ with the space complexity $\boldsymbol{O}(\log N)$.

Shor's algorithm is significant in nature and tells that public-key cryptography can be easily broken by attackers. Shor's algorithm was demonstrated by IBM in 2001, which factor the number 15 into 3 and 5 with 7-qubit using a quantum computer.

Now Lets take an integer N and find an integer $\boldsymbol{P}$ between 1 to $\boldsymbol{N}$ that divides $\boldsymbol{N}$. It can be done in two parts—(i) Classical Part (ii) Quantum Part

i. **Classical part**: In this part factoring problem involves finding the period (order) of a function that can be implemented on a classical computer also.
ii. **Quantum part**: Quantum part is used to find the period (order) of a function using quantum Fourier transform and it is responsible for the quantum speedup.

4.2.1 How Does the Shor's Algorithm Work?

Shor's algorithm consists of the following steps:

Step 1: Take a random integer $a < N$.

Step 2: Find the $gcd(a, N)$; it can be determined using the Euclidean algorithm.

Step 3: If $gcd(a, N)$ **6 = 1** then, as known, there is not any non-trivial factor of N possible and exit.

Step 4: Otherwise, by using the quantum period-finding (order finding) algorithm find $\boldsymbol{r}$ that represents the period or order of the function.

The order or period function is

$$f(X) = a^x \ (mod\ N) \tag{3}$$

Since $f(x) = f(x + r)$ then
$a^{x+r}\ (mod\ N) \equiv a^x\ (mod\ N)$

where $\boldsymbol{r}$ = order or period of $\boldsymbol{a}$ in the group $(Z_N)^X$ represents the smallest positive integer.

Step 5: If $\boldsymbol{r}$ happens to be an odd integer, then go to **step 1**. Or

Fig. 4 Flowchart

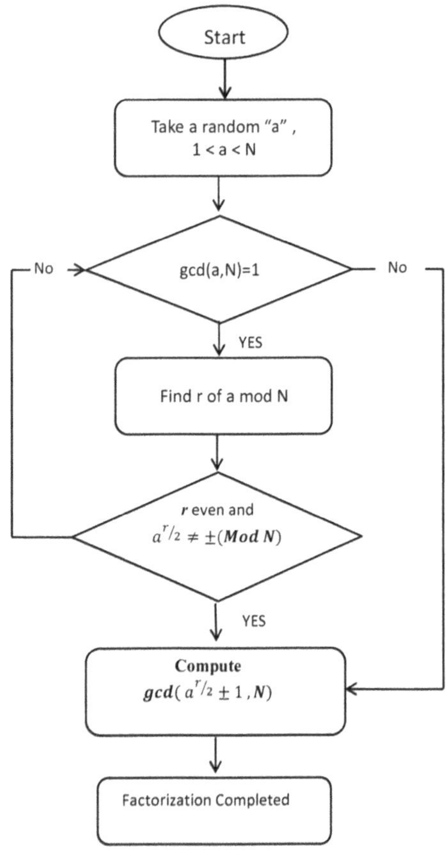

Step 6: If $(a)^{r/2} \equiv -1 (\textbf{mod N})$, then goto **step 1**.

Step 7: Otherwise, find the $gcd((a)^{r/2}+1, N)$ and $gcd((a)^{r/2}-1, N)$, as both are a non-trivial factor of N and exit.

For better understanding, the flowchart of Shor's algorithm is given in Fig. 4.

4.2.2 Example of Shor's Algorithm

lets take an example to understand the Shor's Algorithm—

Find the factor of an odd integer using Shor's Algorithm—

1. Let an odd integer $N = 15$ and choose an integer q between $N^2 < q < 2 * N^2$. Let us take $q = 256$.
2. Take an integer x such that $GCD(x, N) = 1$ using Euclidean algorithm that is $GCD(x, 15) = 1$, so $x = 7$ using Euclidean algorithm.
3. Take two quantum registers that is

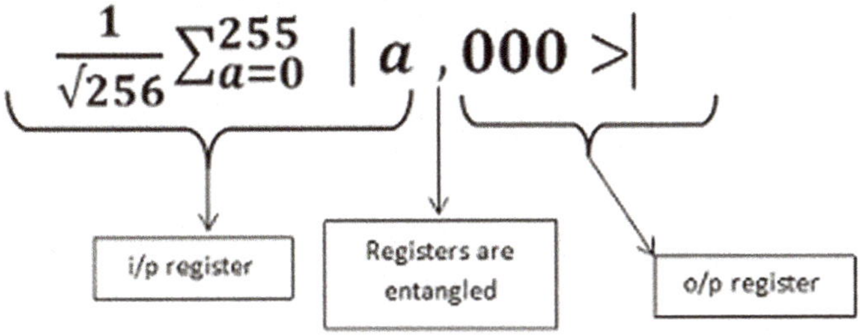

Fig. 5 Example of Shor's algorithm

a. **Input register**: Input register contains $q - 1$ qubits in size. Since it is up to **255**, so **8**-qubit is required.
b. **Output register**: Output register contains $N - 1$ qubits in size. Since it is **14**, so **4**-qubit is required.

4. Initialize the i/p register with weighted superposition; all the integers range from **0** to ($q - 1$), that is **0–255**.
5. Clear the output register that is filled with **0***s*. Now the total state of the system.
6. Now apply the period or order function to all numbers present in the i/p register and store the intermediate result obtained in the corresponding o/p register (Fig. 5).

Now the table is given below (Fig. 6):

7. Now o/p register will collapse because of one of the following: $|1\rangle$, $|7\rangle$, $|4\rangle$ and $|13\rangle$ now lets take $c = |1\rangle$.

Fig. 6 Example of Shor's algorithm

I/P Register	7^a mod 15	O/P Register
|0 >	7^0 mod 15	1
|1 >	7^1 mod 15	7
|2 >	7^2 mod 15	4
|3 >	7^3 mod 15	13
|4 >	7^4 mod 15	1
|5 >	7^5 mod 15	7
|6 >	7^6 mod 15	4
|7 >	7^7 mod 15	13
|8 >	7^8 mod 15	1
|9 >	7^9 mod 15	7
|10 >	7^{10} mod 15	4
⋮	⋮	⋮

8. Since the registers are in an equal superposition state of **64** value, that is, **(0, 4, 8, 12 ... 252)**, hence the probability in this case is: $\frac{1}{64}$. Now the value of collapsed output register is:
$\frac{1}{64}|0\rangle + \frac{1}{64}|4\rangle + \frac{1}{64}|8\rangle + \frac{1}{64}|12\rangle \cdots + \frac{1}{64}|252\rangle$.
9. Apply quantum Fourier transform (QFT) on the partially collapsed i/p register.

Take state |a⟩ and transform it into a state by

$$\frac{\mathbf{1}}{\sqrt{\mathbf{64}}}\sum_{\mathrm{a}\in\mathrm{A}}, |\mathbf{1}\rangle$$

$$\text{where a} = \frac{\mathbf{1}}{\sqrt{\mathbf{256}}}\sum_{\mathrm{c}=0}^{\mathbf{255}} |\boldsymbol{c}\rangle * \boldsymbol{e}^{\frac{2\pi iac}{256}}$$

Now the final value of the input register is

$$\frac{\mathbf{1}}{\sqrt{\mathbf{64}}}\sum_{\mathrm{a}\in\mathrm{A}} * \frac{\mathbf{1}}{\sqrt{\mathbf{256}}}\sum_{\mathrm{c}=0}^{\mathbf{255}} |\boldsymbol{c}\rangle * \boldsymbol{e}^{\frac{2\pi iac}{256}}, |\mathbf{1}\rangle$$

where A is the set of all values (7^a mod ***15***) produces **1** and ***A***= **{0, 4, 8, 12 ... 252}**.

Now QFT will be the maximum probability with the amplitude of integer in the multiples of $\frac{q}{4}$, that is $\frac{256}{4}$ **= 64**, now no longer an equal superposition of states.

Next, find the register that will collapse with a high probability and multiple of **64**. Take $\boldsymbol{p} = \mathbf{4}$

So $|\mathbf{0}\rangle$, $|\mathbf{64}\rangle$, $|\mathbf{128}\rangle$, ... determined by
$\boldsymbol{gcd}(\boldsymbol{x}^{(\frac{p}{2}+1)}, \boldsymbol{N})$ and $\boldsymbol{gcd}(\boldsymbol{x}^{(\frac{p}{2}-1)}, \boldsymbol{N})$.
Now finally,
$\boldsymbol{gcd}(7^{(\frac{4}{2}+1)}, \boldsymbol{15}) = \mathbf{5}$ and $\boldsymbol{gcd}(7^{(\frac{4}{2}-1)}, \mathbf{15}) = \mathbf{3}$.
That is the factor of an odd integer ***N***= **15** is **(5, 3)**.

5 Conclusion

The current research concludes by presenting an in-depth analysis of Grover's algorithm and Shor's algorithm by discussing their problems and applications. The problems/limitations of Shor's algorithm are that the probability is dependent on choosing ***'q'***, so the larger the ***q***, the higher is the probability of finding the correct value. In future, quantum computing has the capability of transforming all the aspects of information security which is vulnerable in a classical way. Quantum computing has the potential to expose computation that is not enough to solve in a classical way.

References

1. Yanofsky NS (2011) An introduction to quantum computing. In: Proof, computation and agency, pp 145–180. Springer, Dordrecht.
2. Kute SS, Desai CG (2017) Quantum cryptography: a review. Ind J Sci Technol 10(3)
3. Sakmann K (2011) Exact quantum dynamics of a bosonic Josephson junction. In: Many-body Schrödinger dynamics of bose-einstein condensates, pp 65–80. Springer, Berlin, Heidelberg
4. Hao L, Li JunLin, Long GuiLu (2010) Eavesdropping in a quantum secret sharing protocol based on Grover algorithm and its solution. Sci China Phys Mech Astron 53(3):491–495
5. Shor PW (1999) Polynomial-time algorithms for prime factorization and discrete logarithms on a quantum computer. SIAM Rev 41(2):303–332
6. Koens T, Ramaekers C, Van Wijk C (2018) Efficient zero-knowledge range proofs in ethereum. ING. blockchain@ing.com

Deviation Settlement Mechanism and Its Implementation in Indian Electricity Grid

Bharti Koul, Kanwardeep Singh, and Y. S. Brar

Abstract India is a big nation that is dependent on electricity to a large extent, like in the agricultural sector, big manufacturing agencies, etc. To achieve electricity at a lesser price is the foremost challenge in such agencies. The incorporation of the Electricity Act has facilitated such agencies to obtain power at a lesser price. This can help the consumers to evade penalties and benefits to acquire incentives by signifying suitable needful arrangements which can aid in maintaining the stability of the power grid. The deviation settlement mechanism helps to maintain the stability of the grid, thus improving the power quality. This paper presents an investigation of the deviation settlement mechanism and its implementation in the Indian electricity grid.

Keywords Deviation settlement mechanism · Availability-based tariff · Demand response

1 Introduction

The Indian power industry is one of the expanded power industries of the world. The installed capacity of the national grid in India is 371.054 GW as per the statistics of 30 June 2020 [1]. This wide area synchronous grid operates economically at 50 Hz and the allowable range of frequency band is 49.9–50.05 Hz [2]. The Indian power sector performs its functioning by being distributed into some main grids which are responsible for its operation as a whole. These main grids, named Regional Load Dispatch Centers (RLDCs), are classified as northern-grid, western-grid, eastern-grid, southern-grid, and north-eastern-grid. All the parts of India are linked with one of these RLDCs subject to its geographical pattern [3]. Demand-side

B. Koul (✉) · K. Singh
Department of Electrical Engineering, Guru Nanak Dev Engineering College, Ludhiana (IKG Punjab Technical University, Kapurthala), India

Y. S. Brar
Department of Electrical Engineering, IKG Punjab Technical University, Kapurthala, India

A. Tomar et al. (eds.), *Machine Learning, Advances in Computing, Renewable Energy and Communication*, Lecture Notes in Electrical Engineering 768,
https://doi.org/10.1007/978-981-16-2354-7_22

management is the forecasting, execution, and monitoring of the job of the utility that is planned to affect consumers' usage of electricity. This results in modifying the load consumption pattern by incentivizing the users to transfer the energy usage to off-peak durations so as to flatten the load curve.

The extensive and quick variations in grid frequency ranging from 48.0 to 52.0 Hz (CERC 1999) on regular basis leads to the problem of grid indiscipline [4]. To sustain the grid discipline, the users of the grid have to follow the forecasted injections or withdrawals, known as grid code, controlled by a commercial mechanism. This mechanism is used for regulating the users of the grid for scheduling and dispatch of power with the least disturbance. So, there is a necessity to improve the efficiency of the grid, grid discipline, capability, and responsibility which is maintained by the unscheduled interchange (UI) [5–8], by imposing charges on those who deviate from their scheduled generation or drawl. The deviation in frequency from its nominal value is caused due to the unbalance or disparity in actual demand and generation. Because of deviation between actual generation (in a time-block (T) means electricity generated or supplied by the seller, measured by the interface meters) and scheduled generation (for a time-block (T) means a schedule of generation in MW or MWh given by the concerned load dispatch center) and also between actual drawl (in a time-block (T) means electricity drawn by a buyer, measured by the interface meters) and scheduled drawl (at any time or for a time-block (T) means a schedule of dispatch in MW or MWh given by the concerned load dispatch center), there is the need of deviation settlement mechanism (DSM). Any utility is permitted to inject/draw power into/from the grid at UI costs till the frequency is stabilized within the specified band [9] and if there is a deviation from the schedule, i.e. the participants withdraw more than the scheduled from the grid they are penalized and if the drawl is less than the actual they are incentivized [10]. The central regulatory electricity commission (CERC) [11] then undertakes amendment if there is a requirement w.r.t. to the working conditions. These amendments are imposed only by taking all the concerned (i.e. producers and beneficiaries) into consideration.

Demand response programs (DRP) incorporate utility commenced incentives to encourage consumers to voluntarily adjust the usage of electricity without any effect on consumer comfort. This is an essential practice for optimized and effective use of electricity which subsequently is the main building block of the future electric grid called a smart grid (SG). Moreover, AI and machine learning-based are some examples listed in [12–19].

Contribution of this paper: This paper builds a study on the introduction of DSC and the structure of the Indian electricity grid to develop a generalized strategy for minimizing the deviation settlement charges (DSC). The importance and significance of demand-side management have been explained to improve the DSM so as to make a ready reference for the researchers who are working in this field.

Organization of the paper: The remaining part of the paper is organized as follows. Section 2 introduces the concept of a grid with the structure of the Indian electricity grid. In Sect. 3, a detailed introduction to the Indian electricity grid is provided. Section 4 explains the deviation settlement charges and the procedure to calculate them. Finally, Sect. 5 concludes the paper by highlighting the major findings.

2 Concept of Grid

The Indian electricity grid is a system of network power lines, transformers along with the related apparatus engaged in allocating the power electricity within a specified topographical zone. It can also be denoted as an interrelated linkage for supplying electricity to customers from the utility. This involves generating stations that generate power, a high power transmission system to transmit electricity from far away stations to the user centers, and distribution lines to link specific consumers as shown in Fig. 1. The generating stations are usually situated nearby to fuel stations, e.g. the dam location, but the electricity generation does not get transmitted in the same form. The generated power is stepped up to a larger voltage to which it joins to the power transmission system (electric grid). The construction of the electric grid can diverge depending on the monetary limitations, need for network consistency, with demand and generation characteristics.

The major assignment done in conventional electric grid involves the major components as:

- *First the generating station for the production of electric power*: These power-producing plants are placed nearby power generation sources, e.g. dams, coal mines, etc.
- *Secondly, the transmission of the generated electric power*: After the electric power is generated it is transmitted to the respective substation from where the voltage is stepped up with the step-up transformers so as to minimize the losses and then is ready for transmission to the several areas. It is then sent to the power grid from where it is then transmitted to different cities.
- *Then finally distribution of the generated power*: The electric grid gets linked within various zones of the entire nation, and the distribution of electricity is done to the different parts of the country or state by the linkage of transmission

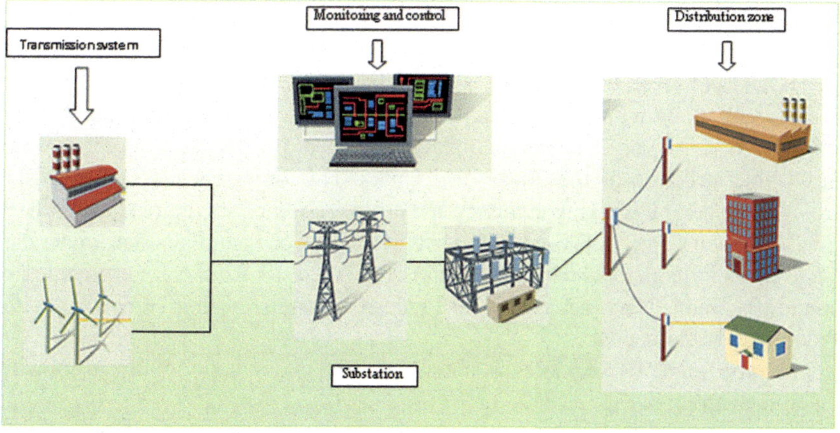

Fig. 1 Basic structure of conventional power grid

network linking various zones. The transmitted power being stepped down at the substations followed by its distribution to the respective consumers.

3 Indian Electricity Grid Structure

The Indian electricity grid structure is grounded on the centralized arrangement of the nation. In the Indian electricity sector, the reforms have been initiated at the beginning of the 2000s, for which the Electricity Act 2003 was one of the revolutionary steps taken by the Government of India. This initiative was persuaded with the aim to impart competition, limpidity, and commercial feasibility in the power sector so as to attain the availability and affordability of electricity. In order to initiate unbundling of unified state electricity board (SEB) into generation, distribution, and transmission, the exposure of transmission and distribution segments for utility and customers was introduced. The creation of an autonomous controller as State Electricity Regulatory Commission (SERC) for the respective states has been envisaged for the states to be an independent entity.

For each of the regions, an exclusive load dispatch center is allocated to further coordinate the regular scheduling practice in the dispatching of centrally generated power. The availability of power collection from the centrally generated system is forecasted day-ahead in 96 timeslots, each of 15 min duration, then the concerned RLDC distributes the power to its corresponding SEBs depending upon the share in the central generation pool. The SEBs perform the corresponding task for meeting the demand of their respective customers for 24 h with the help of their own consumers over the day, from their own generation stations along with the entitlement in the central generation pool. The SEBs give the requests to the RLDC, and then resolve the schedule of dispatch and drawing for the respective SEBs.

The SEB is possessed by every state entity and is accountable for generation, transmission, and distribution. It works under the monitoring of the respective state government and is also partly governed by the central government as far as the financial powers are concerned.

The tasks of several systems are associated with a network of power grid working and control, along with the structural linkages, in order to expedite expansion and smooth working of regional and national grids. There is a hierarchal order in which this whole structure of the power grid works. This order consists of the NLDC which is the central working authority and it further comprises their respective State Load Dispatch Centers. These SLDCs handle respective states that come under their geographical vicinity. As per the regional centers, i.e. RLDCs, e.g. northern region comprises Delhi, Haryana, Himachal Pradesh, Jammu Kashmir, Rajasthan, Uttar Pradesh and Uttaranchal.

Likewise, all SLDCs handle the states that come under their vicinity, as shown in the flow diagram in Fig. 2.

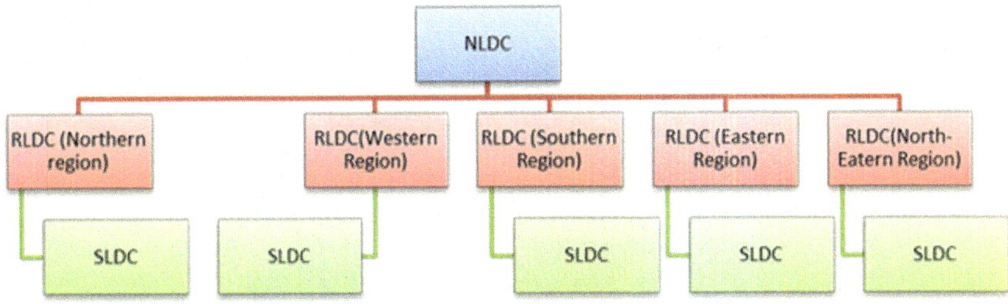

Fig. 2 Hierarchy of various centers in power grid in India

The role of the National Load Dispatch Center (NLDC) that has been framed rendering to the notification Dt. 2 March 2005, from the Ministry of Power, Government of India, based on Section 26(2) of the Act in which NLDC performs various functions for overall regulation of the grid operations, supervising the control and coordination interlink between various regional grids is shown in Fig. 3. The NLDC also takes into account the information regarding the exchange of power between various regions or several RLDCs. The NLDC supervises the scheduling and dispatch of power within the regional links for the safe and efficient operation of the grid.

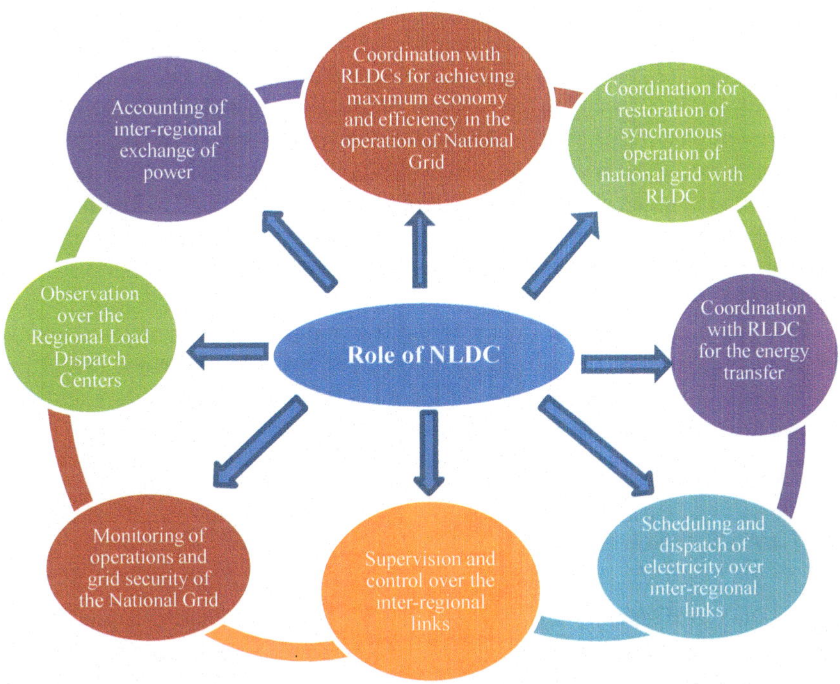

Fig. 3 Role of NLDC

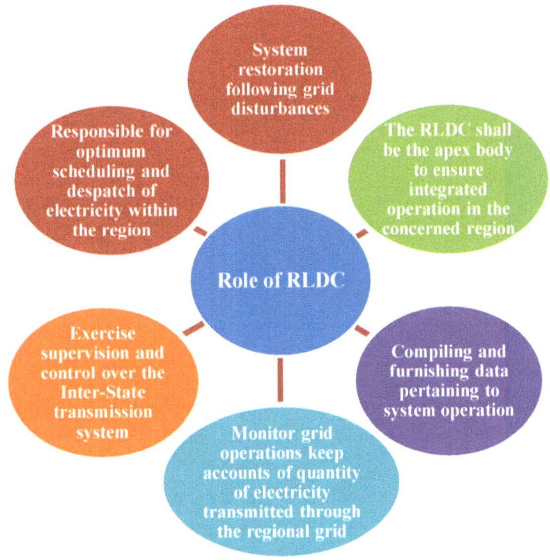

Fig. 4 Role of the Regional Load Dispatch Center

The Regional Load Dispatch Centre (RLDC) is the main body to confirm the combined process of power operation for the specific area, which is framed based on Sections 28 and 29 of the Electricity Act, 2003. The role of RLDCs is shown in Fig. 4. The monitoring and regulation of interstate transmission networks based on the information about the capacity of electricity transferred through regional grids is controlled by the RLDCs. The restoration of the network that is followed by disturbances of the grid is also regulated by the RLDC that makes it responsible for the optimized scheduling and withdrawal of power in a specific region. The integrated work related to the compilation and modification of data for system operation is controlled by the respective regional centers. This leads to the control and systematic operation of the interstate transmission network.

The State Load Dispatch Centre (SLDC) is the prime system to confirm the cohesive task of optimum planning for dispatch of power in a state and that is also accountable for the indentures that are arrived in the generating stations working in the respective state. The functions that SLDC performs are shown in Fig. 5.

4 Deviation Settlement Charges (DSC)

To maintain grid stability CERC introduced various schedule deviation charges according to the frequency of the grid to discourage schedule deviations. The deviation means the structure for managing the deviation in energy, maybe less than or more than the scheduled value. The guidelines for assessing the deviation(s) in these

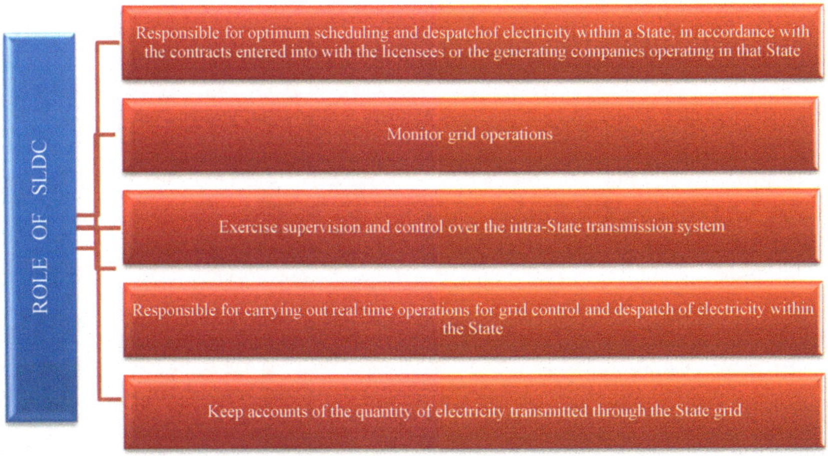

Fig. 5 Role of State Load Dispatch Center

scheduled and actual values which are payable or received by the state utilities and its other parameters are set by the regulatory commissions and the amendments.

Due to the insufficient arrangement for short-term attainment of energy that is supported by the consistent short-term load requirement, renewable generation, and insufficiency of the ancillary services, there are deviations from the scheduled values. So, it becomes necessary to focus on DSM which is considered to report the above reasons, while the ancillary services are responsible for the remaining deviations which are not sufficiently taken into consideration. Ideally, the deviations from the schedule must become inappropriate, with the practicing ancillary services that deal with the matters afar from the realistic control of network parameters.

The deviations settlement mechanism [5] is explained from the consumer's point of view as it depends on the withdrawal by the consumer less or more than actual. The mechanism is fixed based on the frequency calculation and band of frequency from which the deviation settlement charges are calculated. Table 1 explains the limits and bands of frequency for the calculation of DSC. The DSC for various frequency bands whose DSC_Rate is given are categorized between 50.05 and 49.7 Hz [12].

Table 1 Existing deviation price vector

Frequency	Rate of deviation (DSC_Rate)
If $f \geq 50.05$ Hz	DSC_Rate = 0
$50.05 > f \geq 50.00$	DSC_Rate increases by 35.6 p/u for each 0.01 Hz step
$50.00 > f \geq 49.7$	DSC_Rate increases by 20.84 p/u for each 0.01 Hz step
$f < 49.7$	DSC_Rate = 824.04 p/u

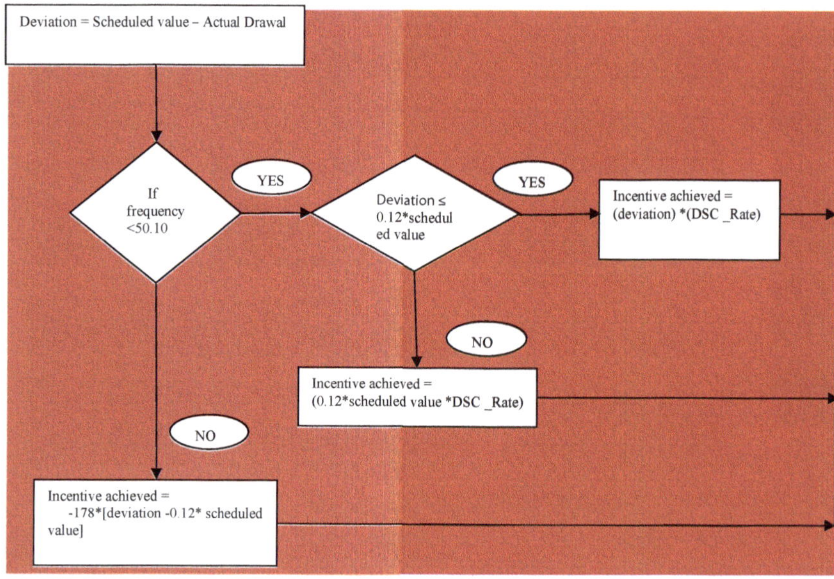

Fig. 6 Flowchart for the incentives achieved on the under-drawl

In the case of a schedule deviation to the value of 12% schedule or to a limit of 150 MW the value of which is lesser in a given time-block of 15 min, the deviation charges are imposed per unit. In the case of over drawl when the deviation exceeds the value of 12% of the scheduled value or greater than 150 MW for a given duration of 15 min time-block, the regular deviation rates need to be given per unit along with additional deviation charges. The schedule deviation in the case of under-drawl that is to the value of 12% of the predefined values, the inducements are given per unit depending on the grid frequency. When the under-drawl deviation becomes greater than 12% of the scheduled value, these inducements to be achieved are nil. The extra deviation charges to be paid become equal to 178 paisa/kWh when the under-drawl is greater than 12% of the scheduled value with the grid frequency larger than 50.10 Hz. The deviation settlement charge rate (DSC_Rate) for deviation charge is calculated depending on the 15-min duration time-block of the grid frequency. The DSC_Rate is taken in paisa/kWh. Fig. 6 shows the flowchart that explains the incentives achieved on the under-drawl by the users/utility.

5 Conclusion

The DSM has transformed from its previous scenario and progressed the power system network in the direction of a modified and reliable stage, thus educating constancy and evading grid disruption. A totally market-motivated setup ensuring significantly upgraded provisions and infrastructure will help to benefit in this direction.

Also, the implementation of availability-based tariffs shall contribute to attaining the objective of transformation of the Indian power sector, thus enhancing the reliability of the power network. It can further augment the grid reliability by a frequency-based pricing mechanism. This paper focuses on the introduction of deviation settlement and its impact on the Indian electricity grid. It also discusses the various procedures to calculate the deviation charges that can help in minimizing the deviation charges to be paid by consumer or state utility.

References

1. All India Installed Capacity of Utility Power Stations. https://www.ibef.org/industry/power-sector-india.aspx
2. https://nrldc.in/guidelines
3. Gupta P, Verma YP (2019) Optimization of deviation settlement charges using residential demand response under frequency linked pricing environment. IET Gener Transm Distrib 13(12):2362–2371
4. Banerjee A, Banerjee S (2004) Availability based tariff: an economic instrument for grid discipline. Econ Polit Weekly 39(35):3939–3946
5. Parkash V, Raj Sood Y (2014)Impact of unscheduled interchange pricing in competitive electricity market. Int J Adv Eng Sci 4(1)
6. http://www.ecodevgroup.com/resources/PageMedia/page45/Reserve%20Power%20Plant.pdf
7. Syadlia H, Abdullaha MP, Faridiansyaha I (2016) An improved load shedding scheduling strategy for solving power supply defisit. Jurnal Teknologi (Sci Eng) 78(5–7):61–66
8. Gelazanskas L, Gamage AA (2014) Demand side management in smart grid: a review and proposal for future direction. Sustain Cities Soc 11:22–30
9. Zhong J, Bhattacharya K (2003) Frequency linked pricing as an instrument for frequency regulation in deregulated electricity markets. In: Proceedings of IEEE PES summer meeting, Toronto, Ontario, Canada, pp 566–571
10. Pujara A, Vilas VG, Bakre SM, Muralidhara V (2017) A novel approach for UI charge reduction using AMI based load prioritization in smart grid. J Electr Syst Inform Technol 4:338–346
11. CERC REPORT, www.cercind.gov.in.
12. Implementation of Deviation Settlement Mechanism and related matters (Fourth Amendment) Regulations (2018). https://nrldc.in/wp-content/uploads/2018/12/DSM-flasher_.pd
13. Aggarwal S et al. (2020) Meta heuristic and evolutionary computation: algorithms and applications. Springer Nature, Berlin, 949 p. https://doi.org/10.1007/978-981-15-7571-6 (ISBN 978-981-15-7571-6)
14. Yadav AK et al. (2020) Soft computing in condition monitoring and diagnostics of electrical and mechanical systems. Springer Nature, Berlin, 496 p. https://doi.org/10.1007/978-981-15-1532-3. ISBN 978-981-15-1532-3
15. Gopal et al. (2021) Digital transformation through advances in artificial intelligence and machine learning. J Intell Fuzzy Syst Pre-press, 1–8. https://doi.org/10.3233/JIFS-189787
16. Fatema N et al. (2021) Intelligent data-analytics for condition monitoring: smart grid applications. Elsevier, 268 p. ISBN: 978-0-323-85511-2. https://www.sciencedirect.com/book/978 0323855105/intelligent-data-analytics-for-condition-monitoring
17. Smriti S et al (2018) Special issue on intelligent tools and techniques for signals, machines and automation. J Intell Fuzzy Syst 35(5):4895–4899. https://doi.org/10.3233/JIFS-169773

18. Jafar A et al. (2021) AI and machine learning paradigms for health monitoring system: intelligent data analytics. Springer Nature, Berlin, 496 p. https://doi.org/10.1007/978-981-33-4412-9. ISBN 978-981-33-4412-9
19. Sood YR et al. (2019) Applications of artificial intelligence techniques in engineering, vol 1. Springer Nature, p 643. https://doi.org/10.1007/978-981-13-1819-1. (ISBN 978-981-13-1819-1)

Integration of Battery Charging and Swapping Using Metaheuristics: A Review

Neha Raj, Manikanta Suri, and K. Deepa

Abstract Electric vehicle (EV) is one of the preferred modes of transportation due to less emission of pollutants. The depleted batteries can be refueled by using a battery charging station (BCS), battery swapping station (BSS), and battery swapping van (BSV). Earlier, the depleted batteries were replenished using the different battery charging modes, but due to less flexibility, battery swapping (BS) was preferred over battery charging (BC). However, battery charging is not completely ruled out as it causes less damage to the battery and the swapped batteries have to be charged using battery charging. The forecasting on the arrival of EVs helps the station owner to serve the customers and in optimizing the various cost(s) associated with BSS. Metaheuristics help to arrive at the solutions at a faster rate when compared to the traditional optimization techniques. BSV is the active mode of replenishing energy which increases the effectiveness and efficiency of the battery swapping process. Further, a case study is carried out to understand the need to serve the customer for an unpredicted situation in the service station.

Keywords EV · BCS · BSS · BSV · Forecasting · Metaheuristics

1 Introduction

The depletion of fossil fuels, which is the primary source in the internal combustion engine (ICE), led to the deployment of EVs. EV is a sustainable solution to reduce greenhouse emissions as it has a very less carbon footprint. EVs have other benefits such as improvement in local air quality, mitigation of global climate, and oil conservation. Thus, EV is considered a clean energy vehicle. Many countries such as India, UK, China, France, and the Netherlands started to reduce ICE vehicles which primarily consume fossil fuels. EV sales have also increased drastically. The global sales hit 2 million at the end of 2016, crossed 3 million by November 2017, reached 5 million by December 2018, and presently it totaled about 7.5 million units. Thus, the

N. Raj (✉) · M. Suri · K. Deepa
Department of Electrical and Electronics Engineering, Amrita School of Engineering, Amrita Vishwa Vidyapeetham, Bengaluru, India

A. Tomar et al. (eds.), *Machine Learning, Advances in Computing, Renewable Energy and Communication*, Lecture Notes in Electrical Engineering 768,
https://doi.org/10.1007/978-981-16-2354-7_23

attention toward EVs is increasing day by day. Lack of charging points, cost, short driving range/charge, long charging time, and a short lifetime with fast charging are the factors that cease the development of EVs. Thus, many countries implement creative solutions to attract the public toward EVs. Some of these are low parking rates, low toll fees, and the ease to obtain license plates. However, these are not long-term solutions for the constraints imposed by EV. BSS is a durable solution that can increase the growth of EV further. Unlike in BCS where the battery is charged using ports, in BSS, the EV battery is swapped with a high SOC battery. This will lower the service time, increase the travel range of vehicles (using high-capacity batteries), and the SOH of the battery is improved if the depleted batteries are charged using a slow-charging method [1, 2].

The depleted EV batteries can be refueled by using three methods: battery swapping (4–12 min to serve), battery swapping van, and battery charging. Battery charging can be further classified into four modes: Mode 1, which is a slow-charging method (230 V, 16 A, AC with 7–15 h to charge); Mode 2, which is a normal charging method (230 V/440 V, 32 A, AC with 3–8 h to charge); Mode 3, which is a fast-charging method (230 V, 63 A, AC with 15–30 min to charge); and Mode 4, which is a super-charging method (fast charging using DC with 8–10 min to charge) [3]. Figure 1 and Table 1 show an overview of EV battery refilling and the key differences between battery charging and battery swapping, respectively.

From Table 1 it can be inferred that the battery swapping method is more flexible than the battery charging method. However, it has a few shortcomings. The drawback of a battery swapping station is its effective battery swapping architecture which can

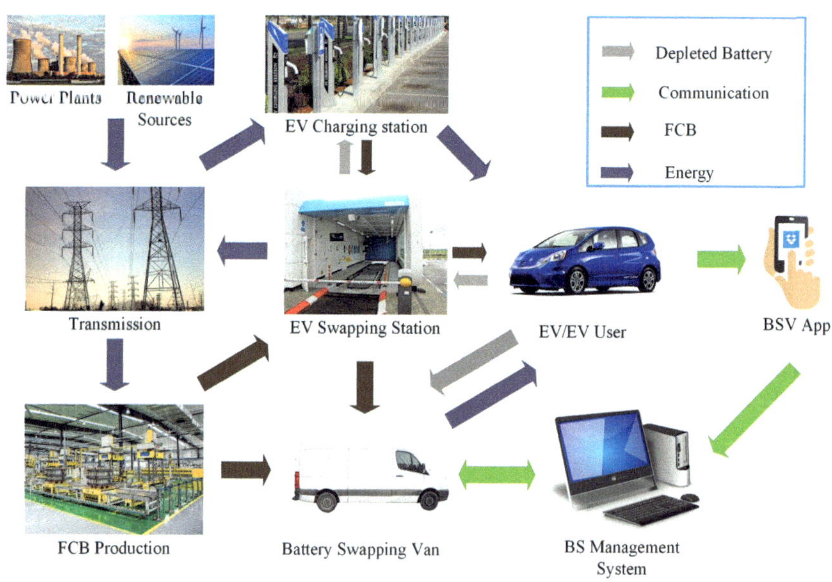

Fig. 1 Overview of EV battery refilling [4]

Table 1 Key differences between battery charging and battery swapping [1, 2, 5]

S. no.	Basis of difference	Battery charging	Battery swapping
1	Service time	Mode 1: 7–15 h Mode 2: 3–8 h Mode 3: 15–30 min Mode 4: 8–10 min	4–12 min
2	SOH	Depends on the method of charging	Improved if depleted battery is charged using slow charging
3	Mileage	Relatively low	High
4	Flexibility to owner of service station	Does not provide flexibility	Flexible
5	Integration with renewables	Relatively tough	Easier
6	Virtual power plant	BCS does not act as a virtual power plant	BSS acts as a virtual power plant
7	Impact on grid	Demands more power from grid	Balances the charging demand

be used for the operation of EV battery swapping stations. Another drawback is because of the usage of a standardized battery with high energy density and high mileage, a high recycling ratio has to be met whenever the battery gets swapped in the swapping station that has flexible battery charging characteristics due to which it has a smoother load profile in comparison to a fast-charging station (FCS) [4].

Though there is a superiority for BSS over BCS, the BCS is not completely ruled out. The depleted batteries swapped with high SOC batteries in a BSS have to be charged for future swapping. A BSS consists of three parts: (1) a BSS, (2) a standalone BCS, and (3) a battery stock. A battery stock is a place where a fully charged battery (FCB) is kept. This acts as a secondary source for a BSS when the BCS is unable to meet the demands of BSS. Initially, when BSS was introduced into the market, there were high expectations on it owing to its advantages. But unfortunately, *Better Place* was recorded for bankruptcy in Israel in May 2013, *Tesla Motor's* battery swap program had to be kept silent for a while, and the State Grid Corporation of China (SGCC) battery-swapping network also faced complications. The reason for this tragedy was the high cost of a BSS. The various costs associated with a BSS are (1) battery degradation cost, (2) electricity charging cost, and (3) cost associated with battery stock. Battery degradation cost is the cost paid for the SOH of the battery. It does not involve money transactions, but it is related to the quality of the battery. SOH is a figure-of-merit of the condition of the battery compared to its ideal conditions. It is expressed as a percentage. It is tempting for the owner to go for FC and Super-charging methods of battery charging but he/she has to pay for a reduction in the SOH of the battery. Electricity charging cost is the cost paid for charging the depleted batteries. Battery stock cost is the cost afforded for purchasing FCB. Thus, the owner of BSS has to carefully plan the charging of depleted batteries without much increase in the costs associated with BSS. To achieve this, he/she should have an account on arrivals of EV at BSS prior. Traditional optimization algorithms (OTA) takes more time to arrive at a solution(s) for the objective functions and the constraints involved in cost(s) associated with BSS. Metaheuristics alleviate this difficulty by logically exploring the search space, thus providing the near-optimal solution in less cost and time.

The energy replenishment techniques for EV discussed until now come under passive mode. In this mode, the user has to drive till BSS/BCS to get served. He/she has to wait in the queue to get served in the service station, as it has limited capacity to hold EVs. The works [6, 7] evaluate the average time spent by a customer in an FCS using queuing theory. For optimal operation of an FCS, the average time spent by a customer in the FCS was found to be 34.638 min. Thus, the average waiting time in the queue will be 4.638 min. For a slow-charging station, the average waiting time in the system will be far more as it takes nearly 7–15 h to charge an EV using a slow-charging method. Though BSS can reduce the waiting time in the queue and the system, the EV users on average have to wait for at least a few minutes before they get served. This calls for an active mode of energy replenishment for EV. For an inactive mode of service, the user need not necessarily have to reach the BSS/BCS to get service, thereby eliminating the chance of the queue. A BSV will improve the efficiency and effectiveness of the battery-swapping service by reducing the average waiting time of an EV user to get served [4]. Moreover, artificial intelligence and machine learning-based are some examples listed in [8–15]. A reader may also refer to these applications.

The paper is organized as follows: Sect. 2 tells about the different forecasting methods. Section 3 introduces the different metaheuristic algorithms; Sect. 4 focuses on BSV, and finally, a case study is given in Sect. 5.

2 Forecasting on the Arrivals of EVs

The forecasting method is used for determining the number of EV batteries that are needed in the BSS. In this paper, the focus is on two methods for forecasting: (backpropagation) neural network and token system.

In the BP neural network model, the EV driving pattern is determined based on a daily pattern for 8 days. The data training in the BP forecasting model is done using historical forecasting and the data obtained from the study pattern is the number of EVs used in 24 h. The BP forecasting is done based on the previous day's study which was carried out for 24 h. It is observed that the number of batteries that are swapped is more during the weekends than the weekdays, as more customers travel through the city and this BSS is situated at the exit tunnels of the city [16].

In the token system, there is a notice issued to the BSS by the customer asking for battery swapping. Four charging methods are adopted in BSS to charge the batteries. Two requirements have to be met to minimize the costs incurred in BSS: to maintain the battery stock and reduce the damage caused due to battery charging methods. Fast charging is preferred as it takes less duration to charge the battery but it causes more damage when compared to normal charging. To avoid this, the slow-charging method is preferred, so there is a need to predetermine the number of batteries to be charged using this method to maintain the SOC of the battery [17–19]. Forecasting the arrival of EVs at BSS helps the owner of the station to schedule FCB for EV battery swapping and battery charging for depleted batteries, thus building up a cost-effective BSS.

3 Metaheuristics

For the viable operation of a BSS, it is necessary to optimize the costs associated with it. There are two ways to approach an optimal solution: (1) using traditional OTA (optimization algorithm) and (2) using heuristics. Traditional OTA uses two methods: (1) direct search methods and (2) gradient-based methods. Direct search methods use function values to arrive at the optimal solution, whereas gradient-based methods use derivatives of the function to achieve the optimal solution. To prove the solution to be global, the OT (optimization) problem should be a convex OT problem. The main disadvantage of traditional OTA was when the derivative of a value in the search space vanishes, gradient-based methods do not provide a solution, computation burden is involved in these algorithms, and the time taken to arrive at a solution is quite high. But most of the time traditional OTA provides the solution. Heuristics are short-cut methods to approach the optimal solution. These provide good enough solutions in a limited frame or deadline. The solutions obtained using heuristics may not be optimum and it will lead to poor decision-making on a limited data set but the speed at which the solution arrives can cope up with the disadvantages. If the search space is too large, it is so difficult to apply traditional OTA. Heuristics eliminate this difficulty by evaluating only the subset of feasible solutions, thus the time taken to arrive at a solution will be faster. The search in heuristic algorithms ends when an optimum occurs or when the algorithm has undergone a specified number of iterations. In most cases, the optimum is a local optimum. In the 1980s, a new generation of metaheuristics was developed which upgraded the quality of heuristics. Metaheuristic algorithms are a form of stochastic-based optimization technique that is mostly drawn from nature and does not depend on the surface gradient and so they are independent of the constraints, allowing the solution to escape entrapment at local minima [20]. Metaheuristics are widely applied in combinatorial OT problems such as travelling salesman problem (TSP). Many authors have used metaheuristic algorithms in the optimization of costs associated with BCS and BSS owing to complexity and nonlinearity associated with objective function and the fastness involved in metaheuristic algorithms. Termination occurs in metaheuristic algorithms when (1) the number of iterations exceeds the given number, (2) the quality of the current solution is adequate, (3) the number of iterations since the last best solution exceeds a specified number, and (4) the neighborhood associated with the current solution is empty or leads to bad solutions. They are independent of the domain of

Table 2 Classification of metaheuristic algorithms [20]

S. no.	Categories	Examples
1	Evolutionary algorithms	GE (genetic algorithm) [21, 17] DE (differential algorithm) [17, 22]
2	Swarm intelligence	PSO (particle swarm optimization) [17, 23–26] Glowworm algorithm [27]
3	Pheromone/stigmergy	ACO (ant colony optimization) [28]
4	Other nature-inspired	Bat algorithm, cuckoo algorithm [29] Artificial bee colony optimization [30] Shuffled frog leaping algorithm (SFLA) [5, 31]
5	Escape from local minima	TS (tabu search) [21, 32] SA (simulated annealing) [21, 32]

the problem [20]. Table 2 is used to depict the categorization of the various metaheuristic algorithms, and the different characteristics of each algorithm are depicted using Table 3.

4 Battery Swapping Van (BSV)

In the passive mode of replenishing energy, the user has to reach the service station to get served. Battery charging and battery swapping methods fall under this category where the EV user needs to travel to BSS/BCS for the service. The user will be dissatisfied if there is a long queue in the service station and this further ceases the development of EV. The other disadvantage is that every time the EV user has to calculate the required SOC such that he/she can reach the service station without any hurdle in the middle of the journey. Though BSS offers less waiting time in the queue compared to BCS, if more flexibility is provided to the customers, this will attract them to buy more EVs and the fellow customers will trade to EV from conventional ICE. This flexibility is achieved through the active mode of replenishing energy [7, 19, 33, 34]. BSV falls under this category which eliminates queues, repetitive checks on the SOC of the battery. BSV converts the passive BSS to active BSS. BSV will improve the efficiency and effectiveness of battery swapping service, reduces latency, missing ratios, and range anxiety. BSV will provide fast, convenient, and flexible BSS service. To understand the architecture of BSV the participants of the entire system, their functions, relationships, and communications must be known. Table 4 lists the functions of various participants involved in BSV. Through the BSV app, the EV user can launch the battery swap request at any time. It is difficult to serve using a single BSV when all customers launch the request at the same time. Thus, the area/city under service has to be divided into n regions and according to the density of EV users in the area, several BSVs should be assigned. If many requests are coming at the same time from a particular region, there is a need of prioritizing the appeals. Since the fundamental requirement for EV users is SOC, the users can

Table 3 Characteristics of each algorithm [32]

Algorithm	GE	PSO	DE
Parameters	Population size, termination criteria crossover, and mutation probability (p_c, p_m) Distribution index (η_c) and (η_m)	Population size, termination criteria Initial weight (w), acceleration coefficients c1 and c2	Population size, termination criteria (F), crossover probability (p_c)
Selection	Survival of the fittest ($\mu + \lambda$)	Always accept new solutions into the population (μ, λ)	Greedy
Number of function evaluations	$N_p + N_p$ T(maximum)	$N_p + N_p$ T	$N_p + N_p$ T
Generation of new solutions	Using other solutions	Using velocity vector p_{best} and g_{best} (they are not a part of the solution)	Using other solutions (the best solution is a part of the population)
Best solution	Part of the population	Need not be a part of the solution	Part of the population
Phases/conditions	Crossover and mutation	Velocity and position update	Mutation and crossover
Applications	TSP, vehicle routing, scheduling	Container terminal scheduling, data mining	Transmission expansion planning (TEP) problem, selection of genes of DNA microarray
Algorithm	TS	SA	ACO
Parameters	Tabu list (L_K), tenure period (τ), neighborhood set ($N(x_k)$), random value (R), no. of iterations (N)	Temperature schedule, inferior move, no. of iterations (N), random value (R)	Pheromone density (τ_{xy}), attractive coefficient (η_{xy}), coefficient of vaporization (ρ), length of arc xy (L_{xy}) The algorithm has n nodes and m arcs
Selection	Next search point x_{k+1} is selected from $N(x_k)$ using R	Next search point x_{k+1} is selected if it satisfies the strategy	Next node y is selected from current node x using probability calculations. The arc is selected based on highest probability
Number of function evaluations	N	N	–
Best solution	From the feasible set of solutions	From the feasible set of solutions	The path which has the least distance

(continued)

Table 3 (continued)

Algorithm	GE	PSO	DE
Phases/conditions	(i) x_{k+1} should be part of $N(x_k)$ (ii) $x_{k+1} \cap L_k = \emptyset$	Accept x_{k+1} if $F(x_{k+1}) < F(x_k)$ otherwise $R_k < e^{-a}$ where, a $= (\lvert F(x_k) - F(x_k + 1)\rvert)/T$	Probability calculation: $P_{xy}{}^{k} = (\tau_{xy}{}^{1})\,(\eta_{xy})$ where:$\tau\frac{1}{xy} =$ $(1-\rho)\left(\tau_{xy}^{0}\right)\left(\sum_{k=1}^{m} \Delta\tau_{xy}^{k}\right)$ Where, $\Delta\tau_{xy}^{k}$ is Amount of pheromone deposited by kth ant on arc xy $= \frac{1}{L_k}$
Applications	Job scheduling problem, TSP, QAP (Quadratic Assignment Problem)	Map coloring problem, TSP, time-table scheduling	TSP, routing vehicles, protein folding

Table 4 Functions of each participant involving battery swapping process based on BSV [4]

S. no.	Participant	Function(s) of the participant
1	BSV	Responsible for battery swapping process
2	EV user	Has to send details such as SOC, speed of EV, the direction of travel through BSV App
3	BSS	(i) To provide FCB to battery swapping van (ii) To satisfy customers reaching BSS
4	BSV App	(i) Provides details such as the distribution of BSS, BSV, real-time battery swapping price to the EV user (ii) Collecting payments from user (iii) Location of the user
5	BS management system	(i) Releases real-time prices (ii) Control of BSV app (iii) Scheduling of BSV/BSS to the user
6	Battery charging facility	A secondary source of FCB to BSV
7	EV battery manufacturers	Manufactures standard battery for different EV brands

be prioritized based on the SOC of their EV battery. If the SOC is less than 15%, then the EV is given priority 1 (highest), and if SOC lies between 15 and 20%, then it is given priority 2; 20–25% of SOC is assigned to priority 3 and greater than 25% SOC is given priority 4 (least).

Since the app does not impose any constraint on the launch of the request, improbable appeals should not be entertained. In such cases, a penalty is issued to the EV user. If the user provides information about the direction of travel other than SOC, it is easy to assign a BSV/BSS to the user. There should be communication between

Table 5 Satisfaction levels for each priority as function of time [4]

Request priority/satisfaction	Very satisfied	Satisfied	Acceptable
1	SOC * d/s	$2T_h$	$3T_h$
2	(SOC-15%) * d/s	max (SOC * d/s, 15%SOC * d/s + T_h)	$3T_h$
3	(SOC-20%) * d/s	(SOC-15%) * d/s	max (SOC * d/s, 20%SOC * d/s + T_h)
4	(SOC-25%) * d/s	(SOC-15%) * d/s	max (SOC * d/s, 25%SOC * d/s + T_h)

different BSVs such that in the case of EV users change in the direction from one region to another is easily notified [4].

Quality of service (QoS) plays a key role in the system, which is described using queuing models. The satisfaction of the customers, once the battery got swapped with a high SOC battery, determines the QoS of the battery swapping process based on BSV. Both (satisfaction and QoS) are related by direction proportionality. EV users will be very satisfied if the service is achieved before a change in his/her priority level. Upon the change in the priority level satisfaction level also decreases. Thus, the BS management system needs to schedule BSV to the user before the satisfaction level crosses the acceptance mark. Table 5 shows different satisfaction levels based on priority. T_h is the threshold level which is constant (7.5% d/s < T_h < 15% d/s) and set according to request density [6, 32]. Energy replenishment using BSV increases the effectiveness and efficiency of the BS process. However, it is difficult to serve EV users only with BSV due to the randomness of requests and more demand during peak hours. This calls for the integration of BSS and BSV when the user is served with BS mode of replenishing energy.

5 Case Study

There are four modes of charging a depleted battery. Mode 1 uses slow charging and takes around 7–15 h to get charged, thus maintaining the SOH of the battery. This mode of charging can be used by just plugging the vehicle for charging in his workplace. It is the cheapest and convenient charging method. Mode 2 is used for normal charging (3–8 h) and is also known as opportunity charging, as the vehicle users prefer this charging method when given a chance without much affecting the SOH of the battery. It is usually found in public places and shopping malls [3]. Figure 2 shows the comparison of average mileage offered by the EV and its refilling time when it is served using BC and BS methods [1]. Consider N EVs are arriving at the charging station to get the battery replenished. The service provider finds it difficult to serve all the EVs at the same time. Based on the SOC, the service is provided to the EV user.

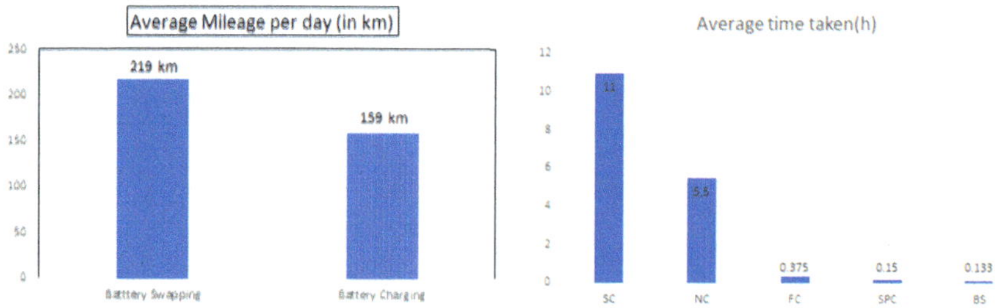

Fig. 2 Comparison of average mileage offered by EV and its refilling time when it is served using BC and BS methods [1, 3]

Case 1:When SOC is between 50 and 80%. Since the SOC lies in the given range, the battery is refueled at BCS. Mode 3 and Mode 4 are used to charge the batteries in a few minutes (4–12 min). Mode 3 is fast charging as it takes a few minutes to get charging but the SOH of the battery gets affected, so it is mostly not preferred by the user. Mode 4 charging can only happen when a match happens for the battery.

Case 2:When the SOC is <50%. Since the SOC is <50%, the battery is refueled at BSS and the depleted battery gets swapped with the battery having the highest SOC present at that time and is done in a few minutes. In case the user is far away from the BSS, then the user is served by BSV which would be moving with the fully charged batteries and the depleted batteries are swapped with fully charged ones.

6 Conclusions

In this paper, the observations made between BS and BC conclude that BS is beneficial to the owner of the station as mentioned in Table 1. The forecasting on the arrival of EVs proves to be an advantage to the owner in terms of pre-determining the number of batteries that have to be made available at the station and it also reduces the cost incurred due to tariff. Optimization techniques are used to determine the number of FCBs from a battery stock, scheduling of depleted batteries to BCS can prove to be a complex approach in terms of time. Metaheuristics is a branch of operations research that helps to obtain solutions at a faster rate. BSV an active mode of replenishing the energy is used to serve the customers who are far away from the station and reduces the waiting time for the users. The case study provides an insight on all the three energy refueling methods and it can be inferred that when the customer is near the station, the customer will be provided with service from BCS or BSS based on the SOC of the battery, and when the user is far from the station, BSV can be used to serve the customer.

References

1. Ban M et al (2019) Battery swapping: an aggressive approach to transportation electrification. IEEE Electr Mag 7:44–54
2. Xu Q et al (2017) Optimal operation of battery swapping-charging systems considering quality-of-service constraints. In: IEEE PES general meeting, Chicago, pp 1–5
3. Mude KN (2018) Battery charging method for electric vehicles: from wired to on-road wireless charging. Chin J Electr Eng 4:1–15
4. Shao S, Guo S, Qiu X (2017) A mobile battery swapping service for electric vehicles based on a battery swapping van. Energies
5. Battapothula G et al (2019) Multi-objective optimal scheduling of electric vehicle batteries in battery swapping station. In: IEEE PES innovative smart grid Technologies, Europe, pp 1–5
6. Suri M, Raj N, Deepa K, Jayan S (2020) Application of aspiration level model in determining QoS for an EV battery charging station. In: International conference on smart technologies in computing, electrical and electronics. IEEE, pp 1–7
7. Thiruvonasundari D, Deepa K (2020) Electric vehicle battery modelling methods based on state of charge—review. J Green Eng 10(1):24–61
8. Aggarwal S et al.(2020) Meta heuristic and evolutionary computation: algorithms and applications. Springer Nature, Berlin, p 949. ISBN 978-981-15-7571-6. https://doi.org/10.1007/978-981-15-7571-6
9. Yadav AK et al (2020) Soft computing in condition monitoring and diagnostics of electrical and mechanical systems. Springer Nature, Berlin, p 496. ISBN 978-981-15-1532-3. https://doi.org/10.1007/978-981-15-1532-3
10. Gopal et al (2021) Digital transformation through advances in artificial intelligence and machine learning. J Intell Fuzzy Syst 1–8. Pre-press. https://doi.org/10.3233/JIFS-189787
11. Fatema N et al (2021) Intelligent data-analytics for condition monitoring: smart grid applications. Elsevier, Amsterdam, p 268. ISBN 978-0-323-85511-2. https://www.sciencedirect.com/book/9780323855105/intelligent-data-analytics-for-condition-monitoring
12. Smriti S et al (2018) Special issue on intelligent tools and techniques for signals, machines and automation. J Intell Fuzzy Syst 35(5):4895–4899. https://doi.org/10.3233/JIFS-169773
13. Jafar A et al (2021) AI and machine learning paradigms for health monitoring system: intelligent data analytics. Springer Nature, Berlin, p 496. ISBN 978-981-33-4412-9. https://doi.org/10.1007/978-981-33-4412-9
14. Sood YR et al (2019) Applications of artificial intelligence techniques in engineering. vol 1. Springer Nature, Basingstoke, p 643. ISBN 978-981-13-1819-1. https://doi.org/10.1007/978-981-13-1819-1
15. Iqbal A et al (2021) Chapter 10—intelligent data analytics for battery health forecasting using semi-supervised and unsupervised extreme learning machines. Intelligent data-analytics for condition monitoring, Academic Press, Cambridge, pp 215–241. ISBN 9780323855105. https://doi.org/10.1016/B978-0-323-85510-5.00010-7
16. Dai Q et al (2014) Stochastic modeling and forecasting of load demand for electric bus battery-swap station. IEEE Trans Power Deliv 29:1909–1917
17. Wu H et al (2017) A charging-scheme decision model for electric vehicle battery swapping station using varied population evolutionary algorithms. Appl Soft Comput J
18. Saxena A, Deepa K (2020) Power quality analysis for electric vehicle charging and its mitigation strategies. Test Eng Manag 5409–5418
19. Sasikumar S, Deepa K (2018) LCL topology based single stage boost rectifier topology for wireless EV charging. J Green Eng 8(8):573–596
20. Wong W et al (2019) A review on metaheuristic algorithms: recent trends, benchmarking and applications. In: International conference on smart computing and communication, Sarawak, pp 1–5
21. Agrawal AP et al (2014) A comparative analysis of memory using and memory less algorithms for quadratic assignment problem. In: 5th international conference-confluence the next generation information technology, pp 815–820

22. Storm R et al (1997) Differential evolution—a simple & efficient heuristic for global optimization over continuous spaces. J Glob Optim 11(4):341–359
23. Sun J, Lai CH, Wu XJ (2012) Particle swarm optimization. CRC Press, Boca Raton
24. Mendes R, Kennedy J, Neves J (2004) The fully informed particle swarm: simpler, maybe better. IEEE Trans Evol Comput 8(3):204–210
25. Hu M, Wu T, Weir JD (2012) An adaptive particle swarm optimization with multiple adaptive methods. IEEE Trans Evol Comput 17(5):705–720
26. Zhan ZH et al (2009) Adaptive particle swarm optimization. IEEE Trans Syst Man Cybern B Cybern 39(6):1362–1381
27. Kaipa KN et al (2009) Glowworm swarm optimisation: a new method for optimizing multi-modal functions. Int J Comput Intell Studies 1:93–119
28. Dorigo M, Birattari M, Stutzle T (2006) Ant colony optimization. IEEE Comput Intell Mag 1(4):28–39
29. Yang XS, Deb S (2009) Cuckoo search via Lévy flights. In: World congress on nature and biologically inspired computing, IEEE Publications, pp 210–214
30. Karaboga D, Basturk B (2007) Artificial Bee Colony (ABC) optimization algorithm for solving constrained optimization problems. In: Foundations of fuzzy logic and soft computing. LNCS, vol 4529. Springer, Berlin, Heidelberg
31. Eusuff M et al (2006) Shuffled frog-leaping algorithm: a memetic meta-heuristic for discrete optimization. Eng Optim 129–154
32. Hamdy AT (2011) Operations research. 9th edn. Pearson publication, London
33. Thiruvonasundari D, Deepa K (2020) Active cell balancing for electric vehicle battery management system. Int J Power Electron Drive Syst 571–579
34. Zhang T et al (2018) A Monte Carlo simulation approach to evaluate service of EV charging & battery swapping stations. IEEE Trans Ind Inform 14(9):3914–3923

A Study of iOS Machine Learning and Artificial Intelligence Frameworks and Libraries for Cotton Plant Disease Detection

Sandeep Kumar, Rajeev Ratan, and J. V. Desai

Abstract Identifying plant disease from leave images using machine learning and artificial intelligence algorithm is a recent trending research area in the technology and agriculture domain. It will fill the gap between hi-tech technology and the old traditional way of doing agriculture. This paper outlines a study of different iOS mobile machine learning and artificial framework and libraries which can be used in plant disease detection in agriculture. This paper also discusses the efficient use of these libraries in the iOS mobile app for better results. Many iOS ML and AI libraries have been studied which can be used directly in the iOS app with offline support, which means without interacting with the server and python language. The main criteria for selecting these libraries are mainly using the libraries directly in the iOS app using either Swift or Objective-C programming language. This will remove the dependency on server-side implementation and the needs of the python programming language. This study is used in designing the architecture of the iOS app for plant leaf disease detection and cotton plant disease classification.

Keywords iOS · Machine learning · Artificial intelligence · Cotton leaf · Disease detection · Offline iOS app · CreateML · Apple · Tensor flow · Swift for TensorFlow · Classification

1 Introduction

In India, we are still dependent on agriculture. Nearly 70% of the Indian population relies on agriculture [1, 2]. In rural India, the main source of income for people is farming. Almost, three main crops are taken by the farmers in a season based on the regional crops. Cotton is one of the most commercial crops in India where farmers get good capital. Different species of cotton is used in India at different parts of the

S. Kumar (✉) · R. Ratan · J. V. Desai
Department of Electronics and Communication Engineering, MVN University, Palwal 121105, India

R. Ratan
e-mail: rajeev.arora@mvn.in

A. Tomar et al. (eds.), *Machine Learning, Advances in Computing, Renewable Energy and Communication*, Lecture Notes in Electrical Engineering 768,
https://doi.org/10.1007/978-981-16-2354-7_24

country at different time. In northern India, the cotton crop is mainly done from April to November month, while on the southern side of India, the cotton crop is done in the winter season.

Four main species of cotton are used in India. These are Gossypium arboreum, G. herbaceum, G. hirsutum, and G. barbadense. The G. hirsutum is the dominant species that contribute to 90% of total cotton harvesting. Even cotton is one of the commercial crops, but disease plays a crucial role in the production of cotton and impacts the income of the farmer. The cotton crops very easily get affected by the diseases. Many times, the farmers take these diseases casually or face issues to identify the disease, which leads to crop damage and loss of money. Every disease has a different diagnosis technique to overcome disease. As of now, there is no technical handy way available for farmers through which they can identify the disease and farmers still depend on the manual way of disease identification or they rely on their own experience which sometimes leads to wrong disease identification, and wrong way of diagnosis again leads to more damage to the crops and loss of more money. Each disease has a different period to affect the plant and farmers need to keep checking the crop from infection. This way of identifying the diseases is a very lengthy process and requires some care while selecting the pesticide.

Keeping all these scenarios in mind, there is a need for some automatic disease detection systems which will help farmers and make their life easy. If this can be done by using the mobile app without internet dependency it would give a huge advantage to the farmers. Nowadays, everyone in the family has a smartphone that will be used for disease identification. This paper gives an overview of the iOS mobile app libraries and frameworks which can be used in cotton leaf disease detection using the iOS app. These libraries will be used for disease classification and detecting the disease from the camera-captured image from the real field, which would help farmers to instantly identify the disease and give the recommended diagnosis pesticide also.

The success of this system will depend on how much precision the system works on the image classification and ML techniques. The huge real image data sets will be required. The significance of this paper is important as it helps in identifying the correct ML and latest libraries that can be used easily in the iOS app. Many iOS open-source libraries have been studied which are available for use. Also listed are some of the best libraries based on the direct use of the ML model in the iOS app. Technology is changing every year and each time there will be more sophisticated and efficient libraries available. In this paper, the regular updates of libraries based on the latest changes and brand of libraries are also kept in consideration. This paper also presented the usage of these libraries for better results of cotton leave disease detection. Moreover, the reader may refer to [3–13] for better understanding and implementation of different AI and ML approaches in a simple way.

2 Related Works

Shah et al. [14] have examined the identification and classification of rice diseases. The paper outlines the rice disease types and their root cause. The paper also displays the identification and measurement operations of image processing in the identification of disease. A comparative study of different segmentation techniques is presented. The paper displays the analysis of machine learning operations applied in rice disease identification.

Shinde and Kulkarni [15] outlined the prediction model using the sensors to collect the field data and then implement the algorithm for disease detection. The alert system also reviews the notification system which alerts the farmers of the unusual condition. The study reviews the IoT and machine learning (ML) techniques that are used in the system.

Shirahatti et al. [16] surveyed the different types of machine learning (ML) techniques that are used in plant disease identification. The paper presented the comparative study of different types of techniques with their advantages and disadvantages. These machine learning (ML) methods will help the system in identifying diseases that happened on plants by processing the images and the system will send the disease information to the farmers.

Shruthi et al. [17] outlined the different stages of plant disease identification system and systematic observation study on machine learning (ML) classification techniques for the detection of plant disease. As per the survey, it perceives that the convolutional neural network (CNN) has high accuracy and identifies more diseases of multiple crops.

Asokan and Anitha [18] present a detailed analysis of different image processing techniques for analyzing satellite images. Some popular machine learning-based image processing techniques are presented with a detailed review. Also, a detailed comparison of various techniques along with limitations is performed.

3 Cotton Leaf Diseases

There are four harvesting species of cotton, viz., Gossypium arboreum, G. herbaceum, G. hirsutum, and G. barbadense. 90% of total production is contributed to the G. hirsutum species. The following seven types of diseases are known in cotton.

3.1 Bacterial Blight

Bacterial blight [19], also known as angular leaf spot, is caused by the bacterium, Xanthomonas citri PV. malvacearum. Bacterial blight starts as small, water-soaked spots on leaves.

3.2 Fungal Leaf Spots

These are purple circular color spots or ring type in shape. These are of Alternaria leaf spot, and the disease caused by phytopathogenic fungi utilizes different mechanisms like synthesis and liberation of cell wall degrading hydrolytic enzymes, toxin production, and synthesis of metabolites which do against the normal growth.

3.3 Grey Mildew

The disease causes premature defoliation and immature bolls in many susceptible genotypes. Irregular to angular pale spot develops on the bottom surface, usually wrapped by veinlets.

3.4 Boll Rot

This is an intricate disease caused by several fungal pathogens like Fusarium moniliforme, Colletotrichum capsici, Aspergillus flavus, A. niger, Rhizopus nigricans, Nematospora nagpuri, and Botryodiplodia sp.

3.5 Root Rot

The disease indication starts at the time when the plant becomes mature. The most important is that the symptoms are quick and patches on the plant appear quickly. The leaves start dying suddenly within few days.

3.6 Leaf Curl

The disease is caused by the CLCuD (cotton leaf curl disease)—begomoviruses are upward or downward leaf curling, vein-thickening, and foliar discoloration. When the leaves are infected in the early stages, the shortened internodes are commonly observed, and the plants have a stunted appearance.

Table 1 Category of cotton diseases

Scientific name	Disease name	Affected area
Xanthomonas citri pv. malvacearum	Bacterial blight	Leaves, lesion ball
Alternaria alternata	Fungal leaf spots	Leaf
Ramularia gossypii/Cercosporella gossypii	Grey Mildew	Leaf
Fusarium moniliforme	Boll rot	Boll
Rhizoctonia solani	Root rot	Boll
Cotton leaf curl virus	Leaf curl	Leaf
anthocyanosis	Leaf reddening	Leaf

3.7 Leaf Reddening

The leaf color is identified by pigment content and concentration. Generally, pigments are present in cotton leaves that include chlorophylls, carotenoids, tannins, and anthocyanins. The different disease categories with scientific names and affected areas are explained in Table 1.

4 Methodology—Usage of Libraries

Every library/framework will have a set of characteristics in terms of usage. Below are the details about the integration and usage of each library/framework.

4.1 CoreML

This is the iOS native framework by Apple Inc. The CoreML [20] is used to integrate ML models into the iOS app. The CoreML feature provides a complete representation of all types of models which can be used in making the prediction, and to train, optimize the models on iOS devices.

The high-level design of the CoreML in the iOS app is shown in Fig. 1. The different layers of artificial intelligence networks are described in Fig. 1.

The CoreML uses the CPU, GPU while optimizing the on-device performance and minimizing the memory and power usages. The requirement of a network connection is omitted when the model runs on iOS devices. The data on iOS devices are securely stored.

CoreML Supports:

- Images analysis, i.e., vision

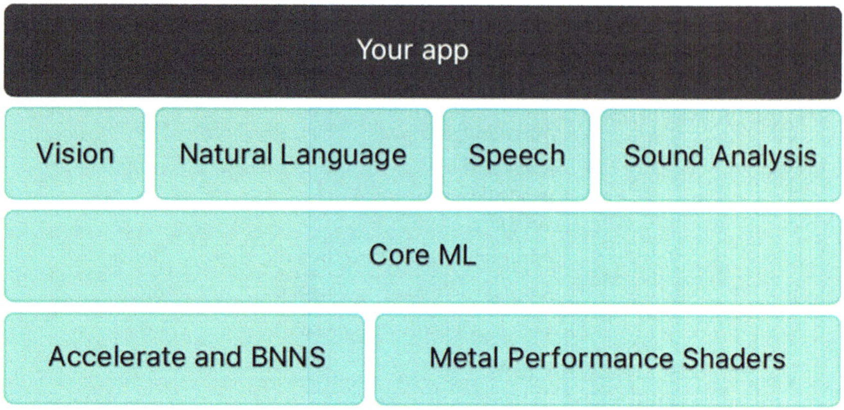

Fig. 1 A high-level overview of CoreML in iOS

- The processing of the text, i.e., natural language
- Converting the audio file into text, i.e., speech
- Sound analysis for sound identification in an audio file.

4.1.1 Usage of the Framework for Plant Diseases Detection

Every library/framework will have a set of characteristics in terms of usage. Below are the details about the integration and usage of each library/framework.

CoreML Request

The instance of VNCoreMLRequest can be created using the model class. CoreML supported model can be created using the CreateML [21] mac application.

The core model instance is created using the VNCoreMLModel class by passing the model name.

Now, once the model instance is created, the CoreML request instance is created by using the VNCoreMLRequest class and passing the model name as an argument. Once, the code ML request instance is created then another property of the instance is set like cropAndScaleOption, etc.

The completion handler call-back is set to catch the response of the request and to further process the request.

Run the Vision Request

An instance of VNImageRequestHandler will be created with an image to be processed. This method will run synchronously by using a background dispatch queue while requests execute to avoid the main queue gets blocked.

The DispatchQueue.global GCD function is used to run the code asynchronously.

Now, with the handler instance, the 'perform' function is called which will operate. The operation will run in the try block, and the error is also caught as part of the error handling process.

Handle Image Classification Result

The completion handler will handle the result either succeeded or error.

The result of the request is processed further to display the result. The operation will be performed in an asynchronous GCD block. The array of the result is fetched using the 'results' property in the request.

The 'result' will always give 'VNClassificationObservation' object which is specified by the CoreML model.

So, using image classification, the sets of healthy and disease images will be used to create the model, and based on the confidence of the model the image result will be shown.

4.2 *Swift for TensorFlow*

Swift for TensorFlow is an ML platform that supports the Swift language. Using this library, the python modules can be imported in Swift and can call the python functions, and the values between python and Swift can also be converted.

4.2.1 Integration of TensorFlow in iOS Project

The different ways of TensorFlow code integration into the iOS project are described in [22].

4.2.2 Import Python Module

The python model is imported into the iOS project by checking availability using the 'if' pragma.

#if canImport(PythonKit)

If this is successful then using the 'import' keyword, the model is imported.

import PythonKit

If PythonKit is not available then in the 'else' case simply import the Python module.

import Python

In Swift, 'PythonObject' gives the reference of a python library object. All APIs related to the python library will use and also return the 'PythonObject' instance.

Usage example:

let nmpy = Python.import("numpy")
NSLog(nmpy)
let someZeros = nmpy.ones([3, 2])
NSLog(someZeros)

4.2.3 Create a Model

```
import TensorFlow

let hiddenSizeCount: Int = 20
struct Model_NAME: Layer {
   var swiftLayer1 = Dense<Float>(inputSize: 8, outputSize: hiddenSizeCount,
activation: relu)
var swiftLayer2 = Dense<Float>(inputSize: hiddenSizeCount, outputSize:
hiddenSizeCount, activation: relu)
var swiftLayer3 = Dense<Float>(inputSize: hiddenSizeCount, outputSize: 6)

@differentiable
   func callAsAFunction(_ input: Tensor<Float>) -> Tensor<Float> {
      return input.sequenced(through: swiftLayer1, swiftLayer2, swiftLayer3)
   }
}

   var modelObject = Model_NAME()
```

The above is a flow of how to create the model, and the parameters of actual model creation will vary and depend on the requirement.

Now, the trained model can make some predictions. The accuracy and performance of the model can be checked with the test data set. An optimizer like SGD is used to optimize the model to overcome the underfit and overfit problems.

4.2.4 Optimizer

To minimize the loss function, the optimizer will apply the computed gradient to the model variable. The loss function is like a curved surface and by walking around, it finds the lowest point. By iterating, it calculates the loss function and gradient for each batch, and it will adjust the model during training. Over time, to minimize the loss, the best combination of the model's weight and bias can be obtained. For the best model prediction, the loss should be minimum.

During the training, the Swift for TensorFlow has many optimization techniques. The stochastic gradient descent (SGD) algorithm will be implemented by the SGD optimizer.

The learning rate sets the step size to take for each iteration down the hill. This is a hyperparameter that will commonly adjust to achieve better results.

```
let optimizer1 = SGD(for: modelObject, learningRate: 0.03)

let (loss1, grads) = valueWithGradient(at: modelObject) { modelObject ->
Tensor<Float> in
    let logits1 = model(firstTrainFeatures1)
    return softmaxCrossEntropy(logits: logits1, labels: firstTrainLabels1)
}

optimizer1.update(&modelObject, along: grads)

let logitsAfterOneStep1 = model(firstTrainFeatures1)
let lossAfterOneStep1 = softmaxCrossEntropy(logits: logitsAfterOneStep1, labels:
firstTrainLabels1)
```

4.2.5 Train the Model

- Importing and parsing the data set
- Model type selection
- Model training
- Evaluation of the model's effectiveness
- For prediction, use the trained model.

For plant disease detection, the sets of images are used to classify healthy and unhealthy leaves. The disease name labeled images set can be used to identify the diseases.

5 Result

In recent years, a lot of development and research has been done in machine learning and the artificial domain. The earlier studies are mostly focused on the backend side where the server is involved and python programming, which requires specialized programming skills. If someone not a python or backend programmer, then it was difficult to work on machine learning and artificial algorithm. If still, someone wants to learn machine learning and artificial intelligence algorithm, then that person needs to learn the special programming language.

Table 2 Comparison of both the libraries/framework

Name	Advantage	Disadvantage
Core ML	• Easy to use and integrate • Less code integration • No dependency on other open-source code/library • Continuous maintenance by Apple • Easy to use the model • Optimized for iOS • High performance and accuracy	• Requires iOS-specific skills • Requires MacBook or iMac for model creation
Swift for TensorFlow	• A lot of option in the library to use • Latest AI algorithm available • Compatible for Swift and python • Easy to integrate with iOS using Swift language • High performance and accuracy • Manage by Google and updated with latest algorithms	• Need to understand the libraries and their functions • Need to learn function call and knowledge about other libraries like pandas, NumPy, etc.

Now, the usage of mobile and mobile app is increasing day by day. So, it is more realistic or demand for time to make machine learning and artificial intelligence algorithm cross-platform and compatible with other programming languages also.

For this paper, the two main libraries/frameworks are presented, which can be used for plant disease detection using image classification or object detection techniques. These two libraries/frameworks are more popular and up to date with the latest development. Also, it is easy to integrate into the iOS project with Swift as a programming language. These libraries are a complete package to achieve plant disease detection. The performance and accuracy of these libraries/frameworks are also high. Table 2 shows the insights into the advantages and disadvantages of both frameworks.

For plant disease detection, the sets of images are used to classify healthy and unhealthy leaves. The disease name labeled images set can be used to identify the diseases.

6 Conclusion

In this paper, the two main libraries/frameworks are presented, which are highly recommended to use in the iOS app for plant disease detection. These libraries/frameworks are a complete package to integrate into the iOS project which can accomplish the task of image classification which is used in plant disease detection. The choice of usage of these libraries/frameworks is also dependent on the programming skills and technical background of the scholar.

However, further work can be carried out in those libraries which have a high popularity and are constantly maintained. Every year there is a new development or enhancement in machine learning algorithms. The algorithm gets enhanced with the latest trend. So, there is always a possibility of better frameworks/libraries that will be available to use.

Acknowledgments The authors would like to thank the Associate. Prof. Dr. Rajeev Ratan and Senior Prof. Dr. J.V. Desai, MVN University, Palwal, India for their guidance and help for our research.

Also, the authors would like to thank Dr. Devvrat Akheriya for the support throughout this work.

References

1. Food and Agriculture Organization in India. http://www.fao.org/india/fao-in-india/india-at-a-glance/en/
2. Vanitha CN, Archana N, Sowmiya R (2019) Agriculture analysis using data mining and machine learning techniques. In: 5th international conference on advanced computing & communication systems (ICACCS), pp. 984–990
3. Aggarwal S et al (2020) Meta heuristic and evolutionary computation: algorithms and applications. Springer Nature, Berlin, 949 pp. https://doi.org/10.1007/978-981-15-7571-6. ISBN 978-981-15-7571-6
4. Yadav AK et al Soft computing in condition monitoring and diagnostics of electrical and mechanical systems. Springer Nature, Berlin, 496 pp. https://doi.org/10.1007/978-981-15-1532-3. ISBN 978-981-15-1532-3
5. Gopal et al (2021) Digital transformation through advances in artificial intelligence and machine learning. J Intell Fuzzy Syst 1–8 (Pre-press). https://doi.org/10.3233/JIFS-189787
6. Fatema N et al (2021) Intelligent data-analytics for condition monitoring: smart grid applications. Elsevier, 268 pp. ISBN: 978-0-323-85511-2. https://www.sciencedirect.com/book/978 0323855105/intelligent-data-analytics-for-condition-monitoring
7. Smriti S et al (2018) Special issue on intelligent tools and techniques for signals, machines and automation. J Intell Fuzzy Syst 35(5):4895–4899. https://doi.org/10.3233/JIFS-169773
8. Jafar A et al (2021) AI and machine learning paradigms for health monitoring system: intelligent data analytics. Springer Nature, Berlin, 496 pp. https://doi.org/10.1007/978-981-33-4412-9. ISBN 978-981-33-4412-9
9. Sood YR et al (2019) Applications of artificial intelligence techniques in engineering, vol 1. Springer Nature, 643 pp. https://doi.org/10.1007/978-981-13-1819-1. ISBN 978-981-13-1819-1)
10. Guruprasad RB, Kumar S, Randhawa S (2019) Machine learning methodologies for paddy yield estimation in India: a case study. In: IGARSS, IEEE, pp 7254–7257
11. Hosseini M, McNairn H, Mitchell S, Davidson A, Robertson LD (2019) Comparison of machine learning algorithms and water cloud model for leaf area index estimation over corn fields. In: IGARSS, IEEE, pp 6267–6270
12. Hasan MdJ, Mahbub S, Alom MdS, Nasim MdA (2019) Rice disease identification and classification by integrating support vector machine with deep convolutional neural network. In: 1st International conference on advances in science, engineering and robotics technology
13. Hasan MdZ, Ahamed MdS, Rakshit A, Hasan KMZ (2019) Recognition of jute diseases by leaf image classification using convolutional neural network. In: 10th ICCCNT, IEEE, Kanpur, India

14. Shah JP, Prajapati HKB, Dabhi VK (2016) A survey on detection and classification of rice plant diseases. In: IEEE
15. Shinde SS, Kulkarni M (2017) Review paper on prediction of crop disease using IoT and machine learning. In: International conference on transforming engineering education, pp 1–4
16. Shirahatti J, Patil R, Akulwar P (2018) A survey paper on plant disease identification using machine learning approach. In: Proceedings of the international conference on communication and electronics systems (ICCES 2018), pp 1171–1174
17. Shruthi U, Nagaveni V, Raghavendra BK (2019) A review on machine learning classification techniques for plant disease detection. In: 5th international conference on advanced computing & communication systems (ICACCS), pp 281–284
18. Asokan A, Anitha J (2019) Machine learning-based image processing techniques for satellite image analysis—a survey. In: International conference on machine learning, big data, cloud and parallel computing (Com-IT-Con), India, pp 119—124
19. Cotton incorporated, identification, and management of bacterial blight of cotton. https://www.cottoninc.com/cotton-production/ag-research/plant-pathology/management-bacterial-blight-cotton/
20. Framework Core ML. https://developer.apple.com/documentation/coreml
21. Framework Create ML. https://developer.apple.com/documentation/createml
22. GitHub TensorFlow swift open source code. https://github.com/tensorflow/swift

Effect of Loss Functions on Language Models in Question Answering-Based Generative Chat-Bots

P. Hemant, Pramod Kumar, and C. R. Nirmala

Abstract Question answering mechanism plays a pivotal role in helping generative chat-bots provide accurate answers to the questions asked by the questioner. Language models are an integral part of question/answer mechanisms where they answer two things, namely classifying the question correctly and also help provide the most suitable answer for the question asked. This paper takes into account the effect of various loss functions on the answer-predicting capability of a language model. Various loss functions were implemented like cross-entropy loss, negative log-likelihood loss, cosine embedding loss, and Kullback–Leibler divergence loss and found that there is a significant impact due to selection of loss functions on accuracy. Based on the research and implementation in this paper, it is found that choosing the multi-label cross-entropy loss for all the general question answering problems has a 0.5–1% raise.

Keywords Loss functions · Language models · Deep learning

1 Introduction

Language models [1] help us to predict the next set of words that need to come up in the contextual part of natural language. Now the sequence of words is the most important aspect when discussing the context. Language models present a way in which the system/architecture provides the probability of occurrence of a series of words $P(w_i, \ldots, w_T)$ given a word w_i. While working on the identification of

P. Hemant (✉)
Department of Computer Science and Engineering, Shri Jagdishprasad Jhabarmal Tibrewala University, Jhunjhunu, India

P. Kumar
Department of Computer Science and Engineering, Krishna Engineering College, Ghaziabad, India

C. R. Nirmala
Department of Computer Science and Engineering, Bapuji Institute of Engineering and Technology, Davangere, India

A. Tomar et al. (eds.), *Machine Learning, Advances in Computing, Renewable Energy and Communication*, Lecture Notes in Electrical Engineering 768,
https://doi.org/10.1007/978-981-16-2354-7_25

words, the position of the word in the entire document plays a very important role in delivering the required context. In various language translation mechanisms, they mainly identify the most possible words that would fit the context, for example: "I have", "I did", "I am" and based on that the scoring of the sequence of translation happens. While considering the bi-grams or tri-grams [2] the usage of counting theory is used for finding the probability of occurrence of two words together or two words with one word is calculated. Formulae for the same is:

$$\Pr(w_1|w_2) = \frac{\text{count}(w_1, w_2)}{\text{count}(w_1)} \tag{1}$$

$$\Pr(w_3|w_2, w_1) = \frac{\text{count}(w_1, w_2, w_3)}{\text{count}(w_1, w_2)} \tag{2}$$

In the above equations, the contextual understanding is not captured and the subsequent word is decided only based on the previous w words. Let us take a simple example of comparing two different books that talk about philosophy, like Plato, The Last Days of Socrates, and Boethius. If we are in a conversation part where it's written "Both the books have a considerable difference in terms of the core concept". Now in this, the system does not have the contextual reference of the said books. Eqs. 1 and 2 considerably fail at this point.

N-gram-based language models have various drawbacks:

1. Huge computational overhead [3] due to the inclusion of n words at the same time. More the words, better the accuracy.
2. The n-gram model typically works on a sparse matrix principle and those words which are not available in the list give out zero probability when encountered in a real-life scenario.

In paper [4], by a neural probabilistic language model, the author showcases the creation of a cost function that takes into account the context of words. Recurrent neural networks make a frame to consider the entire sequence of words. Each hidden layer has a subsequent set of neurons that perform linear calculations on its inputs, and later a nonlinear valuation is used in terms of tan(h). Once the output is acquired, it goes in as an input to the hidden layer with the corresponding word vector x_t, to produce 'y'. Moreover, several advanced-level examples are mentioned in Refs. [12–22].

1.1 Loss Functions

While dealing with supervised machine learning algorithms, we limit the error for every preparation model during the training cycle. We use various optimization strategies like the gradient to do that. These errors originate from the loss function.

Deep learning neural systems are prepared with the stochastic gradient descent algorithm. The loss function is also called the error function sometimes.

Loss functions are used by machines to learn. This technique assesses how nicely explicit calculation models will achieve on certain information. In the event that forecasts strays away a lot from actual outcomes (outliers), the loss function would be huge. So, we use some optimization functions, by which the loss function learns to reduce the error in predictions.

The cost function or a loss function's main objective is to reduce the error and the calculation of the loss function is said to be actual "loss".

1.2 Types of Loss Functions

Loss functions are mainly classified as regression loss [5] and classification loss [6]. In regression, we predict a value based on the given input; for example, given the level and position of a person in a company the model has to predict the salary of that person, where in classification the model has to predict the output from a finite set of categories.

Regression losses are as follows:

1. Mean Square Error (MSE) [7]

It is the mean of the squared difference between the predicted value and real observations. The values are squared so that positive and negative ones should not cancel each other. MSE is the most preferred for regression tasks. It is differential which makes it optimized better. Since it squares the difference of predicted and actual values, it penalizes even a minor error that may lead to over-estimation of how bad the model performs. It is preferred as the loss function for the data whose distribution is Gaussian. MSE is also called L2 loss and quadratic loss.

Formula to calculate MSE:

$$\text{MSE} = \left(\frac{1}{n}\right)\sum_{i=1}^{n}(y_i - x_i)^2 \tag{3}$$

Here
y_i = Predicted value x_i = Real value
x_i = Real value
n = Number of predictions.

2. Mean absolute error (MAE) [8]

It is the average of the difference between the absolute predicted value and absolute actual value. MAE is also called L1 loss. While MSE squares the difference but MAE does not, which makes MAE more robust to outliers. Moreover, MAE requires some complicated tools for calculating the gradients. Following is the formula for MAE:

$$\text{MAE} = \left(\frac{1}{n}\right)\sum_{i=1}^{n}|y_i - x_i| \quad (4)$$

Here
y_i = Predicted value
x_i = Actual value
n = Number of predictions.

2 Classification Loss

1. Binary Classification Problem

This is a problem where we need to classify the given examples into one class or the other. Binary means two classes. The issue is anticipating the probability of a model belonging to class one or class two.

(a) Binary Cross-Entropy [9]

Entropy is used to indicate some kind of disorder or uncertainty. The probability distribution of P(X) is generated for a random variable(X): The –ve sign makes the overall equation positive. The greater the entropy, the greater is the uncertainty. For binary classification problems, we use cross-entropy as the default loss function. Cross-entropy will figure a score that sums up the average difference between the actual and anticipated probability distributions for class 1. The score is limited and an ideal cross-entropy esteem is 0. Cross-entropy can be determined in Keras by indicating 'binary cross entropy' when we compile the model.

$$S = \left\{-\int p(x)\cdot\log p(x)\ \cdot dx, \text{ if } x \text{ is continuous} - \sum_{x} p(x)\cdot\log p(x), \text{ if } x \text{ is discrete}\right.$$

(b) Hinge Loss [10]

The hinge loss function is used as a substitute for binary cross-entropy on support vector machines (SVM). Its target set is $\{-1, 1\}$. In the hinge loss function, the examples are set to the proper sign, proposing more error whenever it finds a discrepancy in the sign amid the real and predicted class values. Sometimes the performance level of the hinge loss function is better than cross-entropy. But it is not always the same case.

Hinge in the compile function is used to specify the hinge loss. The mathematical formulation is

$$\text{Loss} = \left(\sum_{j \neq y_i} \max\left(0, s_j - s_{y_i} + 1\right) \right) \tag{6}$$

The hinge loss detects an inappropriate prediction as well as the predictions that are not sure or might be wrong.

Output layer configuration: hyperbolic tangent activation function.

(c) Squared Hinge Loss [11]

This is an extension of the hinge function. It just computes the square of the score hinge loss. It also brings a certain type of smoothness to the curve of the error function and thus it becomes easy to work on numerical data. Whichever hinge loss method gives better performance on a given binary classification problem, it will probably work on squared hinge loss as well. Its target is set to $\{-1, 1\}$ and squared hinge in the compile () function is used to specify the square hinge loss. r Output layer configuration: hyperbolic tangent activation.

2. Multi-Class Classification Loss Problem

This is a problem where we need to categorize the given examples into two or more classes. Multi-class means two or more optional classes. The issue is surrounded by foreseeing the probability of a model having a place with each class.

The output layer configuration is defined by one node per class by activating SoftMax.

Loss function: Cross-entropy, also called logarithmic loss.

(a) **Multi-Class Cross-Entropy Loss** [12]

Cross-entropy which is used for multiple classes is called multi-class cross-entropy, and this is nothing but the generalization of the binary cross-entropy. It is suggested to use where there is a set of classes that are assigned with unique integer values.

(b) **Sparse Multi-class Cross-Entropy Loss** [13]

Sparse multi-class cross-entropy is used in keras for multi-class classification. It is similar to cross-entropy but here there is no need for one-hot encoding in the data. In cross-entropy, it becomes difficult to perform calculations after applying one-hot encoding in classification problems. Suppose, for example, predicting words in any document may have a minimum of hundred categories. However, it is necessary to perform one-hot encoding to that which forms thousands of zero values to those hundred categories. This unnecessarily consumes a lot of memory and it becomes difficult to apply cross-entropy. So in these cases, sparse multi-class loss is applied.

(c) **Kullback Leibler Divergence Loss** [14]

Divergence means different from each other and KL divergence is basically calculated to identify whether there is any difference in distributions or not. If the KL divergence value is zero then there is no difference in distributions and they are identical

$$D_{KL}(P\|Q) = \begin{cases} -\sum P(x)\cdot \log\frac{Q(x)}{P(x)} & = \sum_x P(x)\cdot \log\frac{P(x)}{Q(x)} \\ -\int P(x)\cdot \log\frac{Q(x)}{P(x)},dx & = \int P(x)\cdot \log\frac{P(x)}{Q(x)}\cdot dx \end{cases} \quad (7)$$

Here we have to note that the divergence function is asymmetric, and so we cannot use KL divergence for the distance matrix.

$$D_{KL}(P\|Q) \neq D_{KL}(Q\|P) \quad (8)$$

In simple terms, P is the right probability distribution of target variables in consideration with input features and Q is an approximate distribution. Minimizing the forward KL-$D_{KL}(P \| Q)$ is used for supervised learning and minimizing the backward KL-$D_{KL}(Q \| P)$ is used for reinforcement learning.

3 Result of Loss Function with Reference to Language Models

Table 1 shows the EM and F1 scores calculated on the benchmark.

SQUAD 2.0 dataset on various language models and pre-trained models.

Here we have evaluated all the loss functions on standard parameters provided below.

Table 1 Exact match and F1 scores for various language models and corresponding loss functions

Model configurations and loss function	Accuracy (in EM)	Accuracy (in F1)
Albert base-Multiclass Cross Entropy Loss	70.58	72.79
Albert base-BCE with Logits	70.58	72.84
Roberta_base-MulticlassCrosseEntropyLoss	65.14	68.30
Roberta_base-Margin Ranking Loss	64.96	68.12
Roberta_base-SmoothL1Loss	64.81	67.84
Roberta_base-SoftMarginLoss	64.56	67.77
Roberta_base-Hinge Embedding Loss	64.33	67.51
Roberta_base-Poisson NLL Loss	64.12	67.30
Roberta_base-BCE Loss	64.10	67.34
Roberta_base-BCE with Logits	63.23	66.09
Distillbert-uncased-MulticlassCrossEntropyLoss	54.94	57.97
Distillbert-uncased-PoissonNLLLoss	52.88	56.30
Distillbert-uncased-KLDivLoss	52.88	56.30
Bert base uncased-CosineEmbeddingLoss	52.81	55.67
Bert_base_uncased-MulticlassCrosseEntropyLoss	50.90	53.73

- Learning Rate: 1.5e−5
- Number of Epochs: 2
- Maximum Sequence Length: 384
- Doc Stride: 128
- Training Batch Size: 24
- Gradient Steps: 6
- Multiprocessing: True—16 Cores
- ADAM Optimizer Epsilon: 1e−5.

LMs are mostly evaluated on less number of EPOCHs as they are computationally expensive.

$$PP = 2^{\tilde{H}_r} \text{ where } \tilde{H}_r = -\frac{1}{T}\log_2 p(w_1, \ldots, w_T)$$

where $\{w_1, \ldots, w_T\}$ is our development data which provides the distribution $q(\cdot)$ in the formulae for multi-class cross-entropy (Fig. 1)

$$\tilde{H} = -\sum_x q(x) \log p(x)$$

and $p(\cdot)$ is probability estimation on the training set.

Interpretations:

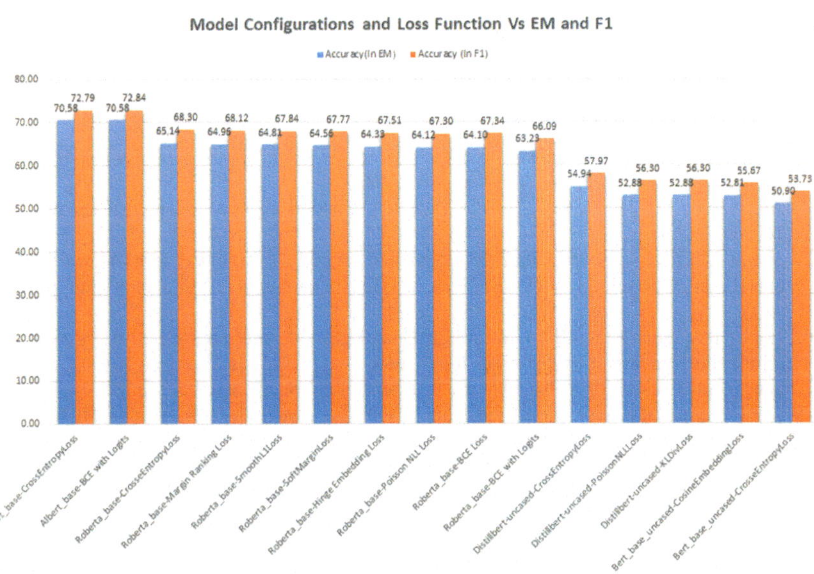

Fig. 1 EM and F1 versus multiple loss functions and language models

- Entropy rate: When the entropy rate is lower in number the estimation of the next word or sentence becomes easier, ruling out other options.
- Average branching factor: In uniform distribution for a larger vocabulary size V, then entropy is log2 V, and perplexity is V.
- Maximum likelihood criterion: Reducing $\tilde{H}$ r is similar to increasing log-likelihood and is a regularly used model selection way.

Caution: Using perplexity makes us realize that it is an average used for each word or words. In some cases, it is easier to find the next set of words.

4 Conclusion

We can clearly see that the multi-class cross-entropy outperforms all the other losses due to the sole reason of it working on classification works as it uses the activation functions in the final output layer for modeling the probability values like sigmoid or softmax for distributions. The fundamental intuitiveness is to use entropy which is an information content when serving probabilities. It has great properties for sigmoid functions, as the question/answering mechanism uses the sigmoid paradigm. The current accuracy numbers are high on the development set.

References

1. Howard J, Ruder S (2018) Universal language model fine-tuning for text classification. arXiv:1801.06146
2. Peng H, Roth D (2016) Two discourse driven language models for semantics. arXiv:1606.05679
3. Pagliardini M, Gupta P, Jaggi M (2017) Unsupervised learning of sentence embeddings using compositional n-gram features. arXiv:1703.02507
4. Ahn S, Choi H, Parnamaa T, Bengio Y (2016) A neural knowledge language model. arXiv:1608.00318
5. Sadhu A, Chen K, Nevatia R (2019) Zero-shot grounding of objects from natural language queries. In: Proceedings of the IEEE international conference on computer vision, pp 4694–4703
6. Conneau A, Lample G (2019) Cross-lingual language model pretraining. In: Advances in neural information processing systems, pp 7059–7069
7. Tang R, Lu Y, Lin J (2019) Natural language generation for effective knowledge distillation. In: Proceedings of the 2nd workshop on deep learning approaches for low-resource NLP (DeepLo 2019), November, pp 202–208
8. Lam MW, Chen X, Hu S, Yu J, Liu X, Meng H (2019) Gaussian process lstm recurrent neural network language models for speech recognition. In: ICASSP 2019–2019 IEEE international conference on acoustics, speech and signal processing (ICASSP), pp 7235–7239, May. IEEE
9. Yang S, Feng D, Qiao L, Kan Z, Li D (2019) Exploring pre-trained language models for event extraction and generation. In: Proceedings of the 57th annual meeting of the Association for Computational Linguistics, pp 5284–5294, July
10. Poostchi H, Piccardi M (2019) BiLSTM-SSVM: training the BiLSTM with a Structured Hinge Loss for Named-Entity Recognition. IEEE Trans Big Data

11. van der Burgh B, Verberne S (2019) The merits of Universal Language Model Finetuning for Small Datasets–a case with Dutch book reviews. arXiv:1910.00896
12. Hsu YC, Lv Z, Schlosser J, Odom P, Kira Z (2019) Multi-class classification without multi-class labels. arXiv:1901.00544
13. Dong Q, Zhu X, Gong S (2019) Single-label multi-class image classification by deep logistic regression. In: Proceedings of the AAAI conference on artificial intelligence, vol 33, pp 3486–3493, July
14. Ji S, Zhang Z, Ying S, Wang L, Zhao X, Gao Y (2020) Kullback-Leibler Divergence Metric Learning. IEEE Trans Cybernet
15. Aggarwal S et al (2020) Meta heuristic and evolutionary computation: algorithms and applications. Springer Nature, Berlin, 949 pp. https://doi.org/10.1007/978-981-15-7571-6. ISBN 978-981-15-7571-6
16. Yadav AK et al (2020) Soft computing in condition monitoring and diagnostics of electrical and mechanical systems. Springer Nature, Berlin, 496 pp. https://doi.org/10.1007/978-981-15-1532-3. ISBN 978-981-15-1532-3
17. Gopal et al (2021) Digital transformation through advances in artificial intelligence and machine learning. J Intell Fuzzy Syst 1–8 (Pre-press). https://doi.org/10.3233/JIFS-189787
18. Fatema N et al (2021) Intelligent data-analytics for condition monitoring: smart grid applications. Elsevier, 268 pp. ISBN: 9780323855112
19. Smriti S et al (2018) Special issue on intelligent tools and techniques for signals, machines and automation. J Intell Fuzzy Syst 35(5):4895–4899. https://doi.org/10.3233/JIFS-169773
20. Jafar A et al (2021) AI and machine learning paradigms for health monitoring system: intelligent data analytics. Springer Nature, Berlin, 496 pp.https://doi.org/10.1007/978-981-33-4412-9. ISBN 978-981-33-4412-9
21. Sood YR et al (2019) Applications of artificial intelligence techniques in engineering, vol. 1. Springer Nature, 643 pp. https://doi.org/10.1007/978-981-13-1819-1. ISBN 978-981-13-1819-1
22. Tsai ST, Kuo EJ, Tiwary P (2020) Learning molecular dynamics with simple language model built upon long short-term memory neural network. Nat Commun 11(1):1–11

Shunt Active Power Filter Based on Synchronous Reference Frame Theory Connected to SPV for Power Quality Enrichment

Dinanath Prasad, Narendra Kumar, and Rakhi Sharma

Abstract As the use of non-linear loads and power electronics (PE) devices are increasing in recent years, the total harmonic distortions and the power quality problems were increased tremendously. Shunt active power filters (SAPF) are proposed for refinement/ enrichment in power quality and reliability of power delivery system. Solar photovoltaic (SPV) system is designed as a renewable energy resource and power electronics converters are used for the interconnection and energy conversion with the grid. The incremental conductance algorithm (Inc) technique is incorporated bring out peak/maximum power from the solar photovoltaic system. The shunt connected compensator is used for current harmonics elimination, balancing and regulation of load terminal voltage and power factor correction. An artificial neural network (ANN) as an artificial intelligence is used for SAPF as a control algorithm for the improvement in the performance of SPV-SAPF, a synchronous reference frame (SRF) based control method has been introduced for generation of switching pulses. The proposed system is implemented in Matlab/Simulink environment and the simulation results are shown in terms of total harmonic distortion (THD) comparison with PI and ANN based techniques.

Keywords Harmonics · SPV · MPPT · SRF · Artificial neural network

1 Introduction

In this modernized world, renewable energy sources create opportunities by providing unprecedented support by various activities [1]. Renewable energy gains much popularities because it is accepted environment friendly sources universally. It can not only provide clean energy but, it also utilized to protecting our health and environment.

D. Prasad (✉) · N. Kumar
Electrical Engineering Department, Delhi Technological University, Delhi, India
e-mail: prasaddinanath@akgec.ac.in

R. Sharma
Indira Gandhi National Open University Delhi, New Delhi, India

A. Tomar et al. (eds.), *Machine Learning, Advances in Computing, Renewable Energy and Communication*, Lecture Notes in Electrical Engineering 768,
https://doi.org/10.1007/978-981-16-2354-7_26

Out of many renewable energy resources SPV system is considered as a sustainable energy resources because of its ubiquity and abundance of photon energy naturally [2]. Moreover, solar system provides solutions in many fields such as industrial, commercial, residential, etc. Also, it can operate in grid connected and isolated mode both. However, the solar intensity and temperature are fluctuating continuously, by virtue of that the power output will vary [3]. Therefore, an efficient MPPT system is required to track maximum power from the solar photovoltaic array and to extract maximum power. There are various MPPT algorithms are reported in many literatures with their merits and demerits [4, 5]. In the proposed work, incremental conductance (INC) based method is utilized due to their ease of working and operation. In the grid connected mode operation role of power electronic based converters has been increased [6, 7]. The quality of power gets affected due to the insertion of power electronics converters and by nonlinear loads. Moreover, various solutions are reported in literatures to overcome power quality (PQ) issues by renewable energy resources connected to grid [8]. Besides, hybrid filter arrangements are adopted as a conditioner to solve PQ problems [9]. The SAPF based on SRF control algorithm is utilized in this paper to make harmonics free source current. The SRF based control algorithm needs current transformation in synchronously d-q frame [10]. Besides, SRF is also well-known technique by another name as a d-q theory. Moreover, the introduced theory is applied for reactive power management and to make grid current operation at unity power factor. Additionally, PI based control technique is applied for stabilizing DC link bus voltage and for loss estimation in DSTATCOM [11, 12]. However, PI controllers having difficulties in tuning the gains of proportional (P) and integral (I) controller and parameters variation due to non-liner loading [13]. Therefore, in the presented work artificial techniques applied as an artificial neural network (ANN) for DC bus voltage regulation. As ANN having excellent learning capability and parallel processing computing nature, this control technique is applied in many recent control applications [14]. By utilizing ANN technique with SRF control method to extract reference current for 3-phase IGBT based voltage source converter (VSC). The complete system is carried out in MATLAB Simulink software to exhibits the performance of the presented system. Moreover, THD of grid current is fall well below 5% according to IEEE-519 slandered.

2 The Proposed System

The proposed system layout is depicted in Fig. 1. The presented system consists of SPV array of 100 kW rating connected to three phase grids via IGBT based VSC. A diode rectifier based nonlinear load is connected at PCC through interfacing inductors (La, Lb, Lc). An MPPT based on INC method is incorporated to utilized maximum power from the 100 kW SPV array. A boost converter is utilized to boosting the voltage level to 700 V from 270 V DC. A SAPF is connected parallelly to the system for injecting compensating currents (Isc, Isb, Isc), to make grid current harmonic free. Moreover, SRF control technique is utilized as a SAPF for generation of reference

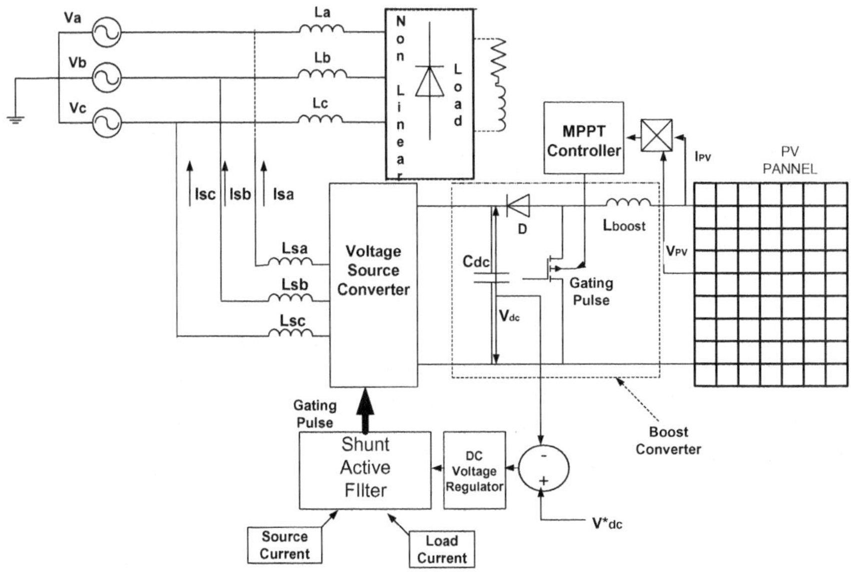

Fig. 1 Proposed system layout

currents for getting pulses to VSC. The main aim is to provides switching pulses to VSC for power factor improvement, DC offset reduction, load balancing, THD reduction and harmonics free source current. An ANN based DC voltage regulator is applied for DC bus voltage regulation.

3 Mathematical Formulation of PV System

For generation of PV system, the solar cell is the fundamental element. Photovoltaic cells are linked at series or parallel for creating solar panels by necessary qualifications. Figure 2 displays the photovoltaic system circuit. At photovoltaic energy subsystem, photovoltaic panels are utilized to change accumulated sunlight into electrical energy. The power created through photovoltaic system is calculated using the Eq. (1)

$$P_{pv} = R_t \eta_{pv} A_{pv} \tag{1}$$

where, P_{pv} denotes PV panels output power, R_t denotes solar radiation in tilted plane module, η_{pv} indicates PV panels efficiency, A_{pv} is PV panel area. The below equation articulates the photovoltaic panel efficiency,

$$\eta_{pv} = \eta_{r-pv} \eta_{PC} \left[1 - N_T \left(T_C - T_{ref}\right)\right] \tag{2}$$

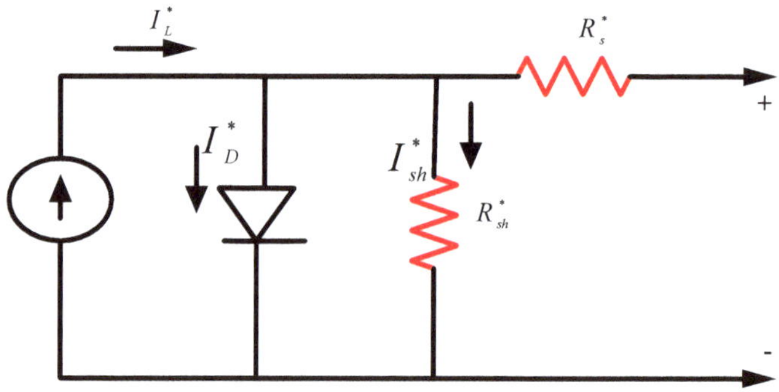

Fig. 2 PV system modeling

where, photovoltaic panel efficiency denotes η_{pv}, reference module efficiency indicates η_{r-pv}, power condition efficiency denotes η_{PC}, photovoltaic panel efficiency temperature coefficient is N_T, cell temperature (°C) is T_C, cell temperature in reference conditions is T_{ref}. If ideal maximal power tracker is utilized, η_{PC} value is equivalent with 1 [15]. Similarly, typically T_{ref} is place with 25 °C and also temperature of mono and polycrystalline silicon N_T as $-3.7 \times 10^{-3\,°C^{-1}}$. Furthermore, T_C articulated as take following,

$$T_C = T_A + \left[\frac{NOCT - 20}{800}\right] R_t \tag{3}$$

Here, T_A denotes ambient air temperature, $NOCT$ indicates nominal cell operating temperature.

The PV module voltage-current relationship is expressed in the Eq. (4)

$$I^* = I_L^* - I_{sat}^* \left[\exp\left\{q^* \frac{V_{pv}^* + I_{pv}^* \cdot R_s^*}{\upsilon k T_c}\right\} - 1\right] - \frac{V_{pv}^* + I_{pv}^* \cdot R_s^*}{R_{sh}^*} \tag{4}$$

where I_L^* implies current generated from Light in A, R_s^* and R_{sh}^* are the shunt and series resistances in Ω, υ indicates ideality factor, I_{pv}^* and V_{pv}^*,I_{sat}^* implies solar output voltage and reverse saturation current in A. Current generated by light I_L^* and reverse saturation current I_{sat}^* are illustrated in Eqs. (5) and (6)

$$I_L^* = \left[I_{sc} + \gamma(T_c - T_c^*)\right] \frac{C_T}{C_T^*} \tag{5}$$

$$I_{sat}^* = \frac{I_{sc}\gamma(T_c - T_c^*)}{\exp\left[\frac{V_{oc}\gamma(T_c - T_c^*)}{V_T}\right] - 1} \tag{6}$$

where T_c ambient temperature of the PV cell, C_T is the irradiance, γ implies circuit current temperature coefficient, and V_T implies thermal voltage.

3.1 *Modeling of Boost Converter*

The PV system maximum consumption can be accomplished through the using boost topology, which regulates the output voltage for a given input voltage. A boost converter is utilized for interfacing the circuit between the inverter and the PV module that attached the PV array to grid. Normally, boost, buck, and buck-boost converter topologies are utilized. To track maximal power point and boosting up PV-cell voltage with appropriate level, boost converter is utilized for grid the connection. Based on switching duty cycle, outcome voltage of the BC is estimated. The modeling of the BC is expressed in Fig. 3.

The chopper switching state S_s is given in the Eq. (7)

$$S_s = \begin{cases} 0, & \textit{switch is open} \\ 1, & \textit{switch is closed} \end{cases} \tag{7}$$

The dynamic model of BC can be attained via applying the Kirchhoff's laws which is expressed as follows

$$\frac{di_{L^*}}{dt} = \frac{V_{pv} - V_O}{L^*} + u\frac{V_O}{L^*} \tag{8}$$

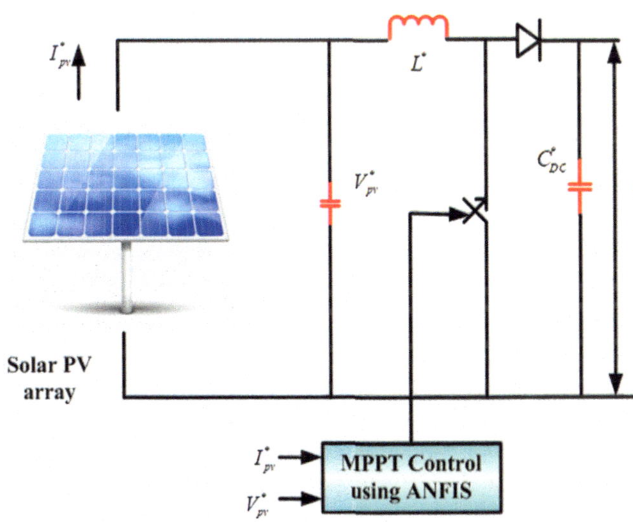

Fig. 3 Modeling of boost converter

$$\frac{dV_O}{dt} = \frac{i_{L^*}}{C_O} - u\frac{i_{L^*}}{C_O} \tag{9}$$

where i_{L^*} implies current injected to inverter V_O is the converter circuit voltage terminals which may be estimated from input voltage function V_{pv} chopper duty cycle D^* in Eq. (10)

$$\frac{V_O}{V_{pv}} = \frac{1}{1 - D^*} \tag{10}$$

The inductor in the BC is estimated using the Eq. (11)

$$L^* = V_{ap}^* D^* \Big/ \Delta i_1^* f_s \tag{11}$$

where Δi_1^* is the input current ripple, f_s implies switching frequency, V_{ap}^* implies outcome voltage ripple.

3.2 Modeling of Grid Connected Inductors

The grid connected inductors are utilized for absorbing switching ripples produced through VSC. Besides, the VSC is linked to the grid through the interfacing inductor. The formula for inductor is expressed by,

$$\Delta L_1^* = \frac{\sqrt{3}m^* V_{dc}^*}{12hf_s \Delta i_1^*} \tag{12}$$

where h is factor of overloading and its value is 1.2.

3.3 DC-Link Capacitor Design

DC link capacitor is connected among photovoltaic array and converter, which diminish the voltage ripple across PV cell terminals. Therefore, there is a ripple of the outcome power. Moreover, the capacitors perform from source and sink each half cycle to produce a power balance in DC bus. Derivation for sizing the DC link capacitor demonstrates that the significance of the component because of the dissimilarity in DC input power and AC output power. The DC-link capacitor for required specifications are achieved via the Eq. (8)

$$C_{dc}^{*} = \frac{P_{dc}^{*}/V_{dc}^{*}}{2\omega V_{dcri}^{*}} \tag{13}$$

where P_{dc}^{*} and V_{dc}^{*} are the dc power and voltage.

4 The Synchronous Reference Frame Theory

Synchronous Reference Frame (SRF) theory is based on currents transformation to rotating d-q frame. In (14) shows the decomposition of load currents (i_{La}, i_{Lb}, i_{Lc}) from three phase to two phase using Park's transformation. Figure 4 shows the basic control algorithm for generating reference current pulses ($I_{\alpha}^{*}\, I_{\beta}^{*}\, I_{\gamma}^{*}$). The i_{La}, i_{Lb}, i_{Lc}, are the load side current and V_{sa}, V_{sb}, V_{sc} are the grid side three phase voltages.

$$\begin{bmatrix} i_{lpd} \\ i_{lqd} \end{bmatrix} = \sqrt{2/3}\begin{bmatrix} \cos\varphi & \cos(\varphi - \frac{2\pi}{3}) & \cos(\phi + \frac{2\pi}{3}) \\ -\sin\varphi & -\sin(\varphi - \frac{2\pi}{3}) & -\sin(\varphi + \frac{2\pi}{3}) \end{bmatrix}\begin{bmatrix} i_{la} \\ i_{lb} \\ i_{lc} \end{bmatrix} \tag{14}$$

The PLL (phase locked loop) is used to synchronization signal as a current along with voltage at PCC. The DC voltage regulation is done by linear conventional PI controller and by ANN based artificial controller. The dynamics of the PI controller over conventional PI controller is evaluated in the proposed work. Moreover, in Fig. 5. Presents the dynamic characteristics with PI and ANN controller is presented. Moreover, low pass filter (LPF) is used to extracting the DC components.

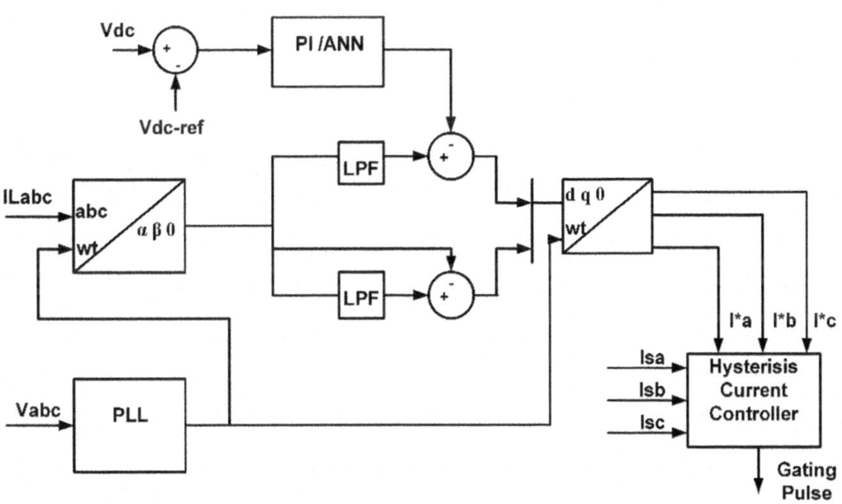

Fig. 4 d-q theory for reference current generation

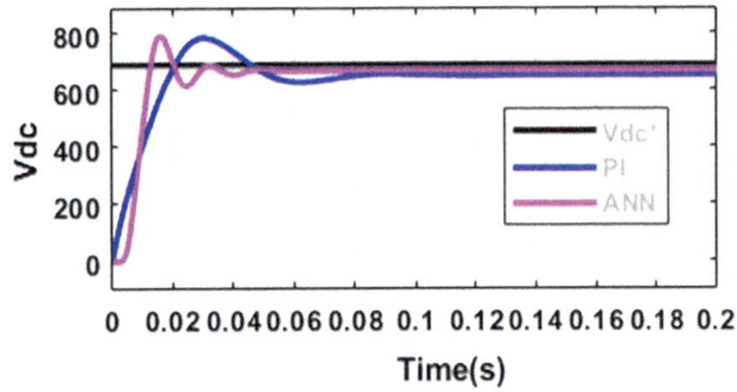

Fig. 5 Dynamic response of PI and ANN controller

$$i_{lpd} = \bar{i}_{Ld} + \bar{i}_{Ld} \tag{15}$$

$$i_{lqd} = \bar{i}_{Lq} + \bar{i}_{Lq} \tag{16}$$

$$\begin{bmatrix} I_\alpha^* \\ I_\beta^* \\ I_\gamma^* \end{bmatrix} = \sqrt{2/3}\begin{bmatrix} \cos\varphi & -\sin\varphi \\ \cos(\varphi - \frac{2\pi}{3}) & -\sin(\varphi - \frac{2\pi}{3}) \\ \cos(\phi + \frac{2\pi}{3}) & -\sin(\varphi + \frac{2\pi}{3}) \end{bmatrix}\begin{bmatrix} i_{lpd} \\ i_{lqd} \end{bmatrix} \tag{17}$$

Finally, converting two phase DC current components in to three phase current components by the help of inverse clark's transformation. The conversion of current is represented in Eq. (17). Additionally, hysteresis current based controller is utilized for the generation of gating pulses to VSC. The complete procedural steps in form of basic block diagram are depicted in Fig. 4.

5 ANN Controllers for DC Bus Voltage Maintenance

The ANN controller is applied to regulate DC link bus voltage and for loss current component estimation. An ANN controller is based on biological neural network as it comprised by involvements of artificial neurons. Therefore, the proposed ANN controllers are more efficient and accurate to maintains the desired constant as well as to minimize the converter losses. Moreover, this control provides better dynamic performance as compared to the FLC and conventional PI controllers. The following steps are adopted by ANN controllers for DC-bus voltage regulation.

Initialize neural network toolbox.

- Setting of input and target parameters
- Import input and target data from workspace
- Network creation
- Define data into the network
- Training of network
- Performance checking.

Table 1 Comparative analysis between Fuzzy logic and conventional controller

Controller	Maximum overshoot	Settling time
ANN	Lower	Less (0.4 s)
PI	Higher	Higher (0.8 s)

6 Simulation Results

The proposed system is analyzed in MATLAB Simulink environment. In the Fig. 4 presents the DC bus voltage regulation by ANN and conventional PI controller. Moreover, Table 1 exhibits the comparative analysis shows that ANN based SRF outperform then conventional PI controller. For the detail information and step-wise-step procedure of implementation of ANN model in a simple, reader may refer [16–25] (Fig. 5).

In Fig. .6 depicts the system behavior under nonlinear load condition, with SPV system operated at constant irradiance. In the depicted Fig. 6 presents the grid voltages, grid currents, irradiance, load currents, DC link voltage (Vdc), active power (Pg) and reactive power (Qg). It is shown that at constant irradiance grid voltage and grid currents are perfectly balanced and sinusoidal with nonlinear loads. Moreover, the Vdc is also maintaining constant at 700V and it is observed that there is no any reactive power is zero in grid side. Therefore, it is concluded that the system is operated at unity power factor as well. Additionally, THD of grid-side current is 3.06% which is below then the IEEE reference standard. In Fig. 7 presented the FFT analysis of source current.

7 Conclusion

In this paper a SRF algorithm is presented for improving the SPV performance. The SAPF is applied for reactive power compensation and harmonic suppression caused by nonlinear load. PI and artificial intelligent technique and ANN are applied for DC link voltage regulation. Moreover, in the presented work shows that DC voltage regulation is better by intelligent ANN based controller then the conventional linear PI controller. The performance of the applied system is simulated in MATLAB/Simulink environment. Moreover, the presented results shows that harmonics of source current have been reduced and system is working with improved power factor. Additionally, ANN based controller effectively regulate DC bus voltage and it maintains the THD of grid current under IEEE-519 limits.

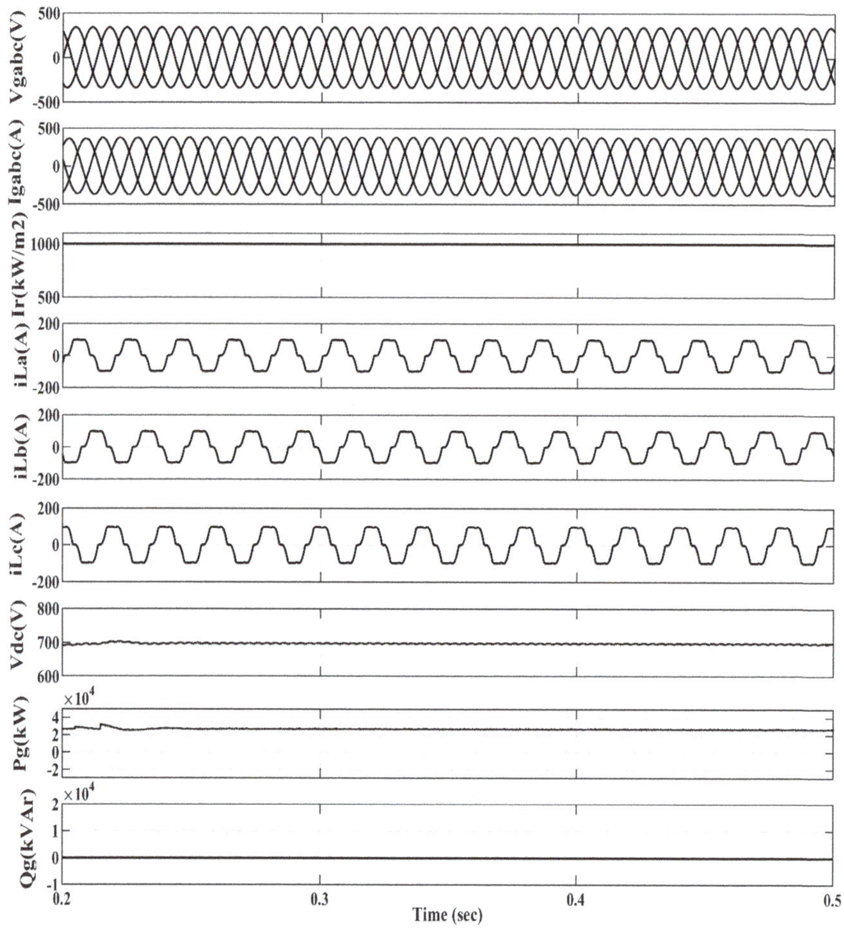

Fig. 6 System behavior under non linear load with fixed solar irradiance

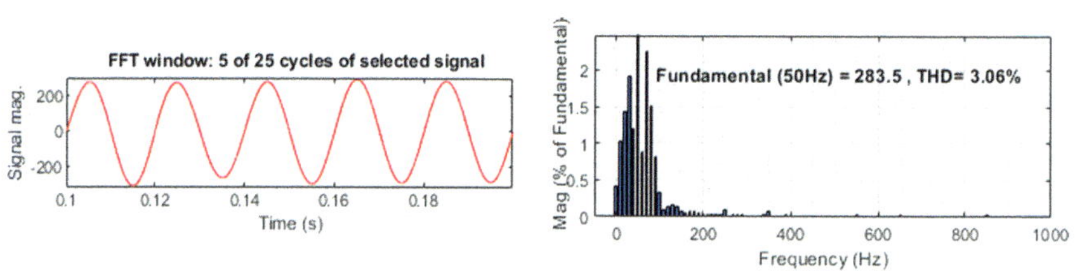

Fig. 7 THD of grid current

Appendix

SPV parameters: irradiance = 1000 w/m^2, V_{oc} = 64.2 V, I_{sc} = 5.96 A, V_{mp} = 54.7 V, I_{mp} = 5.58 A, Ns = 5 and Np = 66, load inductance = 0.1 mH, shunt inductance = 1 mH, DC link voltage 700 V, DC bus capacitor voltage = 3000 μF.

References

1. Renewable energy policy network for the 21st century, global status report 2014. http://www.ren21.net/ren21activities/globalstatusreport.aspx
2. Malinowski M, Leon JI, Abu-Rub H (2017) Solar photovoltaic and thermal energy systems: current technology and future trends. Proc IEEE 105(11):2132–2146. https://doi.org/10.1109/JPROC.2017.2690343
3. Ellabban O, Abu-Rub H, Blaabjerg F (2014) Renewable energy resources: current status, future prospects and their enabling technology. Renew Sustain Energy Rev 39:748–764
4. Subudhi B, Pradhan R (2013) A comparative study on maximum power point tracking techniques for photovoltaic power systems. IEEE Trans Sustain Energy 4(1):89–98. https://doi.org/10.1109/TSTE.2012.2202294
5. Esram T, Chapman PL (2007) Comparison of photovoltaic array maximum power point tracking techniques. IEEE Trans Energy Convers 22(2):439–449. https://doi.org/10.1109/TEC.2006.874230
6. Carrasco JM et al (2006) Power-electronic systems for the grid integration of renewable energy sources: a survey. IEEE Trans Ind Electron 53(4):1002–1016. https://doi.org/10.1109/TIE.2006.878356
7. Liang X (2017) Emerging power quality challenges due to integration of renewable energy sources. IEEE Trans Industr Appl 53(2):855–866. https://doi.org/10.1109/TIA.2016.2626253
8. Singh B, Chandra A, Al-Haddad K (2015) Power quality: problems and mitigation techniques. Wiley, UK
9. Rivas D, Moran L, Dixon JW, Espinoza JR (2003) Improving passive filter compensation performance with active techniques. IEEE Trans Ind Electron 50(1):161–170. https://doi.org/10.1109/TIE.2002.807658
10. Singh B, Solanki J (2009) A comparison of control algorithms for DSTATCOM. IEEE Trans Ind Electron 56(7):2738–2745. https://doi.org/10.1109/TIE.2009.2021596
11. Singh B, Shahani DT, Verma AK (2012) Power balance theory based control of grid interfaced solar photovoltaic power generating system with improved power quality. In: 2012 IEEE international conference on power electronics, drives and energy systems (PEDES), Bengaluru, pp 1–7. https://doi.org/10.1109/PEDES.2012.6484359
12. Singh B, Kumar S, Jain C (2017) Damped-SOGI-based control algorithm for solar PV power generating system. IEEE Trans Ind Appl 53(3):1780–1788. https://doi.org/10.1109/TIA.2017.2677358
13. Bhattacharya A, Chakraborty C (2011) A shunt active power filter with enhanced performance using ANN-based predictive and adaptive controllers. IEEE Trans Ind Electron 58(2):421–428. https://doi.org/10.1109/TIE.2010.2070770
14. Bin L, Minyong T (2012) Control method of the three-phase four-leg shunt active power filter, energy procedia. In: 2011 2nd international conference on advances in energy engineering (ICAEE), vol 14, pp 1825–1830
15. de Oliveira F, Durand F, Bacon V, Oliveira da Silva S, Sampaio L, Campanhol L (2016) Grid-tied photovoltaic system based on PSO MPPT technique with active power line conditioning. IET Power Electron 9(6):1180–1191

16. Farkoush SG et al (2021) Deterministic and probabilistic occupancy detection with a novel heuristic optimization and back-propagation (BP) based algorithm. J Intell Fuzzy Syst Preprint 1–13. https://doi.org/10.3233/JIFS-189748
17. Azeem A et al (2021) Real-time harmonics analysis of digital substation equipment based on IEC-61850 using hybrid intelligent approach. J Intell Fuzzy Syst Preprint, 1–14. https://doi.org/10.3233/JIFS-189745
18. Sanaullah A et al (2021) R analyzing impact of relationship benefit and commitment on developing loyalty using machine intelligence approach. J Intell Fuzzy Syst Preprint 1–14. https://doi.org/10.3233/JIFS-189742
19. Gopal et al (2021) Digital transformation through advances in artificial intelligence and machine learning. J Intell Fuzzy Syst Pre-press 1–8. https://doi.org/10.3233/JIFS-189787
20. Fatema N, Malik H (2020) Data-driven occupancy detection hybrid model using particle swarm optimization based artificial neural network, metaheuristic and evolutionary computation: algorithms and applications, under book series studies in computational intelligence. Springer Nature, pp 283–297. https://doi.org/10.1007/978-981-15-7571-6_13
21. Iqbal A et al (2020) Metaheurestic algorithm based hybrid model for identification of building sale prices, metaheuristic and evolutionary computation: algorithms and applications. Studies in computational intelligence. Springer Nature, pp 689–704. https://doi.org/10.1007/978-981-15-7571-6_32
22. Shah AK et al (2018) EMD and ANN based intelligent model for bearing fault diagnosis. J Intell Fuzzy Syst 35(5):5391–5402. https://doi.org/10.3233/JIFS-169821
23. Azeem A, Fatema N, Malik H (2018) k-NN and ANN based deterministic and probabilistic wind speed forecasting intelligent approach. J Intell Fuzzy Syst 35(5):5021–5031. https://doi.org/10.3233/JIFS-169786
24. Malik H, Sharma R (2017) EMD and ANN based intelligent fault diagnosis model for transmission line. J Intell Fuzzy Syst 32(4):3043–3050. https://doi.org/10.3233/JIFS-169247
25. Fatema N et al (2021) Intelligent data-analytics for condition monitoring: smart grid applications. Elsevier, p 268. ISBN 978-0-323-85511-2. https://www.sciencedirect.com/book/9780323855105/intelligent-data-analytics-for-condition-monitoring

Modeling and Implementation of Statechart for MPPT Control of Photovoltaic System in FPGA

Venkat Pankaj Lahari Molleti, Ramasudha Kasibhatla, and Vijayasanthi Rajamahanthi

Abstract Statecharts, which are an effective modeling and specifications tool for reactive systems, are being used in this work for maximum power point tracking (MPPT) control of the photovoltaic systems. Perturb and observe (P&O) algorithm, which is the most widely used MPPT algorithm is being improvised by implementing via field-programmable gate array (FPGA)-based statechart model developed as the MPPT statechart controller. Statechart controller model is developed for an FPGA target using LabVIEW FPGA and its performance is verified in the simulation. It is then implemented in a Xilinx FPGA and the results are presented. Along with providing better modularity, statecharts also provide determinism and accuracy in their control actions, and the same can be observed from the statechart controller model as well as results.

Keywords Statechart · MPPT · Photovoltaic · FSM · FPGA · LabVIEW

1 Introduction

With the pressing need to utilize renewable energy sources increasing day by day, new and accurate methods to improve the performance of photovoltaic systems, wind energy systems etc. are equally necessary. For these systems, a significant aspect is to ensure operation at the maximum power point. For photovoltaic systems, there are many MPPT algorithms [1–3] already tried, tested and experimented. However, the most widely used algorithm is the perturb and observe algorithm [1, 2, 4].

In this work, an improved P&O algorithm is implemented by a novel methodology, called the statecharts. Statecharts are visual formalisms that offer effective modeling, determinism, modularity in complex and reactive systems [5, 6]. Statecharts are an extension to finite state machines (FSMs) as they possess the unique additional features of depth (hierarchy), orthogonality (parallelism) and broadcast communication (variables being accessed by all states at any point of execution)

V. P. L. Molleti (✉) · R. Kasibhatla · V. Rajamahanthi
Department of Electrical Engineering, Andhra University, Visakhapatnam, Andhra Pradesh 530003, India

A. Tomar et al. (eds.), *Machine Learning, Advances in Computing, Renewable Energy and Communication*, Lecture Notes in Electrical Engineering 768,
https://doi.org/10.1007/978-981-16-2354-7_27

which are very effective attributes for accurate modeling of real-time systems [5–7]. Apart from being highly apt for modeling systems, statecharts are also capable of executing control actions [6]. Hence, they can also be used to implement MPPT control needed by photovoltaic systems and this is being tested in this work by developing a statechart controller, verifying it in simulation and then implementing it in a target embedded controller.

Digital controllers include a range of embedded devices like microcontrollers, digital signal processors (DSPs), field-programmable gate arrays (FPGAs), application-specific integrated circuits (ASICs) and programmable logic controllers (PLCs) [8, 9]. Among these, FPGAs are digital controllers with advanced features like parallelism, hardware execution resulting in fast execution speeds and architecture independence [8, 9]. These features are suitable for statecharts which inherently have the feature of orthogonality. Hence, in this work, for achieving good execution speeds of the MPPT control and thereby better MPPT control with flexibility and ease of development, FPGA is chosen as the target embedded controller for executing the developed MPPT statechart controller.

For implementing the MPPT statechart controller in an FPGA, it is required that statecharts are capable of being transformed into hardware description language (HDL) since FPGA can be programmed only by HDLs. The platform used to develop the MPPT statechart controller is LabVIEW Statechart [10] in conjunction with LabVIEW FPGA [11], through which the developed statechart can be converted into an equivalent HDL code and hence can be deployed into the target FPGA. Automatic HDL-code generation for the statechart controller developed largely simplifies the implementation procedure and enables higher scope for debugging the developed MPPT statechart controller.

Previously, stateflow, which is an FSM-alike, has been used to model MPPT control in [12–14] using MATLAB. In [13, 14], the developed stateflow-based MPPT controller has been implemented in DSP. Some other examples for MMPT control are given in [15–21] using AI and machine learning applications. However, DSP is a digital controller which is not capable of parallel processing and hence it cannot incorporate the feature of orthogonality of statecharts. Hence, in this work, implementation of the statecharts, which are an advanced tool than FSMs, is being performed in FPGA which is capable of parallel processing. In [22], statechart models for MPPT control are developed, but they are not appropriate for FPGA implementation as they were modeled for desktop targets (pure offline simulation), whereas in this work, they are modeled and developed in such a way that they are FPGA-deployable. Simulation and hardware implementation of the MPPT statechart controller in the FPGA target, NI myRIO Xilinx FPGA, is performed in this work and the results are presented.

2 Modeling of FPGA—Deployable Statechart for MPPT Control

In this work, a statechart controller is developed for MPPT control of a photovoltaic system. The control algorithm in the statechart is based on the widely used P&O algorithm. However, the implementation methodology is not conventional, but via statecharts, which are more modular and deterministic. A statechart can be defined as a set of states, transitions, transition conditions, state-actions, inputs, outputs and internal variables [5, 6, 23]. Configuring and defining each of these attributes aptly, a statechart controller is developed using LabVIEW statechart software as shown in Fig. 1. In [22], the P&O-based statechart model is developed, but it is for offline simulation and is not directly deployable into a target embedded controller. In this work, a statechart controller is developed for deployment into the FPGA target. Also, the number of transitions in the present model is lesser by 4, as compared to the model developed in [22]; hence, the proposed model in this work is more efficient. Table 1 shows the specifications of the developed statechart controller.

Figures 1 and 2 show the aspects of the developed statechart controller in LabVIEW Statechart software—1. The statechart project window where the statechart parameters like input variables, output variables, internal variables, their datatypes and limits, and the statechart diagram are all defined—2. The statechart diagram itself, which is the statechart controller performing the MPPT control by updating the duty ratio—3. The transition code of one of the transitions (T1—4. The state-action code of a state, which adds the feature of 'control action' possible in statecharts.

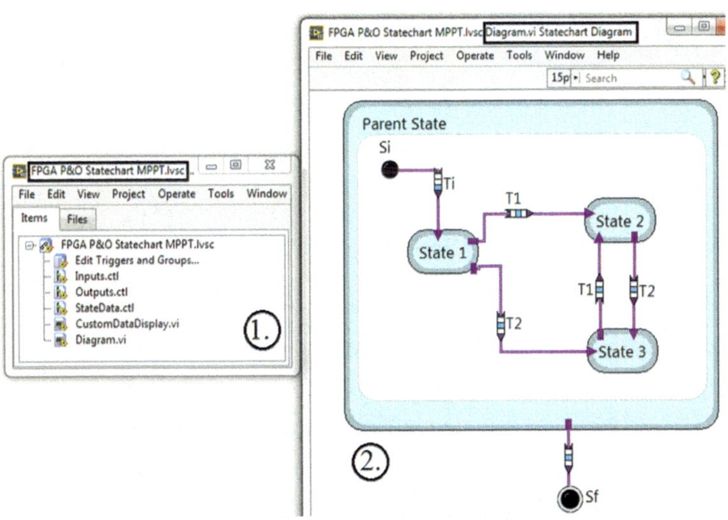

Fig. 1 Development of statechart controller for MPPT—project window and diagram

Table 1 States and transitions of the developed statechart controller

Aspect of statechart	Number	Description
States	5	Si[a], S1, S2, S3, Sf[a]
States with depth	1	Child states—S1, S2, S3
Transitions	4	Ti[a], T1, T2, Tf[a]
Input variables	5	PV voltage, PV current present and previous iterations, a Boolean control for stop/run statechart
Output variables	1	Duty ratio
Internal variables	1	Duty ratio

[a]i and f indicate initial and final, respectively

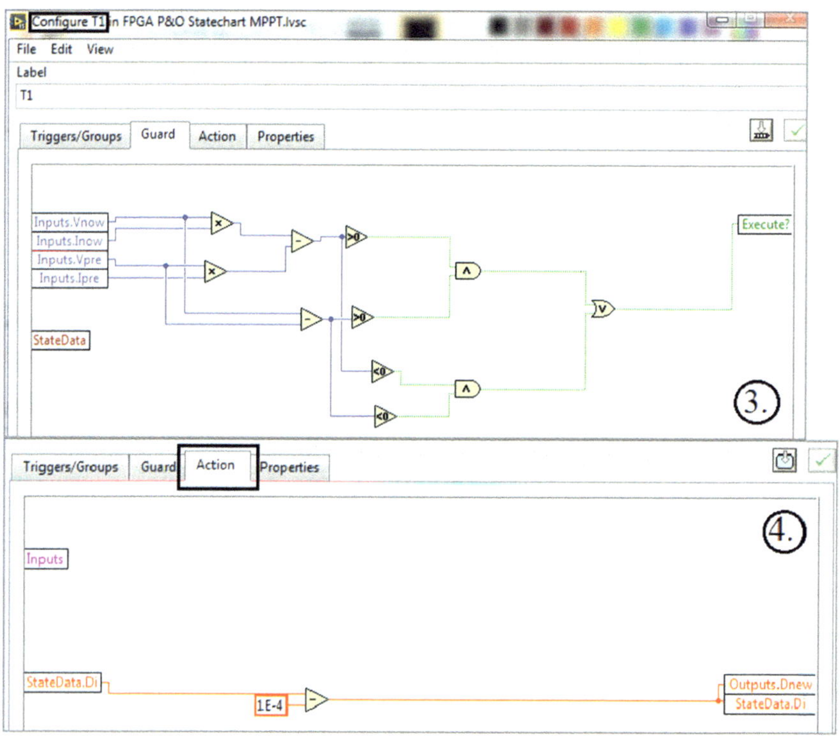

Fig. 2 Development of statechart controller for MPPT—transition and state-action codes

The statechart controller should also be configured for deployment into a target FPGA. To make a statechart FPGA-ready, the first step is to modify its configuration settings by appropriately defining the statechart properties. Also, only the datatypes of the fixed point (FXP) or single-precision (SGL) floating point are supported by the target FPGA of this work. Hence, datatype configuration is to be done for all the defined variables, according to the target FPGA, which is the second step for making the statechart controller FPGA-ready. The third and final step is to build the final FPGA program code such that the necessary inputs to the statecharts controller are communicated to it and the output from the same is accessible via the output channels of the FPGA target. The final FPGA code is developed in LabVIEW FPGA software and is deployed in the target FPGA. However, in this work, before deploying the statechart controller in the target FPGA, the FPGA program is initially verified in simulation. This is explained in the next section.

3 Simulation of MPPT Statechart Controller for Target FPGA

To verify the developed MPPT statechart controller, it is required to communicate the necessary input parameters to the statechart. The input parameters are voltage and current values of the PV panel for which the MPPT statechart controller is being developed. The PV panel model of BP MSX 120 based on [24, 25] is simulated in NI Multisim and the necessary parameters are obtained. These voltage and current values so obtained are used in the FPGA code for passing inputs to the MPPT statechart controller. The FPGA code is explained further.

3.1 Development of FPGA Program and Host Program for MPPT Statechart Controller Verification

Simulation of MPPT statechart controller requires developing two programs, namely the FPGA program developed in LabVIEW FPGA and the host program developed in LabVIEW (host version). The FPGA program is the actual code containing the statechart controller, its inputs and outputs, while the host program is required to run the FPGA program and obtain plots of the output from the FPGA program.

Figure 3 shows the FPGA program developed for the verification of MPPT statechart controller. It has a while loop, the inputs to which are the PV voltage and current values as array constants, obtained from the NI Multisim simulation of the PV panel. Inside the loop in the FPGA program is a cluster defined with five input nodes which are the terminals to communicate the five input variables to the statechart. The statechart is accessed in the FPGA program via a function in LabVIEW FPGA called 'run statechart'. This function receives inputs for the statechart in the form of clusters and also provides output from the statechart as an output cluster. With the PV voltage and current values of the present iteration and previous iteration as inputs, the output from the MPPT statechart controller is the updated duty ratio based on the operating point of the PV panel. With every iteration of the loop, the new values of voltage and current are communicated to the statechart, and the statechart, after assessing the appropriate state of the PV panel, increases or decreases the duty

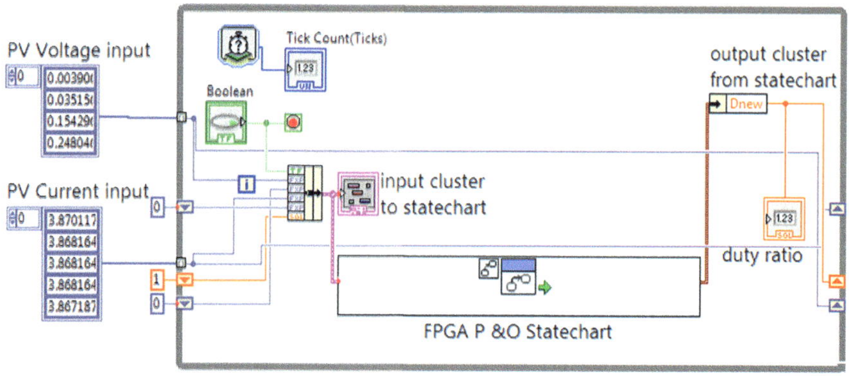

Fig. 3 FPGA program for simulation of MPPT statechart controller

ratio so as to attain MPP. This is the output of the statechart controller which is to be verified, i.e., whether or not the duty ratio generated by it is corresponding to the MPP of the PV panel.

Figure 4 shows the host program developed. The role of the host program is to execute/stop the FPGA program and access all the nodes (input, output) in the FPGA program, including the output of the statechart controller, and enable us to verify the statechart controller by plotting the necessary variables. To access the nodes of the FPGA program in the host program, a function named 'Read/ Write' control is to be used. This function, if the reference of the specific FPGA program is given as input using 'open FPGA-VI reference' function, can access all the input/output nodes of the FPGA program.

Hence, the host program and FPGA program are both developed for verification of the statechart controller for MPPT of a photovoltaic system in simulation.

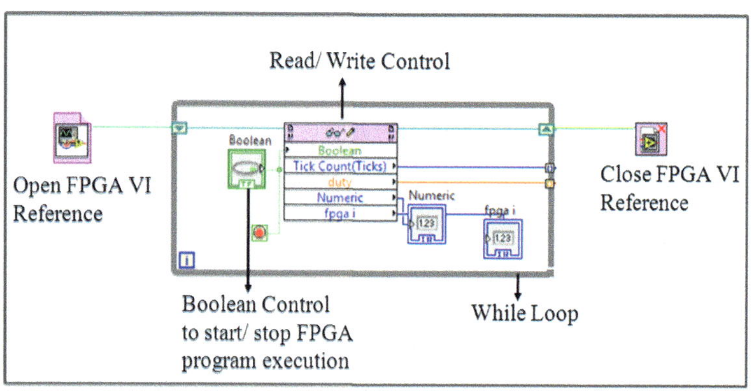

Fig. 4 Host program for simulation of MPPT statechart controller

3.2 *Results Obtained from Simulation of FPGA Statechart Controller*

As mentioned, the inputs to the MPPT statechart controller are PV panel voltage and current values of the present as well as previous iterations. These values are given as inputs in the form of arrays to the statechart controller in the FPGA program. The values are obtained from the current–voltage (I-V) and power–voltage (P-V) curves, as shown in Fig. 5 from the Multisim simulation model. Considering that the PV system is a PV panel interfaced to a resistive load of 1 kΩ via a DC/DC boost converter operating at a duty cycle D [1, 2, 22]. The duty cycle corresponding to the MPP (118.7 W, 33.41 V, 3.551 A) of the PV panel can be calculated mathematically [22] and it is found to be 0.904 for an irradiance 1000 W/m^2 and temperature 25 °C. Figure 6 shows the plot of duty ratio, altered by the MPPT statechart controller such that it corresponds to the MPP of the PV panel. It can be observed that the duty ratio is settled at 0.904 as desired, by the MPPT statechart controller. Hence, the MPP tracking capability of the statechart controller is verified by the simulation result.

4 Implementation of Statechart Controller in FPGA Target

The developed MPPT statechart controller in simulation is to be also verified in the real target FPGA. Hence, in this work, the MPPT statechart controller is deployed in the target FPGA, myRIO Xilinx Z7010 FPGA. This target FPGA can be programmed and compiled using LabVIEW-FPGA and Xilinx-compilation tools [26]. Two analog inputs (one each for PV current and voltage), one analog output (to observe a change in duty ratio caused by the statechart MPPT controller) and one digital output line (to generate the control signal based on the updated duty ratio from statechart MPPT controller) are used for hardware implementation of MPPT statechart controller in the myRIO-FPGA. The PV voltage and current values for irradiance 1000 W/m^2 and

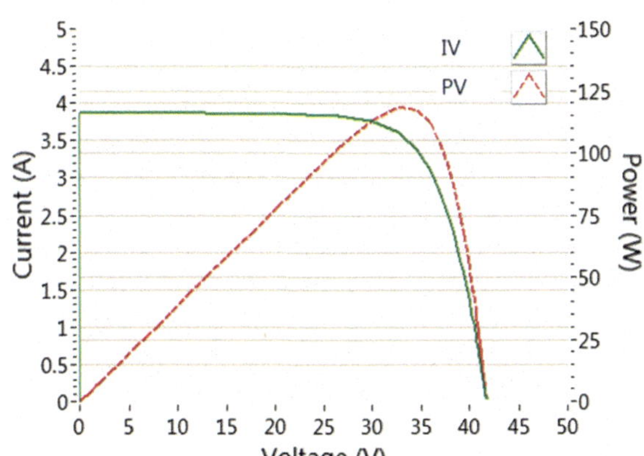

Fig. 5 IV and PV plots of the photovoltaic panel

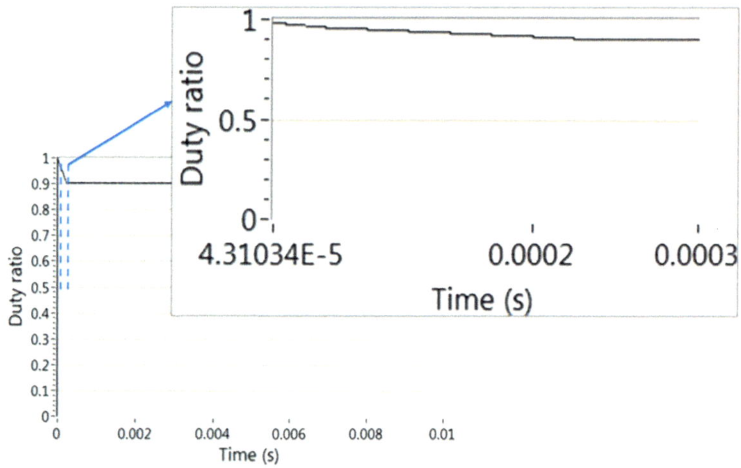

Fig. 6 Output plot of duty ratio generated by MPPT statechart controller simulation

load 1 kΩ are entered as array values and are given as inputs to MPPT statechart in FPGA. Two parallel loops are executed in the target FPGA (the additional loop for digital channel control based on output from the statechart controller). This is a feature of parallelism unique to FPGAs, absent in other digital controllers.

4.1 Specifications for Implementation of MPPT Statechart Controller in myRIO-FPGA

The target FPGA is connected via a USB interface cable to a personal computer (PC) (called host-PC) installed with LabVIEW and LabVIEW-FPGA. The target FPGA is to be externally powered up by the AC/DC adaptor that is compatible with the power specifications of the device. Once the target FPGA and host PC are connected, the myRIO FPGA is to be configured with compatible versions of software versions in it. The host PC used has LabVIEW 2015, LabVIEW-FPGA 2015 and NI Multisim 2014. The myRIO FPGA should hence be configured with 2015 versions of the necessary software. This can be done and verified in NI-MAX (NI ‘Measurement & Automation Explorer’). The FPGA program for the MPPT statechart controller can now be developed for hardware deployment. Compatible Xilinx compilation tool (Vivado 2014.4) is to be installed in the host PC and it generates the necessary HDL code for the deployment of the statechart controller into the target FPGA.

4.2 Results from Implementation of MPPT Statechart Controller

The hardware setup for implementing statechart MPPT controller in FPGA is as shown in Fig. 7 and Fig. 8 shows the output from FPGA digital output channel observed in an oscilloscope (Tektronix-make). The digital output-0 (DO0) of

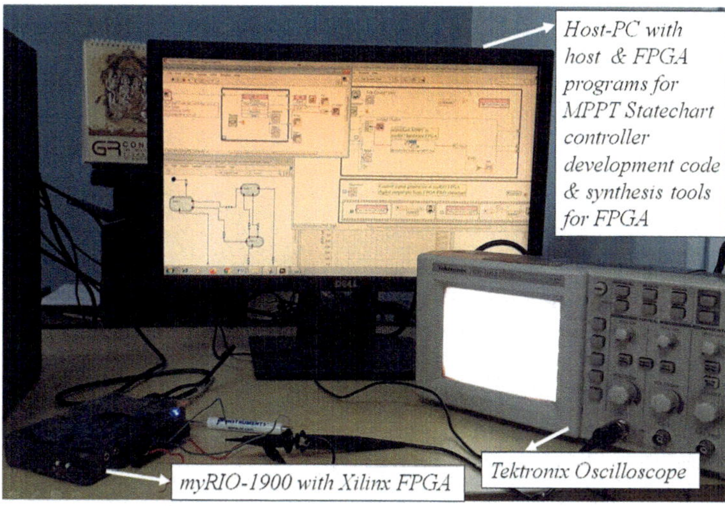

Fig. 7 Hardware setup for executing MPPT statechart controller in FPGA

Fig. 8 Control signal generated by FPGA with MPPT statechart controller deployed (observed at digital output)

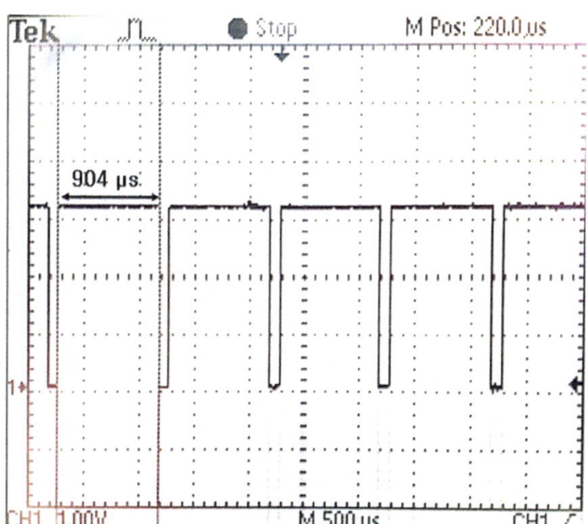

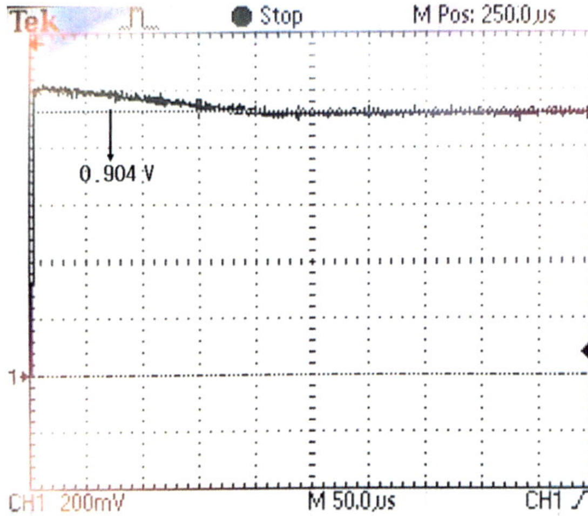

Fig. 9 Duty ratio updated by FPGA with MPPT statechart controller deployed (observed at analog output)

connector-C is programmed to generate a PWM control signal of 1 kHz for a boost-converter, based on the duty ratio updated by the MPPT statechart controller. The on-time of this generated control signal by the FPGA MPPT statechart controller is 0.904 ms, which is the value corresponding to MPP of the PV panel for irradiance of 1000 W/m^2 and load of 1 kΩ. Figure 9 also shows how the duty ratio value is getting updated by the FPGA's MPPT statechart control. Connector-C analog output-0 (AO0) is used to view this parameter computed by the myRIO FPGA. Hence, the FPGA-based MPPT statechart controller is capable of generating the required control signal so as to attain optimum operating point by the PV panel.

5 Conclusion

In this work, a statechart controller is developed for tracking of MPP of a photovoltaic system. With the necessary inputs of PV panel voltage and currents being communicated as arrays, it has been verified in simulation using LabVIEW FPGA. It could be observed that the MPPT statechart controller could update the control variable, i.e., the duty ratio required by the PV system, to the value corresponding to the MPP. After the initial verification, it is then verified by implementing the actual target FPGA, the NI myRIO Xilinx FPGA.

Results observed from the FPGA in the oscilloscope verify the correctness of the developed controller, as it appropriately updated the control variable corresponding to MPP of the PV panel under consideration. Hence, this MPPT statechart controller could be an effective methodology for MPPT control of PV systems, which in future work would be verified with the interface of the PV system.

References

1. Rebei N, Hmidet A, Gammoudi R, Othman H (2015) Implementation of photovoltaic water pumping system with MPPT controls. Front Energy 9(2):187–198
2. Deshpande AS, Patil SL (2020) Robust observer-based sliding mode control for maximum power point tracking. J Control Autom Electr Syst 31(5):1210–1220
3. Esram T, Chapman PL (2007) Comparison of photovoltaic array maximum power point tracking techniques. IEEE Trans Energy Convers 22(2):439–449
4. Amit KP, Naruttam KR, Hemanshu RP (2019) MPPT methods for solar PV systems: a critical review based on tracking nature. IET Renew Power Gener 13(10):1–19
5. Graics B, Molnar V, Voros A, Istvan M, Daniel V (2020) Mixed-semantics composition of statecharts for the component-based design of reactive systems. Softw Syst Model 19(10):1483–1517
6. Harel D (1987) Statecharts: a visual formalism for complex systems. Sci Comput Program 8(3):231–274
7. Said W, Quante J, Koschke R (2020) Mining understandable state machine models from embedded code. Empir Softw Eng 6:1–46. https://doi.org/10.1007/s10664-020-09865-0
8. Patel J, Sood VK (2018) Review of digital controllers in power converters. In: IEEE electrical power and energy conference (EPEC). IEEE, Toronto, pp 2381–2842
9. Choosing the right architecture for real-time signal processing designs. http://www.ti.com/lit/wp/spra879/spra879.pdf. Accessed 11 Nov 2020
10. LabVIEW statechart module tutorial. http://www.ni.com/tutorial/7425/en/. Accessed 11 Nov 2020
11. What is the LabVIEW FPGA module? https://www.ni.com/en-in/shop/electronic-test-instrumentation/add-ons-for-electronic-test-and-instrumentation/what-is-labview-fpga-module.html
12. Shourov EC, Sajid A, Sabzehgar R, Roshan YM (2018) A novel implementation of the state flow approach for maximum power point tracking of photovoltaic systems. In: IEEE international symposium on power electronics for distributed generation systems. IEEE, North Carolina, pp 1–6
13. Ahmed R, Abdelsalam AK, Namaan A, Dessouky YG, M'Sirdi NK (2014) Improved performance state-flow based photovoltaic maximum power point tracking technique. In: 3rd IET renewable power generation conference. IET, Italy, pp 1–5
14. Maher RA, Abdelsalam AK, Dessouky YG, Nouman A (2019) High performance state-flow based MPPT technique for micro WECS. IET Renew Power Gener 13(16):3009–3021
15. Aggarwal et al (2020) Meta heuristic and evolutionary computation: algorithms and applications. Springer Nature, Berlin, 949 pp. https://doi.org/10.1007/978-981-15-7571-6. ISBN 978-981-15-7571-6
16. Yadav AK et al (2020) Soft computing in condition monitoring and diagnostics of electrical and mechanical systems. Springer Nature, Berlin, 496 pp. https://doi.org/10.1007/978-981-15-1532-3. ISBN 978-981-15-1532-3
17. Gopal et al (2021) Digital transformation through advances in artificial intelligence and machine learning. J Intell Fuzzy Syst 1–8. Pre-press. https://doi.org/10.3233/JIFS-189787
18. Fatema N et al (2021) Intelligent data-analytics for condition monitoring: smart grid applications. Elsevier, 268 pp. ISBN: 978-0-323-85511-2. https://www.sciencedirect.com/book/9780323855105/intelligent-data-analytics-for-condition-monitoring
19. Smriti S et al (2018) Special issue on intelligent tools and techniques for signals, machines and automation. J Intell Fuzzy Syst 35(5):4895–4899. https://doi.org/10.3233/JIFS-169773
20. Jafar A et al (2021) AI and machine learning paradigms for health monitoring system: intelligent data analytics. Springer Nature, Berlin, 496 pp. https://doi.org/10.1007/978-981-33-4412-9. ISBN 978-981-33-4412-9
21. Sood YR et al (2019) Applications of artificial intelligence techniques in engineering, vol 1. Springer Nature, 643 pp. https://doi.org/10.1007/978-981-13-1819-1. ISBN 978-981-13-1819-1

22. Lahari MVP, Sudha KR, Santhi RV (2019) Statechart models of MPPT controller for a photovoltaic system in co-simulation environment. In: TENCON 2019 IEEE region 10 conference. IEEE, Kochi, pp 581–586
23. Introduction to UML terminology in the LabVIEW statechart module. https://www.ni.com/en-in/support/documentation/supplemental/08/archived--introduction-to-uml-terminology-in-the-labview-statech.html. Accessed 11 Nov 2020
24. BP MSX 120 datasheet. http://www.soltec-solar.com/html/cms/bp/product_msx_120.pdf. Accessed 11 Nov 2020
25. New models for photovoltaic cells in multisim. https://forums.ni.com/t5/National-Instruments-Circuit/New-Models-for-Photovoltaic-Cells-in-Multisim/ba-p/3473652?profile.language=en. Accessed 11 Nov 2020
26. User guide and specifications-NI myRIO 1900. http://www.ni.com/pdf/manuals/376047c.pdf. Accessed 11 Nov 2020

The Economic Viability of Battery Storage: Revenue from Arbitrage Opportunity in Indian Electricity Exchange (IEX) and NYISO

Asif Nazar and Naqui Anwer

Abstract This paper estimates potential revenue from Arbitrage services in the Indian Electricity Exchange (IEX) for battery energy storage systems. It further compares the potential estimated revenue from battery energy storage for NYISO. It also estimates revenue for both arbitrage services and frequency regulation for NYISO. The estimated revenue from participation in both services in NYISO shows a substantial increase in revenue and hence insight into regulators to make policy for the economic viability of battery energy storage.

Keywords Energy arbitrage · Regulation · Battery storage · IEX · NYISO

1 Introduction

The share of variable renewable energy sources (VRES) like wind and solar photovoltaic (PV) has increased substantially in the energy mix of many developed and developing economies. The glimpses of evidence of the changing energy mix can be had from the present share of VRES in countries like Germany, Denmark, Spain, China, India. At present, the share of VRES in Germany, Denmark, Spain, China and India is 45, 47, 28, 19, and 18% respectively [1–5]. With this changing energy mix, the reliability of the grid comes at its fore. This intermittency of wind and solar PV will also increase the procurement costs of various reliability services. This variability due to intermittency in the generation of wind energy and solar PV can be addressed using storage technologies. These variations can be smoothened using

A. Nazar (✉)
Research Scholar, Department of Energy and Environment, TERI School of Advanced Studies, New Delhi 110070, India
e-mail: asif.nazar2@terisas.ac.in

N. Anwer
Associate Professor, Department of Energy and Environment, TERI School of Advanced Studies, New Delhi 110070, India
e-mail: naqui.anwer@terisas.ac.in

A. Tomar et al. (eds.), *Machine Learning, Advances in Computing, Renewable Energy and Communication*, Lecture Notes in Electrical Engineering 768,
https://doi.org/10.1007/978-981-16-2354-7_28

storage technologies [6]. The economic viability of electrical energy storage technology is dependent upon their participation in various energy market services. Electrical Energy Storage Technologies, Ancillary Services, Energy market and Regulatory Framework form an ecosystem and are interdependent [7]. Revenues for battery energy storage systems may come from participation in services like energy arbitrage, regulation and a combination of energy arbitrage and regulation. Purchasing power when demand is less, storing it, and selling power when demand is high depicts energy arbitrage opportunity. Frequency regulation, an ancillary service, is used for maintaining frequency stability. Further, a battery energy storage can avail both approaches of revenue generation—energy arbitrage and frequency regulation.

The estimation of potential revenues for energy arbitrage and frequency regulation has been dealt with [6, 8, 9]. The estimation for potential revenue from energy arbitrage services in New England ISO Power Market has been dealt with [6]. The estimation of maximum potential revenue in CAISO for 2010–11 has been discussed in [8]. The assessment of potential revenue from battery storage in PJM has been explored in [9].

The estimation of potential revenue from participation in energy arbitrage in Indian Electricity Exchange (IEX), subject to Regulatory policies, has been done for battery energy storage. The estimated potential revenue from IEX for the same period has been compared with NYISO. Further, potential revenue from participation in both energy arbitrage services and frequency regulation services have been estimated for NYISO to derive the point for policymakers, as combined estimated revenues are appreciably and significantly higher than participation in only energy arbitrage services. Moreover, some AI and machine learning-based examples are listed in [10–16].

The paper has been arranged in this way. Section 2 portrays the methods. Section 3 discusses results for a hypothetical 4 MW/32 MWh battery energy system. Section 4 is the conclusion.

2 Methods

2.1 Battery Storage Model for Energy Arbitrage

State of charge (SOC) of a storage battery at time t for energy arbitrage service is the function of SOC of battery at a time prior to t, power recharged for Δt interval and power discharged for Δt interval.

Mathematically, SOC of a battery is stated as:

$$SOC(t) = (1 - d)SOC(t - \Delta t) + [\eta P_R(t) - P_D(t)]\Delta t \tag{1}$$

where, d is the self-discharge of the battery, η is the roundtrip efficiency of the storage battery, $P_R(t)$ is the power recharged at time t, $P_D(t)$ is the power discharged at time t and Δt is the time period between time t and a time prior to t.

2.2 Battery Storage Model for Energy Arbitrage and Frequency Regulation

SOC of a battery at time t for both arbitrage and regulation is the function of SOC of battery at a time prior to t, power recharged for Δt interval, power discharged for Δt interval and power during regulation (regulation up power and regulation down power) for Δt interval.

SOC at time t for storage battery taking part in both energy arbitrage and frequency regulation is represented as:

$$SOC(t) = (1 - d)SOC(t - \Delta t) + [\eta P_R(t) - P_D(t)]\Delta t + \left[\eta \gamma_{rd} P_{RD}(t) - \gamma_{ru} P_{RU}(t)\right]\Delta t \quad (2)$$

where, γ_{rd} and γ_{ru} are the dispatch to contract ratio, and P_{RD} and P_{RU} are power during regulation up and regulation down respectively.

2.3 Optimization Model for Revenue from Energy Arbitrage

Linear program for optimization consists of objective function and constraints.

The objective function of linear program for optimization of revenue from arbitrage is given by:

$$\text{Maximize} \sum_{t=1}^{T} \left[\left(E_{price}(t)(P_D(t)\Delta t)\right) - \left(E_{price}(t)(P_R(t)\Delta t)\right)\right] \quad (3)$$

i.e. selling power when demand is high and purchasing power when demand is low.

Subject to the following constraints:

$$0 \leq SOC(t) \leq SOC_{\max} \quad (4)$$

i.e. state of charge should vary between 0 and rated capacity of the storage battery.

$$0 \leq (P_D(t)\Delta t) \leq (P_{D\max}\Delta t) \quad (5)$$

i.e. discharging power for interval Δt should vary between 0 and rated storage battery power for discharging for Δt interval.

$$0 \leq (P_R(t)\Delta t) \leq (P_{R\max}\Delta t) \tag{6}$$

i.e. recharging power for interval Δt should vary between 0 and rated storage battery power for recharging for Δt interval.

Where, Eprice(t) is the price of electricity at time t, SOC_{max} is the rated capacity of the storage battery, P_{Dmax} and P_{Rmax} are the rated storage battery power for discharging and recharging.

2.4 *Optimization Model for Revenue from Both Arbitrage and Regulation*

The Objective Function of a linear program for optimization of revenue from arbitrage and regulation both is given by:

$$\begin{aligned} \text{Maximize} \sum_{t=1}^{T} [(E_{price}(t)(P_D(t)\Delta t)) - E_{price}(t)(P_R(t)\Delta t)) + (E_{\text{Reg}priceup}(t) \\ + \gamma_{ru} E_{price}(t))(P_{RU}(t)\Delta t) + (E_{pricedown}(t) - \gamma_{rd} E_{price}(t))(P_{RD}(t)\Delta t)]. \end{aligned} \tag{7}$$

The first two terms of the objective function (7) denote energy price arbitrage, i.e. battery will be discharged or sold at that time slot when the electricity price is high and will be recharged or purchased when the electricity price is low. The third and the fourth terms are revenues from frequency regulation up service and frequency regulation down service. The revenue for frequency regulation service consists of capacity payment and energy payment. Energy payment for a resource (battery storage here) is for the period resource is deployed, and it is taken care by dispatch to contract ratio denoted by γ_{rd} and γ_{ru} for regulation down and regulation up services respectively.

Subject to the following constraints:

$$0 \leq SOC(t) \leq SOC_{\max} \tag{8}$$

i.e. state of charge should vary between 0 and rated capacity of storage battery.

$$0 \leq ((P_D(t) + P_{RD}(t))\Delta t \leq (P_{D\max}\Delta t) \tag{9}$$

i.e. the sum of discharging power and power during regulation down for interval Δt should vary between 0 and rated storage battery power for discharging for Δt interval.

$$0 \le (P_R(t) + P_{RU}(t))\Delta t \le P_{R\max}\Delta t \tag{10}$$

i.e. the sum of recharging power and power during regulation up for interval Δt should vary between 0 and rated storage battery power for discharging for Δt interval.

Where, $E_{Regpriceup}(t)$ and $E_{Regpricedown}(t)$ is the electricity price for regulation up and regulation down services.

2.5 Electricity Price Data

The day ahead hourly market clearing price for energy has been taken from Indian Electricity Exchange (IEX) for the month of July 2019 [17]. The hourly average energy price (Rs/MWh) for the month of July 2019 for Indian Energy Exchange (IEX) has been plotted in Fig. 1.

The day ahead hourly market energy price has been taken for Millwood zone in NYISO for the month of July 2019 [18]. The average hourly energy price ($/MWh) for the month of July 2019 for NYISO has been plotted in Fig. 2. The day ahead hourly market regulation price has been taken for the month of July 2019 for Millwood zone in NYISO [19]. The average hourly regulation price ($/MWh) for the month of July 2019 for NYISO has been plotted in Fig. 3.

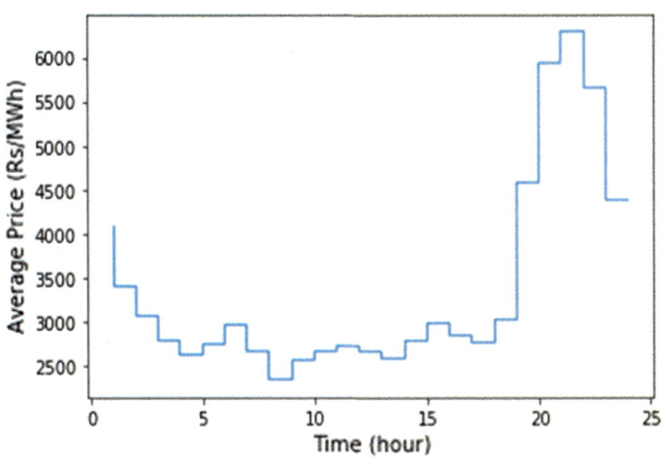

Fig. 1 An hourly average day ahead energy price (Rs/MWh) for the month of July 2019 for IEX, India

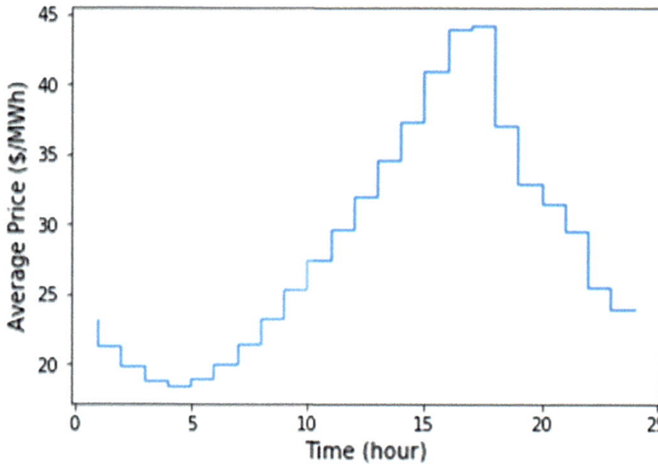

Fig. 2 An hourly average day ahead energy price for the month of July 2019 for NYISO

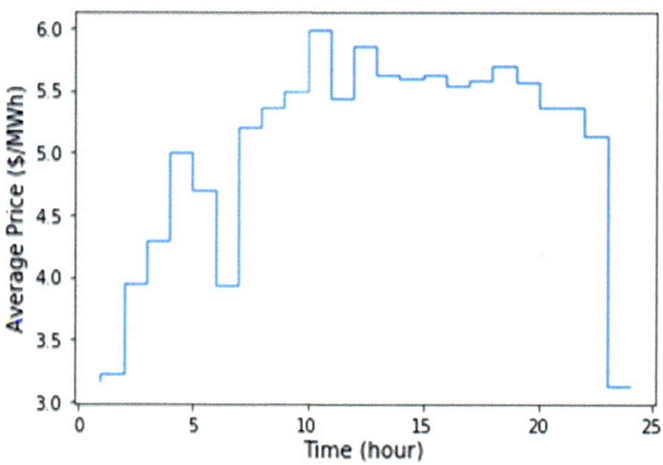

Fig. 3 An hourly average day ahead regulation price for the month of July 2019 for NYISO

2.6 Estimating Dispatch to Contract Ratio

For estimating the value of dispatch to contract ratio, γ_{rd} and γ_{ru}, for regulation down and regulation up services, frequency regulation signal of PJM market has been used as a proxy. γ_{rd} and γ_{ru} show the actual deployment of the reserved capacity of resource (here storage battery). Figure 4 shows the frequency signal versus time plot for an hour for July 2019 [20]. The value of dispatch to contract ratio γ_{ru} has been obtained by the piece-wise integration of area above time axis, and this value comes to be 0.32. The value of dispatch to contract ratio γ_{rd} has been obtained by the piecewise

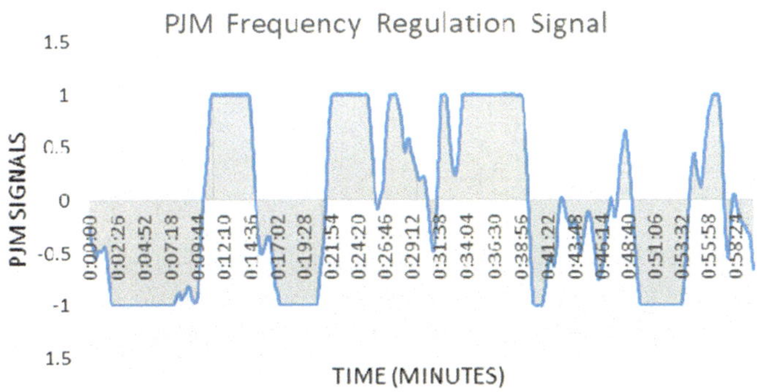

Fig. 4 Plot between regulation signal and time (minutes)

integration of area below time axis, and this value comes to be 0.46. Figure 5 shows the hourly values of γ_{rd} and γ_{ru}. The values have been calculated by the piecewise integration of hourly frequency signal for every hour. This has been done for 24 h. Average hourly deployment of regulation up service is 0.28, whereas for regulation down deployment this value is 0.31.

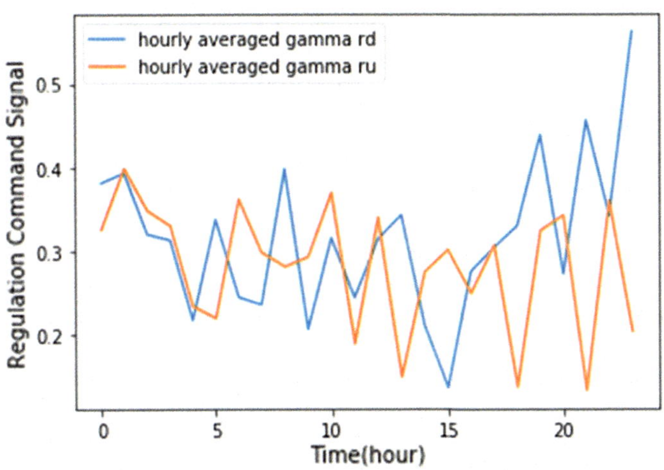

Fig. 5 Plot between regulation signal and time (hour) for 24 h showing hourly averaged γ_{rd} and γ_{ru}

Table 1 Storage battery parameter (NaS 4 MW/32 MWh)

Parameter	Value
$P_{Dmax}\ \Delta t$	4 MWh
$P_{Rmax}\ \Delta t$	4 MWh
SOC_{max}	32 MWh
η	0.80
γ_{ru}	0.28
γ_{rd}	0.31

2.7 Battery Storage Parameter

To illustrate revenue from energy arbitrage service and combination of energy arbitrage and regulation, a hypothetical case of NaS battery of 4 MW/32 MWh has been taken. The battery storage parameter has been shown in Table 1.

3 Results

The optimization of revenue for arbitrage for IEX and Millwood zone in NYISO for NaS battery of 4 MW/32 MWh has been done by generalized algebraic modeling system (GAMS) coding. Decision variables in GAMS coding for estimating arbitrage potential (objective function given by Eq. 3, constraints represented by Eqs. 4–6, and State of Charge given by Eq. 1) are: $P_d(t)$, $P_r(t)$ The result of GAMS coding for linear program optimization has estimated energy arbitrage revenue potential for the month of July 2019 in Indian Electricity Exchange, and the revenue is $20,863 (taking exchange rate $1 = Rs 75). Likewise estimated revenue potential from arbitrage service in Millwood zone in NYISO for the month of July 2019 is $14,849.

Decision variables in GAMS coding for estimating combined arbitrage and regulation potential (objective function given by Eq. 7, constraints represented by Eqs. 8–10, and state of charge by Eq. 2) are: $P_d(t)$, $P_r(t)$, $P_{rd}(t)$ and $P_{ru}(t)$. The combined revenue potential for energy arbitrage service and regulation service for Millwood zone in NYISO is $29,977. This combined service revenue for Millwood zone in NYISO is double of energy arbitrage only revenue. These revenues have been summarized in Tables 2 and 3, respectively.

Table 2 Revenue from arbitrage only

	Revenue IEX (India)	Revenue NYISO
Arbirage only	$20,863 (taking $1 = Rs 75)	$14,849

Table 3 Revenue from arbitrage and regulation both in NYISO

	Revenue NYISO	Remarks
Arbirage and regulation combined	$29,977	Revenue doubles combining both services

4 Conclusion

The economic viability of battery energy storage systems is linked to their participation in the energy market, and revenue generated out of it. Two main sources of revenue for BESS are its participation in energy arbitrage service and regulation service. The revenue generated from the energy arbitrage service in Indian Electricity Exchange is substantial, and it is more than Millwood zone in NYISO. Further revenue generated from combined services of arbitrage and regulation in Millwood zone in NYISO is around double than energy arbitrage service. So regulators and policymakers may think in terms of providing access to BESS for both arbitrage and regulation services for the economic viability of battery energy storage systems.

References

1. Federal Ministry for Economic Affairs and Energy (2020) Electricity market of the future. https://www.bmwi.de/Redaktion/EN/Dossier/electricity-market-of-the-future.html
2. Danish Energy Agency (2018) Energy statistics 2018. https://ens.dk/sites/ens.dk/files/Statistik/energy_statistics_2018.pdf
3. RED Electrica de Espana (2019) The Spanish Electricity System: preliminary report 2018. https://www.ree.es/sites/default/files/11_PUBLICACIONES/Documentos/InformesSistemaElectrico/2019/Avance_ISE_2018_en.pdf
4. Reuters. China's 2018 renewable power capacity up 12 percent on year 2019. https://www.reuters.com/article/us-china-renewables/chinas-2018-power-capacity-up-12-percent-on-year-idUSKCN1PM0HM. Accessed 28 Jan 2019
5. CEA (Central Electricity Authority) (2018) All India installed capacity of power stations as on 31.12.2018. http://www.cea.nic.in/reports/monthly/installedcapacity/2018/installed_capacity-12.pdf
6. Mokrian A, Stephen M (2006) A stochastic programming framework for the valuation of electricity storage. In: Proceedings of the 26th USAEE/IAEE North American conference, Michigan
7. Nazar A, Anwer N (2020) Accommodative energy market for battery energy storage and grid balancing. In: International conference on emerging frontiers in electrical and electronic technologies, Patna
8. Byrne RH, Silva-Monroy CA (2012) Estimating the maximum potential revenue for grid connected electricity storage: arbitrage and regulation. Sandia report, SAND 2012-3863
9. Byrne RH, Concepcion RJ, Silva-Monroy CA (2016) Estimating potential revenue from electrical energy storage in PJM. Sandia National Laboratories
10. Yadav AK et al (2020) Soft computing in condition monitoring and diagnostics of electrical and mechanical systems. Springer Nature, Berlin, 496 pp. https://doi.org/10.1007/978-981-15-1532-3.ISBN 978-981-15-1532-3

11. Fatema N et al (2021) Intelligent data-analytics for condition monitoring: smart grid applications. Elsevier, 268 pp. https://www.sciencedirect.com/book/9780323855105/intelligent-data-analytics-for-condition-monitoring. ISBN 978-0-323-85511-2
12. Smriti S et al (2018) Special issue on intelligent tools and techniques for signals, machines and automation. J Intell Fuzzy Syst 35(5):4895–4899. https://doi.org/10.3233/JIFS-169773
13. Jafar A et al (2021) AI and machine learning paradigms for health monitoring system: intelligent data analytics. Springer Nature, Berlin, 496 pp. https://doi.org/10.1007/978-981-33-4412-9. ISBN 978-981-33-4412-9
14. Sood YR et al (2019) Applications of artificial intelligence techniques in engineering, vol 1. Springer Nature, 643 pp. https://doi.org/10.1007/978-981-13-1819-1. ISBN 978-981-13-1819-1
15. Gopal et al (2021) Digital transformation through advances in artificial intelligence and machine learning. J Intell Fuzzy Syst (Pre-Press) 1–8. https://doi.org/10.3233/JIFS-189787
16. Aggarwal S et al (2020) Meta heuristic and evolutionary computation: algorithms and applications. Springer Nature, Berlin, 949 pp. https://doi.org/10.1007/978-981-15-7571-6.ISBN 978-981-15-7571-6
17. Indian Electricity Exchange, Area Prices. https://www.iexindia.com/marketdata/areaprice.aspx
18. NYISO, LBMP-Zonal. http://mis.nyiso.com/public/P-2Alist.htm
19. NYISO, Ancillary Service Prices-Day-Ahead. http://mis.nyiso.com/public/P-5list.htm
20. PJM, Ancillary Services, RTO Signal Data for 7.19.2019 and 7.20.2019. https://www.pjm.com/markets-and-operations/ancillary-services.aspx

A Technique to Detect Fake News Using Machine Learning

Pritee Yadav and Muzammil Hasan

Abstract The impacts of data spread happen at such a quick pace on social networks and so enhanced that that distorted, mistaken data or false data gets an enormous potential to cause certifiable effects. We examine the fake news issue, and the present specialized concerns related to the need for the robustness of automated fake news detection frameworks, and also examine the opportunity that is appropriately developed. In this paper, we have proposed a systematic framework for fake news identification and gives it's the confusion matrix with a classification report of the accuracy of their perspective model. The proposed approach with comparisons accomplishes classification accuracy 94% and 97% recall. We collect the dataset of real and fake news that is changed from document-based corpus into an event and title-based representation.

Keywords Context analysis · Detection · Fake news · Assessment · Content analysis

1 Introduction

Data veracity is an issue influencing society for advanced media. The sensationalism of not precise eye-getting and features that are intriguingly planned for holding the consideration of audiences to sell data has persevered all since the commencement of a wide range of data communications.

As reported, youths are to be technically knowledgeable when contrasted with their parents [1], but as it may, with respect to the capacity to tell if a news piece is phony or not, they show up as bewildered as the overall public's leftover portion and the examination directed by the Common Sense Media have 44% affirmed of the assessment. Comparative research similarly shows that 31% of kids developed 10–18 have shared online, at any rate, one report that they later found was phony or off base. The present situation expands an altogether unique component of concerns

P. Yadav (✉) · M. Hasan
Department of Computer Science and Engineering, Madan Mohan Malaviya University of Technology, Gorakhpur, India

A. Tomar et al. (eds.), *Machine Learning, Advances in Computing, Renewable Energy and Communication*, Lecture Notes in Electrical Engineering 768,
https://doi.org/10.1007/978-981-16-2354-7_29

recognized with computerized education that goes past the simple capacity to get to and oversee innovation.

Along with cultural difficulties, there is an inconspicuous and dramatic situation happening in the media scene, news-casting industry, and the public sphere that requires assessment and discussion, bringing up two fundamental perspectives. The first relies upon the way that news distributors have lost order over the news circulation, which is introduced to users on social media by algorithms and unpredictable. Also, the news advertises beginners, (for example, Vox, Fusion, and Buzz Feed) have built their quality by grasping these techniques, sabotaging the drawn-out positions involved by distributors of traditional news increasingly. The subsequent viewpoint depends on the expanding power that online networking media, e.g. Amazon, Facebook, Google and, Apple, have picked up in controlling that how the distributions are adapted and who distributes what to whom.

In the above context, building up the unwavering quality of online data daunting but basic current test [2], requesting the guideline, attention, and dynamic observing of computerized content spread by the significant gatherings including web indexes and person to person communication stages, in supporting how the data is introduced and shared among individuals over the Internet. For the Commons Culture, The subject "Fake news" has become so predominant. Sport and Media Committee is right now examining worries about the general population being influenced by untruths and propaganda [3]. The prediction for the odds is defined as detection of fake news of a specific article of news (editorial, expose, news report, etc.) to be deliberately deceiving [2, 4].

2 Literature Review

Fake news detection: Fake news is made by creating non-existent news or altering genuine news. The fake news credibility is supported by (1). Writing styles or well-known authors who are imitating. (2). Communicating assessments with a tone, in many cases, used in genuine news. As of late, an expanding number of strategies for detection have been created. All current identification schemes may be collected into two unmistakable classes, to be specific, network-based methods, and linguistic-based methods [5]. Network-based methodologies for the detection of fake news apply to organize properties as a supporting segment for different approaches that are linguistic-based analysis. Usually utilized network properties that include, yet not restricted to, site data, creators/supporters information, like, and time stamps, such as for decreasing the falsehood by performing the user analysis in online social media gathering identified with Parkinson's sickness. This report analysis that deception embedded inside the conversation string relies upon its substance and user's qualities of the creator. [6] Another examination proposes a model that centers on exploring the crowd-sourced health, the thread question clearness, and potential of the clients for making valuable commitments of the nature of the response in an online source.

Syntax analysis and the existing assumption are redone for exceptional information types, in this manner being insufficient for the system for identifying the fake news.

A CNT model that is rumor detection model [7] receives a features assortment, for example, content-based features (part of speech, segments, and word appearance), features of network-based approach (i.e., tweets propagation or re-tweets), twitter-specific Memes (for example, shared URLs or Hashtag), CNT coordinates a variety of methodologies to choose features to identify falsehood in microblogs [8]. Another examination [9] deep-network- models—CNN for news veracity investigation. Sometimes phony-article shows outrageous conduct for an ideological group.

Knowledge-Based-Detection It is the clearest method for distinguishing and check the truthfulness of significant cases in the article to choose the veracity of the news. This approach plans to utilize outside sources for checking the proposed claims in news content. [10, 11], whereas this approach is a key segment of fake news detection as far as the given viewpoints. Information diagram is utilized to check whether the cases can be deduced from realities, which are existing in the chart or not [12–14].

Visual-Based Detection—Digitally adjusted pictures are wherever flowing a wildfire on social media. These days' applications (likewise Photoshop) are being utilized uninhibitedly for modifying pictures satisfactorily to trick individuals into speculation they are seeing the genuine image. The media criminology field has created a significant number of strategies for tampering detection in videos [15].

Wang (2017), Pros: In this paper, the fake news content is classifying by the CNN method. CNN is additionally used for analyzing the assortment of meta-information. Standard content classifiers, for example, SVM models got significant improvement. *Cons*: Bi-LSTMs did not perform well cause of overfitting [16].

Shu et al. [17], *Pros*: This framework could be implemented for detecting the fake news in real time. *Cons:* This framework using only social engagement information is defeated by using only textual data [18] (Fig. 1).

3 Research Methodology

As shown in figure [19], we initially present the framework of methodology that is used for detecting the fake news and this framework presents the training-method's consistent integration and, testing method and includes these modules: Collection of training data, the pre-processing of data, using TF-IDF and count features trained algorithm and model training and results. Pre-processing of data module coordinates a variety of content procedure strategies to extract topics and from news datasets that are collected from sites that are for the datasets.

Text preparation. The data are highly unstructured of social media—the larger part of them are casual correspondence with grammatical errors, slangs, and typos, etc. For predictive modeling important to clean the information prior to utilizing it. For this reason, fundamental pre-processing was done on the News preparing information.

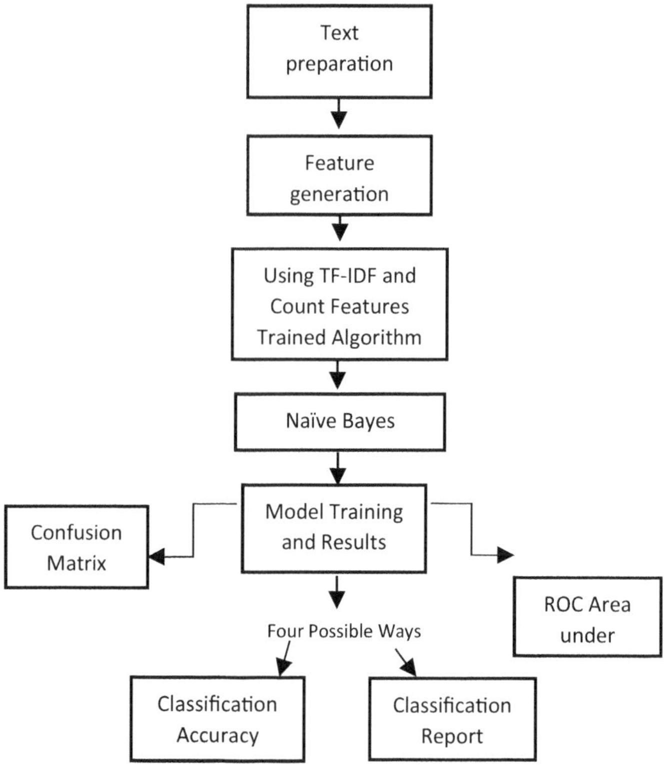

Fig. 1 Flow diagram of the fake news detection method

Feature generation for generating many features we can use the text data such as word count, unique word frequency, large word frequency, n-grams, etc. The words capture their meanings by creating a representation of the words, semantic relationships, they are used in various sorts of context.

TF-IDF vectors (feature): The document also in the corpus for the general significance of a term is represented by the TF-IDF weight.

(a) TF (term frequency): A term may show up additional the short document in a long size document. Thus, the term frequency is divided by the length of the document frequently.

$$TF(t, d) = \frac{no.\ of\ time\ (t)\ occurs\ in\ an\ individual\ document\ (d)}{Total\ word\ count\ of\ d}$$

(b) IDF (inverse document frequency): Certain terms such as "on", "an", "a", "the", "of", etc. seem commonly in the document, however, are of small significance. IDF weighs down the significance of uncommon ones, the significance of these terms and it increments. The more the estimation of IDF, the more one of a kind is the word.

$$IDF\ (t) = \log\ e\left(\frac{Total\ no.\ of\ d\ document}{no.\ of\ document\ with\ t\ (term)\ in\ it}\right)$$

(c) TF-IDF: By assigning them less weightage, it is working by punishing the most-normally words that occurs while putting weightage to terms. It has a high occurrence in a specific document.

$$TFIDF(t, d) = TF(t, d) * IDF(t).$$

TF-IDF is a generally utilized feature for content classification (text Classification). Also,

TF-IDF vectors can be determined at various levels have used, for example, N-gram and Word level.

4 Algorithm

This area manages training the classifier. We used explicitly three algorithms of machine learning, which are Naive Bayes Algorithm, Logistic Regression Algorithm, and Passive-Aggressive Algorithm. For detailed information on the algorithm implementation, reader may refer to the example available in [20–25].

Naive Bayes: This is one of the classification techniques; Naive Bayes Algorithm is based on the Bayesian theorem. It is a set of the supervised learning algorithm, expects that the presence of a specific segment in a class is autonomous of the presence of some other element and gives the way for the posterior-probability calculation.

$$P\left(y|x_{1,\ldots}x_n\right) = \frac{P(y)P\left(x_{1,\ldots}x_n|y\right)}{P(x_{1,\ldots}x_n)}$$

$P\left(y|x_{1,\ldots}x_n\right)$= likelihood (posterior probability) of class given predicator.
$P\left(x_{1,\ldots}x_n|y\right)$= likelihood (probability) of predictor given class.
$P(y)$= prior-probability of class.
$P\left(x_{1,\ldots}x_n\right)$ = prior-probability of predicator.
The independence assumption by using the Naive condition

$$P(x_i|y, x_1, \ldots, x_{i-1}, x_{i+1}, \ldots x_n) = P(x_i|y)$$

The simplified relationship for all i

$$P(y|x_1, \ldots, x_n) = \frac{P(y)\prod_{i=1}^{n} P(x_i|y)}{P(x_1, \ldots, x_n)}$$

Since $P(x_1, \ldots x_n)$ is consistent given the value (input), we use the classification rule which is given:

$$P(y|x_1, \ldots x_n) \propto P(y) \prod_{i=1}^{n} P(x_i|y)$$

$$\Downarrow$$

$$y = \arg maxP(y) \prod_{i=1}^{n} P(x_i|y)$$

also, for estimation, we can use MAP (Maximum A Posteriori) $P(y)$ to $P(x_i|y)$; in the training set, the previous is then the frequency (relative frequency) of class.

- Multinomial NB: The multinomial NB is to implement the Naive Bayes algorithm for multinomial conveyed information (distribution of the data), in-text classification, these classic algorithms Naive Bayes and multinomial variants used for text classification. Where the information is commonly represented as word vector count, even though that tf-idf vectors.
- Passive-Aggressive Classifier: The algorithm is perfect for grouping huge massive of data (e.g. Facebook, Twitter). The Passive-Aggressive Algorithm is easy to implement and very fast.
- Logistic Regression: Logistic Regression is used for the prediction of the event occurrence (true/false, 1/0, yes/no). It is a sigmoid function used for probability estimation. It is the classification algorithm.

5 Evaluation Metrics for Accessing the Performance

In this area, have analyzed probably the significant metrics measurements by which the performance is estimated of ML model. These measurements quantify how well model can evaluate predictions or classify. In this project, the introduction of metrics was used.

Classification Accuracy: It is characterized as the quantity of right predication as against the quantity of all-out prediction. In any case, this metric alone can not give enough data to choose whether the model is a decent one or not. *Confusion Matrix:* Confusion Matrix shows the model performance also called an error matrix that is a type of Contingency table. The table has two dimensions: label "predicted," on the y-axis and label "actual" on the x-axis. The quantity of predictions that are the cells of the table is made by the algorithm.

Classification Accuracy: When working on classification issues, the Scikit-learn gives a report also call classification report, which supports each class that has the outcomes Precision, Recall, and f-score.

$$ClassificationAccuracy = \frac{True\ Positive + True\ Nagative}{True\ Positive + True\ Nagative + False\ Positive + False\ Nagative}.$$

Precision: Precision is the ratio of correctly predicted to total positive instances which have been also predicted. High accuracy (Precision) implies low False Positive rate.

$$Precision = \frac{TP}{TP + FP}$$

Recall: It is the positive instance ratio that is accurately predicted to all cases in genuine class—Yes.

$$Recall = \frac{TP}{TP + FN}$$

F-score: F-score is the weighted average of both Precision and Recall. Their consideration is the combination of both FP and FN. It is generally more helpful than the accuracy, particularly for lopsided class dissemination. In the event that FP and FN having the same cost or instances then accuracy performs best.

$$F1 = 2 * \frac{Precision * Recall}{Precision + Recall}$$

If their instances are different widely, at that point, it is better to look both Recall and Precision.

Tf-Idf vector and Count Vector are used for classifying the responses at two levels that are N-gram level and Word level. N-gram level: The range of 1–3 words will be taken. Word level: The range of single words will be taken. Words will be considered as a token (bigram, trigram). The Word level classification accuracy is better than the performance of the N-gram level. At the N-gram level while passive aggressive classifier stochastic gradient descent, using Tf-Idf vectors performance at both is better above 90% accuracy. Since, the accuracy of classification alone is not sufficient to decide the viability of the model, for these algorithms using Tf-Idf Vectors at word level other metrics explored.

6 Experimental Result

We are using Vector features Tf-Idf vectors and Count Vectors and by using algorithms at word level. Accuracy will be noted for the model. We applied content (text) classification on the body of the articles in the distinctive freely accessible datasets of UCI Machine Learning.

Confusion Matrix:

- *Linear-TFIDF:* Fig. 2. *True Positive = 56, False Positive = 80, True Negative = 1003, False Negative = 952*
- *Multinomial NB-TFIDF:* Fig. 3. *True Positive = 48, False Positive = 76, True Negative = 1007, False Negative = 960,*
- *PAC-TFIDF*: Fig. 4. *True Positive = 52, False Positive = 82, True Negative = 1001, False Negative = 956* (Fig. 5).

Classification Report: Linear-TFIDF (Fig. 6).
PAC-TFIDF (Fig. 7).
Multinomial NB-TFIDF (Fig. 8).

Total instances		Predicted class		Description
		Yes	No	
Actual class	Yes	True-Positive	False-Positive	TP: It is positive values, correctly predicted FP: Negative observation, but predicted positive. TN: It is negative values, correctly predicted FN: Positive observation, but predicted negative.
	No	False-Negative	True-Negative	

Fig. 2 Confusion matrix

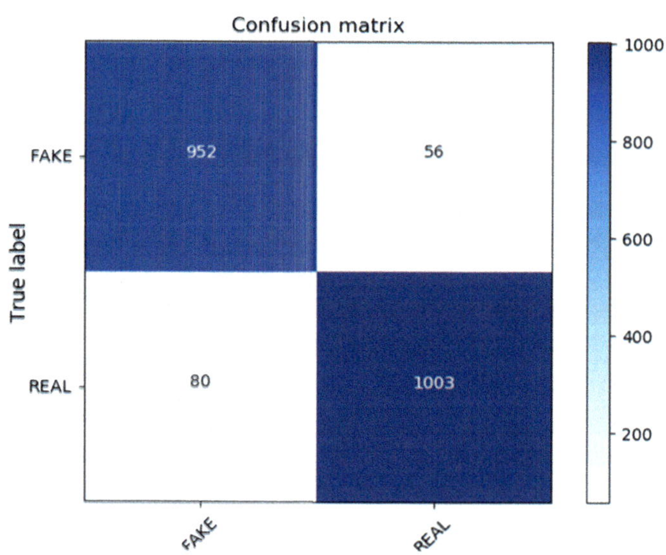

Fig. 3 Linear-TFIDF Confusion Matrix

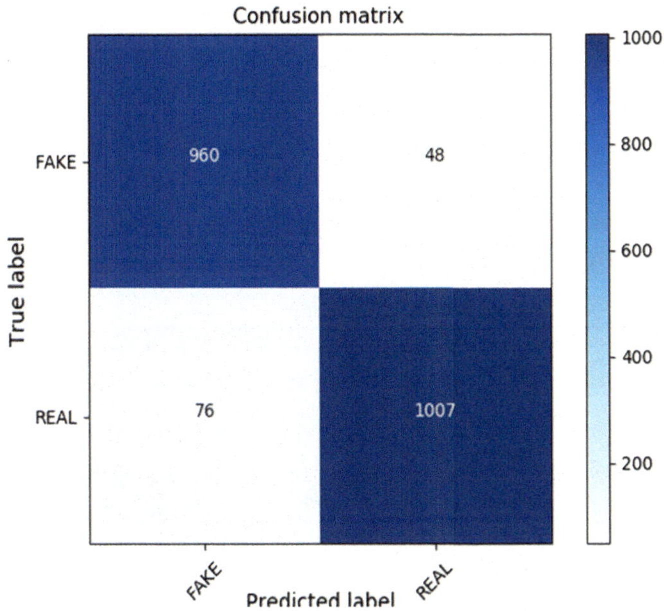

Fig. 4 Multinomial NB-TFIDF Confusion Matrix

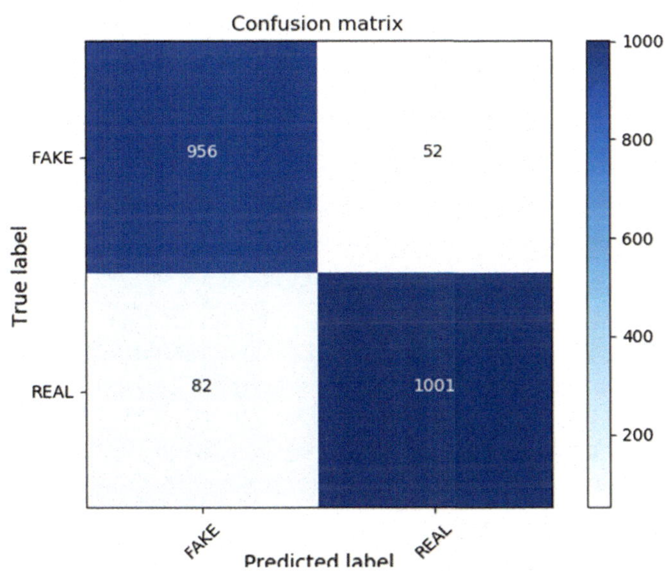

Fig. 5 PAC-TFIDF confusion matrix

	precision	recall	f1-score	support
FAKE	0.96	0.73	0.83	1008
REAL	0.80	0.97	0.88	1083
avg / total	0.88	0.86	0.85	2091

Fig. 6 Linear-TFIDF classification report

	precision	recall	f1-score	support
FAKE	0.92	0.86	0.89	1008
REAL	0.88	0.93	0.90	1083
avg / total	0.89	0.89	0.89	2091

Fig. 7 PAC-TFIDF classification report

	precision	recall	f1-score	support
FAKE	0.92	0.95	0.93	1008
REAL	0.95	0.92	0.94	1083
avg / total	0.94	0.94	0.94	2091

Fig. 8 Multinomial NB-TFIDF classification report

Precision value of linear-TFIDF is 96% and is higher than other models have precision value 92%. Multinomial NB-TFIDF model has the avg/total is 94% that is higher than another model. For fake news detection, the sensitivity advises how sensitive is the classifier, while in the genuine news specificity tells how specific or selective the model is for prediction. Along these, sensitivity ought to be higher, because FP is more adequate than FN in classification issues of such applications.

7 Conclusion and Future Scope

The misinformation spread by counterfeit news presents a genuine hazard for its target consumers, which could be enterprises and also individuals. While an individual expending the fake news creates misinterpreted or distorted the real-world perception, which impacts their decision-making and beliefs, enterprises experience the effects of fake news because of loss of damaging impact or competitive advantage on their image. In this paper, the proposed work for fake news identification and gives it's the confusion matrix with a classification report of the accuracy of their perspective model. The sensitivity is high for both the models and is having comparable to esteem. By greater sensitivity streamlining, we can give indications of progress results. We can increase the classifier's sensitivity by a decrement of the threshold for the prediction of fake news. So it would help for incrementing the TP (true positive) rate in this work. This analysis also required the improvement and the complete repository of fake and real news, which might be used for future analysis in this development of the significant area of research.

References

1. Anderson J (2017) Even social media-savvy teens can't spot a fake news story
2. Conroy N, Rubin V, Chen Y (2015) Automatic deception detection: methods for finding fake news. Proc Assoc Inf Sci Technol 52(1):1–4
3. BBC N (2016) The rise and rise of fake news. BBC Trending. http://www.bbc.com/news/blogs-trending-37846860
4. Rubin V, Conroy N, Chen Y (2015) Towards news verification: deception detection methods for news discourse
5. Yadav P, Hasan M (2020) A review: issues and challenges in various fake news detection mechanism. In: Proceedings of international conference on advances in computational technologies in science and engineering ACTSE
6. Conroy NJ, Victoria LR, Yimin C (2015) Automatic deception detection: methods for finding fake news. Proc Assoc Inf Sci Technol 52(1)
7. Qazvinian V et al (2011) Rumor has it: identifying misinformation in microblogs. In: Proceedings of the conference on empirical methods in natural language processing. Association for Computational Linguistics
8. Aggarwal A, Kumar M (2020) Image surface texture analysis and classification using deep learning. Multimed Tools Appl
9. Aggarwal A et al (2020) Landslide-data-analysis using various time-series forecasting-models. Comput Electr Eng 88
10. Banko M, Cafarella MJ, Soder-land S, Broadhead M, Oren Etzioni O Open information extraction from the web
11. Magdy A, Wanas N (2010) Web-based, istatistical, fact ichecking, of, itextual documents. In: Proceedings of the 2nd international workshop on Search and mining user-generated contents
12. Ciampaglia GL, Shiralkar P, Rocha LM, Bollen J, Menczer F, Flammini A (2015) Computational fact checking from knowledge networks. PloS one
13. Wu Y, Agarwal PK, Li C, Yang J, Yu C (2014) Toward computational fact-checking. In: Proceedings of the VLDB endowment
14. Shi B, Weninger T Fact checking in heterogeneous-information-networks. In: WWW'16
15. Kumar M, Srivastava S (2019) Image authentication by assessing manipulations using illumination. Multimed Tools Appl 78(9):12451–12463
16. Shu K, Wang S, Liu H (2017) Exploiting tri-relationship for fake news detection (2017). arXiv: 1712.07709
17. Shu K, Wang S, Liu H (2017) Exploitingitri-relationshipifor fake inews idetection
18. Shu K, Wang S, Tang J, Zafarani R, Liu H (2017) User identity linkage across online social networks: a review. ACMSIGKDD Explorations Newsletter 18(2)

19. Ahmad I, Yousaf M, Yousaf S, Ahmad MO (2020) Fake news detection using machine learning ensemble methods. Complexity 2020:1–11. Article ID 8885861
20. Aggarwal S et al (2020) Meta heuristic and evolutionary computation: algorithms and applications. Springer Nature, Berlin, 949 p https://doi.org/10.1007/978-981-15-7571-6. ISBN 978-981-15-7571-6)
21. Yadav AK et al (2020) Soft computing in condition monitoring and diagnostics of electrical and mechanical systems. Springer Nature, Berlin, 496 p https://doi.org/10.1007/978-981-15-1532-3. ISBN 978-981-15-1532-3
22. Gopal et al (2021) Digital transformation through advances in artificial intelligence and machine learning. J Intell Fuzzy Syst, Pre-press, 1–8. https://doi.org/10.3233/JIFS-189787
23. Smriti S et al (2018) Special issue on intelligent tools and techniques for signals, machines and automation. J Intell Fuzzy Syst 35(5):4895–4899. https://doi.org/10.3233/JIFS-169773
24. Jafar A et al (2021) AI and machine learning paradigms for health monitoring system: intelligent data analytics. Springer Nature, Berlin, 496 p https://doi.org/10.1007/978-981-33-4412-9. ISBN 978-981-33-4412-9
25. Sood YR et al (2019) Applications of artificial intelligence techniques in engineering, vol 1, Springer Nature, 643 p https://doi.org/10.1007/978-981-13-1819-1. ISBN 978-981-13-1819-1

Efficiency Improvement for Regenerative Braking System for a Vehicular Model Using Supercapacitor

Sangeeta Singh, Kushagra Pani Tiwari, Ananya Shahi, Marut Nandan Singh, and Shivam Tripathi

Abstract This hybrid model, developed in Simscape, is to maximize the use of electric vehicles where each component of an HEV is developed keeping in mind the basic principle of approach in a commutable trend, authenticated by available experimental data. To achieve it, a hybrid energy storage system is used, which will increase the efficiency of regenerative braking to convert the heat energy lost in braking to electric energy. Battery with supercapacitors stores this energy and provides it to the motor during acceleration thereby increasing the efficiency of battery and also the life of vehicle. Simulations with several drive cycles are used to observe the vehicle behavior under different conditions. The result indicates the authenticity of the model for realistic prediction of the vehicle's efficiency as well as the battery life.

Keywords Energy storage system · H-bridge · Hybrid energy storage system · Hybrid electric vehicle · State of charge · Supercapacitor

1 Introduction

As the awareness toward the environment is increasing, various alternatives in the field of transportation are sought and the best option available for today's world considering different problems has emerged as hybrid electric vehicle (HEV) [1]. Almost every transport system chemical batteries have been used as main energy storage system (ESS) [2]. Though there are many flaws of using chemical batteries such as limited life cycle, limited power density, higher cost, etc. [3], they are used in large scale. So, to overcome these demerits, an electric double-layer capacitor (ELDC) also known as ultra-capacitor/SC is used with battery. SC has many advantages, which overpower the shortcoming of batteries but, in spite of this, they can

S. Singh (✉)
Department of Electrical Engineering, NIT Kurukshetra, Kurukshetra, Haryana, India

K. P. Tiwari · A. Shahi · M. N. Singh · S. Tripathi
Department of Electrical Engineering, JSSATE, Noida, U.P, India

A. Tomar et al. (eds.), *Machine Learning, Advances in Computing, Renewable Energy and Communication*, Lecture Notes in Electrical Engineering 768,
https://doi.org/10.1007/978-981-16-2354-7_30

not be used individually in electric vehicles because of their low energy density and high self-discharge rate. Hence, SC is used in parallel with batteries, which leads to hybrid energy storage system (HESS).

Battery is a constant source of energy, which has a fixed rate of charging/discharging. If more energy is demanded from battery, it develops stress on it. High stress causes heat loss that affects the life cycle of battery and electric vehicle. Hence during acceleration, battery performance gets deteriorated. To minimize this effect, SC is added that checks for the high energy demand and supplies it to electric vehicle in combination with battery thus avoiding the possibility of developing stress and improves the performance. At the time of deceleration/braking, SC uses the energy dissipated by HEV to charge itself and the battery which in turn improves the battery's source of charge [4, 5].

2 Modeling

The hybrid electric vehicle considered in this work utilizes a vehicle body available is further connected with a DC motor drive and also H-Bridge converter as shown in Fig. 1. Since the vehicle body is connected as shown in the simulink model in Fig. 4, which is further controlled by PWM inverter as a drive controller [6]. Thus, the interconnection is operated physically with the help of a longitudinal driver, which is responsible for the acceleration and deceleration of the vehicle [7]. To do so several drive cycles are there. The given block diagram represents the flow of energy of a vehicular system using SC in parallel with the battery.

2.1 Drive Controller

Pulse width modulation (PWM) technique is used to control the DC motor as it provides the benefit of small power losses in switching transistors. By reducing the power dissipation control becomes linear, which provides better speed stability to

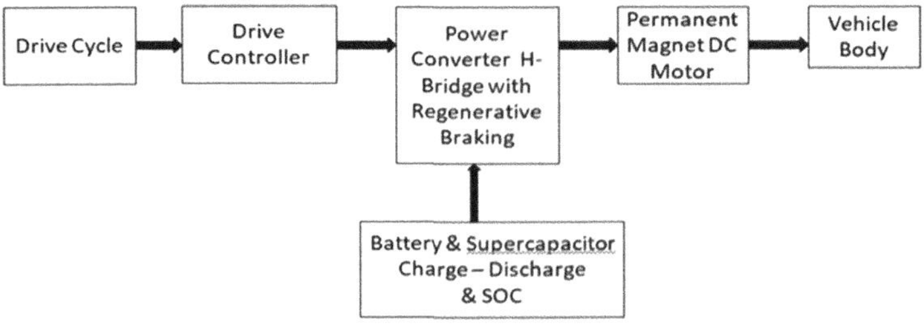

Fig. 1 Drive cycle

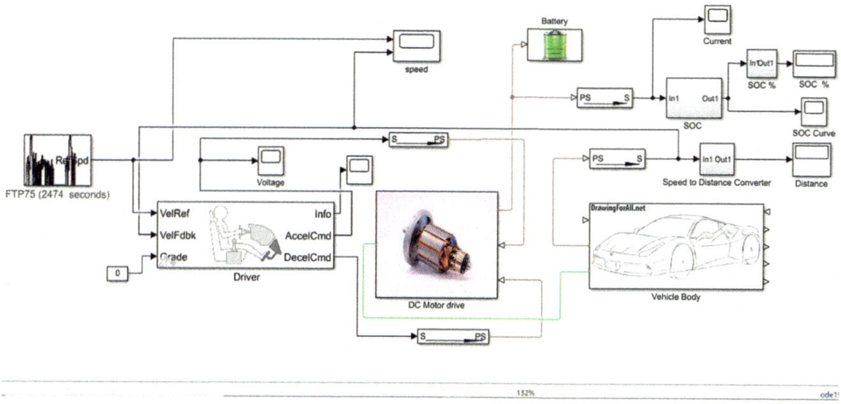

Fig. 2 Simulink model of HEV

motor. When pulse is low (0), output voltage becomes zero and it became equal to output voltage amplitude when pulse is high (1). PWM is the simplest and most efficient method for driving small motors.

2.2 *H-Bridge*

To control the direction of motor either forward or backward H-bridge is connected to it. In this modeling of electric vehicle, H-bridge is controlled by PWM converter. The working principle of H-bridge is very simple comprising of four switches, which can be transistor or MOSFET with motor adjusted at center giving it H like formation. By operating two particular switches at a time, direction of current flowing in the motor gets altered thus changing its direction of rotation from forward to backward or vice versa. Changing the direction of motor is crucial as regenerative braking works on this to store charge at the time of deceleration/braking. In a regenerative braking system when brakes are applied then energy released during the process is again utilized by running the motor in opposite direction to work as generator and capture the energy for charging of battery or SC at the time of need [8].

2.3 *Battery*

A battery package consists of 12 cells per cell row. Lithium-ion battery is widely used because of its high energy density as compared with other batteries. SOC—source of charge of battery denotes the amount of power available in battery at any particular instant of time. It varies 0–100%. When SOC is 100% then battery is said to be fully charged and 0% SOC is the state of complete discharge. In practical usage, it is not suitable to let SOC go below 50% and it is recharged immediately after that. With time maximum capacity of SOC decreases around 75–80%, which requires frequent charging of batteries and degradation of battery takes place. The average life of lithium-ion battery in electric vehicles is within 8 years and in that life span, it can complete a distance of 100,000 miles. SC can last for 10–15 years.

2.4 Supercapacitor

Based on their charge, storing mechanism SC is characterized in two different kinds, which are electric double-layer capacitors (ELDC) and pseudo-capacitors. SC in a hybrid electric vehicle enhances the SOC of battery considerably. One of the greatest superiorities of SC is that it increases the life expectancy of battery as well as vehicles. Nowadays SC is used in the braking system and stop-go hybrid electric vehicles with the idea that its application can be seen in a wide perspective near future [9]. Some important features that make SC beneficial in HEVs are:

- High power density than traditional batteries.
- Fast charging.
- Wide working temperature.
- Better life cycle as compared with batteries.
- Safety factor of SC is more reliable.

3 Electric Vehicle Dynamics

3.1 Vehicle Body

The vehicle body block is a normal chassis of the vehicle with one set of axel per two tyres, i.e., rear and front sets [3–5, 10]. The total force applied on any vehicular system is obtained by following equation:

$$F_{total} = F_a + F_r + F_c + F_{linear-acc} + F_{angular-acc}$$

where,

F_a	Aerodynamic force.
F_r	Rolling resistance force.
F_c	Hill climbing force.
$F_{linear-acc}$	Linear acceleration force.
$F_{angular-acc}$	Angular acceleration force.

Rolling Resistance Loss

Rolling resistance loss is calculated to overcome the friction between the tires and the road. Equation associated with this is as follow:

$$F_r = K_r mg \cos\theta \quad (1)$$

K_r Coefficient of rolling resistance.
m Mass of the vehicle body (Kg).
θ Slope angle (degrees).
g Gravitational constant (m/s^2).

Aerodynamic Resistance Loss

Aerodynamics loss results from the friction between the airstream and the body of the vehicle. Aerodynamic resistance is disabled in the model. When the aerodynamic resistance appears then the aerodynamic force is obtained as:

$$F_a = 0.5 \times \rho A C_d v^2 \quad (2)$$

where,

ρ air density (Kg/m^3).
A Frontal area of the vehicle (m^2).
C_d Aerodynamic drag coefficient.
v Vehicle linear velocity (m/sec^2).

Hill Climbing Loss

Hill climbing force can be exerted on any vehicle system only when it is in running or moving mode with the upside steam pattern. Since here the hill climbing losses are considered to be negligible but can be observed by having expression as,

$$F_c = mg \sin(\alpha) \quad (3)$$

where,

α Hill climbing angle.

3.2 H-Bridge

There are two modes of operation that can be performed in H-bridge, one is average mode and another one is PWM mode [8, 11, 12]. In PWM, the H-Bridge output is a controlled voltage that depends on the input signal at the PWM port. If the input signal has a value greater than the enable threshold voltage parameter value, the H-Bridge output is on and has a value equal to the value of the output voltage amplitude parameter. If it has a value less than the enable threshold voltage parameter value, the block maintains the load circuit using one of the following three Freewheeling mode options: First, using one semi-conductor and one freewheeling diode and second, two freewheeling diodes and with two semi-conductor and one freewheeling diode. Since the output of the H-Bridge controller is given by,

$$\left(\frac{V_o \times V_{pwm}}{A_{pwm}}\right) - I_{out} \times R_{on} \tag{4}$$

V_o output voltage amplitude (Volts).
V_{pwm} voltage at PWM port (Volts).
A_{pwm} PWM signal amplitude.
I_{out} value of output current (A).
R_{on} bridge on resistance (Ohms).

3.3 PWM Controller H-Bridge

The controlled voltage PWM is used to give power to H-bridge separately and is used as pulse width modulation of voltage source [6, 11]. This can be tuned to reduce switching losses in the circuitry by observing the desired duty cycle.

$$Duty\,Cycle = \frac{v_{ref} - v_{\min}}{v_{\max} - v_{\min}} \times 100\% \tag{5}$$

where,

v_{ref} reference voltage across ports.
$v_{\min}$ minimum reference voltage.
$v_{\max}$ maximum reference voltage.

3.4 DC Motor

Permanent magnet DC (PMDC) motor acts as an actuator in electric vehicle modeling. Speed of PMDC motor can be varied by PI or PID controller with some specific values [11].

$$V = R_a i(t) + L\frac{di(t)}{dt} + V_D \tag{6}$$

where,

V_D developed voltage.
V voltage at high side.
$i(t)$ current at high side.
R_a armature resistance and
L Inductance

The permanent magnet in motor induces back EMF in armature by,

$$v_b = K' \times \omega \tag{7}$$

K' constant of back-EMF.
ω ANgular velocity.

Also, torque produced is given by,

$$\tau \times \omega = v_b \times i. \tag{8}$$

So, value of one parameter is sufficient. Rated speed, rated power, and no-load speed are calculated by,

$$i = \frac{V - v_b}{R} = \frac{V - K' \times \omega}{R}. \tag{9}$$

Finally, the developed torque will become

$$\tau = \frac{K}{R} \times (V - K' \times \omega). \tag{10}$$

3.5 Longitudinal Driver

The longitudinal driver block is used to maintain equilibrium between the generated speed and reference speed. This can be done by varying the signal from 0 to 1. The nominal control output is,

$$y = \frac{K_{ff}}{v_{nom}} + \frac{K_p}{v_{nom}} + \left(\frac{K_i}{v_{nom}} + K_{aw}e_{out}\right) \int e_{reg}dt + K_g\theta \tag{11}$$

K_{ff} & K_g velocity and grade feed-forward gain.
v_{nom} nominal Vehicle Speed.
K_p and K_i proportional and integral Gain.
K_{aw} Anti-Windup Gain.
θ Grade Angle.

3.6 Battery

A battery package consists of 12 cells per cell row. Lithium-ion battery is widely used because of its high energy density as compared with other batteries including lead-acid battery. So, lithium-ion batteries can be made as small as possible retaining the same storage capacity. Battery used for power electric vehicles is rechargeable battery and is designed for ampere hour capacity [3, 5, 10].

Battery voltage is defined as:

$$V = V_O\left(\frac{SOC}{(1-\theta)\times(1-SOC)}\right) \tag{12}$$

where,

V_O nominal voltage parameter and
θ constant.

Since, battery SOC and energy can be obtained by,

$$SOC(t) = SOC(t-1) + \int_0^t \frac{1}{C_b}.dt \tag{13}$$

$$E_b = C_b \times U_b \tag{14}$$

$SOC(t)$ battery state of charge at time 't' (%) and
$SOC(t-1)$ Battery initial state of charge (%).
C_b Battery capacity (Ah).

Also, the number of cells connected in series per string, energy of a string and total number of strings in a battery pack is:

$$N_{sc} = \frac{U_{bp}}{U_b}. \tag{15}$$

$$E_s = N_{sc} \times E_b \tag{16}$$

$$N_p = \frac{E_p}{E_s} \tag{17}$$

N_{SC} & N_p battery cell and Strings in series and
U_b & U_{bp} Voltage of battery and battery pack.

3.7 Supercapacitor

SC is capable of handling millions of charge and discharge cycles [8, 11, 12]. Thus, the charging voltage and current across the nth RC branch of the SC as follow:

$$V_{Cn} = \frac{V}{N_{Series}} - i_n R_n \tag{18}$$

$$i_n = C_n \times \frac{dV_{Cn}}{dt}. \tag{19}$$

And if, $V_{C1} > 0$ Then,

$$\begin{aligned} & i_1 = (C_1 + K_C \times V_{C1}) \\ & Else, i_1 = C_1 \times \frac{dV_{C1}}{dt} \end{aligned} \tag{20}$$

where,

V	Voltage across the block and
V_{C1}	Voltage across the first branch of capacitor.
N_{Series}	No of cells in series.
i_n	Current in nth branch.
i_1	Current of first branch.
R_n	Resistance through nth branch.
C_n & C_1	Capacitance of nth & 1st branch.
K_C	Voltage-dependent capacitance gain.

Number of cells in series and parallel of SC is always 1. Here, for HEV, SC is modified according to the need of battery and capacity of charge require by driver. Following data have been prescribed, i.e., number of cells in series 150 and number of cells in parallel is 1 (Fig. 2 and Table 1).

4 Methodology Proposed

4.1 Drive Cycle

A battery package consists of 12 cells per cell row. Lithium-ion battery is widely used because of its high energy density as compared with other batteries including lead-acid battery. So, lithium-ion batteries can be made as small as possible retaining the same storage capacity [10]. Drive cycle used here is FTP 75. The maximum speed of this drive cycle is 90 Kmph. This cycle runs for 2474 s.

4.2 Hybrid Energy Storage System (HESS) Drive Cycle

In electric vehicle during regenerative braking and also at the time of acceleration/deceleration, there are unwanted spikes that harm the battery and also decrease its life expectancy [11]. To overcome this power, electronic controller called as Hybrid Energy Storage System (HESS) [1, 3–5, 10] is introduced as shown in Fig. 3 (Fig. 4).

The HESS is basically used to increase the efficiency of regenerative braking system and also improve the SOC of battery, which in turn increases the life of the vehicle [7]. HESS is designed such that the battery used in vehicle supplies the continuous energy demand and SC takes care of the demand during instant loading, Buck/Boost converter is used in the HESS model for necessary controlling required in the vehicle [5]. In the above model, Conn 1 is the physical signal, i.e., physical current signal. SC is connected in shunt with battery to improve the load balancing [7]. In the block of SC, cells receive equal amount of voltage to keep the HESS in equilibrium thus keeping the vehicle connections stable. There are two modes of working with HESS:

Normal Mode: During the normal, running condition of vehicle battery is sufficient to deliver the supply hence; SC module is inactive during that period.

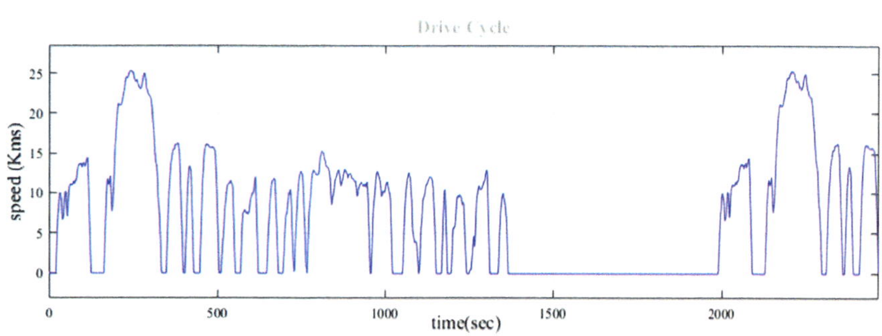

Fig. 3 Drive cycle

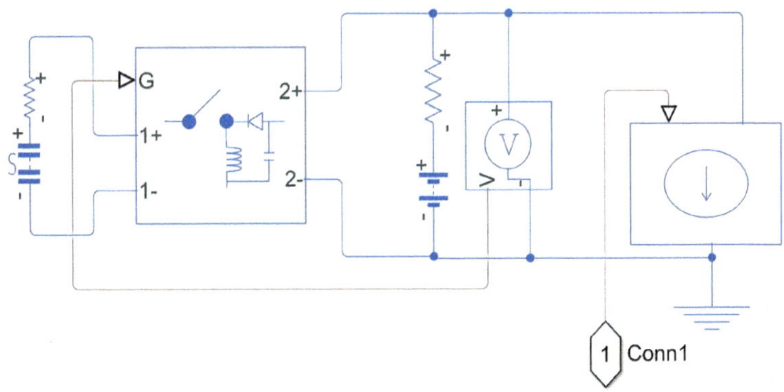

Fig. 4 Simscape HESS

Acceleration Mode: In this mode, battery needs to go deep discharging, which can harm battery life. In such cases, SC assists the battery by buck/boost converter to maintain inside equilibrium. If at any instance, SC voltage decreases by battery voltage then it will increase battery stress as anyhow SC will start charging through battery. To avoid this, SC only acts as a back-up for battery module.

5 Results and Discussion

Different simulation results have been observed, which include SOC of battery without SC, SOC of battery with SC, and graph of the speed comparing the longitudinal drive cycle with actual drive cycle. Model is made to run for 2474 s and battery's SOC without HESS is observed in Fig. 5 and with HESS in Fig. 6. With SC, energy captured by regenerative braking is more often than electric vehicle with

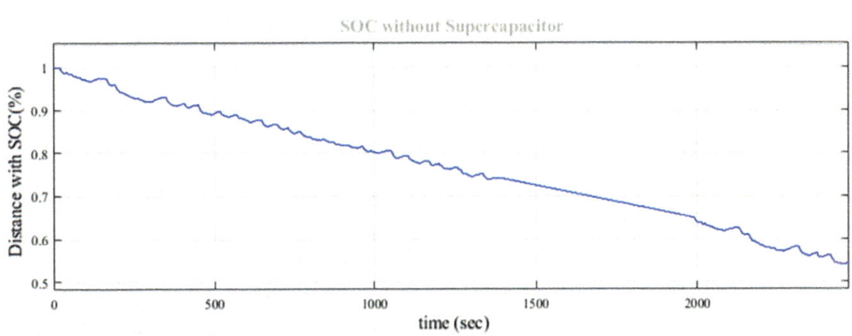

Fig. 5 State of charge for HEV without HESS

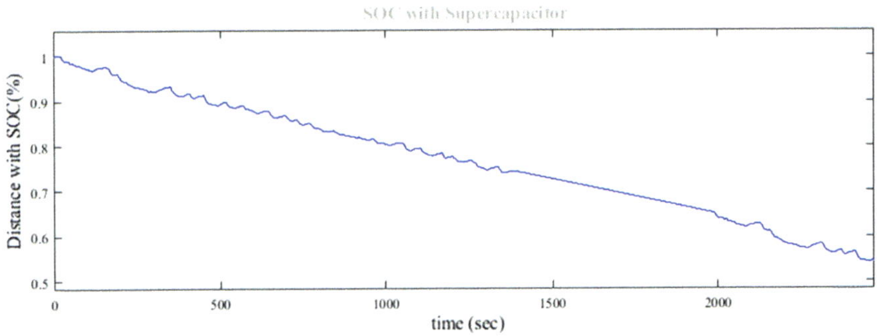

Fig. 6 State of charge for HEV with HESS

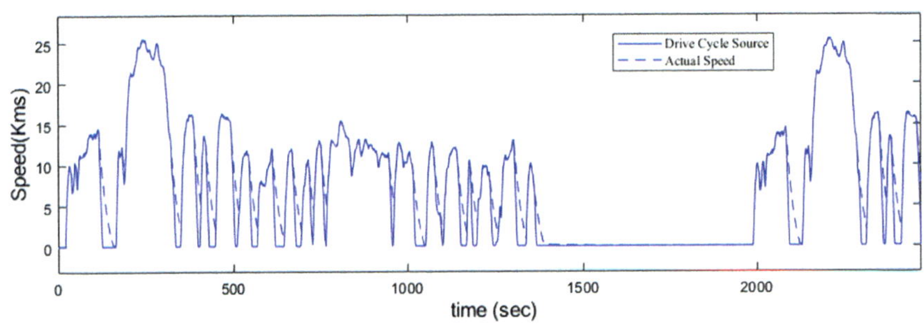

Fig. 7 Actual speed versus drive cycle speed

battery only, i.e., for the same interval of drive cycle, charge available in battery with SC is more than charge in battery without SC after the completion of drive cycle.

Where the SOC for the same distance of 5.59 Km is obtained, 54.49% with SC and 54.44% without SC with a limit of 0.05%, which can be observed and a comparative result for SOC of battery with or without SC is obtained by comparing Figs. 5 and 6 in which it is clear that longitudinal driver is closely following the actual drive cycle and is given in Fig. 7 (Table 2)

6 Conclusion

With the introduction of SC along with appropriate parameters in hybrid electric vehicle, the SOC of battery improves with significant amount, which has been observed in the above two graphs Figs. 6 and 7. The increased percentage of SOC in battery with SC signifies that losses in the battery have been decreased, which mainly takes place in the form of heat at the time of acceleration/deceleration. One of the major problems of using hybrid electric vehicles is frequent charging, which can be resolved by this alternative. The rate of discharge of battery is defined by the application of SC in such a way that during normal speed, vehicle is driven by battery and at the time of instant demands SC takes the control and maintain the continuous flow of energy smoothly without any excess burden on battery. With this improvisation of SOC in battery, the life expectancy of battery gets better than before, which contributes in efficient regenerative braking.

7 Future Scope

Advancement of adding an SC with a battery in an electric vehicle shows makeable progress in the improvement of SOC of the battery for the same time reference. With high energy density of SC, its self-discharge rate can be optimized to a considerable level that again increases the efficiency of an electric vehicle. Furthermore, an increase in the voltage of individual cells of SC is another field to be explored and work accordingly.

Table 1 EV modeling parameter

Component used	Significance	Rating
Vehicle Body	Mass	1200 kg
	Drag coefficient	0.4
	Air density	0.28 kg/m3
Tire (magic formula)	Wheels per axle	2
	Weight per wheel	8000 N
	Rolling radius	301 mm
H-bridge	Bridge resistance	0.1
	Output voltage amplitude	320 V
PWM controller	Frequency	10 K Hz
DC motor	No load speed	8000 RPM
	Rated load	50KW
Longitudinal driver	Nominal speed	25.352 m/s
Drive Cycles	FTP75	2474 s
Battery	Nominal voltage	320 V
	Rating	50Ah
Supercapacitor	Self-discharge resistance	300 Ω
	R1	0.2 Ω
	R2	90 Ω
	R3	1 KΩ
Buck-boost converter	Threshold voltage	6 V

Table 2 Parametric comparison of EV with battery and supercapacitor

Parameters	Battery	Supercapacitor
Cell voltage	3.60 V/cell	2.75 V/cell
Life cycle	2–3 years	10–15 years
Self-discharge	2–3% per month	5% per month
Price (per KWh)	$175 per KWh	$100 per KWh
Recyclability	25–96%	20–70%
Safety	Moderate	Safer than Li-ion battery
Temperature range	+10 °C– + 55 °C	−40°C to + 70°C
Power density	1.6 KW/L	10 KW/L

References

1. Mutoh N (2012) Driving and braking torque distribution methods for front-and rear-wheel-independent drive-type electric vehicles on roads with low friction coefficient. IEEE Trans Ind Electron 59(10):3919–3933
2. Future Electric Cars (2007). http://www.future-car.ghnet/future-electric-cars.html. Accessed 29 January 29 2010
3. Blanes JM, Gutierrez R, Garrigós A, Lizán L, Cuadrado JM (2013) Electric vehicle battery life extension usingultracapacitors and an FPGA controlled interleaved buck–boostconverter. IEEE Trans Power Electron 28(12):5940–5948
4. Nasheri F, Frajah E, Ghanbari T (2016) An efficient regenerative braking system based on battery or supercapacitor for electric, hybrid and plug-in hybrid electric vehicles with BLDC motor. IEEE
5. Kim Y, Chang N (2014) Design and management of energy-efficienthybrid electric energy storage systems. Springer, pp 19–25
6. Khaligh, Li Z (2010) Battery, ultra capacitor, fuel cell, and hybrid energy storage systems for electric, hybrid electric, fuel cell, and plug-in hybrid electric vehicles: state of the art. IEEE Trans Veh Technol 59(6):2806–2814
7. Montazeri M, Soleymani M (2009) Investigation of the energy regeneration of active suspension system in hybrid electric vehicles. IEEE Trans Ind Electron 57(3):918–925
8. Jiang Z, Dougal RA (2006) A compact digitally controlled fuel cell/battery hybrid power source. IEEE Trans Industr Electron 53(4):1094–1104
9. Arnet BJ, Haines LP (2010) Hig-power DC-to-DC converter for supercapacitors. Proc IEEE Power Convers Conf, Osaka, Japan 2002:1160–1165
10. Yang SM, Chen JY (2012) Controlled dynamic braking for switched reluctance motor drives with a rectifier front end. IEEE Trans Ind Electron 60(11):4913–4919
11. Li Z, Onar O, Khaligh A, Schaltz E (2009) Design and control of a multiple input DC/DC converter for battery/ultracapacitor based electric vehicle power system. In: IEEE conference on applied power electronics and exposition (APEC), Washington DC, pp 591–596
12. Brain M (2002) How electric cars work. http://auto.howstuffworks.com/electric-car2.html

Formulation of C++ program for Quine–McCluskey Method of Boolean Function Minimization

Mayank Joshi, Sandeep Kumar Sunori, Naveen Tewari, Sudhanshu Maurya, Mayank Joshi, and Pradeep Kumar Juneja

Abstract There are a number of techniques available for the minimization of Boolean functions. Quine–McCluskey (Q-M) method, which is a very efficient technique, is one of them. However, while doing it manually, as a large number of comparisons are to be done in various tables, there are great chances of human error. Moreover, it takes a long time to simplify a function using this technique manually. This method is fruitful when it is performed fast within seconds. For this, we can rely on computer. By using the programming, we can make computer do all the long comparisons and get the minimized Boolean expression instantly. This paper presents a report on the work that we have done for developing the C++ program for Q-M method for Boolean functions of 3, 4, and 5 variables. Finally, the outputs of this program have also been showcased for various considered Boolean functions.

Keywords Boolean function · Minimization · Program · Q-M method · Prime implicant

1 Introduction

Any logical problem can be easily solved by expressing it in form of a Boolean function and then minimizing it. A Boolean function is comprised of the Boolean variables and logical operations. The primary logical operations on which the Boolean algebra is based are 'OR', 'AND', and 'NOT' [1]. The logical operations can also be expressed in the mathematical form [2].

The minimization of a Boolean function is highly desirable as after minimization, the resulting logic circuit is less complex and contains a minimum number of logic gates, which in turn results in a reduction in its cost, increase in speed of operation, and decrease in the power consumed [3, 4]. A given SOP or POS form of the

M. Joshi · S. K. Sunori (✉) · N. Tewari · S. Maurya · M. Joshi
Graphic Era Hill University, Bhimtal Campus, Dehradun, India

P. K. Juneja
Graphic Era University, Dehradun, India

A. Tomar et al. (eds.), *Machine Learning, Advances in Computing, Renewable Energy and Communication*, Lecture Notes in Electrical Engineering 768,
https://doi.org/10.1007/978-981-16-2354-7_31

Boolean function gets easily minimized manually using laws of Boolean algebra and Karnaugh map (K-map) [5]. However, with a Boolean function containing a large number of variables, typically more than four, Boolean algebra and K-map techniques become very tedious for finding the minimal form of the given function.

A third very popular and efficient technique for minimization is Quine–McCluskey (Q-M) method. This method is based on searching the prime implicants (PIs) [6]. It can very effectively deal with the Boolean functions in which the number of variables is very large [7]. But when this technique is used manually, there are great chances of errors as a large number of comparisons are to be performed at every step [8]. However, as this method is a tabular method based on an algorithm, a very efficient computer program can be developed for it which can minimize the given Boolean function instantly.

2 Developing the C++ Language Program for Q-M Method

To make a successful C++ program for this requires a little deep knowledge of C++ language, different functions to compare the terms of each variable, to print the tables in an ordered way, to take input in an organized way, to find the prime implicants (PIs), to create PI table, to find out essential prime implicants (EPIs), to derive the minimized Boolean expression and to print them. To handle different expression in a uniform way, to get unique output expression, requires a complex programming with the concept of classes, arrays, functions, and Quine–McCluskey method.

The algorithm to show the working of Quine–McCluskey method for obtaining minimized SOP expression for a particular Boolean function consists of some basic steps that are shown in the flowchart presented in Fig. 1.

3 Difficulties and Solutions

(i) *The uniformity of table heading and data*: To maintain the gap between each column and to set column width the ***setw*** function is used.

(ii) *To minimize the expression having all the terms*: To print 1 when all the possible terms (minterms and do not care terms) are present, the 'if' condition is used.

(iii) *To decide the result according to the number of variables (3,4,5)*: A menu-driven program is attached to make it user-dependent that asks user for choosing among 3 variable, 4 variable, and 5 variable programs.

(iv) *To print the correct expression when few or all the minterms are not in EPIs*: To check whether all the terms are covered or not, a separate function is created that gives the product terms, which covers the uncovered minterms.

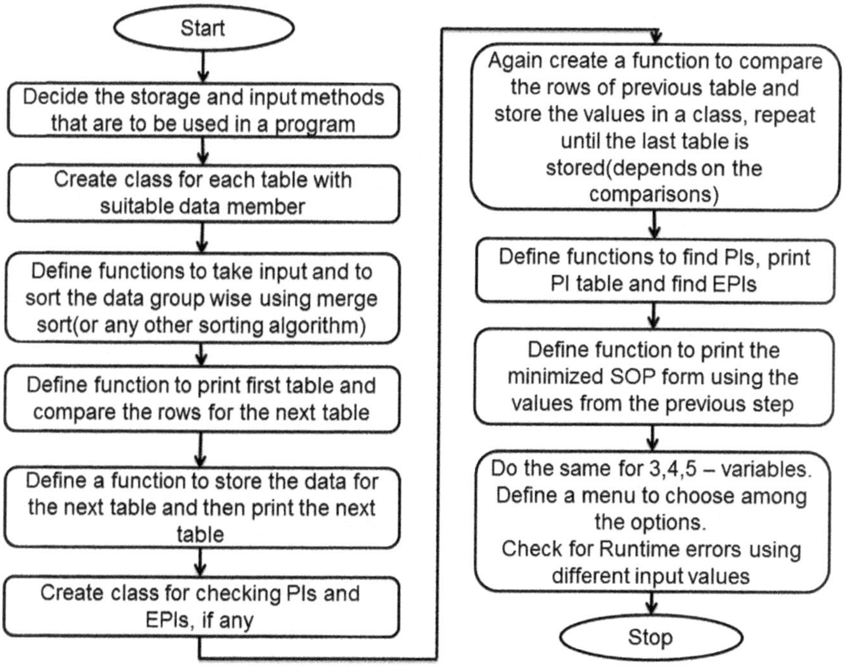

Fig. 1 Flowchart of the C++ program for Q-M method

(v) *Repetition of product terms*: Multiple product terms having the same value may repeat, especially from the last table. For this, a few lines of code are added to destroy the uniqueness of the repeating PIs.

(vi) *The plus + sign at the end*: The sign at the end due to recursion is not necessary and may be confusion. A zero at the end of SOP expression is inserted, which clearly shows that the minterm expression is finished.

(vii) *Handling the Don't care terms*: If the user enters any value in Don't care terms then code will execute as usual but if the user enters—1 as Don't care terms (in case the function has no Don't care term) then code having evaluation part of Don't care terms will be skipped.

4 Results

The developed C++ program is now executed on some Boolean functions of 3, 4, and 5 variables. The snapshots of the outputs generated by this program are presented in Figs. 2, 3, and 4. The Red Cross signs ('**X**') in PI tables indicate the essential prime implicants (EPIs). In every snapshot, the last term that is shown to be added to the final minimal SOP expression is **zero**, which indicates the end of expression.

(i) Minimization of 3 variable function:

```
Enter Min terms : 1,2,3
Enter Dont care if any else enter -1: 6,7

GROUP          MINTERM          ABC
1                 1             001
1                 2             010

2                 3             011
2                 6             110

3                 7             111


GROUP          MINTERM          ABC
1              (1,3)            0_1
1              (2,3)            01_
1              (2,6)            _10

2              (3,7)            _11
2              (6,7)            11_
```

```
GROUP          MINTERM          ABC
1              (2,3,6,7)        _1_
1              (2,6,3,7)        _1_


Prime Implicants:
A'C, B,


Minterms -->     1  2  3
PIs
A'C              X     X
B                   X  X

Minimized SOP Expression:
A'C + B + 0_
```

Fig. 2 Output of program for function of 3 variables

(i) Minimization of 3 variable function:
(ii) Minimization of 4 variable function:
(iii) Minimization of 5 variable function:

5 Conclusion

In this paper, a C++ program has been developed for Quine–McCluskey (Q-M) method of Boolean function minimization for Boolean functions with 3, 4, and 5 variables. This program has been tested on a number of Boolean functions and the desired minimal SOP form is obtained for each. So, this validates the developed program. This program can be further modified for 6, 7, 8, and more variables.

(ii) Minimization of 4 variable function:

```
Enter Min terms : 2,3,6,9,10
Enter Dont care if any else enter -1: 0,13,14

GROUP              MINTERM                  ABCD
0                     0                     0000

1                     2                     0010

2                     3                     0011
2                     6                     0110
2                     9                     1001
2                     10                    1010

3                     13                    1101
3                     14                    1110
```

```
GROUP              MINTERM                  ABCD
0                  (0,2)                    00_0

1                  (2,3)                    001_
1                  (2,6)                    0_10
1                  (2,10)                   _010

2                  (6,14)                   _110
2                  (9,13)                   1_01
2                  (10,14)                  1_10
GROUP              MINTERM                  ABCD
1                  (2,6,10,14)              __10
1                  (2,10,6,14)              __10
Prime Implicants:
A'B'D', A'B'C, AC'D, CD',

Minterms -->         2  3  6  9  10
PIs
A'B'D'               X
A'B'C                X  X
AC'D                          X
CD'                  X     X     X

Minimized SOP Expression:
A'B'C + AC'D + CD' + 0_
```

Fig. 3 Output of program for function of 4 variables

(iii) Minimization of 5 variable function:

```
Enter Min terms : 1,2,3,4,5,6,30,29
Enter Dont care if any else enter -1: -1

GROUP              MINTERM            ABCDE
1                  1                  00001
1                  2                  00010
1                  4                  00100

2                  3                  00011
2                  5                  00101
2                  6                  00110

4                  29                 11101
4                  30                 11110

GROUP              MINTERM            ABCDE
1                  (1,3)              000_1
1                  (1,5)              00_01
1                  (2,3)              0001_
1                  (2,6)              00_10
1                  (4,5)              0010_
1                  (4,6)              001_0


Prime Implicants:
ABCD'E, ABCDE', A'B'C'E, A'B'D'E, A'B'C'D, A'B'DE', A'B'CD', A'B'CE',

Minterms -->    1  2  3  4  5  6  29 30
PIs
ABCD'E                            X
ABCDE'                               X
A'B'C'E         X     X
A'B'D'E         X           X
A'B'C'D            X  X
A'B'DE'            X           X
A'B'CD'                  X  X
A'B'CE'                  X     X
```

Fig. 4 Output of program for function of 5 variables

References

1. Yang Y, Jung I (2017) Boolean algebra application in simplifying fault tree analysis. Int J Safety Sci 01(01):12–19
2. Azram M, Daoud JI, Elfaki FAM (2009) Arithmetic version of Boolean algebra. Adv Appl Discrete Math 4(2)
3. Ashfak Habib AHM, Abdus Salam M, Nadir Z, Goswami H (2004) A new approach to simplifying Boolean functions. J Eng Res 39–45
4. El-Bakry HM, Atwan A (2010) Simplification and implementation of Boolean functions. Int J Univ Comput Sci 1(1):41–50
5. Nosrati M, Karimi R, Aziztabar R (2011) Minimization of Boolean functions which include don't-care statements, using graph data structure. WiMoA 2011/ICCSEA 2011, CCIS 154, pp 212–220
6. Morris Mano M (2004) Digital logic and computer design. Prentice-Hall of India Private Limited
7. Petrík M (2008) Quine–McCluskey method for many-valued logical functions. Soft Comput
8. Majumder A, Chowdhury B, Mondal AJ, Jain K (2015) Investigation on Quine McCluskey method: a decimal manipulation based novel approach for the minimization of Boolean function. In: International conference on electronic design, computer networks & automated verification (EDCAV), pp 18–22

Survey of Security Issues in Cyber-Physical Systems

Aditya Tandon

Abstract The curious case of "a sense of security" showed clear weaknesses in the existing Cyber-Physical Systems or widely known as the Internet of Things (IoT) and is not close to that of the existing computers. The Internet of Things unveils a variety of new issues and challenges that need security and privacy in a different way that are usual in the existing information systems. This work lays the groundwork for how protection within the Internet of Things will be treated, both in current and future systems. Current devices have been studied in various domains and with various technologies to create a simple, concrete image of the problems and solutions that exist in today's IoT. Within the IoT framework, computation, resources and bandwidth, the three key constraints are defined and used to establish the basis for the challenges posed. Further discussed the numerous potential futures for the IoT, and what challenges would involve in the successful operations of IoT. The approaches to the problems will vary from what resources are available and an in-depth overview of potential solutions will be discussed depending on available resources. To increase developers' attention on important IoT security issues, complete recommendations are described chronologically from start to finish of design, creating and maintaining devices on the Internet of Things, aimed at developers with in-depth information security expertise. It poses potential solutions and alternatives, as well as major issues that will make programmers think about the implications of the choices, they make during the overall process in IoT.

Keywords Privacy · Cyber-Physical Systems · Security · Smart City · Challenges

A. Tandon (✉)
Department of Computer Science and Engineering, Krishna Engineering College, Ghaziabad, UP, India
e-mail: aditya.tandon@krishnacollege.ac.in

A. Tomar et al. (eds.), *Machine Learning, Advances in Computing, Renewable Energy and Communication*, Lecture Notes in Electrical Engineering 768,
https://doi.org/10.1007/978-981-16-2354-7_32

1 Introduction

Kevin Ashton originally gave the term Internet of Things (IoT) in 1998 in a presentation [1]. The IoT allows us to connect people and things with Anywhere, Anyplace, Everything and Anything [2]. The communication, configuration and management of these devices are not feasible if it is not done automatically because of the huge number of internet-based devices. The sensor networks are key components of IoT. The sensor networks consist of one or more sensing nodes that use wired or wireless means to communicate with each other. Every other sensor node has the capability to locally or remotely sense, communicate and process data. Sensor nodes in sensor networks can be homogenous or heterogeneous. These sensor nodes are built with the phenomenon that we would like to sense the environment activities [3, 4]. Through the development of hardware and software innovations, mixture of real-world entities wireless sensor networks and smart objects have become a practical solution through the Internet capabilities. The IoT is the introduction into the Internet of clearly identifiable Smart Objects, manufactured systems with actual objects and their realizations. IoT engineers operate in tandem with WSN hardware, but the deployment and analysis procedures for sensor devices to act as smart objects are not insignificant. The expansion of applications for the IoT relates in particular to various usability characteristics of WSN. The design and implementation of IoT systems need the basic issues that are explained in [5] to be addressed:

- Hardware and Software heterogeneity linked networking and compatibility problems.
- Flexibility and scalability of application.
- Standardized communication and descriptions of the services.
- Procedures for the automation.
- Handling the Big Data

The IoTs have the capability to deploy billions of very low-cost, internet protocol (IP)-enabled wireless sensor nodes, allowing sensors to sense and track any object or individual in real world. The integration of sensing objects allows us to communicate easily with others in the world [6]. An IP address is provided for any device that connects to the Internet. The existing IPv4 has an address space of 32 bits (i.e. approximately 4.3 billion distinctive IP's, which is less than today's population of the world). IPv6 is the latest version active to solve the limitation of 32-bit address space issue and plays a significant role in IoT implementation. IPv6 can handle more than 340 undecillion distinct IP addresses (128-bit). So even, the new IPv6 will recognize trillions of WSN nodes [7]. Internet technologies and WSN are realizing a new trend in the age of ubiquity. A rapid growth in internet usage and improvements to various networking technologies allow the internetworking of daily objects [8]. IoT has always been about real objects communicate with each other, a machine-to-machine contact would be applied to things [9]. The phase of convergence supports the core concept of the IoT, two-way communication is possible between two network devices. The tools may be overwhelming in nature, such as a high-speed server system

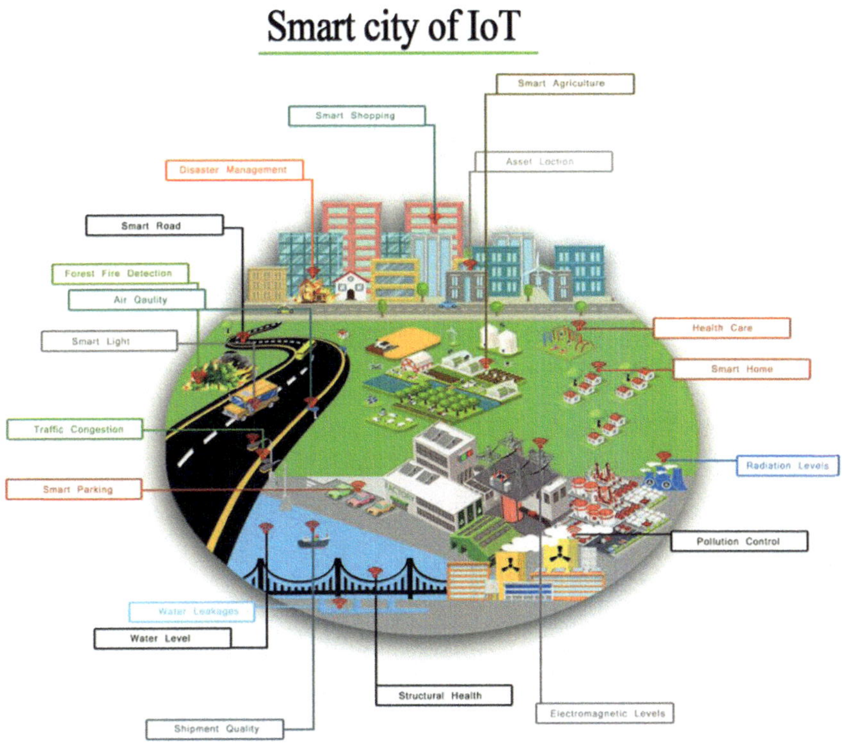

Fig. 1 IoT-based Smart City

tracking out a weather monitor, or a smartphone user manipulating bulbs. This correspondence is made possible by the presence of a universal communication network, using structured protocols. This extensive-scale integration is expected to enhance many of the existing systems such as logistics, transport and different automated systems. However, it will also allow the implementation of novel applications such as smart towns (Fig. 1).

2 Motivation for Security in IoT

IoT protection is a very popular area of study that draws researchers from science, business and government domains. Threats to IoT systems are fast and easy to enforce. An attacker can breach a residential alarm device by infringing the RF signal that used to turn the alarm on and off [10]. Security problems are discussed in [11], such as basic system security, network safety in the IoT. The internet will be connected with more than 200 million devices by year 2020, with a large portion of such devices being phones, appliances and a huge opportunity will also be there for

Table 1 Types of industries that will use IoT widely

Category of industry	2014	2015	2020
Domestic consumer	2,244.5	2,874.9	13,172.5
Vertical business	836.5	1,009.4	3,164.4
Standard business	479.4	623.9	5,158.6
Automotive	189.6	372.3	3,511.1
Total	3750.0	4,880.6	25,006.6

the hackers to attempt "DoS" attacks, malicious emails from other hazardous Trojans or Worms. A well-known organization's (Hewlett Packard) study report states that approximately 80% of IoT devices infringed the privacy of user's personal data such as name, address, date of birth, contact number, etc. on commercialized IoT deployments, more than 80% system did not require passwords having sufficient length, and 60% had security flaws on their user application interfaces [12, 13].

2.1 *Need for Security in IoT*

A wide range of protocols and algorithms are available in today's internet environment to solve the wireless network security issues; however, the latest methodologies comprise with a restriction on their implementation in Internet of Things (IoT) domain due to the hardware system and WSN constraints in IoT. Another major factor related to security is that traditional security protocols consume a huge quantity of computing and memory resources. The IoT devices generally have to operate in harsh environment, unpredictable and dangerous environmental conditions around them, where they are vulnerable to a variety of security breach. A huge number of IoT devices (about 25 billion) will be used worldwide by the year 2020. Table 1 shows the various types of industries, which will use IoT below [14].

2.2 *IoT Malware Attacks Skyrocket in 2018, Continuing to 2019*

SonicWALL estimates that IoT malware attacks in 2018 (up from 10.3 million in 2017) soared 215.7% to 32.7 million. The initial two quarters of the year 2019 already outperformed 55% in the initial two quarters of 2018. If this frequency continues, this will be another record-making year for attacks on IoT malware (Fig. 2) .

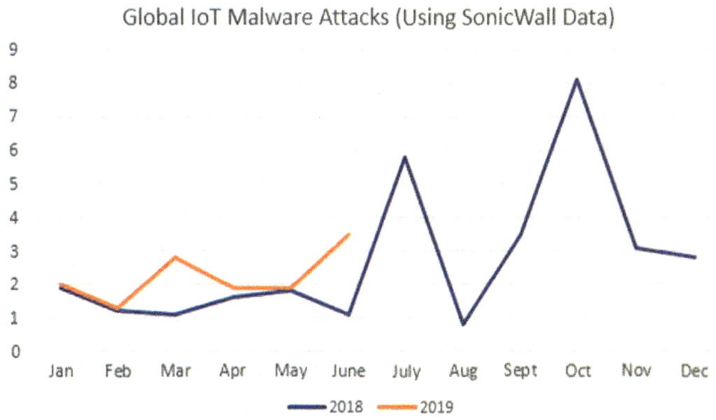

Fig. 2 Global Malware attacks in IoT [15]

2.3 The Role of IoT

IoT can help to grow a vast range of industrial domains; this section gives a brief overview of its significance and how its impact on daily life. The Internet of Things should not only be seen as an evolution of today's internet but as a collection of advance autonomous networks operating their own services and infrastructures' [15]. Various sectors where IoT helps to grow are mentioned below: (1) Healthcare [16–19], (2) Smart Home [20, 21], (3) Automated Vehicles [22], (4) Big Data [23] and (5) others [24–30].

3 Current Challenges IoT

Compared with what is theoretically feasible, the present Internet of Things is found relatively easy. Many devices connect to a phone that acts as a hub to a central server, communicate to a static home hub or directly connect to a central server. Some devices can communicate with a central server using a mesh network linked to a router. In this section, the numerous current and future challenges in Internet of Things (IoT) are presented.

The IoT technical challenges [31] are identifiable in several areas which are mentioned below: (1) Connectivity, (2) Power Consumption, (3) Architecture, (4) Interoperability and Integration, (5) Storage and Computation Complexity, (6) Authorization.

4 Attack Types on the IoT System

Many attack types will affect the IoT framework. IoT attacks are generally classified into four major categories: software attacks, network attacks, physical attacks and crypto attacks. Physical aggression happens when the attacker is close to IoT. When the intruder accesses the IoT network, a network attack occurs, manipulating a computer to inflict harm. A software attack happens when the IoT program has bugs that allow the hacker to access the IoT devices and damage the system. Eventually, an assault occurs when the hacker cracks the IoT encryption to trigger an intrusion. IoT is suggested to take additional steps to enhance its security, such as authentication, safe booting using digital certificates, secure applications and data encryption, so that only approved users can access and track IoT device data. Some researchers identified other types of attacks other than the typical ones are shown in Fig. 3.

As shown, there are six primary categories that are namely: physical attacks; side-channel attacks; environmental attacks; attacks by cryptanalysis. However, a network attack may lead to multiple IoT device problems and information sharing between IoT devices and servers is the most dangerous type of attack. Table 2 compares some existing solutions for these attacks and their role in optimizing the basic security features and in most of the types of attacks listed above. As shown in the table, all solutions should avoid software-related attacks. And moreover, there is no way

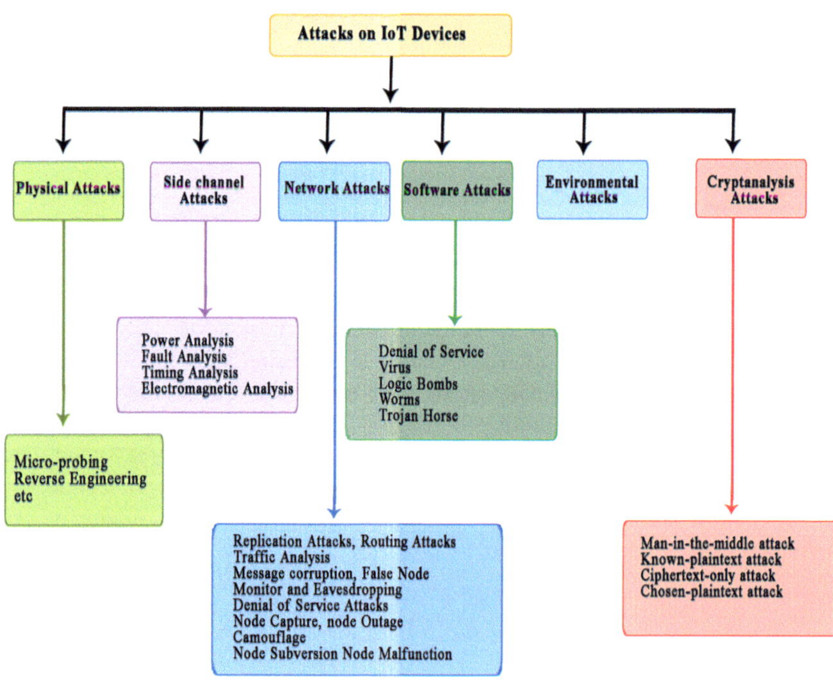

Fig. 3 Classification of attacks in IoT

Table 2 Various aspects of security challenges in IoT and proposed solutions

Challenge	Result	Constraints	Available solution
Authentication	Incorrect data can be considered as correct	Power, bandwidth, processing	Authenticated end-cryption
Authorization	Adversary access to important data and functions	Processing, bandwidth, power	Authorization techniques
Availability	Deferred updates and management	Power and bandwidth	Work with presented restrictions
Lacking multi-layer security	No added security if one fails	Processing, bandwidth, power	Use encryption on multiple levels
Key distribution	Key can be snapped up during transit	Processing, bandwidth, power	Use-case specific different solutions
Privacy	Ability to track users	–	Privacy
DoS	Render device unusable, loss of data	Power, bandwidth, processing	Detection, network design
Unintended uses	Device is insecure because the use case was not predicted	–	Always prioritize security
Usability before security	Easy to use, but in-secure	–	User-friendly security. Always include security
Local storage	Large amounts of local data can get in wrong hands	Bandwidth	Use case-specific. Offloading possible?
Local processing	Local processing will impact the ability to do other things	Power, processing	Enhanced processing. Offloading possible!
Interoperability	No devices talk with each other, extra layer needed (added breakpoint)	Bandwidth, processing	Adaption, corporate unity

to get rid of side-channel assaults. Therefore, all devices are configured to operate basic security activities except for the latter, an embedded data-powered security technique. This table shows that a defensive solution is needed in IoT, as most of them concentrate on software attacks without worrying about device hardware elements.

5 IoT Security

5.1 Information Security (IS) and Privacy Threats

The way an individual or organization protects assets from unauthorized access, alteration, disturbance or disruption is handled by Information Security (IS). Companies have invested a large sum of revenue for many years in getting high-end firewalls, virus protection software, spam email filters and robust cyber security networks as well. Kumar and Srivastava [32] performed some work on the development of an optimal amount of capital that businesses can invest to build IT safety systems. Kumar and Srivastava [31] published a report on ‘Costs of Data Breach’ the report found that a company’s total data breach costs increased from $3.520 million in the year 2014 to $3.790 million in the year 2015. Hacker groups like anonymous are focusing on major corporations and risking a very huge amount (millions of $) in damages for businesses. PayPal had to pay 3.50 million loss due to this in 2012. Cybercrime is at the forefront of several EU states’ National Security Policy—for example, France, Netherlands and Great Britain. Several independent studies have recently found that in the past year between 36 and 90% of companies experienced security breaches. IT incident volume is rising at an impressive pace. It is not just companies that need to be cautious, but regular people do likewise. While companies may make invest wealth to train workers to take adequate safety measures while using technology, individual users need to take the steps to educate themselves for such concern.

5.2 IoT Security Vulnerabilities

A quick description of the several security flaws in IoT systems and eco-systems is outlined below. A device’s level of safety is the possibility or threat it would be compromised, the loss it would cause, as well as the time and resources required to secure a degree of protection OWASP is a firm involved that emphasizes IS security concerns and seeks to raise standards. Several of those security issues are mentioned in brief below:

- Attack Vectors
- Attacks Spread Quicker
- Limited Resources
- Data Integrity
- Lack of Encryption
- Eavesdropping

6 Conclusion

The (IoT) was the world's biggest breakthrough and is a promising innovation to transform our lives. IoT is facing a lot of challenges at the same time. Threats to privacy and protection, which include safety standards, are the greatest challenge. The main aim of the current study was to establish safety criteria that could enhance the output rate of the IoT. The advent of IoT, including context, resources, links and IoT applications, was briefly presented in this paper. Security specifications play the most important role in formulating security solutions and maintaining the IoT network. Awareness of the meaning and types of risks, visibility, vulnerability and attacks is one of the most significant safety topics to be derived from this report. The goal of this research was to concentrate on the IoT layers and features to tackle IoT security issues. For each layer of IoT protection and its security measures, this survey paper identified different attacks and problems. This paper proposed contrasts between security measures required to determine the impact of security mechanisms on the use of power and time for each IoT layer of security. There is still a need to find a method to avoid some form of IoT threats, such as side channels, that present a challenge that researchers need to solve. This paper describes and addresses those problems. The IoT network fails due to heterogeneous network attacks on IoT nodes. While there are several technology-based privacy and security issues, developers and researchers need to work together to overcome these risks, as they have achieved with many other related technologies.

References

1. Ashton K (2013) That 'Internet of Things' thing. [Online]. http://www.rfidjournal.com/articles/view?4986. Accessed 20 May 2013
2. Friess P, Guillemin P (2009) Internet of things strategic research roadmap. The cluster of European Research Projects
3. Corcho O, Garcia-Castro R (2010) Five challenges for the semantic sensor web. Semantic Web 1(1–2):121–125
4. Brush AJ et al (2011) Home automation in the wild: challenges and opportunities. In: Proceedings of the SIGCHI conference on human factors in computing systems. ACM, pp 2115–2124
5. Ning H, Liu H, Yang LT (2013) Cyberentity security in the internet of things, vol 46, no 4, pp 46–53
6. Suoa H, Wan J, Zoua C, Liu J (2012) Security in the internet of things: a review. In: Proceedings of the 2012 international conference on computer science and electronics engineering (ICCSEE), Hangzhou, China, 23–25 March 2012, pp 648–651
7. Farooq MU, Waseem M, Khairi A, Mazhar S (2015) A critical analysis on the security concerns of Internet of Things (IoT). Int J Comput Appl 111(7):1–6
8. Stepanova, Zegzhda (2015) Achieving internet of things security via providing topological sustainability. In: Science and information, London
9. Raza, Seitz L, Sitenkov D, Selander G (2016) S3K: scalable security with symmetric keys—DTLS key establishment for the internet of things. IEEE Trans Autom Sci Eng 13(3)
10. Frey S, Rashid A, Anthonysamy P, Pinto-Albuquerque M, Naqvi SA (2008) The good, the bad and the ugly: a study of security decisions in a cyber-physical systems game. In: IEEE/ACM

40th international conference on software engineering (ICSE), Gothenburg, Sweden, pp 496–496. https://doi.org/10.1145/3180155.3182549
11. Sarigiannidis P, Karapistoli E, Economides AA (2015) Detecting Sybil attacks in wireless sensor networks using UWB ranging-based information. Expert Syst Appl 42(21):7560–7572
12. Sharma M, Tandon A, Narayan S, Bhushan B (2017) Classification and analysis of security attacks in WSNs and IEEE 802.15.4 standards: a survey. In: 2017 3rd international conference on advances in computing, communication and automation (ICACCA) (Fall), Dehradun, India, pp 1–5. https://doi.org/10.1109/ICACCAF.2017.8344727
13. Tandon A, Srivastava P (2019) Trust-based enhanced secure routing against rank and Sybil attacks in IoT. In: 2019 twelfth international conference on contemporary computing (IC3), Noida, India, pp 1–7. https://doi.org/10.1109/IC3.2019.8844935
14. Pongle P, Chavan G (2015) A survey: attacks on RPL and 6LoWPAN in IoT. In: Proceedings of international conference on pervasive computing (ICPC)
15. Yadav EP, Mittal EA, Yadav H (2018) IoT: challenges and issues in Indian perspective. In: 2018 3rd international conference on internet of things: smart innovation and usages (IoT-SIU), Bhimtal, India, pp 1–5. https://doi.org/10.1109/IoT-SIU.2018.8519869
16. Hameed S, Ali U (2018) HADEC: hadoop-based live DDoS detection framework. EURASIP J Inf Secur 2018(1):11
17. Khan FI, Hameed S (2016) Software defined security service provisioning framework for internet of things. Int J Adv Comput Sci Appl 7(12)
18. Hameed S, Khan HA (2018) SDN based collaborative scheme for mitigation of DDoS attacks. Future Internet 10(3):23
19. Aggarwal A, Alshehrii M, Kumar M, Alfarraj O, Sharma P, Pardasani KR (2020) Landslide data analysis using various time-series forecasting models. Comput Electr Eng 88. https://doi.org/10.1016/j.compeleceng.2020.106858
20. Bhardwaj A, Shah SBH, Shankar A, Alazab M, Kumar M, Reddy GT (2020) Penetration testing framework for smart contract blockchain. Peer Peer Netw Appl 2020. https://doi.org/10.1007/s12083-020-00991-6
21. Chithaluru P, Al-Turjman F, Kumar M, Stephan T (2020) I-AREOR: an energy-balanced clustering protocol for implementing green IoT for smart cities. Sustain Cities Soc 61:102–254. https://doi.org/10.1016/j.scs.2020.102254
22. Aggarwal A, Rani A, Sharma P, Kumar M, Shankar A, Alazab M (2020) Prediction of landsliding using univariate forecasting models. Internet Technol Lett e209. https://doi.org/10.1002/itl2.209
23. Aggarwal A, Rani A, Kumar M (2019) A robust method to authenticate license plates using segmentation and ROI based approach. Smart Sustain Built Environ. https://doi.org/10.1108/SASBE-07-2019-0083
24. Aggarwal S et al (2020) Meta heuristic and evolutionary computation: algorithms and applications. Springer Nature, Berlin, 949 pp. https://doi.org/10.1007/978-981-15-7571-6. ISBN 978-981-15-7571-6
25. Yadav AK et al (2020) Soft computing in condition monitoring and diagnostics of electrical and mechanical systems. Springer Nature, Berlin, 496 pp. https://doi.org/10.1007/978-981-15-1532-3. ISBN 978-981-15-1532-3
26. Gopal et al (2021) Digital transformation through advances in artificial intelligence and machine learning. J Intell Fuzzy Syst (Pre-press) 1–8. https://doi.org/10.3233/JIFS-189787
27. Fatema N et al (2021) Intelligent data-analytics for condition monitoring: smart grid applications. Elsevier, 268 pp. ISBN 978-0-323-85511-2. https://www.sciencedirect.com/book/9780323855105/intelligent-data-analytics-for-condition-monitoring
28. Smriti S et al (2018) Special issue on intelligent tools and techniques for signals, machines and automation. J Intell Fuzzy Syst 35(5):4895–4899. https://doi.org/10.3233/JIFS-169773
29. Jafar A et al (2021) AI and machine learning paradigms for health monitoring system: intelligent data analytics. Springer Nature, Berlin, 496 pp. https://doi.org/10.1007/978-981-33-4412-9. ISBN 978-981-33-4412-9

30. Sood YR et al (2019) Applications of artificial intelligence techniques in engineering, vol 1. Springer Nature, 643 pp. https://doi.org/10.1007/978-981-13-1819-1. ISBN 978-981-13-1819-1
31. Kumar M, Srivastava S (2017) Image forgery detection based on physics and pixels: a study. Aust J Fornesic Sci 51(2):119–134
32. Kumar M, Srivastava S (2018) Image authentication by assessing manipulations using illumination. Multimed Tools Appl 78(9):12451–21246

A Detailed Analysis of Adaptive Kernel Density-Based Outlier Detection in Volatile Time Series

Kumar Gaurav Ranjan and B. Rajanarayan Prusty

Abstract Preprocessing of historical observations is an essential step before any data-based analysis. Most power system studies using statistical methods use historical data of multiple variables collected from different places and time instants. A preprocessing method that detects outliers in a multivariate dataset in a single step is generally preferred. This paper presents a comprehensive analysis of a popular density-based preprocessing approach, i.e., the adaptive kernel density-based method applied to multivariate datasets. An attempt is made to explain the aforesaid method's algorithm in detail and the significance of its crucial parameters. The applicability of the approach is elucidated by applying it to various kinds of datasets and evaluating the performance using different measures. Furthermore, the analysis demonstrates the impact of crucial parameters on the method's performance. This paper will help a novice researcher apply and evaluate the aforesaid method according to the dataset type.

Keywords False positive · Outlier detection · Preprocessing · True positive · Volatile time series

1 Introduction

Probabilistic load flow (PLF) is an essential tool used to characterize the power system steady-state in a realistic way compared with deterministic load flow [1, 2]. PLF of a power system integrated with renewable generations includes numerous input random variables [3]. Forecasting of the above multivariate inputs is essential to perform the steady-state analysis of the power system. The effectiveness of the probabilistic forecasts required for PLF depends on input data quality. That is to

K. G. Ranjan
National Institute of Science and Technology, Brahmapur, India
e-mail: gaurav17998@gmail.com

B. R. Prusty (✉)
Vellore Institute of Technology, Vellore, India
e-mail: b.r.prusty@ieee.org

A. Tomar et al. (eds.), *Machine Learning, Advances in Computing, Renewable Energy and Communication*, Lecture Notes in Electrical Engineering 768,
https://doi.org/10.1007/978-981-16-2354-7_33

say; the input time series must be free from outliers to avoid the creeping of errors in modeling. Therefore, data cleaning/preprocessing of the input time series plays a vital role in detecting outliers and correcting them aptly. While much research attention has been given toward data preprocessing, most of them apply to univariate time series only. Multivariate data preprocessing can preprocess a multivariate time series and be considered an effective preprocessing compared with when each univariate time series of the same stochastic system is preprocessed. In the literature, the adaptive kernel density (AKD)-based approach is mostly appreciated for detecting outliers in multivariate data [4]. This paper aims to critically analyze the approach's applicability in a detailed manner by applying it to both clean and raw data.

In the literature, various categorizations of data cleaning approaches are available [5–9]. Out of the different categories, density-based and cluster-based approaches deal extensively with multivariate data. However, increased interest in research in the density-based category over cluster-based methods is because the former, unlike the latter, are not binary, i.e., they give a quantitative indication of outlierness. Further, nearby dense regions can be effectively preprocessed, and optimum performance can be attained by significantly less parameter optimization. The AKD-based approach is one such efficient density-based method [4]. It is a recent robust approach that has the capability of detecting outliers in multivariate time series. This approach is adaptive, eliminating over-smoothing and noisy estimates in regions of high and low densities, respectively. Despite possessing these qualities, the AKD-based method, like all other approaches, does not perform outlier correction.

The AKD-based approach being a recent discovery, many aspects of the method are yet not critically analyzed. The approach was tested with limited datasets. Therefore, the applicability of the approach to different datasets was not elucidated. The importance of the crucial parameters and the performance dependence on these parameters were not elaborated. These pitfalls led to the foundations of this paper. The paper's contributions are listed underneath.

- A wide range of density-based outlier detection approaches applicable to multivariate data are categorized (refer to Table 1).
- The AKD-based outlier detection approach is explained in detail with appropriate algorithmic steps.
- The AKD-based method's performance is critically analyzed considering various kinds of volatile time series.
- The significance of the crucial parameters associated with the AKD-based method is elaborated, and their impact is studied in detail.

In Sect. 2, the AKD-based approach is explained with its algorithmic steps. The result analysis, including a comprehensive description of different datasets, indices for performance evaluation, application of AKD-based approach to different datasets, and the effect of crucial parameters, is included in Sect. 3, followed by conclusions in Sect. 4.

Table 1 Existing density-based outlier detection methods for multivariate data

Year/original method	Current version(s)
2000/local outlier factor (LOF) [10]	Distributed LOF computing [11]
2004/relative density factor [12]	–
2007/Kernel density functions-based [13]	AKD-based [4]
	Scalable kernel density estimation-based [14]
2009/local outlier probabilities [15]	–
2009/resolution-based outlier factor [16]	Dynamic window outlier factor [17]
2017/relative density-based outlier score [18]	–

2 Adaptive Kernel Density-Based Outlier Detection Approach

It is an unsupervised approach that can be effectively applied to nonlinear multivariate time series [4]. This method gives a binary output (data point is an outlier or not) and calculates each data point's degree of outlierness. This approach's primary purpose is to compute the data points' extent of deviation in a local sense. To compute a data point's local measure of outlierness, this approach follows three necessary steps: define the reference set, i.e., the k-nearest neighbors, compute the local density (primary metric) and then compute the local outlierness (secondary metric). This approach uses a kernel function to estimate a data point's local density and attain smoothness in the final local outlierness measure.

Consider a $m \times n$ multivariate time series matrix $Y = \{y_1, y_2, \ldots, y_m\}$, where y_i denotes the data point at i^{th} instant, having n attributes. The very first step that this approach adopts is data normalization. Normalization standardizes different features' numeric ranges, which prevents the greater numeric ranges from dominating the lower ones in calculations. For this purpose, Z-score normalization is used that converts the matrix Y into normalized matrix $\widehat{Y}$. The ith data point y_i can be normalized as:

$$\widehat{y}_i = \frac{(y_i - \overline{y})}{\sigma} \tag{1}$$

where $\overline{y}$ is the column-wise mean vector of Y and σ is the column-wise standard deviation vector of Y.

Once the normalization is done, pair-wise distance between a particular data point of interest and the remaining data points $(d(\widehat{y}_i, \widehat{y}_j), \forall i, j \in \{1, 2, \ldots, m\}, i \neq j)$ are calculated using L_2 norm, and then the distances are arranged in ascending order to choose the k-nearest neighbors, i.e., the reference set of a data point. Each data point's local density is calculated by applying the Gaussian kernel, which is otherwise known as the radial basis function (RBF). The local density of $\hat{y}_i$ is given as

$$\rho_i = \frac{1}{m-1}\sum_{j=1}^{m} \exp\left\{-\left(\frac{\hat{y}_i - \hat{y}_j}{h_i}\right)^2\right\} \tag{2}$$

where h_i is the adaptive kernel width.

The term "adaptive kernel" comes from the adaptive selection of width parameter, which is not fixed, unlike classical density estimation. The selection of width parameter in this approach is based on the fact that small width and large width are, respectively, preferred in high-density and low-density regions. This avoids over-smoothing in high-density regions and noisy estimates in regions of low density.

Let $d_k(\widehat{y}_i)$ be the average distance of $\widehat{y}_i$ to its k-nearest neighbors and $d_{k-\min}$, and $d_{k-\max}$ be the smallest and the largest quantities in the set $\{d_k(\widehat{y}_i)|i = 1, 2, \ldots, m\}$. The kernel width is given as

$$h_i = C\left[d_{k-\max} + d_{k-\min} + \varepsilon + d_k(\hat{y}_i)\right] \tag{3}$$

where $C(C > 0)$ is the scaling factor. It controls the overall smoothing effect. ε is chosen to be a significantly small positive value for non-zero width.

In this approach, the term "local density" does not mean local outlierness measure; instead, it defines a relative measure of local outlierness, i.e., local outlier score (LOS). LOS of $\widehat{y}_i$ is computed as

$$\mathrm{LOS}(\hat{y}_i) = \log\left[\frac{\sum_{j\in k\mathrm{NN}(\hat{y}_i)} \rho(\hat{y}_j)}{k\rho(\hat{y}_j)}\right]. \tag{4}$$

Once the LOS of all the data points is calculated, outliers can be filtered out following two different ways, depending upon the user's choice, as follows:

i. Mark the top-most data points with largest LOS as outliers, with number of outliers known a priori.
ii. Set a threshold value of LOS, and mark the points as outliers whose LOS is larger than the set threshold.

The flowchart for outlier detection using AKD-based approach is shown in Fig. 1. The accuracy of the method depends on certain crucial parameters and Table 2 elucidated the significance of those parameters.

Begin: Select a $m \times n$ time-series matrix Y.

- Initialize the parameters C and k.
- Set the threshold value of outlier score LOS^{thr}.

for $i = 1$ until $i = m$

do

- Normalize the data point y_i using (1) and save it as $\hat{y}_i$ in normalization matrix $\hat{Y}$.

end for

for $i = 1$ until $i = m$

do

for $j = 1$ until $j = m$ and $j \neq i$

do

- Calculate the distance $d(\hat{y}_i, \hat{y}_j)$ between $\hat{y}_i$ and $\hat{y}_j$ using L_2 norm.

end for

- Obtain the k-nearest neighbors $k\text{NN}(\hat{y}_i)$ by sorting the distances calculate above.
- Calculate the average distance $d_k(\hat{y}_i)$ of $\hat{y}_i$ to its k nearest neighbors.

end for

- Obtain $d_{k-\min}$ and $d_{k-\max}$ from all the quantities $d_k(\hat{y}_i)$ where $i = 1,2,\ldots,m$.

for $i = 1$ until $i = m$

do

- Calculate the kernel width of $\hat{y}_i$ i.e., h_i, using (3).
- Calculate the local density of $\hat{y}_i$ i.e., ρ_i, using (2).

end for

for $i = 1$ until $i = m$

do

- Calculate the local outlier score of $\hat{y}_i$ i.e., $\text{LOS}(\hat{y}_i)$, using (4).

if $\text{LOS}(\hat{y}_i) > \text{LOS}^{\text{thr}}$

Mark y_i as an outlier.

end if

end for

Fig. 1 Algorithm for AKD-based preprocessing approach

Table 2 Significance of crucial parameters of AKD-based approach

Crucial parameters	Significance
Scaling factor (C)	It is a factor selected by the user to control the smoothing effect
k	It decides the number of data points to be included in the reference set. Based on the reference set of a data point, its kernel width, local density and local outlier score are determined
Threshold outlier score (LOS^{thr})	Since every data point has a local outlier score, a data point is treated as an outlier only if its outlier score is greater than the threshold outlier score

3 Result Analysis

3.1 Datasets

In this paper, the AKD-based outlier detection approach is tested with two different types of multivariate datasets (clean and raw). A dataset is considered clean if it is free from outliers. To test the performance of a method on such a dataset, outliers are inserted manually; therefore, the outliers' information is available a priori. On the contrary, a raw dataset is the practical data collected from different solar panels, substations, etc. Information about the outliers present in the data is not available a priori [19].

A bivariate synthetic data are created manually to analyze the performance of the adaptive kernel density-based approach (applicable to multivariate data) on clean data. The bivariate wind speed data are collected from two sites at Desert Knowledge Australia Solar Centre (DKSAC), Australia [20]. Table 3 describes the sample size and time information. The various cases listed in Table 3 are used for result analysis.

Table 3 Comprehensive portrayal of the data used for result analysis

Case no	Data information	Location	Notation	Data variety	Time information	Time step	Sample size
1	Synthetic	–	SD_1	Clean	–	–	1500
			SD_2				
2	Wind speed	DKASC, Australia	WS_1	Raw	August 2013	5 min	1000
			WS_2				

3.2 Indices for Performance Evaluation

For evaluating the method's performance for outlier detection in clean data, indicators as frequently used by many authors [4, 7] are adopted and elaborated underneath:

(a) TP—It stands for true positive. It refers to a data point, which is an outlier and is identified correctly by the preprocessing method.
(b) FP—It stands for false positive. It refers to a genuine data point, which is identified as outlier by the preprocessing method.
(c) FN—It stands for false negative. It refers to an outlier, which is overlooked by the preprocessing method.

Using the above indicators, the bases for evaluation are given as.

(1) Precision (P): $P = \frac{\text{No. of TP}}{\text{No. of TP+No. of FP}}$
(2) Recall (R): $R = \frac{\text{No. of TP}}{\text{No. of TP+No. of FN}}$
(3) F-measure (F): $F = \frac{2\times P\times R}{P+R}$.

Any obtained higher values for the above bases indicate better performance of the preprocessing method. Since preliminary information about the outliers is not available in a practical scenario, the bases used for evaluating a method's performance when applied to clean data cannot be applicable for raw data. So, an index-based performance evaluation is not possible for raw data. Hence, the preprocessing method's effectiveness is evaluated by critically analyzing the approach as well as its data-specific performance [21].

3.3 Application to Clean Data

The AKD-based approach's performance is analyzed considering clean multivariate data (Case 1 of Table 3). One hundred outliers are intentionally created in the clean data by increasing/decreasing the value of randomly selected data points by 30%. The resulting dataset obtained by inserting outliers manually is termed as "polluted data." The AKD-based approach is applied to the polluted data, and the results are shown in Fig. 2 and Table 4. It is inferred from Fig. 2 and Table 4 that the AKD-based approach performs above average when applied to clean data. Though it detects a small number of FP, a large section of outliers are not detected, even when they fall in the low-density region (evident from Fig. 2). Also, if a data point is so altered that it falls in the different areas of high density, it is treated as a genuine data point, which should not be the case.

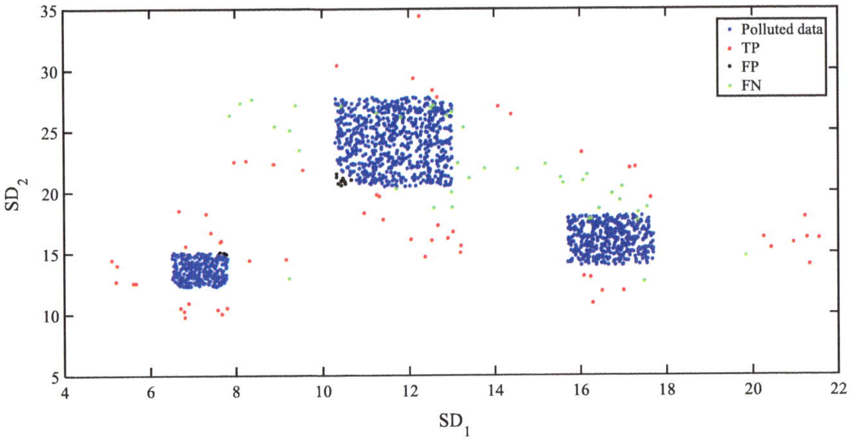

Fig. 2 Application of AKD-based preprocessing to Case 1 of Table 3

Table 4 AKD-based approach's performance analysis considering polluted data (bivariate)

Indices	AKD-based
TP count	60
FP count	13
FN count	40
P	0.8219
R	0.6
F-measure	0.6936

3.4 Application to Raw Data

This section aims to analyze the AKD-based approach's performance when applied to raw data. Case 2 of Table 3 is used for the analysis. The outlier detection capability of the AKD-based approach is elucidated in Fig. 3.

In Fig. 3, the AKD-based approach is applied to volatile multivariate raw data (Case 2 of Table 3) for outlier detection. It can be seen from the figure that the method detects many real outliers in the data without noticing a vast number of genuine data points as outliers. This shows the effective outlier detection performance of the approach. Further, the approach also reduces the over-smoothing effect in high-density regions. The method eliminates the noisy estimates in the low-density areas as the kernel width used in this approach is adaptive. But this approach cannot detect outliers in the high-density region, as this method assumes that outliers occur only in the low-density areas. The crucial parameters of AKD-based data have a significant impact on its outlier detection performance. These effects are elaborated in Sect. 3.5.

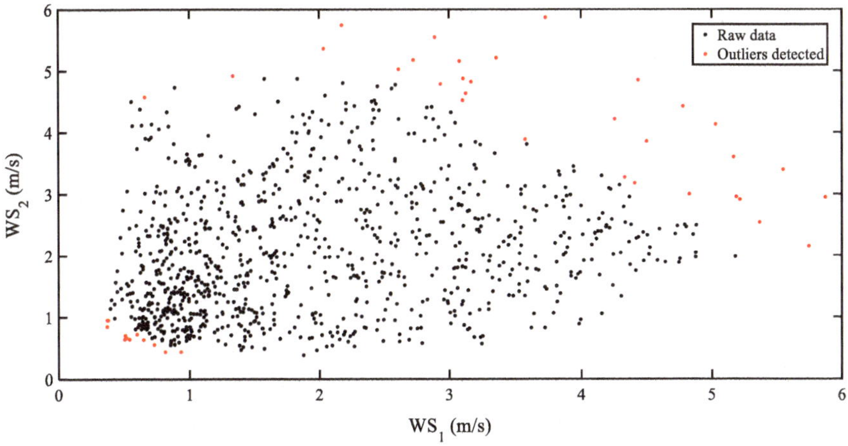

Fig. 3 Application of AKD-based outlier detection approach to Case 2 of Table 3

3.5 Impact of Crucial Parameter Values on AKD-Based Approach' Performance

Case 1 of Table 3 is used to show the effect of the crucial parameters on the method's performance. The outlier detection using AKD-based approach is performed by selecting various sets of values for the parameters (C and k), and the obtained results are shown in Table 5. The threshold outlier score is kept constant (0.2) in all the cases. It is inferred from Table 5 that the scaling factor C should be kept low to avoid the over-smoothing effect, whereas the value of k should be chosen high for the proper calculation of the data points' outlier score. Only then, the performance of the method will be optimum.

It is inferred from the analyses done in Section 3.3 through 3.5 that, with a proper set of crucial parameters value AKD-based approach is capable of giving good outlier detection results, irrespective of the type of data to which it is applied. The aforesaid method's final remarks, along with its merits, lacunae, and scope of improvisation, are summarized in the preceding section.

4 Conclusion

A compendious analysis of a well-established outlier detection method applicable to multivariate data, i.e., AKD-based approach, is presented in this paper. The method's algorithmic steps are explained lucidly, along with the significance of the associated crucial parameters. The approach is then applied to two different types of bivariate data, and its performance is evaluated using data-specific measures. The crucial parameters' impact on the outlier detection performance of the AKD-based

Table 5 Impact of crucial parameter values on the AKD-based approach's performance

Sl. no	Scaling factor (C)	k	TP count	FP count	FN count	P	R	F-measure
1	0.1	10	55	38	45	0.5914	0.55	0.5699
2	0.1	30	57	35	43	0.6196	0.57	0.5938
3	0.1	50	59	39	41	0.6020	0.59	0.5960
4	0.25	15	53	16	47	0.7681	0.53	0.6272
5	0.25	35	54	9	46	0.8571	0.54	0.6626
6	0.25	45	56	8	44	0.8750	0.56	0.6829
7	0.5	20	42	1	58	0.9767	0.42	0.5874
8	0.5	40	42	0	58	1	0.42	0.5915
9	0.5	50	43	0	57	1	0.43	0.6014
10	0.7	15	33	0	67	1	0.33	0.4962
11	0.7	25	36	0	64	1	0.36	0.5294
12	0.7	35	34	0	66	1	0.34	0.5075
13	0.9	15	24	0	76	1	0.24	0.3871
14	0.9	35	26	0	74	1	0.26	0.4127
15	0.9	65	28	0	72	1	0.28	0.4375

approach is also explicated. Through the detailed study of the AKD-based approach in Section 2 and the rigorous result analyses in Section 3, it is concluded that the AKD-based approach is an acceptable way for outlier detection in any multivariate data. It gives satisfactory outlier detection performance when applied with an optimized set of parameter values. It does not necessitate any prior information on the data set's probability distribution. It can overcome over-smoothing and noisy estimates in regions of high and low densities, respectively, because of its adaptive nature. However, simultaneously the AKD-based approach does not perform outlier correction; therefore, it does only half preprocessing. Furthermore, the need for generalized rules for optimizing the crucial parameters and detecting outliers in dense regions instigates scopes for future improvisation.

References

1. Bhat NG, Prusty BR, Jena D (2017) Cumulant-based correlated probabilistic load flow considering photovoltaic generation and electric vehicle charging demand. Front Energy 11(2):184–196
2. Prusty BR, Jena D (2017) A detailed formulation of sensitivity matrices for probabilistic load flow assessment considering electro-thermal coupling effect. In: 2017 IEEE PES Asia-Pacific power and energy engineering conference (APPEEC), pp 1–6
3. Prusty BR, Jena D (2019) A spatiotemporal probabilistic model-based temperature-augmented probabilistic load flow considering PV generations. Int Trans Electr Energy Syst 29(5):e2819

4. Zhang L, Lin J, Karim R (2018) Adaptive kernel density-based anomaly detection for nonlinear systems. Knowl-Based Syst 139:50–63
5. Zhang J (2013) Advancements of outlier detection: a survey. ICST Trans Scalable Inform Syst 13(1):1–26
6. Bhosale SV (2014) A survey: outlier detection in streaming data using clustering approached (IJCSIT) Int J Comput Sci Inform Technol 5:6050--6053
7. Ranjan KG, Prusty BR, Jena D (2020) Review of preprocessing methods for univariate volatile time-series in power system applications. Electr Power Syst Res 191:106885
8. Bakar ZA, Mohemad R, Ahmad A, Deris MM (2006) A comparative study for outlier detection techniques in data mining. In: 2006 IEEE conference on cybernetics and intelligent systems, pp 1--6
9. Goldstein M, Uchida S (2016) A comparative evaluation of unsupervised anomaly detection algorithms for multivariate data. PLoS ONE 11(4):e0152173
10. Breunig M, Kriegel H, Ng R, Sander J (2000) LOF: Identifying density based local outliers. In: Proceedings of the 2000 ACM SIGMOD international conference on management of data, Dallas, vol 29, pp 93--104
11. Bai M, Wang X, Xin J, Wang G (2016) An efficient algorithm for distributed density-based outlier detection on big data. Neurocomputing 181:19–28
12. Ren D, Wang B, Perrizo W (2004) RDF: a density-based outlier detection method using vertical data representation. In: International conference on data mining (ICDM'04), pp 503--506
13. Latecki LJ, Lazarevic A, Pokrajac D (2007) Outlier detection with kernel density functions. In: Proceedings of the 5th international conference on machine learning and data mining in pattern recognition, pp 61--75
14. Qin X, Cao L, Rundensteiner EA, Madden S (2019) Scalable kernel density estimation-based local outlier detection over large data streams. EDBT, pp 421--432
15. Kriegel H, Kröger P, Schubert E, Zimek A (2009) LoOP: Local outlier probabilities. In: Proceedings of the 18th ACM conference on information and knowledge management (CIKM'09), Hong Kong, China, pp 1649--1652
16. Fan H, Zaıane OR, Foss A, Wu J (2009) Resolution-based outlier factor: detecting the top-n most outlying data points in engineering data. Knowl Inform Syst 19(1):31--51
17. Momtaz R, Mohssen N, Gowayyed MA (2013) DWOF: a robust density-based outlier detection approach. In: Iberian conference on pattern recognition and image analysis, pp 517--525
18. Tang B, He H (2017) A local density-based approach for outlier detection. Neurocomputing 241:171–180
19. Ranjan KG, Prusty BR, Jena D (2019) Comparison of two data cleaning methods as applied to volatile time-series. In: 2019 International conference on power electronics applications and technology in present energy scenario (PETPES), pp 1--6
20. Wind speed data. http://dkasolarcentre.com.au
21. Ranjan KG, Tripathy DS, Prusty BR, Jena D (2020) An improved sliding window prediction-based outlier detection and correction for volatile time-series. Int J Numer Model e2816. https://doi.org/10.1002/jnm.2816

Vehicle Health Monitoring System Using IoT

Sharmila, Annu Bhardwaj, Madhu Bala, Priyanka Mishra, Lavita, and Janvi Gautam

Abstract Nowadays, as we all know that automobile industry is one of the growing industries. In every house, there is a vehicle and also there is necessary. Vehicle Health Monitoring System (VHMS) is designed to make our journey more suitable. In this, we use multiple sensors, which are used to fetch the activity of vehicle. It is hard to know any failure in advance due to sensors that are present in the vehicle. By this project, we are able to know about any damage, fuel level, temperature of engine, condition of tires of vehicle, level of carbon dioxide emitted from vehicle and give feedback to avoid loss. Using Internet of Things (IOT), in this project, is remarkable. IOT is changing the structure of technologies and allows us to implement various complex systems.

Keywords IOT · VHMS · Automobile industry · Sensor

1 Introduction

Vehicles are now a major part of our comfortable life or we can say that it is necessary too. Automobile industries are most advanced and also developing drastically. Automobiles are having complex system, both hardware as well as software, and need the best maintenance strategy. We can not avoid technology and advancement in our daily life, and at the same time, we are facing problems too. It is very necessary for human to be healthy, similarly, we have to be very careful toward our vehicle too. Between all our necessity and glamorous life, we are also harming nature.

As we all know, the production and sale of automobiles increase rapidly. Domestic automobile production increases at 2.36% CAGR between FY16 and 20 with 26.36 million vehicles being manufactured in the country. By this increment, it is approaching to solve the risk of air pollution. Vehicles are powered by fuels, which are major causes of air pollution. After the combustion of fuel, it releases CO_2 directly into our atmosphere and also in very big amounts. Components from vehicle are

Sharmila (✉) · A. Bhardwaj · M. Bala · P. Mishra · Lavita · J. Gautam
Krishna Engineering College, Ghaziabad, Uttar Pradesh, India

A. Tomar et al. (eds.), *Machine Learning, Advances in Computing, Renewable Energy and Communication*, Lecture Notes in Electrical Engineering 768,
https://doi.org/10.1007/978-981-16-2354-7_34

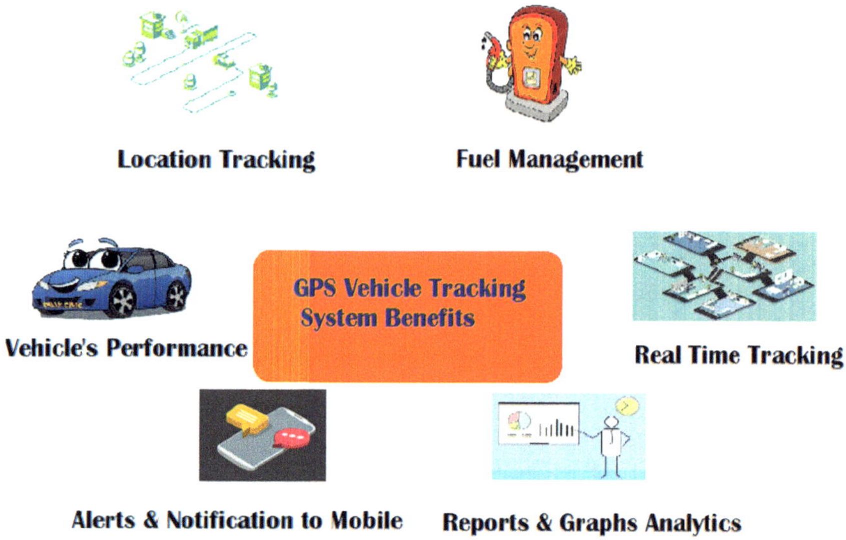

Fig. 1 Benefits of vehicle health monitoring system

harmful to our lungs and also global warming is a very big issue. Figure 1 shows the benefits of vehicle monitoring system.

The main contribution of our paper is to give solution to all your problems. After a well maintenance, some faults take place in vehicle sometimes. This project is about IOT and vehicle health monitoring system (VHMS). A vehicle itself is having many sensor IoT effective monitorings. It is necessary that our vehicle is at good condition.

The objective of this paper is to inform the owner about the health of vehicle to avoid any loss.

Innovation in automobile is endless, now there are so many vehicles with advance technology. Our "HEALTH MONITORING SYSTEM" gives the overall health report of any vehicle in this project when we are using IOT. We are using GSM and GPS systems and sensor connected with microcontroller for monitoring vehicles. The controller sends message to the owner about any unusual activity of the vehicle. This system sends message about any problems with tires, engine failure and engine temperature. As we mention air pollution, it also informs the amount of co_2 and harmful components that are released in air. We believe that by using IOT, our project is tremendous.

2 Related Works

The vehicle health monitoring system is being approached by the industry with various existing technologies, which are used in vehicle industry to improve the functioning of vehicles using machine learning, algorithms, different data and operators to make it connected remotely for communication [2]. Machine learning system is proposed to the vehicle industry to automatically improve and being able to perform its intended function as it measures the system reliability with the existing AI (artificial intelligence) transforming industry into more automotive industry by the learning approaches, computing and designing [1], while new algorithms are presented as "Sequential Pattern Mining" to mining the sequential pattern between data examples and delivered sequence. By the increasing crowd in traffic caused dangerous situations frequently to overcome those of situation new intelligence system and technologies being used to detect the problem in advance [3]. Presenting an existing model that is diagnosed the automatic term and having reporting capabilities as the function of OBD2 that is remotely used by the system interface as supervised techniques to understand the data that systems gather as it can be diagnosed the real-time data, while OBD1 was introduced as a consoler or connected through wireless medium [4–6]. In the other way, VHMS may be informed of health monitoring of energy storage system of the EV [7, 8].

The proposed vehicle remote health monitoring system includes the subsystem of Arduino with Wireless System, which is introducing the wireless communication service in the era of advancement in technologies where the remotely connected vehicle will be monitored by its user or server and can predict its fault functioning in advance by sensing its signal processing and directly alert its user, as it is also used as an integrated temperature sensor that operates the temperature and reduces the fuel emission of the engine as it detects and integrates the thermal element.

3 Proposed System and Methodology

The proposed vehicle remote health monitoring system, using Arduino, sensors and GSM module, for sensing the concentration of gas, remotely operates the temperature and transmits the detected level of information to the users or service provider whenever the limit exceeds for the safe access of the vehicle.

Figure 2 shows the diagram of the vehicle health monitoring system that we used as the hardware in our project for the implementation while explaining the required functions of the given blocks as all of them sequentially related to each other and responsible for the health monitoring system of the vehicles as they all are interconnected to each other and accessing the given functions while informing the user/server, and it will be able to conduct the required task digitally.

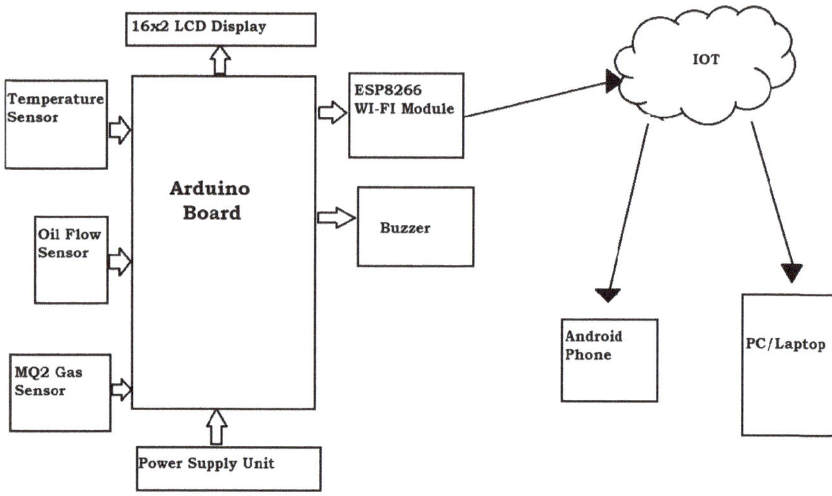

Fig. 2 Block diagram of vehicle health monitoring system

3.1 Proposed Functional Pattern

The functional pattern stated as:

- **MQ2 gas sensor**. It is used to detect combustible gases and smoke as it has an electrochemical sensor, which is able to detect any toxic and combustible gas from the vehicle that is causing the excessive level of pollution.
- **Arduino Nano**. It is the type of microcontroller that is used to transmit the information of the vehicle to its user and service provider, which is processed with the help of sensors.
- **GSM Module**. It is the device that stands as global system for mobile communication and used as a communication device, which helps to make communication between computing machine and microcontroller and able to transmit the data through mobile communication system. As the above-mentioned device helps to detect the level when the level exceeds its limit, device sent message about the level to its user. Figure 3 shows the functional pattern of the proposed system.

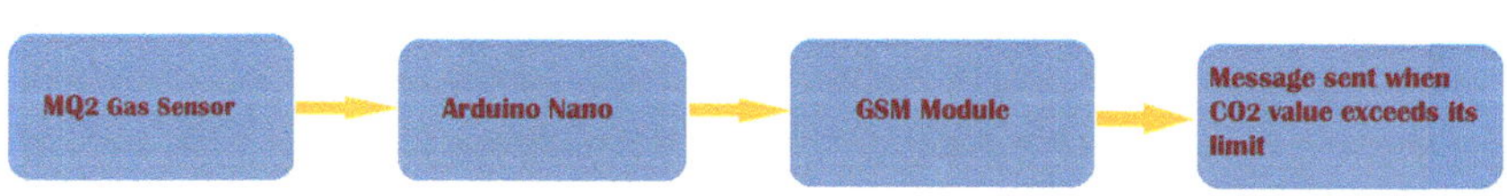

Fig. 3 Functional pattern of the proposed system

3.2 *Construction of Vehicle Health Monitoring System*

Power Supply. This supplies the power to connect the device of microcontroller for the transmission of data.

MCU (Microcontroller unit). It is used to process the data and loaded the data to provide its user in an accessed form.

Gas Sensor. It is used to detect any type of combustible and toxic gas that is causing pollution using electrochemical sensors.

GSM Module. System that is used for mobile communication helps to communicate between the computing machine and microcontroller to transmit the information.

SMS. It is the easiest way to alert the users about the exceeded limit of the level of temperature, fuel emissions of the engine and causing the pollution through message.

The above-mentioned method is to remotely connect the users who are able to access important information about their vehicles. Figure 4 shows the construction of vehicle engine monitoring system.

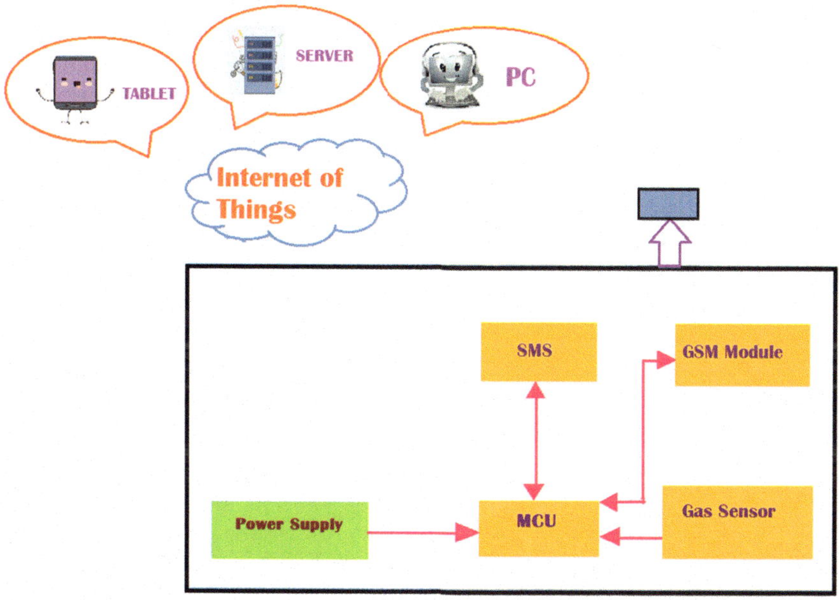

Fig. 4 Construction of vehicle engine health monitoring system

4 Observation

In India, pollution has become one of the major problems due to which many health problems have been recorded so far. Among all, air pollution has the highest contribution to deaths. So, to control this problem, we have made a device that can give information about the pollution level of vehicles and according to which we can control or maintain it. This pie chart shows the sources of air pollution in which various factors are responsible:

•38% of pollution in the air is caused by dust and construction.
•21% of pollution in the air is caused by transport.
•17% of pollution in the air is caused by water boring.
•9% of pollution in the air is caused by diesel generators.
•8% of pollution in the air is caused by industries.
•7% of pollution in the air is caused by domestic cooking.

Figure 5 shows the major sources of air pollution in India. There is about 2 million premature deaths of Indians every year due to air pollution as per the records. This increase in air pollution is due to the exhaust emissions from diesel vehicles, smoke from factories, etc., which is responsible for the problem of asthma and breathlessness, cancer, eye infection and many others as shown in the graph.

- The bar graph shows the various reasons for concern due to air pollution in which Asthma and Breathlessness are at the highest and Global warming is at the lowest.
- These eight problems are the major consequences of air pollution.

Figure 6 shows the major health problems faced by people due to air pollution produced by vehicles.

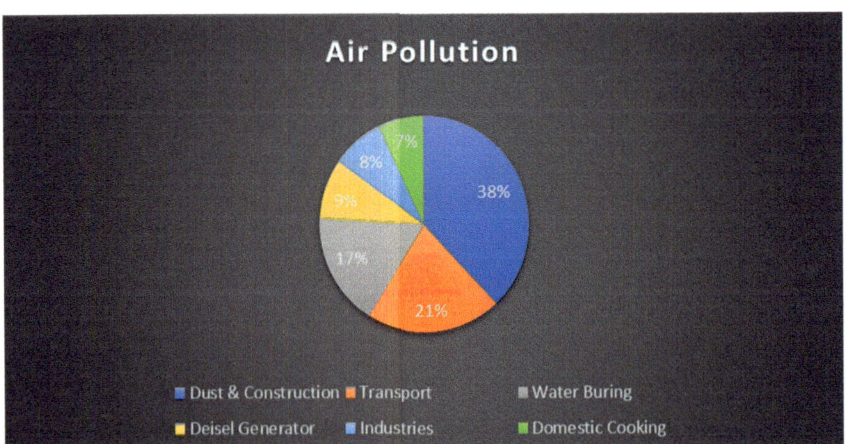

Fig. 5 Pie chart shows the major sources of Air Pollution in India

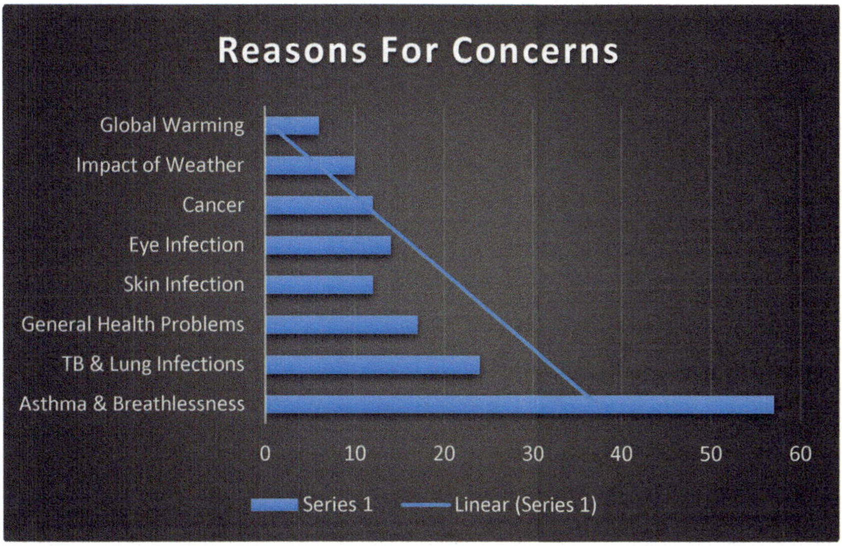

Fig. 6 Major health problems faced by the people

•The increase in the PM 2.5 level is due to the emission of CO_2 per passenger when they travel via solo vehicle, public transport, etc.

5 Conclusion

The paper is concerned with the health of vehicle and their monitoring. The information given by the device includes vehicles' performance on various parameters of their subsystems. In this paper, we have performed various tests for the vehicle to monitor and analyze the health and obtained their real-time results. In this era of advancement and development in technologies and with the great usage of smartphones, laptops, tablets and wireless communication, it has become possible to use the technology along with various tools, languages and devices to solve the problems of industries, companies and the mankind. The main motive of the proposed system is to reduce the problems and fault in systems and make it a user-friendly vehicle. This system is designed in such a way that can frequently and rapidly give information about the fault in any system, i.e. vehicle can respond faster than manual methods. All the possible scenarios of vehicle crashes are studied in this paper. A clear conclusion that emerged from the work is that a good approach must be taken for the development of such a system with better use of technologies, applications and tools to make it a successful device.

6 Future Work

The WiFi module is used to connect it with WiFi network to make it more user friendly, so, as the future work, it can be extended to more signal processing by advance GPS, which is able to keep track of its functioning and fault and will be able to provide advance information to the users and regular service station to save from dangerous situations. With the oil flow sensors, we can keep the information about the fluid that is flowing in the pipe and can keep the access of the surrounding with the help of temperature sensor that is able to keep the access to the present energy. By effective monitoring system with modified technologies about its functioning and data can record the route of the vehicle by sensors, it will also keep track about its pollution level and severe accident caused by the negligence of its owner and can be able to take action against. While the given functions, its wireless medium which is able to connect both user/server and vehicle, are all user friendly and keep the pattern of the vehicle digitally.

References

1. Desai M, Phadke A (2017) Internet of things based vehicle monitoring system. In: Proceedings of the 2017 fourteenth international conference on wireless and optical communications networks (WOCN). Mumbai, pp 1–3. https://doi.org/10.1109/WOCN.2017.8065840.
2. Bojan TM, Kumar UR, Bojan VM (2014) Designing vehicle tracking system an open source approach. In: Proceedings of the IEEE international conference on vehicular electronics and safety (ICVES), pp 16–17
3. Türk E, Challenger M (2018) An android-based IoT system for vehicle monitoring and diagnostic. In: Proceedings of the 2018 26th signal processing and communications applications conference (SIU). Izmir, Turkey, pp 1–4. https://doi.org/10.1109/SIU.2018.8404378
4. Patel R, Dabhi VK, Prajapati HB (2017) A survey on IoT based road traffic surveillance and accident detection system (A smart way to handle traffic and concerned problems). In Proceedings of the 2017 innovations in power and advanced computing technologies (i-PACT), Vellore, pp 1–7
5. Hamid AFA, Rahman MTA, Khan SF, Adom AH, Rahim MA, Rahim NA (2017) Connected car: engines diagnostic via Internet of Things (IoT). J Phys Conf Ser
6. Li Q, Cheng H, Zhou Y, Huo G (2015) Road vehicle monitoring system based on intelligent visual internet of things. J Sens 2015:720308. https://doi.org/10.1155/2015/720308
7. Fatema N et al (2021) Intelligent data-analytics for condition monitoring: smart grid applications, Elsevier, Amsterdam, p 268. ISBN: 978-0-323-85511-2. https://www.sciencedirect.com/book/9780323855105/intelligent-data-analytics-for-condition-monitoring
8. Iqbal A et al (2021) Chapter 10–intelligent data analytics for battery health forecasting using semi-supervised and unsupervised extreme learning machines. Intelligent data-analytics for condition monitoring, Academic Press, Cambridge, MA, pp 215–241. ISBN 9780323855105. https://doi.org/10.1016/B978-0-323-85510-5.00010-7

Analyzing the Attacks on Blockchain Technologies

Vinay Kumar Vats and Rahul Katarya

Abstract Blockchain technology has gained significant interest because of a wide variety of possible uses. It was first developed as a Bitcoin cryptocurrency but has since been used in many other business and non-business applications. There are various innovations available from Bitcoin to financial innovations, risk management, IoT, and public and social services. The blockchain infrastructure, instead of most current systems based on center-based architectures, combines peer-to-peer networks and distributed platforms that use blockchain registers to store transactions. But, like the Internet in 1990, the blockchain is regarded as still growing. In the future, it can alter so many aspects of technology. However, there are many setbacks, mainly due to the defense area, as a new undeveloped sector. So, it becomes important to analyze their success in multiple use cases and scenarios as more and more diverse blockchain technologies have appeared. We address the security concerns of blockchain technology in this survey. In this paper, with a focus on shared blockchains, we systematically analyze the attack surface of blockchain technologies.

Keywords Blockchain · Vulnerability · Smart contract · Consensus algorithms · Ethereum · Hyperledger

1 Introduction

Since Bitcoin's debut in 2009, its fundamental methodology, blockchain, has demonstrated exciting implementations and drawn much academic and industrial knowledge. The blockchain is Bitcoin's central system. In 2008 and 2009, blockchain was first suggested by Satoshi Nakamoto [1]. In 2015 [2] and the most powerful asset in 2016 [3], Bitcoin was ranked as the first crypto-currency and checked in May 2017 more than 300 k transactions [4] per day. The theoretical framework of blockchain's crypto-monetary technology has been investigated in previous studies

V. K. Vats (✉) · R. Katarya
Department of Computer Science & Engineering, Delhi Technological University, Shahbad Daulatpur, Delhi 110042, India

A. Tomar et al. (eds.), *Machine Learning, Advances in Computing, Renewable Energy and Communication*, Lecture Notes in Electrical Engineering 768,
https://doi.org/10.1007/978-981-16-2354-7_35

[5]. While some studies focused on blockchain's security concerns, because of rising demands for crypto-monetary and security issues, they did not concentrate on the cyber vulnerability of blockchain. The study in [6] focused on the basic sense of Bitcoin's crypto-currency, its use, and functions, and its aspects of privacy are one of these studies. Smart contracts in Ethereum [7], with general programming glitches and bugs associated with taxonomy blockchain vulnerabilities. Many blockchains are designed to increase their efficiency by improving the setup of the protocol and the creation of new consensus algorithms, for example, the performance benefits/drawbacks of the new update in comparison with the Hyperledger Fabric versions such as HLF v0.6 and HLFR v1.0 are in the same sense of calculation [8]. Besides, bottlenecks can be defined and used through performance measurement and analysis to facilitate further improvement ideas. Therefore, in the field of blockchain science, performance assessment plays an important role.

Blockchain technology is a decentralized technology for data and transaction management that offers data security, privacy, and accountability without a third-party entity. The technology can be interpreted as a shared directory of all transactions in a blockchain. This chain is constantly extended by the addition of new blocks. Blocks are arranged according to a sequence, of which the stack base is the bottom element. The previous block of the chain is connected to each block. Any block is defined by a hash created by cryptographic hash algorithms. A block contains a header forming a chain that links it to its parent blocks, consisting of a certain hatch of its parent blocks. The first block is known as a genesis block.

Blockchain technology also incorporates the technique of Hashing for security. There are some properties in cryptographic hashes that are very helpful in blockchain operations. The property's hiding property should be well built to be crash-proof and help to make puzzles easier. The hiding property can be difficult to find when a hash output is given. It is hard to find two plaintext items producing the same Hash performance because of the collision resistance function.

While there are some recent studies on blockchain performance, none of them conducts a comprehensive review of the threats to blockchain systems, the related actual attacks, and the security improvements. Some of the recent studies related to attacks and blockchain applications are also available in [9–13]. Based on these factors, our research provides a thorough analysis of security vulnerabilities in blockchain technologies by exploring attack vectors that concentrate on user security and their vulnerabilities. We also analyze the types of attacks that pose both realistic and theoretical threats to blockchain technology at different levels. The rest of the paper is organized in the following order:

I. Section 2 presents the structure and some key features of blockchain technology.
II. In Section 4 different blockchain generations and there, vulnerabilities towards the attacks are discussed.
III. In Section 5 risks related to blockchain technology are classified and discussed.

2 Structure and Overview of Blockchain Technology

The key technologies used in blockchain are presented in this section.

2.1 *Consensus*

At the heart of a blockchain system is a consensus protocol. The regulations are established and all nodes can be used to agree on blockchain content, for instance. Two primary modes of consensus [14] are usually focused on facts and vote. The most popular proof of work (PoW), used by many blockchain systems, is consensus-based on evidence. PoW is a very computer-based consensus. The nodes will solve an appalling puzzle to fight for the right to record. The first node (called the winner) was the only way to solve the jigsaw. In comparison to PoW, voting solutions produce a deterministic outcome and produce relatively high output in general. For each block order to achieve agreement, they depend on normal message transitions between different functions on a network. Two members of this consensus type are the Raft and Byzantine Fault Tolerance algorithms (e.g. pBFT and BFT-SMaRt). Raft tolerates the crash defects only, while the byzantine erroneous defects are fixed by pBFT and BFT-SMaRt.

2.2 *Block Structure*

The blockchain is a block series, including traditional public records, with an exhaustive list of transaction records. Any block points to a link called the parent block, which is mostly a previous block hash value. A block is a header and the block body. In specific Block editions, Parent Block Hash and Merkle Tree Root have block header, timestamp, and nuncio. The block body creates a counter and a transaction. The maximum number of transactions per block depends on each transaction's package size and size (Fig. 1).

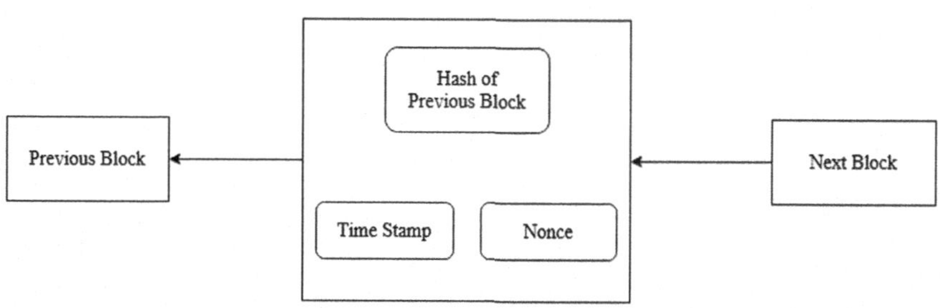

Fig. 1 Block structure of blockchain

2.3 *Key Cryptography*

Every entity owns two basic keys, one is a private key for its encryption and the other for verification used by other users. For signing transactions, a private key is used. Digital transactions signed are distributed across the network and can only be accessed through public keys to any network user. The first step is to establish the hash value derived from the contract when a user wants to enter into a contract. Then he uses his private key to encrypt this hash value and sends the encrypted hash to another user with original information. Another user tracks the transaction obtained by matching the decrypted hash with the hash extracted from the information received (Fig. 2).

3 Blockchain Generations

Relating blockchain technology operations are structured and accessible as follows: (a) public blockchain in the first generation, (1.0), (b) public blockchain in the second generation, (2.0), and (c) blockchain in the third generation. In this section, we will classify the security threats faced by different blockchain generations [15] (Fig. 3).

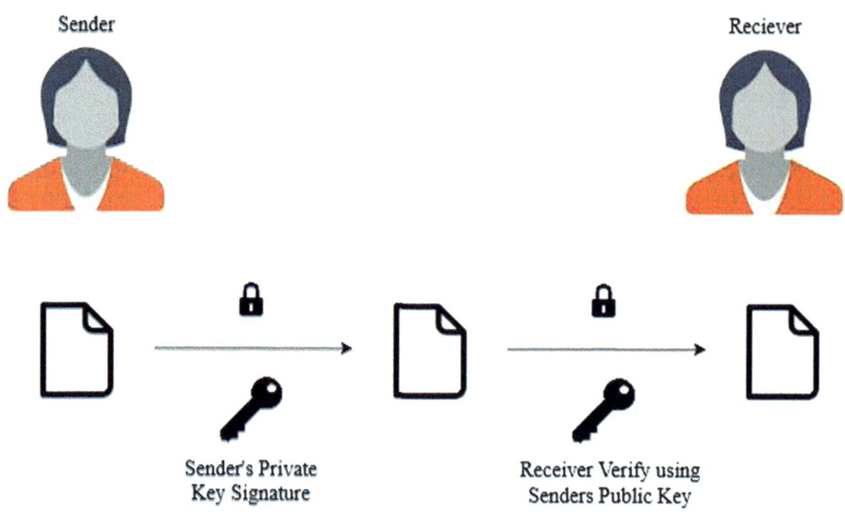

Fig. 2 Illustration of key cryptography

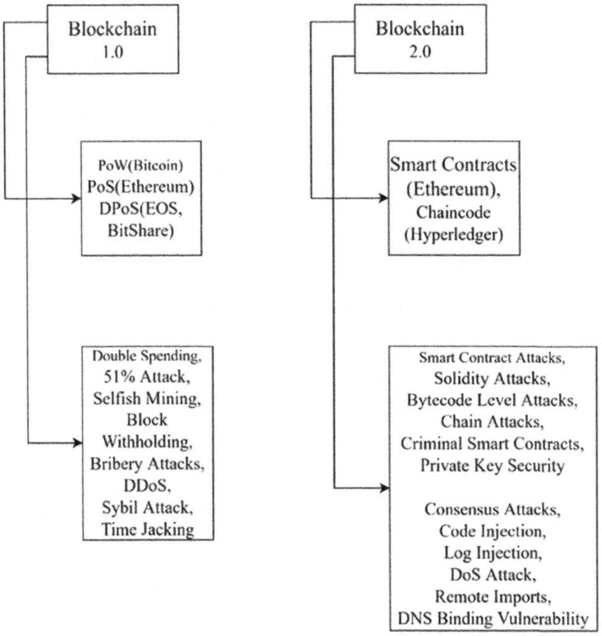

Fig. 3 Generation wise attack vulnerability of blockchain system

3.1 *First Generation of Blockchain System*

Blockchain 1.0 was the first blockchain technology implementation in 2009 and is thus v1.0.0. Cryptocurrency is an exchange medium that, using encryption techniques for monitoring money generation, is manufactured and stored electronically in the blockchain and allows for verifying the transfer of money. When blockchain and cryptocurrency technology first came to the market, one of the key roles was to eliminate third-party interaction in all forms of currency movement.

3.2 *Second Generation of Blockchain System*

In effect, Blockchain 2.0 is a system that enables programmable transactions (a condition-modified transaction or collection of conditions) introduced in the year 2010, that is when Ethereum [16], a platform where the developer community can create distributed blockchain network applications, also known as "smart contracts", was born. These are an autonomous computer program that is automatically executed and conditions specified in advance, such as facilitating, checking, or implementing the execution of a contract.

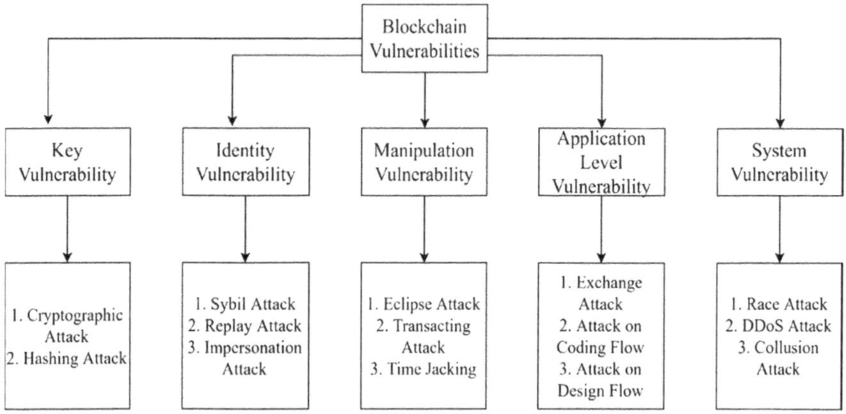

Fig. 4 Classification of blockchain attacks

3.3 Third Generation of Blockchain System

Blockchain 3.0 is a summary of attempts in blockchain industries to solve current problems, particularly scalability, interoperability, and privacy problems. Here is a new concept for Decentralized Apps (DApp). Decentralized storage and decentralized network are used because most DApps operate a blockchain, decentralized peer-to-peer network with their backend code. However, a DApp can host its portfolio in decentralized storages like Ethereum, Swarm [17], and Hyperledger [18] (Fig. .4).

4 Risks and Vulnerability

4.1 Key Vulnerability

In most blockchain processes, cryptographic hashes are used to protect chain integrity and transactional property [19]. The most common algorithms for blockchain implementations are digital signature (DSA) algorithms based on asymmetric key encryption and cryptographic hacking.

4.1.1 Cryptographic Attacks

A cryptographic attack is a method of discovering a loophole in a document, cipher, cryptographic protocol, or key management system to bypass the security of a cryptographic system. The attacker's objective in cryptography is to crack the confidentiality of encryption and to learn the hidden message and the secret key.

4.1.2 Hashing Attacks

There occurs a condition when the attacker attacks the hashing algorithms used by the blockchain system. A pass-the-hash attack (PtH) was a technique used by an attacker to collect the hash key, which it transmits to other networked systems for authentication and possibly lateral access, as opposed to key functionality. To obtain a plain password, the threat actor needs to decode the hash.

4.2 Identity Vulnerability

When stored over a blockchain, information is cryptographically encrypted and cannot be changed or erased, rendering large breaches of information extremely difficult, if not technically impossible. With the digital identities of users cryptographically stored directly on a blockchain inside an internet browser, technically, users will no longer need to provide any third party with sensitive data. Some of the attacks are listed below.

4.2.1 Sybil Attack

These types of attacks are common and are used by a single rogue group to recognize and monitor multiple false identities. To isolate the target node from the rest of an honest network in the blockchain network, this method of attack is used.

4.2.2 Replay Attack

This kind of attack spooks the attacker and gives the two legitimate parties access to correspondence. Stealing and reuse the hash key make the attacker a legitimate user to block the ecosystems.

4.2.3 Assault Impersonation

The impersonation of a legit consumer is commonly used to get entry. An ECDSA algorithm can also be extended to some other approaches suggested for using a distributed incentive-based approach. In comparison, BSeiN [20] used the users to validate an attribute-based signature.

4.3 Manipulation Vulnerability

If the rest of the network can partition one or more nodes in the blockchain system, different routing attacks may occur for malicious purposes. The use of such attacks, DoS attacks, and large portions of network mining resources can be postponed for a significant period, and other attacks can be carried out. The three major forms of attacks are eclipse, time jack, and transaction malleability-based assault.

4.3.1 Eclipse Attack

The eclipse attacks [15] constitute a kind of attack and the assailant tries to isolate a target by monopolizing all incoming and incoming links. This lets the attacker damage the blockchain's goal vision, waste his computing energy, or weaken the target's computer power for malicious purposes.

4.3.2 Transactional Attacks

The malleability of transactions faults the design of Bitcoin, which may change transactions before being added to a block after generation. It is not possible to change the source/government addresses or transaction numbers, but another component of the transaction can be modified.

4.3.3 Time Jacking

The Time Jack is an attempt by communicating with several people to skew the time signature for the target node and reporting the time to the target. To validate fresh blocks, the node network time is used. When a network time view from a node is skewed, a timestamp greater than a given time frame is rejected for new blocks. It makes it easy to isolate a target node from the rest of the network. Betrayed transactions can be created and transferred to the target by isolating a target node.

4.4 *Application-Level Vulnerability*

Utilities that operate on or connect with the blockchain are by far the poorest link to blockchain protection. While the underlying blockchain protocol is robust, monetary loss and the people affected, of course, have all been the target of a variety of attacks that have fallen prey to trade, wallet, and decentralized apps (DApps).

4.4.1 Design Flow Attacks

The Ethereum blockchain is a strong framework for intelligent contracts and DApps to be built and run. A decentralized independent organization (DAO) is a clever contract-based system in which citizens can fund voting by proposals.

4.4.2 Code Flow Attack

Some wallet applications were also targeted because of code errors. By accident in November 2017, the 513,753 ether "kill" bug in Parity's wallet program, which at the time of this writing was worth some 355 million. This is possible because Parity has concluded every multi-signature agreement in a library.

4.4.3 Exchange Attack

The term exchange refers to the place where the transaction and exchange of currencies take place for keeping the wealth moving. So, sometimes the attackers directly try to attack the exchange where payments and transactions, and parties take place. This happens because most exchanges are relatively small enterprises with less money to invest in cybersecurity and are (or were) start-up companies.

4.5 *System Vulnerability*

Blockchain technology systems are based on a data structure that is only connected to network storage and transaction. The system-based attacks attack the network architecture and physical hardware, servers, and access terminals of the working network. The attacker tries to overload the whole connected terminals with service requests and this overflow of requests degrades the system performance and ultimately the attacker tries to crash or manipulate the network.

4.5.1 Race Attack

The race attack enables the attacker to produce two transactions, one real and one false. The purpose is to allow any node accepting 0 unconfirmed status transactions that show that the transaction is not yet visible in a block. As a network peer, the attacker links the target directly.

4.5.2 DDoS Attack

The term DDoS refers to Distributed Denial of Service. A hijacked computer in this type of attack is normally used by an attacker to overload a network with disproportionate requests to impede the network's ability to support the supply of high traffic [9].

4.5.3 Double Spending Attack

The attacker here tries to spend the same currency multiple times. After waiting for some trust, the assailant created and put him into a new fraudulent block and a fraudulent dispute settlement. The attacker then eliminates or leases a large part of the network's mining power.

4.5.4 Collusion Attack

The 51% attack is perhaps the most established attack due to its ability to fully subvert the blockchain. The mining force of more than 50% of the network is dominated by one entity or group in this form of attack (Tables 1 and 2).

Table 1 Vulnerabilities and their precautionary measures

Vulnerability	Flaws	Precautionary measures
Key	Vulnerable cryptographic algorithm	A strong and powerful cryptographic algorithm, password-protected secret sharing, hardware wallets
Identity	A weak network system	Stronger authentication
Manipulation	Network process	Network encryption
Application	Weak development process	Better design and development
System	Service speed vs. Time tradeoff	Trained user base

Table 2 Classification of attacks based on blockchain surface

Blockchain surface	Attacks
Blockchain structure attack	Forks, orphans
Peer to peer system attacks	Eclipse attacks, selfish mining, DDoS attacks, DNS hijacks, BGP hijacks, time jacking, block withholding
Attacks over blockchain applications	Double spending, wallet theft, crypto-jacking, blockchain ingestion, overflow attacks, replay attacks

5 Conclusion

The blockchain's decentralized platform and peer-to-peer design are highly regarded and endorsed. Through storing data across the network, blockchain has removed the threats that come with data centralization. The use of encryption technology in blockchain security systems improves defense. In this paper, we present a thorough investigation into the blockchain. First, we give an overview of blockchain architecture and main blockchain functionality. Then we discuss the different generations of blockchain development over the years of work and risks related to them. Also, many risks can complicate the implementation of a blockchain that is classified and compared in the survey.

For future work, we are planning toward an in-depth investigation and analysis of blockchain 2.0 and 3.0 generation-based applications including Smart contracts, Hyperledger, and some blockchain-based Decentralized Applications (DApps).

References

1. Nakamoto S (2008) A peer-to-peer electronic cash system. https://bitcoin.org/bitcoin.pdf. Accessed 12 August 2020
2. Desjardins J (2016) It's official: bitcoin was the top-performing currency of 2015. http://money.visualcapitalist.com/its-official-bitcoin-wasthe-top-performing-currency-of-2015/. Accessed 06 Aug 2020
3. Adinolfi J (2016) 2016's best-performing commodity is ... bitcoin? http://www.marketwatch.com/story/and-2016s-best-performing-commodity-isbitcoin-2016-12-22. Accessed 16 Aug 2020
4. Blockchain.info (2017) Confirmed transactions per day. https://blockchain.info/charts/n-transactions?timespan=all/
5. Tschorsch F, Scheuermann B (2016) Bitcoin and beyond: a technical survey on decentralized digital currencies. IEEE Commun Surv Tutor 18:2084–2123
6. Conti M, Sandeep Kumar E, Lal C, Ruj S (2018) A survey on security and privacy issues of bitcoin. IEEE Commun Surv Tutor 20(4):3416–3452
7. Atzei N, Bartoletti M, and Cimoli T (2017) A survey of attacks on ethereum smart contracts (SoK). Lect Notes Comput Sci 164–186
8. Saad, Spaulding J, Njilla L, Kamhoua C, Shetty S, Nyang D, Mohaisen D (2020) Exploring the attack surface of blockchain: a comprehensive survey. IEEE Commun Surv Tutor 22(3):1977–2008

9. Aggarwal S et al (2020) Meta heuristic and evolutionary computation: algorithms and applications. Springer Nature, Berlin, p 949. https://doi.org/10.1007/978-981-15-7571-6. ISBN 978-981-15-7571-6
10. Yadav AK et al (2020) Soft computing in condition monitoring and diagnostics of electrical and mechanical systems. Springer Nature, Berlin, p 496. https://doi.org/10.1007/978-981-15-1532-3. ISBN 978-981-15-1532-3
11. Gopal et al (2021) Digital transformation through advances in artificial intelligence and machine learning. J Intell Fuzzy Syst 1–8. https://doi.org/10.3233/JIFS-189787
12. Smriti S et al (2018) Special issue on intelligent tools and techniques for signals, machines and automation. J Intell Fuzzy Syst 35(5):4895–4899. https://doi.org/10.3233/JIFS-169773
13. Sood YR et al (2019) Applications of artificial intelligence techniques in engineering, vol 1. Springer Nature, London, p 643. https://doi.org/10.1007/978-981-13-1819-1. ISBN 978–981-13-1819-1
14. Zheng Z, Xie S, Dai H, Chen X, Wang H (2018) Blockchain challenges and opportunities: a survey. Int J Web Grid Serv 14(4):352
15. Heilman E, Zohar A, Goldberg S (2015) Eclipse attacks on bitcoin's peer-to peer network. In: Proceedings of the USENIX conference on security symposium, pp 129–144
16. Gavin W (2014) Ethereum: a secure decentralised generalised transaction ledger. In: Proceedings of the ethereum project yellow paper. https://ethereum.github.io/yellowpaper/paper.pdf
17. Trón V, Fischer A, Nagy (2016) State channels on swap networks: claims and obligations on and off the blockchain (tentative title). Ethersphere orange papers, ethersphere, Tech. Rep.
18. Hyperledger Fabric. Available https://hyperledger-fabric.readthedocs.io. Accessed 06 July 2018
19. Bodkhe U, Tanwar S, Parekh K, Khanpara P, Tyagi S, Kumar N, Alazab M (2020) Blockchain for Industry 4.0: a comprehensive review. IEEE Access 8:79764–79800
20. Lin C, He D, Huang X, Choo KR, Vasilakos A (2018) BSeIn: a blockchain-based secure mutual authentication with fine-grained access control system for industry 4.0. J Netw Comput Appl 116:42–52

Comparative Analysis of Classifiers Based on Spam Data in Twitter Sentiments

Santosh Kumar, Ravi Kumar, and Sidhu

Abstract In the present scenario, sentiment analysis has gained much attention in the field of text mining. As social media have a huge impact on one's life, people use social media as a tool to express their feelings, thoughts, opinions, emotions and ideology. With the help of sentiment analysis, we can provide computational treatment for sentiments, opinions and subjectivity of text. Detecting spams from various social media sites is a challenging task as the messages contain the short informal text. In this paper, we have tested different classifiers on spam data of users' tweets based on spammer and non-spammers. The classifiers used for the purpose are Naives Bayes, Simple Logistic, J48 pruned tree, Bayes network classifier and Random Forest. Analysis results showed that random forest is proving the highest accuracy among all the algorithms under consideration.

Keywords Social media · Spam detection · Naives Bayes · Random forest · Classifiers

1 Introduction

Sentiment analysis is one of the extremely popular and fastest grown research areas in recent times. In sentiment analysis, identification and mining of subjective information like the persons' opinions, their thoughts, emotions and attitude are done by applying various techniques and procedures. Nowadays, social media has come up as a very important tool through which people can put their thoughts, discuss over blogs, comment and share their opinions on various subjects. Basically through sentiment analysis, we can easily review the polarity of any opinion that on any issue what is one's opinion like neutral, positive or it may be negative. Sentiment analysis

S. Kumar (✉) · R. Kumar
Computer Science & Engineering Department, ABES Engineering College, Ghaziabad, UP 201009, India

Sidhu
Shri Venkateshwara University, Gajraula, UP, India

A. Tomar et al. (eds.), *Machine Learning, Advances in Computing, Renewable Energy and Communication*, Lecture Notes in Electrical Engineering 768,
https://doi.org/10.1007/978-981-16-2354-7_36

has been applied to very vast areas like marketing of products, websites' reviews, consumer-related informations and many more, and this helps the users in effective decision-making [1]. The sub-task of sentiment analysis is the subjectivity detection. The different tools that are used to differentiate among negative and positive text are optimized for doing so. We need to ensure that the facts and data should not be filtered out while data are passed to the polarity classifier, as it becomes tedious to interpret products' reviews that are neutral [2]. Various challenges are faced during the process of analysis of the sentiments like ambiguity in words, multipolarity, noise, selection of appropriate algorithms, etc. If these are overlooked, it becomes difficult to extract accurate information. As in sentiment analysis, a large set of information are required on various topics, which are combined. Subjectivity is an important subtask for analysis of sentiments, identification of subjectivity is needed. We have to remove the comments that are factual or neutral or which are lacking sentiments. The analysis of sentiments can be categorized into three levels of classification namely aspect level, sentence-level and document-level sentiment analysis. Aspect-level sentiment analysis is a detailed analysis where opinions about various aspects of the entity are extracted. For example, in the review: "from last two months I am using this apple Laptop. It's wonderful. I just love this laptop; LED display is awsome", here laptop is entity whereas LED display is the aspect wonderful, awesome is sentiment with respect to its aspects through aspect level analysis of sentiments which can gain information regarding the perception of the user [3]. At document level, the classification of document is done on the basis of opinions framed or expressed it may be negative or positive opinion. In sentence-level sentiment analysis, we used to analyze the sentences in order to classify sentiment and to find that a sentence is objective or subjective [4]. If there is some subjective sentence then sentence level SA will derive the opinion that may be positive or negative. In the sentence level SA, the detailed information is not produced of all opinions. Fundamentally, there is no difference between the sentence and document level classification as the document is composed of small sentences [5].

2 Related Work

In [6], the author has worked on the domain of infectious diseases they have used the ideas like the transmission procedures, symptoms and risks. The connection between these concepts is analyzed to find how one concept impacts the other. The information that is produced is applied and a model has been framed that relies on linguistic and statistical features by analyzing the sentiments based on various aspects using the models of deep learning. In [7], the author has discussed about various challenges faced during the process of sentiment evaluation and analysis. Due to these challenges, it is very difficult to analyze the exact sentiments and finding out the accurate sentiment polarity. They have surveyed different challenges of sentiment analysis and the relevant techniques and approaches. In [8], the author has presented the development and limitations saddling Urdu sentiment analysis they have also proposed the corrections. The whole study has been classified into three dimensions namely sentiment classification, text pre-processing and lexical resources. The pre-processing phase is divided into parts of speech tagging, segmentation of word, cleaning and spell checking. The lexical resources like lexicons and corpus have been evaluated and examinations were carried out on constructs of sentiment analysis like modifiers, words and negations. In [2], the author has reviewed various models that are used to detect the subjectivity of sentiments automatically. They have talked about key assumption and challenges that need to be traversed comparison of advantages and shortcomings have been discussed. These procedures have been classified as automatic, handcrafted and multi-modal. Handcrafted templates are applied on the sentences having strong sentiments they do not work well on sentences having low subjectivity. Automatic techniques like deep learning help to represent meta information which can further be generalized on new domains and languages. Using many kernels, the multi-modal mechanisms are applicable to social data, which integrate the video and audio forms. In [3], the author has surveyed different approaches of sentiment analysis opinions aspect levels. The aspects can be specified explicitly by using the words or it may be derived from the text data. They have talked about implicit aspect mining techniques. The classification of studies as per approaches and the limitation have been stated. In [4], the author has explored recent algorithms that have been implemented in the field of sentiment analysis. They have discussed the connected fields like building resources, transfer learning and detection of emotions. They have done the classification on recent papers and their implementation in the related field. In [9], the author has proposed a framework that can be applied in multilingual analysis of sentiments and textual resources. They have taken two languages namely Greek and English, which are high inflection and low inflection languages respectively. A hybrid vectorization procedure has been implemented that works on lexicons (with emotional words and polarized) and different encoding learning algorithms and the effectiveness has been demonstrated and achieved high accuracy of classification. This can be applied to multilingual sentiment analysis applications, which are quick exact and flexible.

Twitter is a very popular platform for most web users to share their opinions, and it has become an attractive platform for Spammers too. Spammers send spam messages and malicious links and cheat many honest users.

3 Classifiers

In the digital domain, there are numerous examples and case studies performed by the researchers for analyzing the performance of the intelligent classifiers [10–16]. Out of them some of the intelligent algorithms have been illustrated below in sub-sequence section.

3.1 Bayesian Network Classifier

It is based on the Bayes theorem and works on the assumption of strong (naive) independence. From a given training set data, we can easily encode distribution Pb (A1, … An, D). The output model can be applied to the given set of attributes a1 … an, the label D, which increases the posterior probability Pb(Dja1, … an) probability can be returned by the classifier.

3.2 Logistic Regression

It is applied to represent data; it is useful in understanding the relationship between ordinal, nominal, ratio independent variables and dependent binary variables. Through logistic regression, we can carry out the predictive analysis.

3.3 Simple Logistic

It is a mechanism through which we can relate the binary outcome to a single predictor, which may be categorical, continuous or binary.

3.4 Smo

It stands for sequential minimal optimization and a useful iterative optimization algorithm applied in support vector machine implementation. It breaks the problem into small sub-problems and analytically these can be solved.

3.5 J48

It is a decision tree that is used for the prediction of new datasets. If we have dataset containing independent variables and predictors then J48 can be applied to anticipate the target variables.

3.6 Random Forest

It contains a lot of decision trees and can be applied as a classification algorithm. The two main parameters are feature randomness and bagging for accurate prediction.

Next, classification has been done in the weka benchmark to classify the instances using supervised learning techniques. Different parameters are evaluated on the data set and steps are shown in Fig. 1.

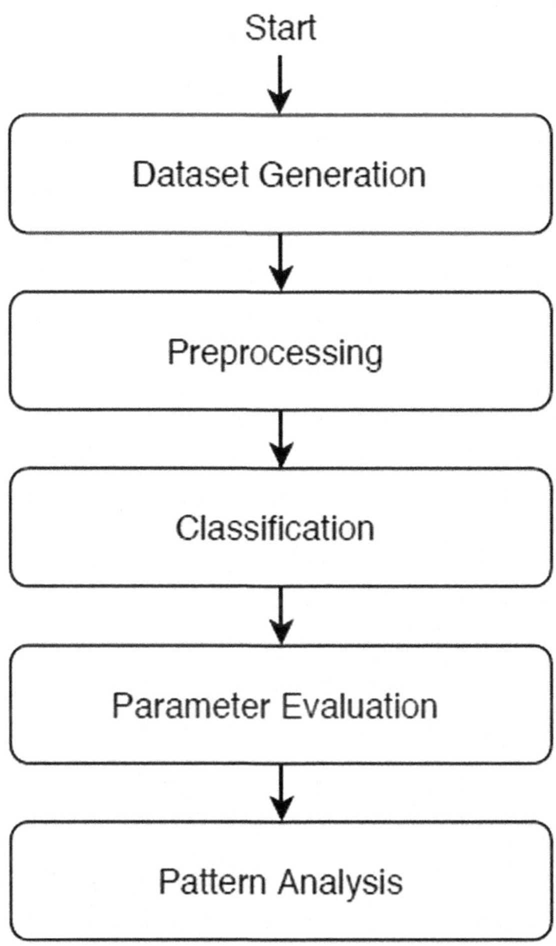

Fig. 1 Process of classification

4 Attribute Analysis

Ten thousand tweets of users have been collected having the features age of user's account, number of followers, number of friends, number of favorites received by user, number of lists added by user, number of sent tweets, number of retweets, hashtags user mentions, URLs included in tweet, length of tweet and number of digits. Some statistical analysis of these data is given in Table 1.

Table 1 Attribute analysis of tweets data

Attribute		Mean	Standard deviation	Weight sum	Precision
account_age	Spammer	570.7222	478.6063	5000	1.3742
	Non-spammer	747.4432	560.4431	5000	1.3742
no_follower	Spammer	1596.538	21,610.26	5000	759.6774
	Non-spammer	3914.922	49,676.13	5000	759.6774
no_following	Spammer	1137.477	13,210.52	5000	332.4595
	Non-spammer	1007.751	6078.379	5000	332.4595
no_userfavourites	Spammer	567.7489	3287.153	5000	92.5215
	Non-spammer	888.5579	5129.138	5000	92.5215
no_lists	Spammer	3.6007	51.2772	5000	94.2587
	Non-spammer	35.1962	490.0219	5000	94.2587
no_tweets	Spammer	8436.321	23,646.08	5000	157.9735
	Non-spammer	20,080.89	46,148.61	5000	157.9735
no_retweets	Spammer	725.3447	6957.689	5000	113.5551
	Non-spammer	273.1682	2090.838	5000	113.5551
no_hashtag	Spammer	0.5708	0.8769	5000	1.0769
	Non-spammer	0.493	1.2251	5000	1.0769
no_usermention	Spammer	0.3716	0.7144	5000	1
	Non-spammer	0.412	0.7135	5000	1
no_urls	Spammer	1.0924	0.3103	5000	1
	Non-spammer	1.0506	0.3436	5000	1
no_char	Spammer	63.378	31.4408	5000	1.0606
	Non-spammer	67.0903	32.2555	5000	1.0606
no_digits	Spammer	3.1781	4.7484	5000	1.697
	Non-spammer	1.4421	2.8436	5000	1.697

5 Experimentation and Results

The above data are analyzed using different classifiers based on parameters given in Table 2, and it has been found that random forest is able to correctly classify the tweets with the highest accuracy of 86.82% in case of these data. Naïve Bayes multinominal text showed the lowest accuracy. Also, the mean absolute error is lowest among all in case of random forest.

The graph in Fig. 2 is plotted for different classifiers showing the accuracy of correctly classified instances. Random forest is having the highest accuracy among all the algorithms under consideration.

Furthermore, detailed accuracy of classifiers based on spammer and non-spammer is given in Table 3.

Graph for detailed accuracy of classifiers for class is plotted in Fig. 3. For this, the weighted average of spammer and non-spammer is considered, and it has been found that the random forest is performing better among all the algorithms under consideration.

Table 2 Comparison of classifiers based on different parameters

Parameters versus classifiers	Bayes network classifier	Naive Bayes classifier	Naive Bayes multinomial	Naïve Bayes multinomial text	Logistic regression	Simple logistic	SMO	J48 pruned tree	Random forest
Correctly classified instances	74.22%	56.58%	60.87%	50%	66.56%	66.61%	65.48%	74.22%	86.82%
Incorrectly classified instances	25.7896	43.42%	39.13%	50%	33.44%	33.39%	34.52%	25.78%	13.18%
Kappa statistic	0.4844	0.1316	0.2174	0	0.3312	0.3322	0.3096	0.4844	0.7364
Mean absolute error	0.3086	0.4317	0.3914	0.5	0.4368	0.4371	0.3452	0.3086	0.2279
Root mean squared error	0.4255	0.5716	0.6252	0.5	0.4656	0.4657	0.5875	0.4255	0.3175
Relative absolute error	61.72%	86.34%	78.29%	100%	87.37%	87.42%	69.04%	61.72%	45.59%
Root relative squared error	85.09%	114.32%	125.04%	100%	93.12%	93.14%	117.51%	85.09%	63.51%
Coverage of cases (0.95 level)	97.53%	93.85%	60.97%	100%	99.71%	99.74%	65.48%	97.53%	99.63%
Mean relative region size (0.95 level)	87.54%	88.14%	50.14%	100%	98.51%	98.53%	50%	87.54%	86.57%
Total number of instances	10,000	10,000	10,000	10,000	10,000	10,000	10,000	10,000	10,000

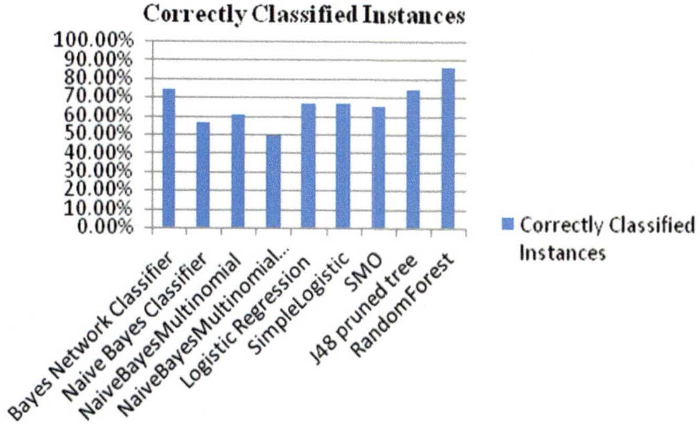

Fig. 2 Classifiers versus accuracy

Table 3 Detailed accuracy of classifiers based on spammer and non-spammer

Classifiers		FP rate	Precision	Recall	F-measure	MCC	ROC area	PRC area	Class
Bayes network classifier	Spammer	0.713	0.228	0.757	0.713	0.734	0.485	0.817	0.811
	Non-spammer	0.772	0.287	0.729	0.772	0.75	0.485	0.817	0.803
	Weighted	0.742	0.258	0.743	0.742	0.742	0.485	0.817	0.807
Naive Bayes classifier	Spammer	0.944	0.813	0.537	0.944	0.685	0.201	0.671	0.649
	Non-spammer	0.187	0.056	0.771	0.187	0.301	0.201	0.671	0.668
	Weighted	0.566	0.434	0.654	0.566	0.493	0.201	0.671	0.658
Naïve Bayes multinomial	Spammer	0.724	0.507	0.588	0.724	0.649	0.223	0.617	0.573
	Non-spammer	0.493	0.276	0.641	0.493	0.558	0.223	0.618	0.598
	Weighted	0.609	0.391	0.615	0.609	0.603	0.223	0.618	0.586
Naïve Bayes multinomial text	Spammer	1	1	0.5	1	0.667	0	0.5	0.5
	Non-Spammer	0	0	0	0	0	0	0.5	0.5
	Weighted	0.5	0.5	0.25	0.5	0.333	0	0.5	0.5
	Spammer	0.687	0.355	0.659	0.687	0.672	0.331	0.718	0.676
Logistic regression	Non-Spammer	0.645	0.313	0.673	0.645	0.658	0.331	0.718	0.726
	Weighted	0.666	0.334	0.666	0.666	0.665	0.331	0.718	0.701
	Spammer	0.688	0.355	0.659	0.688	0.673	0.333	0.718	0.676
Simple logistic	Non-Spammer	0.645	0.312	0.674	0.645	0.659	0.333	0.718	0.725
	Weighted	0.666	0.334	0.666	0.666	0.666	0.333	0.718	0.701
	Spammer	0.728	0.418	0.635	0.728	0.678	0.313	0.655	0.598
SMO	Non-Spammer	0.582	0.272	0.681	0.582	0.628	0.313	0.655	0.605
	Weighted	0.655	0.345	0.658	0.655	0.653	0.313	0.655	0.602

(continued)

Table 3 (continued)

Classifiers		FP rate	Precision	Recall	F-measure	MCC	ROC area	PRC area	Class
	Spammer	0.713	0.228	0.757	0.713	0.734	0.485	0.817	0.811
J48 pruned tree	Non-Spammer	0.772	0.287	0.729	0.772	0.75	0.485	0.817	0.803
	Weighted	0.742	0.258	0.743	0.742	0.742	0.485	0.817	0.807
	Spammer	0.854	0.118	0.879	0.854	0.866	0.737	0.937	0.942
Random forest	Non-spammer	0.882	0.146	0.858	0.882	0.87	0.737	0.937	0.926
	Weighted	0.868	0.132	0.868	0.868	0.868	0.737	0.937	0.934

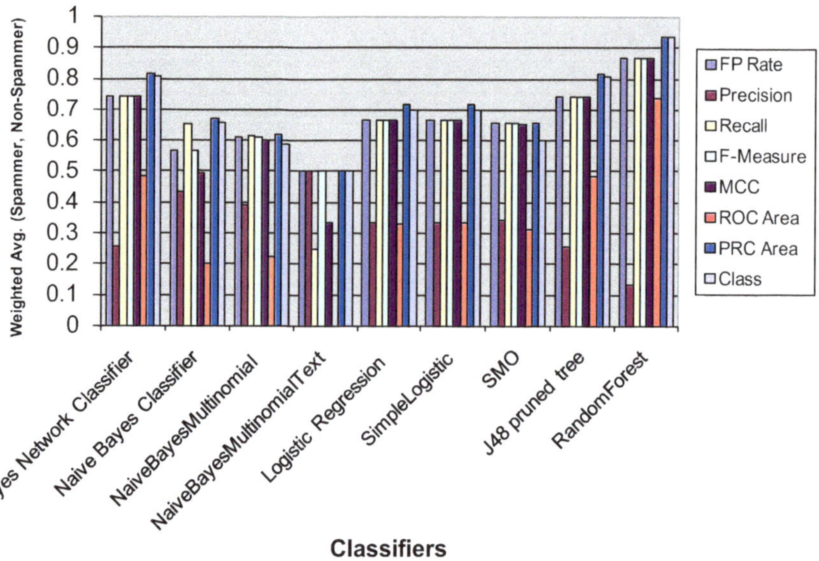

Fig. 3 Classifiers versus weighted average for spammer and non-spammers

6 Conclusion

Tweets data of 10,000 tweets of spammer and non-spammers are taken for analysis of classifiers. The attributes of importance are age of user's account, number of followers, number of friends, number of favorites received by user, number of lists added by user, number of sent tweets, number of retweets, hashtags user mentions, URLs included in tweet, length of tweet and number of digits. The data are analyzed using different classifiers based on different parameters listed in text, and it has been found that random forest is able to correctly classify the tweets with the highest accuracy of 86.82% in case of this data. Naïve Bayes multinominal text showed the lowest accuracy. Also, the mean absolute error is lowest among all in case of the random forest.

References

1. Johnson F, Gupta SK (2012) Web content minings techniques: a survey. Int J Comput Appl 47(11):44
2. Chaturvedi I, Cambria E, Welsch R, Herrera F (2018) Distinguishing between facts and opinions for sentiment analysis. Surv Chall Inf Fusion 44:65–77
3. Ganganwar V, Rajalakshmi R (2019) Implicit aspect extraction for sentiment analysis: a survey of recent approaches. In: ICRTAC-disruptive innovation, 2019 November 11–12, 2019, vol 165, pp 485–491
4. Medhat W, Hassan A, Korashy H (2014) Sentiment analysis algorithms and applications: a survey. Ain Shams Eng J 5(4):1093–1113
5. Liu B (2012) Sentence analysis and opinion mining. Synth Lect Human Lang Technol
6. García-Díaz JA, Cánovas-García M, Valencia-García R (2020) Ontology-driven aspect-based sentiment analysis classification: an infodemiological case study regarding infectious diseases in Latin America. Future Generat Comput Syst: FGCS 112:641–657. https://doi.org/10.1016/j.future.2020.06.019
7. Hussein DME-DM (2016) A survey on sentiment analysis challenges. J King Saud Univ—Eng Sci
8. Khattak A, Asghar MZ, Saeed A, Hameed IA, Hassan SA et al (2020) A survey on sentiment analysis in urdu: a resource-poor language. Egypt Inf J. https://doi.org/10.1016/j.eij.2020.04.003
9. Giatsoglou M, Vozalis MG, Diamantaras K, Vakali A, Sarigiannidis G, Chatzisavvas KC (2017) Sentiment analysis leveraging emotions and word embeddings. Expert Syst Appl 69:214–224
10. Aggarwal S et al (2020) Meta heuristic and evolutionary computation: algorithms and applications. Springer Nature, Berlin, 949 pp. https://doi.org/10.1007/978-981-15-7571-6. ISBN 978–981–15–7571–6
11. Yadav AK et al (2020) Soft computing in condition monitoring and diagnostics of electrical and mechanical systems. Springer Nature, Berlin, 496 pp. https://doi.org/10.1007/978-981-15-1532-3. ISBN 978–981–15–1532–3
12. Gopal et al (2021) Digital transformation through advances in artificial intelligence and machine learning. J Intell Fuzzy Syst 1–8 (Pre-press). https://doi.org/10.3233/JIFS-189787
13. Fatema N et al (2021) Intelligent data-analytics for condition monitoring: smart grid applications. Elsevier, 268 pp. ISBN: 978–0–323–85511–2. https://www.sciencedirect.com/book/978 0323855105/intelligent-data-analytics-for-condition-monitoring
14. Smriti S et al (2018) Special issue on intelligent tools and techniques for signals, machines and automation. J Intell Fuzzy Syst 35(5):4895–4899. https://doi.org/10.3233/JIFS-169773

15. Jafar A et al (2021) AI and machine learning paradigms for health monitoring system: intelligent data analytics. Springer Nature, Berlin, 496 pp. https://doi.org/10.1007/978-981-33-4412-9. ISBN 978–981–33–4412–9
16. Sood YR et al (2019) Applications of artificial intelligence techniques in engineering, vol 1. Springer Nature, 643 pp. https://doi.org/10.1007/978-981-13-1819-1. ISBN 978–981–13–1819–1

Hybrid Solar Wind Charger

Sharmila, Malti Gautam, Neha Raheja, and Bhawna Tiwari

Abstract Renewable energy resources like wind, sun, hydropower, geothermal, and biomass are better alternatives for conventional non-renewable energy resources such as fossil fuel reserves. Renewable energy resources are the better technological option to generate clean energy and overcome the depletion of non-renewable energy resources. This paper presents the complete system design of hybrid solar wind charger. The main contribution is to develop a compact system, which utilizes the eternal solar and wind power to solve the major crisis of pollution as well as the scarcity of fossil fuels. The functionality of the proposed system allows a reliable source of power generation for human beings in the energy crisis.

Keywords Solar power · Wind power · Renewable energy

1 Introduction

Electricity plays a very important in our day-to-day life. Generally, it is generated by either conventional method or non-conventional method. The demand and usage of electricity increase day by day. At present, electricity is generated by conventional energy resources such as coal, diesel, nuclear, etc. The conventional energy resources will deplete in the future due to the increased demand for fossil fuels [1]. The downside of conventional energy resource is that it produces a lot of waste and pollutes the environment also. Another method of generating electricity is non-conventional method, which is free from pollution, economical, and more reliable as compared with the conventional energy resources. It is good alternative energy resource for conventional energy resources. The different sources of non-conventional energy resources are tidal, wind, solar, biogas, geothermal, etc.

The limitations of non-conventional energy resources are as follows: the tidal energy powered by the ocean tides, which is implemented on the sea shores only, the geothermal energy is location-specific, and in extreme cases, it causes earthquake,

Sharmila (✉) · M. Gautam · N. Raheja · B. Tiwari
Krishna Engineering College, Uttar Pradesh, Ghaziabad, India

A. Tomar et al. (eds.), *Machine Learning, Advances in Computing, Renewable Energy and Communication*, Lecture Notes in Electrical Engineering 768,
https://doi.org/10.1007/978-981-16-2354-7_37

solar energy could not power the electrical energy during rainy and cloudy seasons [2]. Earlier and even nowadays, the hybridization process was done by separately placing the windmills and solar panels. This required large land areas and could be used only for commercial purposes. They cannot be shifted to any other place [3]. In this proposed design, we have come up with a compact device, which can be placed on the rooftop of the house making it possible to be used both for personal as well as commercial purposes [4]. Each conventional energy resource has its own pros and cons, the hybrid combination of solar and wind energy will keep generating the energy in case of any one of the sources fails [5]. Moreover, some examples are listed in Springer's books and journals [6–12]. Whereas, the proposed system consists of solar panel, vertical windmill, which helps generate electricity with the help of wind and the sun. Since these two resources are renewable energy resources, it helps controlling pollution. In this proposed system, the DC–DC boost converter and DC–DC buck converter are used to regularize the voltage levels of the solar panels and the windmills. Since the solar panel has a higher voltage and the windmill has lower voltage, their voltage variation is high. There is a need to regularize their voltage levels.

The main contribution of the proposed system is as follows:

There are many problems that can be tackled using the concept of hybrid solar wind charger, such as:

1. Hybridization increases the output,
2. Provides uninterrupted power supply,
3. Provides clean and pure energy,
4. Efficient and easy installation,
5. Long-term sustainability,
6. Low installation and maintenance cost.
7. Compact size.

The organization of the paper is as follows: Sect. 2 reviews the related work; Sect. 3 describes the proposed hybrid solar wind charge system. Section 3 discussed the implementation of the proposed system. The conclusion is given in Sect. 4.

2 Design of the Proposed System

The methodology of making this proposed system includes three main parts. The first part is the construction of the vertical windmill. The second part is the construction of a body on which the solar panels as well as the windmill can be attached. The third part is the final furnishing part that has to be done to make the project presentable.

To begin with the construction, the waste materials are collected that could be used. This project is made from waste iron parts and wooden parts. The first step is headed toward the construction of a vertical wind turbine. A vertical wind turbine proves to be more advantageous than a horizontal wind turbine. The horizontal wind turbine needs the wind to flow at the right angle to the blades. By contrast, a vertical turbine runs well regardless of wind direction, making it better suited to urban areas. The blades of the wind turbine are made up of hollow PVC pipes. The blades then have to be attached to a wheel. This is done through rectangular-shaped wooden sticks. The dynamos can be attached to the wheel with the help of any frictionless rubber band or a tire tube. The wind turbine can generate a voltage of 3 to 6 V. Then we headed toward the construction of the body on which the wind turbine and the solar panels need to be attached. The solar panels are of 21 V each. Two solar panels are required. The body is made up of wood. Since wood is lighter and will make the system compact enough to install.

To place the body on the ground, a stand made up of iron is made.

The block diagram with the detailed description of the project is shown in Fig. 1. The windmill generates a voltage of about 3 V–6 V. While the solar panels generate a voltage of about 21 V. To combine and regularize their voltage, we use IC7812 and DC–DC buck converter for the solar panel side. For the windmill, we use DC–DC booster.

The block diagram above consists of:

- Solar panels.
- Control circuit (IC7812, DC-DC buck converter).
- Vertical windmill.
- Control circuit (DC–DC booster).
- Hybrid charger.

Solar Panels. Generally, the solar modules are fabricated from crystalline silicon. More than 90% of solar cells are made up of mono and multi-crystalline crystal, and the rest of the solar cells in the market are made up of cadmium telluride and amorphous silicon. In the third generation, thin film cells are used in advanced solar technologies.

Vertical Windmills. In vertical axis wind turbines, the rotor shaft is placed transverse to the wind direction. The main components are placed at the base of the turbine and the generator and gearbox are placed near the ground, simplifying service and

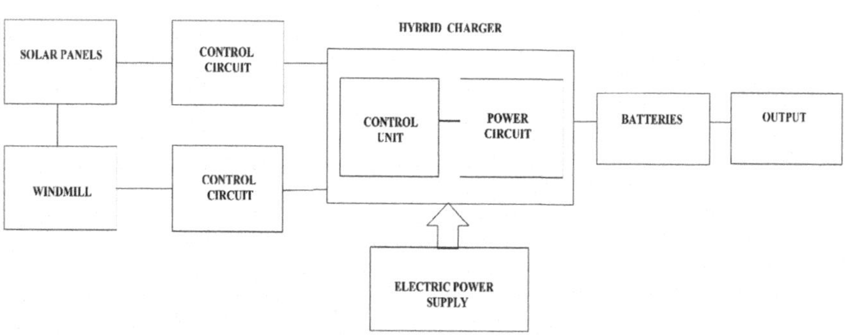

Fig. 1 Block diagram of the proposed hybrid solar wind charger

repair. Vertical axis wind turbines (VAWTs) need not to be directed toward the wind, and it has an automatic wind sensing and oriented mechanisms.

DC–DC Booster. A DC to DC booster converter is a step-up converter, which steps up the voltage from its input to its output.

DC–DC Buck Converter. A DC–DC buck converter is a step-down converter, which reduces the voltage from its input to output.

IC 7812. IC 7812 is a self-contained linear voltage regulator Integrated circuit. It is easy to use and low cost.

3 Implementation of Hybrid Solar Wind Charger

- Solar panels of voltage 21 V and vertical windmill of 6 V are used.
- To charge a battery or to light any device of about 11 V, we need to bring the voltage levels of the solar panels as well as the windmills to the same level of about 12 V.
- So, for this, we used a voltage limiter i.e. a DC–DC buck converter (step-down converter) to bring down the voltage of the solar panels from 21 to 12 V (Fig. 2).

Also, IC 7812 is used to regulate the voltage up to 12 V and do not let it drop down below.

- In the windmill, we used a DC boost converter (step-up converter) to raise the voltage from 6 to 12 V.
- Now they are at the same levels so batteries can be charged.
- The hybrid charger gives us three kinds of voltage variations.
- A power of about 30 watts with voltages 5 V, 12 V and 18 V choose whichever one is needed.
- To generate electricity from the windmill, a dynamo is used.

4 Conclusion

The conventional energy resources will deplete in the future due to the increased demand for fossil fuels. Therefore, the solar and wind energy are the best substitute for the future. The solar panel does not offer power during the night, but the wind picks up contrariwise. Due to the frequent power loss, a regular unit power supply (UPS)/battery backup gets more time to charge the battery from the supply. The proposed hybrid model combines solar and wind energy, which provides an excellent solution. This paper mainly focuses on integrated solar and wind energy due to the large quantity, availability of resource and ease of energy utilize for power generation. The main advantage of the proposed system is cost-effective and highly safe as it does not emit any radiation, and it generates clean energy. It needs initial investment

Fig. 2 The proposed hybrid solar wind charger

and affordable for electricity generation. Further research work can be done to make it proper for military uses too.

References

1. Arjun A, Athul S, Ayub M, Neethu R, Anith K (2014) Micro-hybrid power systems—a feasibility study. J Clean Energy Technol 3(1):27–32
2. Balaji TS, Damodhar, Sethil Kumar A (2015) Design of high step up modified for hybrid solar/wind energy system. Middle-East J. Sci Res 23(6):1041–1046
3. Jenkins P, Elmnifi M, Younis A, Emhamed A (2019) Hybrid power generation by using solar and wind energy: case study. World J Mech 9:81–93
4. Ataei A, Biglari M, Nedaei M, Assareh E, Choi JK, Yoo C, Adaramola MS (2015) Techno-economic feasibility study of autonomous hybrid wind and solar power systems for rural areas in Iran, a case study in Moheydar village. Environ Prog Sustain Energy 34:1521–1527
5. Mohamed A, Al-Habaibeh H (2013) An investigation into the current utilization and prospective of renewable energy resources and technologies in Libya. Renew Energy 50:732–740
6. Fatema N et al. (2021) Intelligent data-analytics for condition monitoring: smart grid applications. Elsevier, p 268. ISBN: 978-0-323-85511-2. https://www.sciencedirect.com/book/978

0323855105/intelligent-data-analytics-for-condition-monitoring
7. Smriti S et al (2018) Special issue on intelligent tools and techniques for signals, machines and automation. J Intell Fuzzy Syst 35(5):4895–4899. https://doi.org/10.3233/JIFS-169773
8. Yadav AK et al. (2020) Soft computing in condition monitoring and diagnostics of electrical and mechanical systems. Springer Nature, Berlin, p 496. https://doi.org/10.1007/978-981-15-1532-3. ISBN 978-981-15-1532-3
9. Gopal et al. (2021) Digital transformation through advances in artificial intelligence and machine learning. J Intell Fuzzy Syst Pre-press 1–8. doi: https://doi.org/10.3233/JIFS-189787
10. Jafar A et al. (2021) AI and machine learning paradigms for health monitoring system: intelligent data analytics. Springer Nature, Berlin, p 496. https://doi.org/10.1007/978-981-33-4412-9. ISBN 978-981-33-4412-9
11. Sood YR et al. (2019) Applications of artificial intelligence techniques in engineering, vol 1. Springer Nature, p 643. doi: https://doi.org/10.1007/978-981-13-1819-1. ISBN 978-981-13-1819-1)
12. Aggarwal S et al. (2020) Meta heuristic and evolutionary computation: algorithms and applications. Springer Nature, Berlin, p 949.doi: https://doi.org/10.1007/978-981-15-7571-6. ISBN 978-981-15-7571-6

Determinants of Artificial Intelligence Systems and Its Impact on the Performance of Accounting Firms

Shelly Oberoi, Sunil Kumar, R. K. Sharma, and Loveleen Gaur

Abstract The study intends to identify and validate the various determinants of artificial intelligence (AI) systems and also aims to investigate its impact on the performance of accounting firms. A sample of 176 accountants working in accounting firms of Delhi-NCR was taken to test the literature-driven conceptual models. The data for the study were collected using a self-administered questionnaire based on 5-point Likert's scale using convenience sampling. Hypotheses were tested using factor analysis and structural equation modeling. Cronbach alpha and KMO and Bartlett test were also being applied to measure internal consistency and to test sample adequacy. The present study discloses that artificial intelligence (AI) will help accountants to work better and smarter and would also contribute to provide better quality accounting work by eliminating errors and frauds. The present study is among few studies, which have empirically investigated the impact of AI on the performance of accounting firms.

Keywords Accounting · Artificial intelligence · Systems · Firms · Performance

1 Introduction

Accounting is defined as a method of collecting and documenting the economic and financial information of a firm, which is further divided into two subordinate areas, i.e. external accounting and internal accounting. External accounting involves the preparation of financial reports containing information to external parties about economic situation of the firm, whereas, internal accounting mainly focuses on the information about company's future. It is the call of the firm whether to conduct

S. Oberoi (✉) · S. Kumar
SOMS, IGNOU, New Delhi, India

R. K. Sharma
Guru Ghasidas University, Bilaspur, India

L. Gaur
Amity University, Noida, India

A. Tomar et al. (eds.), *Machine Learning, Advances in Computing, Renewable Energy and Communication*, Lecture Notes in Electrical Engineering 768,
https://doi.org/10.1007/978-981-16-2354-7_38

in-house accounting or to outsource it to third party who has knowledge, expertise and also the specialized technology for all such situations [42, 53]. In the last few years, there has been enormous development of technology in accounting sector. In 1980s, computers were available in markets and companies were able to afford them. It is being estimated that Microsoft Corporation produced and sold millions of hard copies to stakeholders for financial records, as it was too difficult to store large data on computers [14, p. 99], but it was also tough to distribute hard copies at fast pace, which was not cost-efficient, hence, digitization seemed to be the simplified mechanism for dispersal of financial information to stakeholders [15, p. 94]. At present, accounting professionals emphasize technological development and its application is also rising comprehensively [27, p. 4].

Nowadays, in the accounting field, automated accounting is a growing concept, due to standardization of accounting reporting and time consuming tasks. Artificial intelligence has really contributed a lot in automated accounting in recent years.

The term '*Artificial intelligence*' was first given by John McCarthy in 1956. It involves the development of computer systems, which performs tasks, which need human intelligence like speech recognition, decision-making, etc. AI helps in performing function like human brains like problem-solving, learning, etc. over the decades, the research on AI systems has increased [18, 30]. Many past studies have reported the widespread use of AI in accounting and integrated audit support system (Dowling and Leech 2014) [23, 25].

The present study objects to identify and validate the determinants of AI in accounting firms and to analyze its impact on accounting firm's performance.

2 Review of Literature

AI is perilous to the imminent of accounting and auditing professions and its application in this profession date back to decade ago [2, 16, 31]. In today's business environment, it has emerged as a research discipline and its operation grips digital technology through the implementation of machine systems [22]. It has been noted that the domain of accounting field has applied many AI technologies in financial reporting and analysis [35]. AI involves the development and use of expert systems in accounting disciplines, which facilitates and advices accounting professionals [39, 55]. It has been said by many researchers that expert systems to be more of value to accountant than technological thrives who painstakingly checks debit and credit entries [6, 21]. Such expert systems and automation of data processes would facilitate business owners and their accountant to have up to date minute information about business enabling them to make quick decisions and eliminating anything, which affects the performance and profitability of business and takes necessary rectification procedures. Application of expert systems in accounting is classified into following: auditing, taxation, financial planning, management accounting, financial accounting, etc. Artificial intelligence systems help accountants in audit planning, identification of audit risk and internal control valuation. Expert systems in internal audit facilitate the screening and verification of transactions susceptible to frauds [10, 26, 36]. AI is utilized in terms of cash flow and authorization and processing of claims. It has vital functionality in financial accounting areas like determination of financial status by ratios, analysis of financial reports and mergers and acquisition (Yang 2008) [24, 34]. Artificial intelligence has brought automation of labor-intensive tasks in auditing, hence, it has become imperative to identify areas likely to be impacted by AI to a great extent [1, 5]. Past studies have argued that the automation of external audit process could lead to extinction of human auditors [37, 41, 47]. AI helps in the performance of analytical review procedure, risk assessment and genetic algorithm [12, 46, 50]. AI helps in materiality assessment, internal control evaluations, going concern judgment (Iss 2015) [48]. Nowadays, AI is mainly prominent in data acquisition, data extraction, data comparison and data validation [17, 43, 45]. It helps in locating relevant information and makes it usable for human auditors. AI facilitates full automation of time-consuming tasks for further substantive testing [17]. Modern artificial intelligence tools enable to identify and extract relevant accounting information from various sources, for example, it can spot company records before the end of the reporting time and also after the end of the reporting time [5, 44]. Artificial intelligence has made audit fully automated resulting in less emphasis on ticking, vouching and more emphasis will be on identifying trends and patterns and a better understanding of inputs and assumptions [3, 17].

3 Theoretical Framework and Hypotheses

Artificial intelligence and factors related to firms

Nowadays, it is difficult to carry out the accounting process without the use of computers and accounting software, hence accounting technology has become very sophisticated, and it has changed the accounting profession to great extent. Artificial intelligence has integrated into accounting profession to make it automated, but there many factors to be kept in mind before integrating AI. The size of the firms and type of the firm are accelerating factors for need of more advanced software to support all aspects of accounting operations [26].

***HA1**: Artificial intelligence system to be adopted is influenced by the factors related to firm*

Artificial intelligence and factors related to technology

Incorporation of AI in accounting profession helps in reducing risks of frauds and errors, but accountants and auditors need to consider the firm's objectives and also they need to compare the costs of developing artificial intelligence technologies with the benefits they would reap [41].

***HA2**: Artificial intelligence system to be adopted is influenced by the factors related to technology*

Artificial intelligence and factors related to personnel

Artificial intelligence technologies in ERP systems in accounting are exposed to major risk causing due to untrained personnel, which may result in inappropriate access and financial misstatements [16, 33].

HA3*: Artificial intelligence system to be adopted is influenced by the factors related to the personnel*

Artificial intelligence and improved data

Artificial intelligence systems and big data provide more evidence than traditional accounting process and audit processes in fraud investigations due to improved data recording and better data analysis [53], but, its integration in accounting processes needs to be taken care of due to a lot of risks involved [16, 33].

HA4*: Artificial intelligence system improves the recording and analysis of data*

Improved data and improved performance of accounting firms

The output given by accounting information system has many concerns as far as data quality like timeliness, accuracy and consistency [51], which requires proper data collection, data storage and data utilization, which is possible only by implementing artificial intelligence techniques in the accounting process, which would improve the efficiency of the accountants and auditors, thereby improving the overall performance of the accounting performance (Emeke-Nwokeji 2012) [9].

HA5*: Improved data due to artificial intelligence system improves the performance of the accounting firms*

Artificial intelligence and transparency

A lot of attention is being given to AI nowadays, which helps in automated decision-making process delegated to machines and systems [7], but, with automated process, transparency has become main the concern [11, 19, 40]. Artificial intelligence allows machines and software to run without human control and also informs users about data processing and prospective transparency [8, 54].

HA6*: Artificial intelligence system leads to transparency in the accounting firms*

Transparency and performance of accounting firms

Past studies have hypothesized and suggested that financial transparency positively and significantly impacts the firm's performance through a reduction in data errors and financial frauds (Ozbay 2009) [29, 49].

HA7*: Transparency improves the performance of the accounting firms*

On the basis of the existing literature, two proposed models were framed, Model 1 validating the determinants of artificial intelligence and Model 2, representing the impact of artificial intelligence on the performance of accounting firms (Figs. 1 and 2).

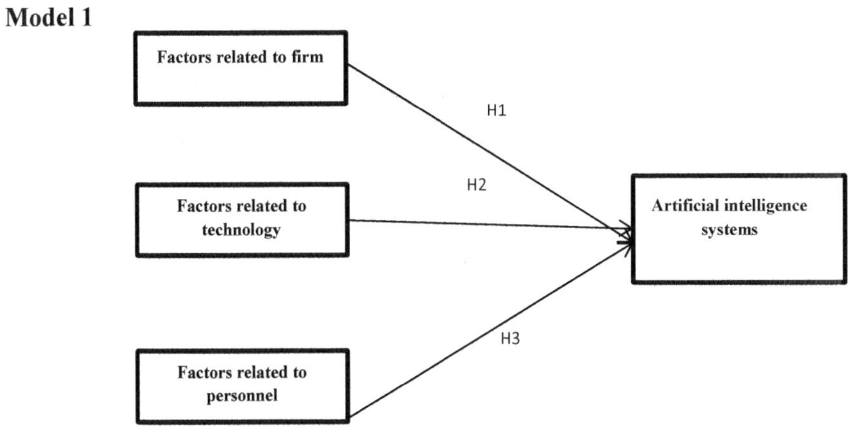

Fig. 1 Determinants of artificial intelligence systems

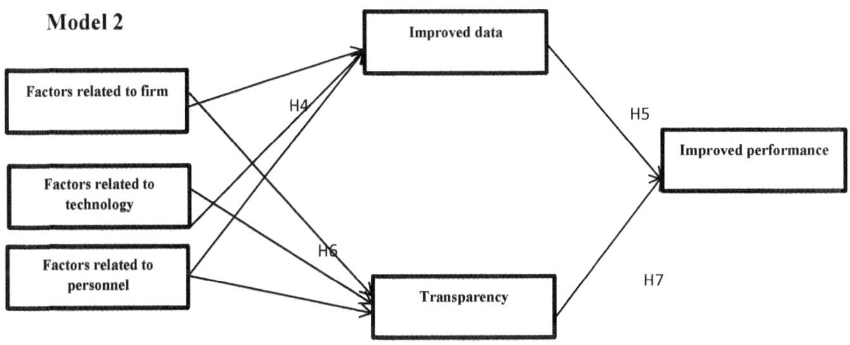

Fig. 2 Artificial intelligence systems and performance of accounting firms

4 Methodology

The main focus of the study is on primary data that were collected using a Google Form. It was decided to follow a self-report type approach to get opinions from respondents towards certain objects. It is one of the best approaches, which provides dominant statistical approaches for data analysis [28]. The questionnaire was first used in conducting pretesting and some vague expressions were altered and then final testing was done for final results based on the responses given by the respondents. The study also highlights the descriptive analysis. A prior permission was taken from the authority concerned to obtain data collection. Quantitative data were collected through two approaches: online survey and in-person administered survey. The potential respondents were contacted through emails and telephonic conversation and they were assured that their opinions will be used only for academic purposes and will remain confidential. All the questions were made compulsory to avoid missing data.

Table 1 Sample profile

Measure	Variable	Frequency	Percentage
Age	21–30 years	68	38.63
	31–40 years	84	47.72
	41–50 years	14	7.95
	51–60 years	5	2.84
	Above 60 years	5	2.84
Gender	Male	97	55.11
	Female	79	44.88
Education	Graduate	38	21.59
	Post-graduate	49	27.84
	Professional Qualification	89	50.56
Status	Married	136	77.27
	Unmarried	40	22.72

4.1 Sample Description

A total of 204 responses were received working in accounting firms from Delhi/NCR and a sample size of 176 accounting professionals was finally taken after rejecting the biased responses. Most of the respondents were men (55.11%) and women were only 44.88%. Most of them had professional certificate such as CPA, certified financial planner and accountants. Only 27.84% possessed master degree or beyond it, majority of them had professional qualification (Table 1). The data were analyzed using various statistical techniques and hypotheses were tested through factor analysis.

4.2 Measurement Tools and Techniques

The data were collected through self-administered questionnaire based on a 5-point Likert's scale ranging from strongly disagree to strongly agree consisting of four constructs, namely, artificial intelligence, improved data, transparency and improved performance. Artificial intelligence systems have three constructs, namely, factors related to firm (size of the firm (F_1), type of the firm (F_2) and investment by firm (F_3), factors related to technology (cost (T_1), knowledge (T_2) and infrastructure (T_3) and factors related to personnel (training, (PER_1), attitude of personnel (PER_2), perception of personnel (PER_3), each of the items has three sub-items. Second, improved data have three items, namely, proper data record (ID1), proper data analysis (ID2) and predictive analysis (ID3). Third, transparency has two items, namely, reduction in errors (T1) and reduction in financial frauds (T2) and fourth, improved performance has four items, namely, saves time (IP1), virtual assistance (IP2), risk assessment (IP3) and improved communication (IP4). A total of four constructs having 18 items were identified and validated and hypotheses were tested using factor analysis. Structural equation modeling (SEM) was applied to bring out a causal relationship between all the constructs, using SPSS and the internal consistency of the survey instrument was determined through Cronbach's alpha.

5 Analysis of Data

5.1 EFA Results of EFA

Table 2 represents KMO results, which measures sample adequacy and scores of >0.70 are satisfactory. It also displays Bartlett's test of sphericity results, which is also significant.

The EFA results of Model 1 are exhibited in Table 3, which depicts that all the three constructs have been extracted as their Eigen value is 1 or higher. The total variance for all the constructs was 75.881%.

The result depicts that factors related to the firm (Eigen value = 4.061 and Variance explained = 27.295%) are the most important determinants of artificial intelligence system to be employed in any accounting firm. After extracting the items using PCA, rotation was done using Varimax with Kaiser Normalization, which showed that three factors had nine items in all. Table 2 represents the internal consistency of various constructs measured through Cronbach's alpha. The reliability was also checked and shown in Table 3, where alpha value of all the constructs is above 0.70, which is considered acceptable for basic research [38].

5.2 Confirmatory Factor Analysis (CFA) Results of Model 1

Validity Results

Convergent Validity

Convergent validity ensures that various methods of measuring give the same results (O'Leary-Kelly and Vokurka 1998). The factor loadings should be looked at first,

Table 2 KMO and Bartlett's results

KMO and Bartlett's test		
Kaiser–Meyer–Olkin measure		0.784
Bartlett's test of sphericity	Approximately chi-square	864.424
	Df	36
	Sig.	0.000

Table 3 EFA and reliability results (model 1)

Constructs	Items	Factor loadings	Eigen value	Variance explained	Cronbach's alpha
Factors related to firm	Size of the firm (F_1)	0.861	4.061	27.295	0.792
	Type of the firm (F_2)	0.894			
	Investment by firm (F_3)	0.834			
Factors related to technology	Cost (T_1)	0.826	1.526	25.455	0.709
	Knowledge (T_2)	0.832			
	Infrastructure (T_3)	0.832			
Factors related to personnel	Training (PER_1)	0.946	1.242	23.130	0.765
	Attitude of personnel (PER_2)	0.798			
	Perception of personnel (PER_3)	0.921			

which shall be atleast 0.5 and preferable 0.7 or higher. All the items showed significant loadings (Table 3). Average variance explained (AVE) and composite reliability (CR) should also be looked into, CR scores should be >0.70 [20], as per rule of thumb, CR >0.70, CR >AVE and AVE >0.50 [32]. The values of CR and AVE for all the constructs are significant (Table 4).

Discriminant validity

AVE values are compared with squared inter-variable correlation estimates (SIC) to measure discriminant validity, AVE > SIC to ensure discriminant validity [32]. All the constructs confirm the discriminant validity.

Path Analysis and Model Fit (Model 1)

For measurement model 1, CFA was conducted, which showed good fit with df = 2.065, GFI = 0.942, NFI = 0.944, CFI = 0.970 and RMSEA = 0.078 (Table 5).

There was no multicollinearity issue as bivariate correlation did not exceed 0.80. Path coefficients were calculated on the basis of model 1 and its results are shown in Fig. 3. Table 6 exhibits the effects from path analysis. Path analysis results revealed that factors related to technology, i.e. T_1 ($\beta = 0.958$), T_2 ($\beta = 0.440$) and T_3 ($\beta = 0.930$), factors related to firm, i.e. F_1 ($\beta = 0.877$), F_2 ($\beta = 0.884$) and F_3 ($\beta = 0.753$) and factors related to personnel, i.e. PER_1 ($\beta = 0.724$), PER_2 ($\beta =$

Table 4 AVE and CR values

Constructs and items	AVE	CR
Artificial intelligence systems		
Factors related to firm	0.745	0.897
F_1		
F_2		
F_3		
Factors related to technology	0.688	0.869
T_1		
T_2		
T_3		
Factors related to personnel	0.792	0.907
PER_1		
PER_2		
PER_3		

Table 5 Model fit (model 1)

Model	CFI	RMSEA	GFI	AGFI	TLI	NFI	Df	*p* value
Model 1	0.970	0.078	0.942	0.892	0.955	0.944	2.065	0.000

0.838) and PER_3 ($\beta = 0.657$), all are correlated significantly and are determinants of artificial intelligence system, hereby, supporting ***HA1***, ***HA2*** and ***HA3***.

5.3 Exploratory Factor Analysis (EFA) Results of Model 2

The EFA results of model 2 are exhibited in Table 7, which depicts that all the three constructs (improved data, transparency and improved performance) have been extracted as their Eigen value is 1 or higher. All the constructs of model cumulatively accounted for 74.457% of the total variance. The result depicts that transparency significantly impacts the improved performance in an accounting performance (Eigen value = 3.586, Variance explained = 28.981%). After extraction of items using PCA, rotation was also done using Varimax with Kasier normalization, which yielded that out of nine items extracted, one of the item, i.e. ID2 is excluded from further analysis, as its factor loading <0.50. The internal consistency of various constructs (Model 2) is represented in Table 7, where, the Cronbach's alpha of all the constructs is above 0.70, which indicates a good internal consistency and considered satisfactory for basic research [38].

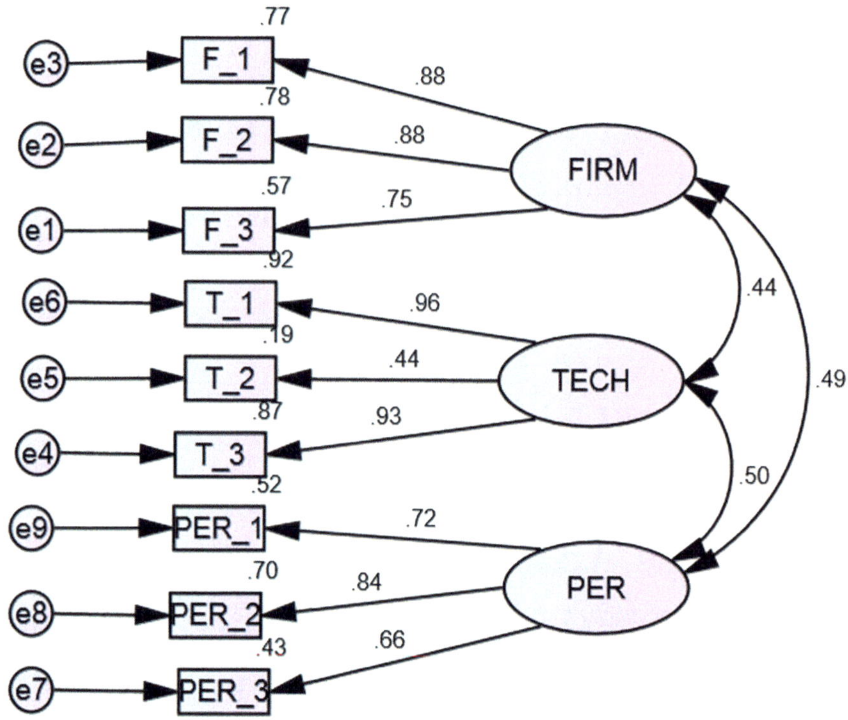

Fig. 3 Standardized regression coefficients and correlation (Path model 1)

Table 6 Path coefficients based on model 1

			Unstandardized coefficient estimate	Standard error	Composite reliability	*p* value	Standardized coefficient estimate (β)
F_3	<---	FIRM	1.000				0.753
F_2	<---	FIRM	1.261	0.109	11.570	***	0.884
F_1	<---	FIRM	1.117	0.097	11.524	***	0.877
T_3	<---	TECH	1.000				0.930
T_2	<---	TECH	0.413	0.067	6.176	***	0.440
T_1	<---	TECH	1.038	0.063	16.522	***	0.958
PER_3	<---	PER	1.000				0.657
PER_2	<---	PER	1.278	0.160	7.984	***	0.838
PER_1	<---	PER	1.027	0.134	7.651	***	0.724

Table 7 EFA and reliability results (model 2)

Constructs		Loadings	Eigen value	Variance explained	Cronbach's alpha
Improved performance	Saves time (IP1)	0.783	1.287	19.829	0.820
	Virtual assistance(IP2)	0.810			
	Risk assessment (IP3)	0.796			
	Improved communication (IP4)	0.756			
Improved data	Proper data record (ID1)	0.843	1.828	25.647	0.845
	Proper data analysis (ID2)	0.471			
	Predictive analysis(ID3)	0.831			
Transparency	Reduction in errors (T1)	0.931	3.586	28.981	0.862
	Reduction in financial frauds (T2)	0.932			

5.4 *Confirmatory Factor Analysis (CFA) Results of Model 2*

Validity Results

Table 8 depicts that the CR and AVE values for all the constructs of model 2 are significant, thereby, confirming the convergent validity [32], all the constructs of the

Table 8 AVE and CR values

Constructs and items	AVE	CR
Improved performance	0.618	0.866
IP1		
IP2		
IP3		
IP4		
Improved data	0.540	0.770
ID1		
ID3		
Transparency	0.867	0.930
T1		
T2		

model 2, even confirms the discriminant validity by satisfying all pre-requisites, i.e. AVE > SIC [32].

Path Analysis and Model Fit (Model 2)

For measurement model 2, CFA was conducted and it showed a good fit with df = 2.065, GFI = 0.923, NFI = 0.921, CFI = 0.987, RMSEA = 0.032 (Table 9).

Path coefficients were calculated on the basis of the hypothesized model 2 and its results are shown in Fig. 4. Table 10 displays the path coefficients. The results of path analysis discovered that transparency has no significant influence on the improved performance ($\beta = -0.062$), hereby not supporting ***HA7***, whereas improved data due to artificial intelligence systems significantly influence the improved performance of the accounting firms ($\beta = 0.661$), hereby supporting ***HA5***. The path analysis also discovered that there is no significant relationship between factors related to the firm and transparency ($\beta = -0.079$) and factors related to personnel and transparency (β

Table 9 Model fit (model 2)

Model	CFI	RMSEA	GFI	AGFI	TLI	NFI	Df	*p* value
Model 2	0.987	0.032	0.923	0.891	0.984	0.921	1.179	0.000

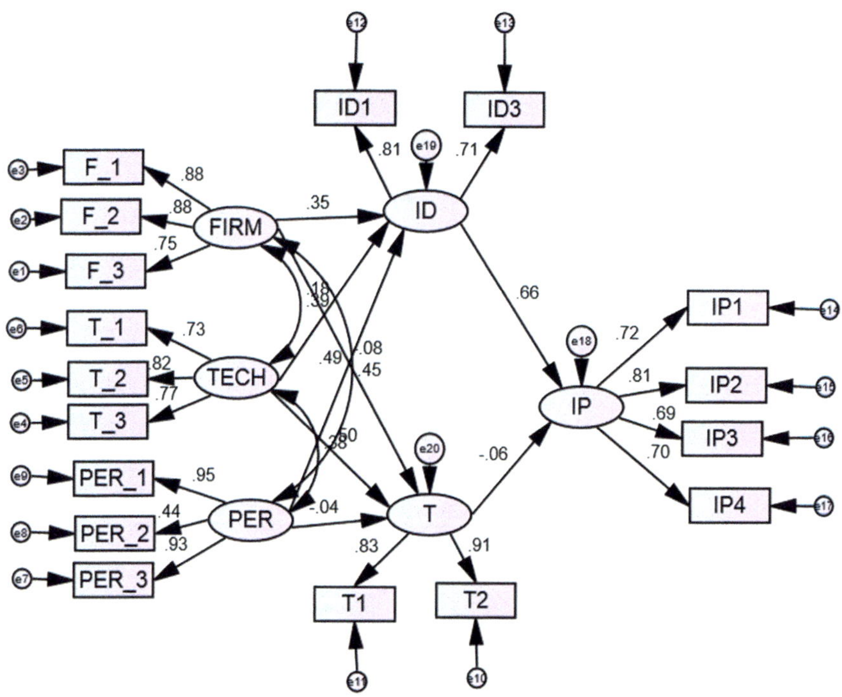

Fig. 4 Path model 2 with standardized regression coefficients and correlation between variables

Table 10 Path coefficients based on model 2

			Unstandardized coefficient estimate	Standard error	Composite reliability	*p* value	Standardized coefficient estimate (β)
ID	<---	FIRM	0.326	0.078	4.167	***	0.346
ID	<---	TECH	0.130	0.059	2.203	0.028	0.177
ID	<---	PER	0.325	0.053	6.149	***	0.491
T	<---	FIRM	−0.098	0.119	−0.826	0.409	−0.079
T	<---	TECH	0.486	0.101	4.801	***	0.498
T	<---	PER	−0.039	0.080	−0.485	0.628	−0.044
IP	<---	ID	0.547	0.084	6.498	***	0.661
IP	<---	T	−0.039	0.048	−0.806	0.420	−0.062
F_3	<---	FIRM	1.000				0.755
F_2	<---	FIRM	1.250	0.108	11.615	***	0.880
F_1	<---	FIRM	1.124	0.097	11.634	***	0.882
T_3	<---	TECH	1.000				0.768
T_2	<---	TECH	1.000	0.103	9.698	***	0.823
T_1	<---	TECH	0.964	0.108	8.954	***	0.727
PER_3	<---	PER	1.000				0.933
PER_2	<---	PER	0.411	0.066	6.227	***	0.444
PER_1	<---	PER	1.032	0.055	18.646	***	0.955
T2	<---	T	1.000				0.912
T1	<---	T	0.941	0.138	6.830	***	0.832
ID1	<---	ID	1.000				0.807
ID3	<---	ID	1.013	0.114	8.925	***	0.708
IP1	<---	IP	1.000				0.725
IP2	<---	IP	1.198	0.129	9.286	***	0.806
IP3	<---	IP	1.030	0.125	8.249	***	0.694
IP4	<---	IP	0.850	0.103	8.270	***	0.696

= −0.044), hereby not supporting ***HA6***. The findings also revealed that the determinants of artificial intelligence (factors related to firm, $\beta = 0.346$, factors related to technology, $\beta = 0.177$, and factors related to personnel, $\beta = 0.491$), significantly influence improved data, hereby, supporting ***HA4***.

6 Discussion and Conclusion

There have been exceptional changes in the accounting profession at present and it is in evolutionary stage. The results of the current study that AI systems in the accounting firms of Delhi/NCR improve their performance. Artificial intelligence system is a very powerful tool in the accounting profession as it helps in providing extremely accurate output superseding human efforts. Past literature has acknowledged a positive influence of AI techniques on the accounting profession. It has been identified in past studies that AI techniques mainly to specific tasks like financial reporting, auditing etc [35]. Many studies have noted that expert system in accounting simplifies the training and advising of accountants with the number of prototype expert systems (O'Leary 2003) [55]. It has also been asserted that AI in accounting has made the job of the accountants and auditors easier with the advent of analytics and cognitive technology [21]. On the basis of the findings and results of the present study, it is concluded that AI not only improves the recording and analysis of data but also leads to more transparency of financial transactions, which ultimately leads to improved performance of accounting firms. The choice of which AI technique to be adopted mainly depends on the factors related to the firms like size of the firm, type of the firm, etc, this decision also depends on the technological factors and the type of personnel hired to do such jobs. Implementation of AI technologies in accounting profession not only saves time but also is also beneficial for detecting errors and frauds and also helps in better understanding of trends and patterns in financial transactions. The findings recommend that auditors and accountants should improve their knowledge regarding AI, which will facilitate the accounting and auditing process, which may involve changes in risk analysis, internal auditing report issuing, etc. Integration of AI in accounting functions helps in reducing the size of the work from both client and accounting firm side. Accountants and auditors can collect useful data with machine learning for identifying the relevant data as well as the risk solving factors. It may bring many opportunities for accountants to provide more insights and value to the business and also to improve their efficiency. The present study examines the determinants of the artificial intelligence systems in accounting firms in Model 1. The result of the path analysis suggests significant results, i.e. all the three majors constructs, namely, factors related to firm, factors related to technology and factors related to personnel, all influence the decision to employ AI techniques, hereby supporting ***HA1***, ***HA2*** and ***HA3***, which is in consistent with the result given by past literature [16, 26, 33, 41]. Furthermore, the paper also examines the influence of AI systems on the performance of the accounting through improved data and transparency. The path results advocate that AI techniques improve the performance of the accounting firms through improved data recording and analysis, which is in line with the past literature (Emeke-Nwokeji 2012) [9], but it has no significant impact on the transparency in the firm, which can lead to improved performance.

7 Scope and Implications of the Study

The present research has theoretical as well as practical implications. Theoretically, the study gives major contribution to the existing literature by identifying and confirming the factors, which determine the adoption of AI techniques in accounting firms and by exploring how such techniques improve the performance of accounting firms through improved data and increased transparency. The current study will also help policymakers, educational institutions and government to initiate toward organizing training sessions and seminars for future accountants and auditors to develop their technical skills.

8 Limitations of the Study and Future Scope

The present study has ignored to highlight complexity in AI applications, which has to be investigated to the fullest extent possible before integrating and implementing AI techniques in accounting firms as they need to make additional investments to redesign the business models. The study has also not highlighted the probability of loss of accounting jobs due to progression of technology. The future research could address the challenges involved in the implementation of AI technologies and can also suggest possible solutions to resolve them. Future researchers can also report the potential biases and highlight more implications for the quality and process of accounting work.

References

1. Abdolmohammadi M, Wright A (1987) An examination of the effects of experience and task complexity on audit judgments. Account Rev 62(1):1–13
2. Abu-Musa A (2004) Auditing E-business: new challenges for external auditors. J Am Acad Bus 4(1):28–41
3. Accounting Today (2016) The audit of the future. Accounting Today. Available at: https://www.accountingtoday.com/news/the-audit-ofthe-future
4. Adebayo A, Lee A, Epps R (2008) Developing a theory of auditing behavior in the electronic business environment. J Theor Account Res 4(1):38–82
5. Agnew H (2016) Auditing: pitch battle. Financial Times. Available at: https://www.ft.com/content/268637f6-15c8-11e6-9d98-00386a18e39d
6. Alex H et al (2014) AI, robotics and the future of jobs. Pew Research Centre. Available at: http://www.pewinternet.org/2014/08/06/futureofjobs
7. Algorithm Watch (2019) Automating society: taking stock of automated decision making in the EU. A report by Algorithm Watch in cooperation with Bertelsmann Stiftung supported by the Open Society Foundations, 1st edn. Available at: https://algorithmwatch.org/. Google Scholar
8. Alpaydin E (2016) Machine learning: the new AI. MIT Press, Cambridge, MA
9. Al Qudah H, Shukeri S (2014) The role of data quality and internal control in raising the effectiveness of Ais in Jordan companies. Int J Technol Enhanc Emerg Eng Res 3(8):298–303

10. Amin H, Mohamed E (2016) Auditors' perceptions of the impact of continuous auditing on the quality of Internet reported financial information in Egypt. Manag Audit J 31(1):111–132
11. Amoore L (2018) Doubtful algorithms: of machine learning truths and partial accounts. Theory Cult Soc (Forthcoming). Available at: http://dro.dur.ac.uk/26913/. Google Scholar
12. Baldwin A (1993) The impact of expert system audit tools on auditing firms in the year 2001: a Delphi investigation. J Inf Syst 7(1):16–34
13. Bierstaker L, Burnaby P, Thibodeau J (2001) The impact of information technology on the audit process: an assessment of the state of the art and implications for the future. Manag Audit J 16(3):159–164
14. Boggs SM (1999) Accounting—the digital way. J Account 187(5):99–108
15. Boylan H, Boylan L (2017) Technology in accounting: social media as effective platform for financial disclosures. Int J Digit Account Res 17:93–109
16. Brazel J (2005) A measure of perceived auditor ERP systems expertise: development, assessment, and uses. Manag Audit J 20(6):619–631
17. Brennan B, Baccala M, Flynn M (2017) Artificial intelligence comes to financial statement audits. CFO.com. Available at: http://ww2.cfo.com/auditing/2017/02/artificial-intelligence-audits/
18. Brynjolfsson E, McAfee A (2014) The second machine age: work, progress, and prosperity in a time of brilliant technologies. W. W. Norton & Company Inc., New York
19. Burrell J (2016) How the machine 'thinks': understanding opacity in machine learning algorithms. Big Data Soc 3(1):1–12
20. Carmines G, Zeller A (1982) Reliability and validity assessment. Sage Publications, Beverly Hills
21. Davenport T (2016) Innovation in audit takes the analytics. AI routes. Audit analytics, cognitive technologies, to set accountants free from grunt work. Dupress Publisher
22. Dilek S, Çakır H, Aydın M (2015) Applications of artificial intelligence techniques to combating cyber-crimes: a review. Int J Artif Intell Appl 6(1):21–39
23. Dowling C, Leech S (2007) Audit support systems and decision aids: current practice and opportunities for future research. Int J Account Inf Syst 8(2):92–116
24. Dzuranin C, Malaeseu I (2016) The current state and future direction of IT audit: challenges and opportunities. J Inf Syst 30(1):7–20
25. Elbashir M, Collier P, Sutton G (2011) The role of organizational absorptive capacity in strategic use of business intelligence to support integrated management control systems. Account Rev 86(1):155–184
26. Enofe A et al (2012) Major changes affecting the accounting profession: empirical investigation. Int J Bus Public Admin 9(2):77–96
27. Ernst & Young (EY) (2016). How big data and analytics are transforming the audit. Available at: http://www.ey.com/gl/en/services/assurance/ey-reporting-issue-9-how-big-data-and-analytics-are-transforming-the-audit
28. Field A (2005) Discovering statistics using SPSS, 1st edn. Sage Publications, London
29. Francis J, Huang S, Khurana K, Pereira R (2009) Does corporate transparency contribute to efficient resource allocation. J Account Res 47(4):943–989
30. Gray G et al (2014) The expert systems life cycle in AIS research: what does it mean for future AIS research? Int J Account Inf Syst 15:423–451
31. Greenman C (2017) Exploring the impact of artificial intelligence on the accounting profession. J Res Bus Econ Manag 8(3):1451–1454
32. Hair F et al (2006) Multivariate data analysis. Pearson International Edition, New Jersey
33. Janvrin D, Bierstaker J, Lowe J (2009) An investigation of factors influencing the use of computer-related audit procedures. J Inf Syst 23(1):97–118
34. Janvrin J, Bierstaker J, Lowe J (2008) An examination of audit information technology use and perceived importance. Account Horizons 22(1):1–21
35. Lam M (2004) Neural network techniques for financial performance prediction: integrating fundamental and technical analysis. Decis Support Syst 37(4):567–581

36. Lin C, Taiwan I, Wang C (2011) A selection model for auditing software. Ind Manag Data Syst 111(5):776–790
37. Majdalawieh M, Zaghloul I (2008) Paradigm shift in information systems auditing. Manag Audit J 24(4):352–367
38. Nunnally JC (1978) Psychometric theory. McGraw-Hill, New York, NY
39. O'Leary D (2008) Gartner's hype cycle and information system research issues. Int J Account Inf Syst 9(4):245–252
40. Pasquale F (2015) The black box society: the secret algorithms that control money and information. Harvard University Press, Cambridge, MA
41. Pathak J (2004) A conceptual risk framework for internal auditing in E-commerce. Manag Audit J 19(2):556–564
42. Pathak J, Lind MR (2002) Integrated information systems, SAS 94 and auditors. J Corp Account Financ 19(1):57–67
43. Petterson M (2005) The keys to effective IT auditing. J Corp Account Financ 16(5):41–46
44. Rapoport M (2016) Auditors count on tech for backup. Wall Str J
45. Razi M, Madani H (2013) An analysis of attributes that impact adoption of audit software: an empirical study in Saudi Arabia. Int J Account Inf Manag 21(2):170–188
46. Rezaee Z, Reinstein A (1998) The impact of emerging information technology on auditing. Manag Audit J 13(8):465–471
47. Srinivasan V (2016) Will financial auditors become extinct? In: The intelligent enterprise in the era of big data, Chapter 7. Wiley, New York, NY, pp 171–183. https://doi.org/10.1002/9781118834725.ch7
48. Sun T, Alles M, Vasarhelyi M (2015) Adopting continuous auditing: a cross-sectional comparison between China and the United States. Manag Audit J 30(2):176–204
49. Tarus DK, Omandi EM (2013) Business case for corporate transparency: evidence from Kenya. Eur J Bus Manag 5(3):113–125
50. Vilsanoiu D, Serban M (2010) Changing methodologies in financial audit and their impact on information systems audit. Informatica Economică 14(1):57–66
51. Xu H (2003) Critical success factors for accounting information systems data quality. Doctor of Philosophy (PhD) thesis. Transferred from ADT 01/12/2006
52. Yang DC, Miklos AV (2016) The application of expert system in accounting. Available at: http://www.sciencejrank.org/pages
53. Yoon K, Hoogduin L, Zhang L (2015) Big Data as complementary audit evidence. Account Horizons 29(2):431–438
54. Zerilli J, Knott A, Maclaurin J (2018) Transparency in algorithmic and human decision-making: is there a double standard? Philosophy & Technology. Springer, pp 1–23
55. Zhao N, Yen DC, Chang I (2004) Auditing in the e-commerce era. Inf Manag Comput Secur 12(5):389–400

Design of a Controller for the Microgrid to Enhance Stability and Synchronization Capability

Suhaib Khan, Naiyyar Iqubal, and Sheetla Prasad

Abstract In this study, a state feedback controller is designed for a linearized model of a diesel generator-based generation plant and a battery storage-based inverter model plant. The diesel generator plant and inverter-based battery storage system constitute as a microgrid. The microgrid is a localized power system network and minimized the mismatch between the generation and loads in a specific region. Thus, to achieve better performance and stability, a proposed controller is designed for microgrid. The stability convergence of the control law is analyzed through Lyapunov stability theorem as well as the Nyquist diagram stability criteria. The proposed controller improves the overall system performance in the presence of initial parameter variation and mechanical shaft power random step variations by reducing over/under shoots, settling time and oscillations. In addition, the proposed controller regulated both diesel generator model and inverter-based DG independently. The performance, stability and ability to keep in synchronism of the proposed control scheme are validated on a diesel generator and inverter-based battery storage model simulated in MATLAB©.

Keywords Diesel generator system · Inverter-based battery storage system · Microgrid · State feedback controller

1 Introduction

Due to green aspects as well as limited conventional resources availability, the power generation is trying to shift toward renewable energy sources. In the decade, air pollution and environmental pollution and same issues have become more prominent. Pollution is mainly caused by burning of coal and other fossil fuels vigorously at the world level, as these fossil fuels will be consumed by humans 1 day inevitably. So, the future focus should be on renewable energy, which will be inexhaustible

S. Khan · N. Iqubal · S. Prasad (✉)
School of Electrical, Electronics and Communication Engineering, Galgotias University, Greater Noida, Uttar Pradesh, India
e-mail: sheetla.prasad@galgotiasuniversity.edu.in

A. Tomar et al. (eds.), *Machine Learning, Advances in Computing, Renewable Energy and Communication*, Lecture Notes in Electrical Engineering 768,
https://doi.org/10.1007/978-981-16-2354-7_39

and can be renewed, and most of the domestic as well as commercialize renewable generation system can be directly connected to either utility grid or not connected to utility grids. Thus, this structure is called as microgrid. In this context, the microgrid is the localized network with at least one distributed generation (DG) and is made to deliver electricity to small community like town, business complex [1, 2]. Thus, the electrical distribution utility needs more a sophisticate control structure to manage power quality issues at end terminals.

In the conventional power system, the whole grid collectively is connected to the larger grid. In this type of conventional power system, the cost is more and there are other operational difficulties. Sometimes, a particular power plant is not able to supply the required energy. This issue was solved by integrating the supply with power grid. But the problem was not solved but later to resolve this issue, the concept of islanded microgrid was introduced [3].

The microgrid is only connected at a single point to utility grid. The basic objective of the microgrid is to generate and control power demand localized way [4]. This connection acts as a switch that makes it possible to disconnect the microgrid from the main grid and can be operated temporarily in islanded mode as well. The microgrid combines distributed power load, control devices, etc. and classified as (a) Institute microgrid: this type of microgrid focuses on site generation of the energy and utilizing within the institutional campus, (b) Community microgrid: it serves thousands living within the community and they also comprise the renewable sources and can supply energy on demand within the community. They have several distributed energy storages but, in a system, at least one should be there. Such type can be in both AC or DC and can operate both as bi-lateral system and (c) Military microgrid: this is used in military base and can also provide security.

The national and centralize microgrids tend to more research and developments to resolve load frequency control, protection in both island as well as in grid-connected modes. The operation and control of microgrids variables can enhance the benefits of the power supplier as well as power consumers [5]. But the output of renewable sources such as the photovoltaic can vary frequently because of environmental conditions due to this serious issue arises on reliability on PV and power quality of utility grid [6]. The microgrid can help in lowering feeder loss and increase the reliability of local power supply and increase in the energy efficiency along this microgrid is dominated by converter network interface also [7–9]. Based on the characteristics of feeder, there are three types of microgrid and are classified as [10]: (a) urban microgrid, (b) rural microgrid and (c) off-set microgrid, and also comprised as a dispersed or distributed generation network.

The distributed generation is a system that uses the number of dispersed energy sources to generate power. The energy could be achieved from fossil fuels and renewable power plants such as solar generation system, wind generation system, etc. [11]. Such grids operate in hundreds of kilowatts and megawatts [12]. This can help reducing carbon emissions as well as increasing the quality of air and giving higher efficiency for the overall system, controller can be used to get proper as well as desirable output hence resulting in higher overall performance of the plant and making the plant more stable. This is positioned closer to the dispersed network, misplacing these can cause network losses and higher operating costs [13]. Controller can also be placed and used to reduce the PV fluctuations and a local controller can be used as protection as well as it can be used as an individual device [14, 15]. Most common type of controller used in feedback control is PID [16]. Controller is used as a bridge between microgrid model and utility grid. It receives desired instructions and after validating the instruction it executes to control accordingly [17]. The tuning PID controller can be done using optimization approach to get a better response [18]. Moreover, some advanced AI and machine learning-based studies are listed in Springer books and journals [20–27].

In this paper, a linearized model of both diesel generator and inverter-based battery storage model is considered for study and farmed as a microgrid system. A simplified state feedback control approach is applied. The present study has the following contributions: (1) a state feedback control scheme designs for the linearized model of microgrid, (2) the convergence of control law analyzes using Lyapunov's stability theorem and Nyquist diagram, (3) the performance of the controller demonstrates in the presence of mechanical shaft power disturbances and (4) closed-loop system responses produce satisfactorily over/under shoots with negligible oscillations and maintains in synchronization.

The remaining sections of this paper are composed as follows: Sect. 2 describes the linearization of both diesel generator dynamic model and inverter-based battery storage dynamic model. The control law design steps followed by its convergence are depicted in Sect. 3. The simulation investigations and results' demonstration are given in Sect. 4. At last, the conclusion is drawn based on simulation demonstration in Sect. 5.

2 System Structure and Dynamic Modeling

The microgrid is a localized power system network and minimizes the mismatch between the generation and loads in a synchronized way. The microgrid can operate to generate maximum revenue as autonomously and also synchronized with power grids. The optimal operation of microgrid needs more sophisticated control structure. As a result, a simplified microgrid structure with supporting equipment is considered in this study and as shown in Fig. 1a. The diesel generation simplified model and an equipped DG such as solar PVs etc. [1, 19] are considered in this study. All simplified microgrid system model is represented as first-order dynamics for small signal analysis.

The dynamic model of diesel generator contains an automatic voltage regulator, synchronous generator and speed governor are considered [1]. The classical diesel generator model acceleration dynamics is described as:

$$\Delta\dot{\omega}_r = \frac{1}{2H}(\Delta \mathrm{T}_m - \Delta \mathrm{T}_e - K_D \Delta\omega_r) \tag{1}$$

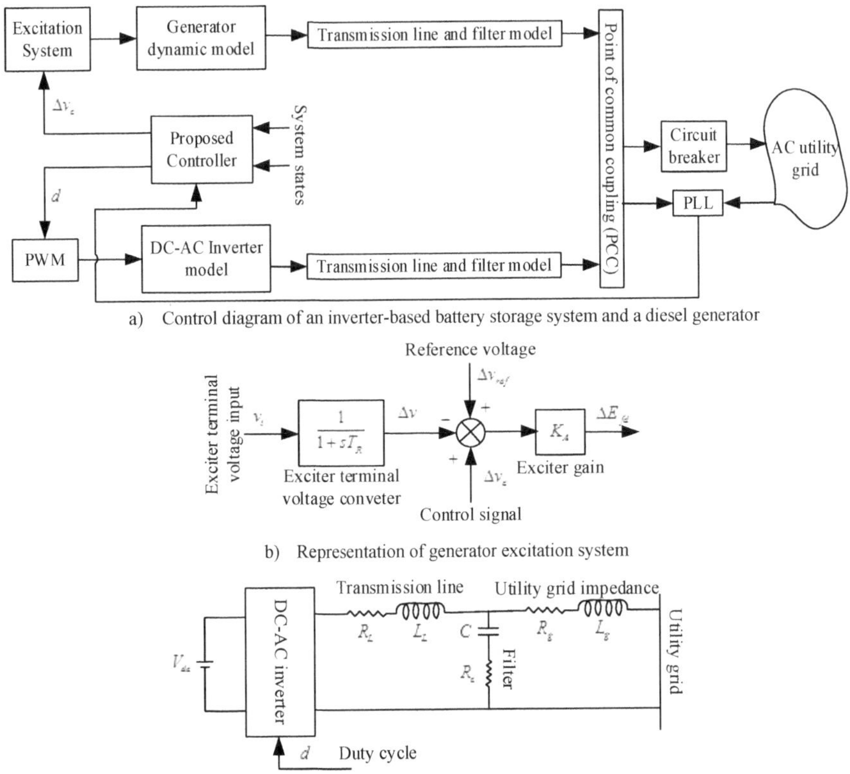

a) Control diagram of an inverter-based battery storage system and a diesel generator

b) Representation of generator excitation system

c) Simplified inverter based solar PV system model

Fig. 1 Schematic diagram of a microgrid

where, the terms H, T_e, K_D, T_m and f_0 are the inertia constant (per MVA), electrical torque (N-m), damping constant (N per m), mechanical torque (N-m) and nominal frequency (Hz) respectively. The electrical torque linearized value is considered [1] and substitutes in Eq. (1) as:

$$\Delta\dot{\omega}_r = \frac{1}{2H}(\Delta \mathrm{T}_m - K_1\Delta\delta - K_2\Delta\psi - K_D\Delta\omega_r) \tag{2}$$

where, K_1 and K_2 are the synchronizing coefficients.

$$\Delta\dot{\delta} = 2\pi f_0 \Delta\omega_r \tag{3}$$

The state variables are $\Delta\omega_r$ and $\Delta\delta$ are generator angular speed (rad/s) and rotor angle (rad) respectively. The generator field circuit system dynamics are written as:

$$\Delta\dot{\psi} = -\frac{2\pi f_0 m_1 R_{fd} L'_{ads}}{L_{fd}}\Delta\delta - \frac{2\pi f_0 R_{fd}}{L_{fd}}\left[1 - \frac{L'_{ads}}{L_{fd}} + m_2 L'_{ads}\right]\Delta\psi$$
$$- \frac{2\pi f_0 R_{fd} K_A}{L_{adu}}\Delta v + \frac{2\pi f_0 R_{fd} K_A}{L_{adu}}\Delta v_c \quad (4)$$

where, terms R_{fd}, L'_{ads}, L_{fd}, L_{adu}, m_1 and m_2 are the generator field system parameters. The term K_A is derived from exciter transfer function and known as exciter gain. The state variables $\Delta\psi$ and Δv are magnetic flux and field excitation voltage (volt) of the diesel generator respectively. The generator automatic voltage regulator system is considered [1] as shown in Fig. 1b and its dynamics is written as:

$$\Delta\dot{v} = \frac{K_5}{T_R}\Delta\delta + \frac{K_6}{T_R}\Delta\psi - \frac{1}{T_R}\Delta v. \quad (5)$$

The terminal voltage error can be obtained using the following equation:

$$\Delta v_t = K_5\Delta\delta + K_6\Delta\psi \quad (6)$$

where, terms K_5 and K_6 are the excitation system parameters. The value of term K_5 is always positive while term K_6 can be either positive or negative due to dependency on the external circuit impedance.

The grid connected inverter-based battery storage system is considered [20] in this study and represented in Fig. 1c. The transmission line parameter can be represented into short transmission line model. The filter parameters are neglected here for small perturbations due to low harmonic effect and inverter switches are considered as ideal. The output voltage of the inverter v_i is varied from V_{dc} to $-V_{dc}$ according to value of duty cycle d. Thus, the duty cycle is defined based on t_{on} and t_{off} time and written as:

$$d = \frac{t_{on} - t_{off}}{T} \quad (7)$$

where, term $T = t_{on} + t_{off}$ is known as switching time interval. The output of the inverter can be written using time-average approach as:

$$v_i = \frac{t_{on} V_{dc} + t_{off}(-V_{dc})}{T} = d V_{dc}. \quad (8)$$

The transmission line impedance can be represented as equivalent impedance between inverter and utility grid as shown in Fig. 1c. The utility grid is also considered here in the presence of negligible harmonics production. The equivalent complete dynamics of transmission line and utility grid can be written as:

$$v_i - v_g = (R_L + R_g) i_0 + (L_L + L_g)\frac{\partial i_0}{\partial t}. \quad (9)$$

Equation (9) can be rewritten for small perturbation analysis as:

$$\Delta \dot{i}_0 = \frac{1}{(L_L + L_g)} v_i - \frac{(R_L + R_g)}{(L_L + L_g)} \Delta i_0. \tag{10}$$

Substitute inverter output voltage from Eq. (8) into Eq. (10), it gives as:

$$\Delta \dot{i}_0 = -\frac{(R_L + R_g)}{(L_L + L_g)} \Delta i_0 + \frac{V_{dc}}{(L_L + L_g)} d. \tag{11}$$

The complete microgrid state space dynamics are represented using Eqs. (2)–(5) and Eq. (11) as follows:

$$\begin{aligned} \dot{x}(t) &= A\,x(t) + B\,u(t) + D\,d(t) \\ y(t) &= C\,x(t) \end{aligned} \tag{12}$$

where, $x(t)$ is the microgrid state variables, $u(t)$ is input, $y(t)$ is output and $d(t)$ is mechanical torque disturbances respectively. The microgrid plant matrices A, B, C and D are given as:

$$A = \begin{bmatrix} -\frac{K_D}{2H} & -\frac{K_1}{2H} & 0 & -\frac{K_2}{2H} & 0 \\ 2\pi f_0 & 0 & 0 & 0 & 0 \\ 0 & \frac{K_5}{T_R} & -\frac{1}{T_R} & \frac{K_6}{T_R} & 0 \\ 0 & -\frac{2\pi f_0 m_1 R_{fd} L'_{ads}}{L_{fd}} & -\frac{2\pi f_0 R_{fd} K_A}{L_{adu}} & -\frac{2\pi f_0 R_{fd}}{L_{fd}}\left[1 - \frac{L'_{ads}}{L_{fd}} + m_2 L'_{ads}\right] & 0 \\ 0 & 0 & 0 & 0 & -\frac{R_L + R_g}{L_L + L_g} \end{bmatrix},$$

$$x(t) = \begin{bmatrix} \Delta\omega_r \\ \Delta\delta \\ \Delta v \\ \Delta\psi \\ \Delta i_0 \end{bmatrix}.$$

$$B = \begin{bmatrix} 0\,0\,0 & 0 & \frac{V_{dc}}{L_L + L_g} \\ 0\,0\,0 & \frac{2\pi f_0 R_{fd} K_A}{L_{adu}} & 0 \end{bmatrix}^T, C = \begin{bmatrix} \frac{1}{2\pi f_0} & 0\;0\;0\;0 \\ 0 & K_5\;0\;K_6\;0 \end{bmatrix},$$

$$D = \begin{bmatrix} \frac{1}{2H} & 0\,0\,0\,0 \end{bmatrix}^T,\ u(t) = \begin{bmatrix} \Delta v_c & d \end{bmatrix}^T \text{ and } d(t) = \Delta \mathrm{T}_m.$$

3 Control Methodology

The solar PV-based battery storage terminal DC voltage of the distributed generator is varied according to environmental conditions. So, the inverter output terminal voltage is regulated using duty cycle as input to the inverter. Similarly, the diesel generator terminal voltage is also regulated using excitation voltage. Thus, to minimize the mismatch between power generation and load, the inverter duty cycle and excitation voltage are needed to regulate in a precise way. As result, the microgrid can withstand synchronization with utility grids in the presence of large disturbances in the system. A state feedback-based pole placement control scheme is applied in this study to achieve robust performance.

Lemma 1 *If microgrid matrices A and B are controllable system then system states can be controlled according to desired response.*

To obtain system state feedback gain F is calculated using the following steps:

- Test the controllability conditions of the state space system dynamics as per given in Lemma 1.
- Find the open-loop dominant pole/poles location with its damping ratio and undamped natural frequency.
- Select closed-loop desired damping ratio and damped natural frequency, and corresponding location of its poles.
- To achieve robust performance of the microgrid system, the desired poles should be located at least three times far away from the open-loop dominant poles.
- Obtain state feedback gain F using "*place*" command in the MATLAB© with system matrices and desired poles location.
- Finally, calculate state feedback duty cycle and exciter voltage control signal using control law.

The overall microgrid system dynamics are calculated using $u(t) = -Fx(t)$ and substituted in Eq. (12) and it gives:

$$\dot{x}(t) = (A - BF)x(t) + Dd(t). \tag{13}$$

The convergence of the controller is evaluated using Lyapunov's stability theorem.

Proof Let the Lyapunov function is defined with positive definite matrix P as:

$$\nu = x^T(t)Px(t). \tag{14}$$

Obtain the first derivative with respect to time and Eq. (14) gives:

$$\dot{\nu} = \dot{x}^T P x + x^T P \dot{x}. \tag{15}$$

Substitutes from Eq. (13) into Eq. (15) and it gives after simplification as:

$$\dot{\nu} = x^T(t)\left[P^T(A - BF) + P(A - BF)^T\right]x(t) + x^T(t)P^T Dd(t). \tag{16}$$

Using Lyapunov's stability conditions, it gives:

$$\dot{\nu} = -\|x\|[\varepsilon\|x\| - \|PD\|\|d(t)\|] \tag{17}$$

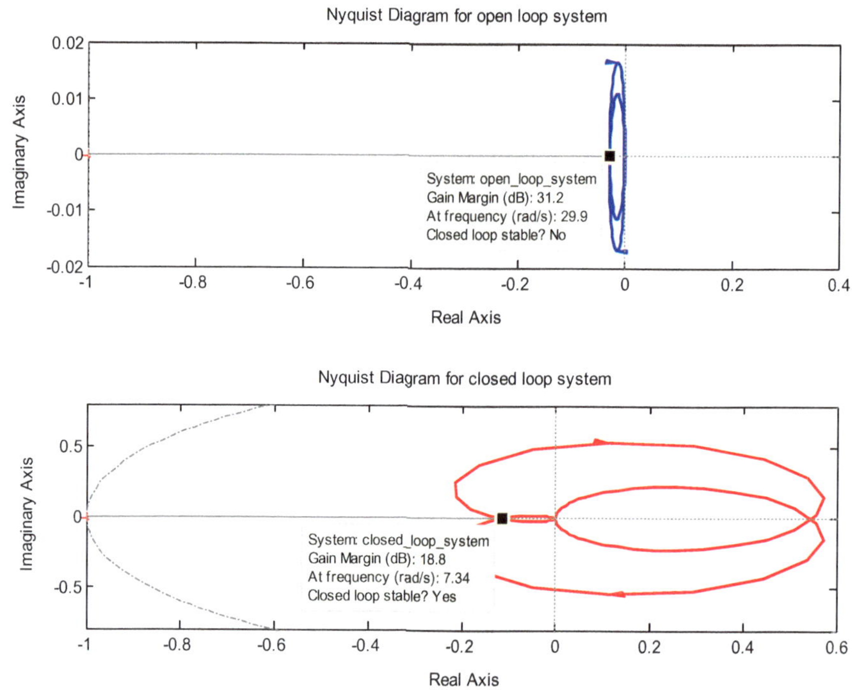

Fig. 2 Nyquist diagram of open- and closed-loop system

where, $\|.\|$ is the matrix norms.

Thus, the microgrid state dynamics enter into bounded space with radius $\frac{\|PD\|\|d(t)\|}{\varepsilon}$. Hence, closed-loop system trajectories converge as asymptotically due to $\dot{v} < 0$. This completes the proof.

Furthermore, the Nyquist plot of open-loop microgrid system and overall microgrid system are considered to demonstrate the convergence of the control law effectively as shown in Fig. 2. The transfer function is derived with respect to duty cycle of the inverter. It is observed from the Nyquist plot, without control law, system is unstable and more sensitive with respect to disturbances compared to overall microgrid system with control law.

4 Responses and Discussions

The state-space dynamics of the microgrid is simulated to demonstrate controller performance using MATLAB© 2014a software as depicted in Fig. 1a. The linearized microgrid parameters are tabulated in Table 1. The microgrid contains diesel generation and solar PV-based battery source with rating 20 kW, three phases with four wires, 60 Hz, 380 by 220 V, 1800 rpm and 6 kW, 400 V, respectively. The generator voltage exciter is considered as brushless and represented using first-order dynamics

Table 1 Microgrid parameters

System parameter	Values (in unit)	System parameter	Values (in unit)
K_D	0	K_a	200
H	3	R_{fd}	0.0006
K_1	1.591	m_1	1.0473
K_2	1.5	L_{fd}	0.153
K_3	0.333	L_{ads}	1.64899
R_L	0.01	L_{adu}	1.65
Rg	0.01	m_2	0.8802
L_L	0.005	L_{as}	1
K_5	0.12	K_a	187
K_6	0.3	L_g	0.001
ω_0	376.8	V_{dc}	400
T_R	0.05	K_r	1

with gain 187. The generator voltage exciter time constant is 0.05 s. All simulations are performed at randomly selected initial condition $x_0 = \left[0.9\ 0\ 0.04\ 0.03\ 0.032\right]^T$.

The open-loop system poles, dominant poles damping ratio and undamped natural frequency of the linearized microgrid are tabulated in Table 2. It is observed that open-loop system has one pole lie on right-hand side and the damping ratio as well as undamped natural frequency oscillations is too low and high, respectively. As result, open-loop system dynamics are unstable. The desired overall system time characteristics are also given in Table 2 and enhanced overall system performance against different disturbances.

The performance of the proposed controller is demonstrated against initial condition and ΔT_m.

Table 2 Microgrid system poles location and time characteristics

Open-loop system poles locations	Open-loop system ∂ and ω_n(rad/s)	Overall microgrid system poles locations	Overall system ∂ and ω_n (rad/s)
Dominant pole: −0.9132 + 29.6993i; −0.9132 – 29.6993i; Other poles: −18.4711, 12.6009, – 3.3333	$\omega_n = 29.7133$ and $\partial = 0.0307$ Remark: open-loop system is unstable	Dominant poles: −2.3089 + 2.2991i, – 2.3089 – 2.2991i Other poles: −16.9789, −3.8203, – 3.3333	$\omega_n = 3.2584$ and $\partial = 0.7086$ Remark: closed loop is stable

4.1 Microgrid System Response in Presence of Initial Condition

The controller performance of the microgrid system is tested in the presence of randomly selected initial condition $x_0 = \begin{bmatrix} 0.9 \ 0 \ 0.04 \ 0.03 \ 0.032 \end{bmatrix}^T$.

The deviations in microgrid state variables and control effort signal are depicted in Fig. 3 and Fig. 4, respectively. Due to initial conditions, the deviation in angular speed, rotor angle, excitation voltage and flux are settled within 3 s as given in Fig. 3 and found within permissible limits. Similarly, the deviations if inverter load current and generator frequency with control signals are converged within 3 s as shown in Fig. 4. The inverter is switched on and off ideally and so that deviation in inverter load current as well as control duty cycle signals are varied with small in magnitude. The proposed controller is capable to adjust generator excitation voltage and inverter duty cycle in accordance to minimize the system disturbances. Thus, the proposed controller settles system dynamics with permissible overshoots and reduced settling time.

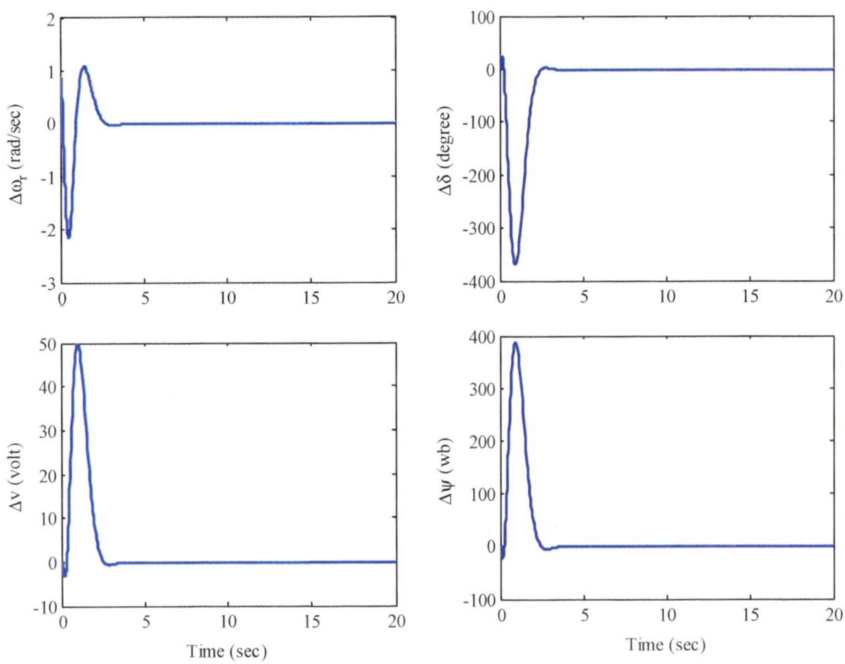

Fig. 3 Deviations of the closed-loop microgrid system variables at initial conditions

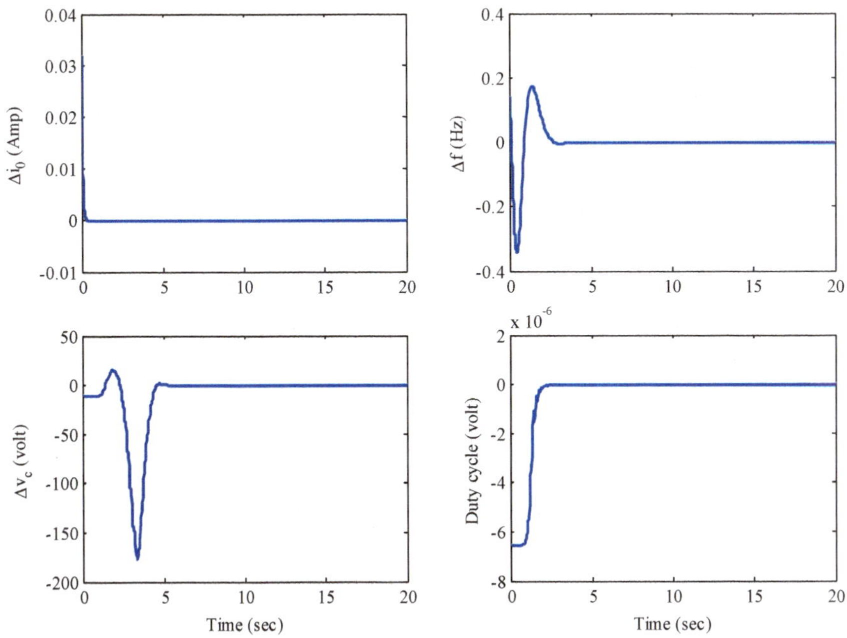

Fig. 4 Deviations in load current, frequency and control signals at initial conditions

4.2 Microgrid System Responses in the Presence of Deviation in ΔT_m

The robustness of the proposed controller is analysed in presence of ΔT_m and initial condition. The deviation patterns in mechanical power in diesel generator shaft are shown in Fig. 5. The said variation is applied in diesel generator model. The system responses are shown in Figs. 6 and 7 in the presence of mechanical power deviation pattern as given in Fig. 5. The inverter-based DG system responses are not affected

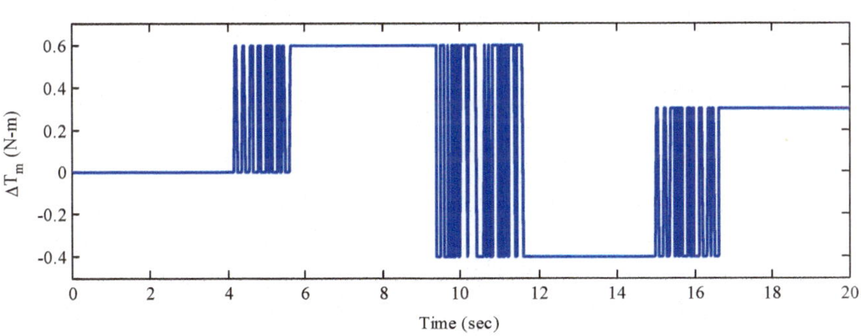

Fig. 5 Deviation pattern in mechanical power in diesel generator shaft

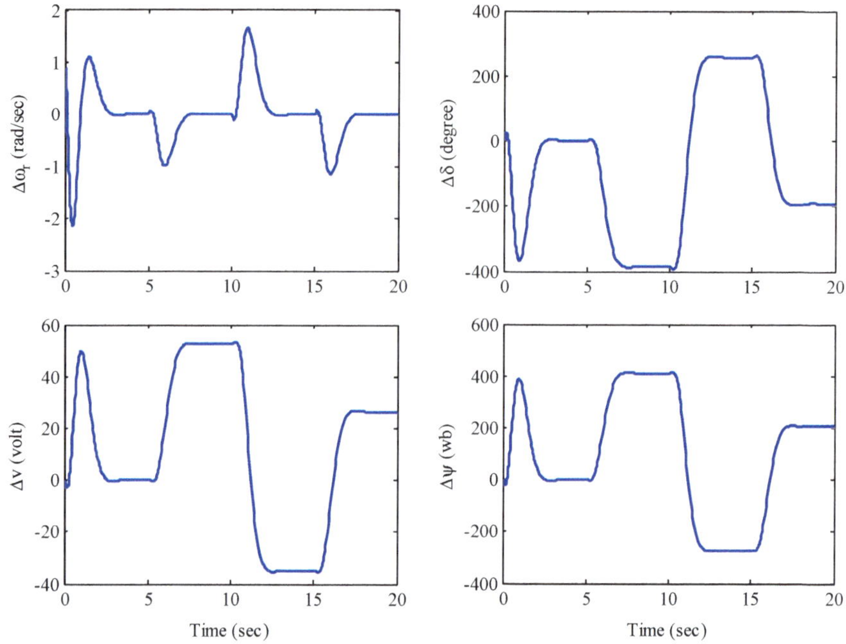

Fig. 6 Microgrid system responses against mechanical power deviations and initial conditions

due to the said variation is applied in diesel generator model. The deviation in angular speed, rotor angle, excitation voltage and flux are settled down within short time as given in Fig. 6 without any oscillations and over/under shoots.

Thus, the proposed controller is capable to sustain system within permissible limits and remains in synchronization with utility grid even in the presence of system disturbances. As observed, the proposed controller sustains the system in synchronization with negligible over/under shoots and oscillations and maintained all variables are lied within permissible limits. Hence, the microgrid system stability and performance are enhanced significantly even in the presence of system disturbances. Also, the proposed controller is regulated both diesel generator-based generation plant and inverter-based energy plant independently.

5 Conclusion

In this study, a state feedback controller designed for a linearized model of a diesel generator-based generation plant and inverter-based energy plant. The diesel generator-based generation plant and inverter-based battery storage system were

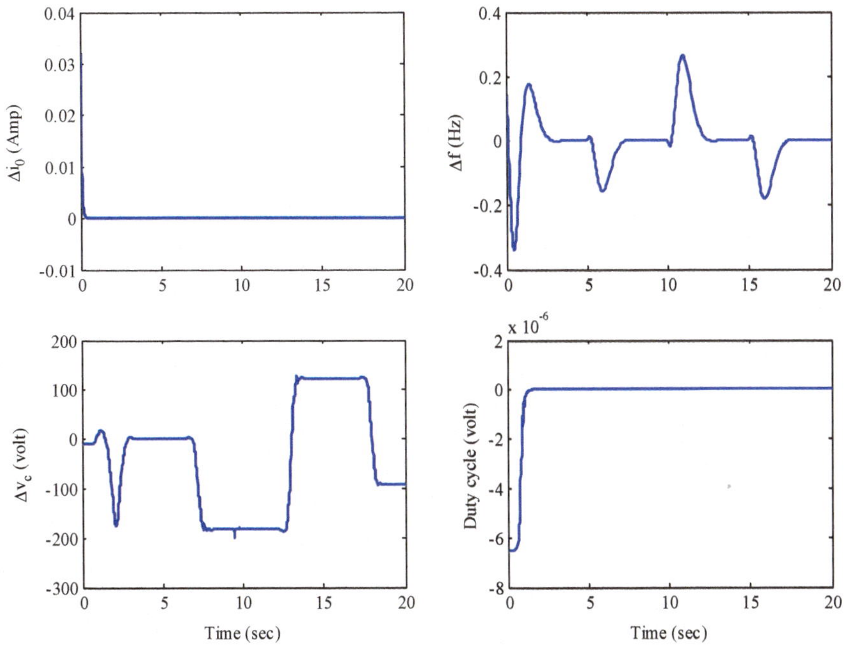

Fig. 7 Deviations in load current, frequency and inputs signal against mechanical power deviations and initial conditions

constituted as a utility grid-connected microgrid. The diesel generator voltage excitation model and inverter-based battery storage model filter characteristics are considered as linear region and low-frequency range in harmonics, respectively. The proposed controller design steps are described in a simplified manner. The convergence of control law is analyzed through Lyapunov stability theorem as well as Nyquist diagram stability condition approach. The proposed controller performance is observed in the presence of initial parameters and mechanical shaft power random step variations. It is evident that the microgrid system performance and stability enhance and maintain system in synchronization with permissible operating limits. The over/under shoots and settling time reduced with negligible oscillations. Also, the proposed controller regulated both diesel generator-based generation plants independently. In the future, the dynamics of solar PV system will be considered in place of battery storage system and designed a robust controller.

References

1. Kundur P (1994) Power system stability and control. McGraw Hill, New York, pp 780–808
2. Olivares DE, Mehrizi-Sani A, Etemadi AH, Cañizares CA, Kazerani RIM, Hajimiragha AH, Gomis-Bellmunt O, Saeedifard M, Palma-Behnke R, Jiménez-Estévez GA, Hatziargyriou ND

(2014) Trends in microgrid control. IEEE Trans Smart Grid 5(4):1905–1919
3. Singh M, Singh O, Kumar A (2019) Renewable energy sources integration in micro-grid including load patterns: In: 2019 3rd international conference on recent developments in control, automation and power engineering (RDCAPE), pp 1–6
4. Gupta P, Ansari MA (2019) Analysis and control of AC and hybrid AC-DC microgrid: a review. In: 2019 2nd international conference on power energy, environment and intelligent control (PEEIC), pp 1–6
5. Lu W, Zhao Y, Li W, Du H (2014) Design and application of microgrid operation control system based on IEC 61850. J Mod Power Syst Clean Energy 2(3):256–263
6. Han Y, Ning X, Yang P, Xu L (2019) Review of power sharing, voltage restoration and stabilization techniques in hierarchical controlled DC microgrids. IEEE Access 7:149202–149223
7. Zhang W, Xu Y (2019) Distributed optimal control for multiple microgrids in a distribution network. IEEE Trans Smart Grid 10(4):3765–3779
8. Kaviri SM, Pahlevani M, Jain P, Bakhshai A (2017) A review of AC microgrid control methods. In: 2017 IEEE 8th international symposium on power electronics for distributed generation systems (PEDG), pp 1–6
9. Qingdong H, Bin X, Wei W, Yueliang C (2019) Operation control strategy of microgrid based on energy storage system. In: IEEE 4th advanced information technology, electronic and automation control conference (IAEAC), pp 1–6
10. Parisio A, Rikos E, Glielmo L (2014) A model predictive control approach to microgrid operation optimization. IEEE Trans Control Syst Technol 22(5):1813–1827
11. de Andrade F, Castilla M, Bonatto BD (2020) Microgrids: operation and control methods. In: Basic tutorial on simulation of microgrids control using MATLAB® and Simulink® software. Springer briefs in energy. Springer, Cham. https://doi.org/10.1007/978-3-030-43013-9_1
12. Xuefeng L, Kaiju L, Weiqiang L, Chaoxu M, Dan W (2018) A brief analysis of distributed generation connected to distribution network. In: 2018 33rd youth academic annual conference of Chinese Association of Automation (YAC), pp 1–6
13. John N, Janamala V, Rodrigues J (2019) Impact of variable distributed generation on distribution system voltage stability. In: 2019 international conference on data science and communication (IconDSC), pp 1–6
14. Rahmanov NR, Karimov OZ (2020) AC and DC combined microgrid, modeling and operation. In: Mahdavi Tabatabaei N, Kabalci E, Bizon N (eds) Microgrid architectures, control and protection methods. Power systems. Springer, Cham. https://doi.org/10.1007/978-3-030-23723-3_3
15. Yazdanian M, Mehrizi-Sani A (2014) Distributed control techniques in microgrids. IEEE Trans Smart Grid 5(6):2901–2909
16. Bordons C, Garcia-Torres F, Ridao MA (2020) Microgrid control issues. In: Model predictive control of microgrids. Advances in industrial control. Springer, Cham. https://doi.org/10.1007/978-3-030-24570-2_1
17. Chowdhury MKI (2016) Pre and post controller based MVC architecture for web application. In: 2016 5th international conference on informatics, electronics and vision (ICIEV), pp 1–6
18. Prainetr S, Phurahong T, Janprom K, Prainetr N (2019) Design tuning PID controller for temperature control using ant colony optimization. In: 2019 IEEE 2nd international conference on power and energy applications (ICPEA), pp 1–6
19. Taher SA, Zolfaghari M, Cho C, Abedi M, Shahidehpou M (2017) A new approach for soft synchronization of microgrid using robust control theory. IEEE Trans Power Deliv 32(3):13701381
20. Taher SA, Zolfaghari M (2014) Designing robust controller to improve current-sharing for parallel-connected inverter-based DGs considering line impedance impact in microgrid networks. Electr Power Energy Syst 63:625–644
21. Gopal et al (2021) Digital transformation through advances in artificial intelligence and machine learning. J Intell Fuzzy Syst (Pre-press) 1–8. https://doi.org/10.3233/JIFS-189787
22. Fatema N et al (2021) Intelligent data-analytics for condition monitoring: smart grid applications. Elsevier, 268 pp. ISBN 978-0-323-85511-2. https://www.sciencedirect.com/book/9780323855105/intelligent-data-analytics-for-condition-monitoring

23. Smriti S et al (2018) Special issue on intelligent tools and techniques for signals, machines and automation. J Intell Fuzzy Syst 35(5):4895–4899. https://doi.org/10.3233/JIFS-169773
24. Jafar A et al (2021) AI and machine learning paradigms for health monitoring system: intelligent data analytics. Springer Nature, Berlin, 496 pp. https://doi.org/10.1007/978-981-33-4412-9. ISBN 978-981-33-4412-9
25. Sood YR et al (2019) Applications of artificial intelligence techniques in engineering, vol 1. Springer Nature, 643 pp. https://doi.org/10.1007/978-981-13-1819-1. ISBN 978-981-13-1819-1
26. Yadav AK et al (2020) Soft computing in condition monitoring and diagnostics of electrical and mechanical systems. Springer Nature, Berlin, 496 pp. https://doi.org/10.1007/978-981-15-1532-3. ISBN 978-981-15-1532-3
27. Aggarwal S et al (2020) Meta heuristic and evolutionary computation: algorithms and applications. Springer Nature, Berlin, 949 pp. https://doi.org/10.1007/978-981-15-7571-6. ISBN 978-981-15-7571-6

Power Loss Reduction in Distribution System Using Wind Power as DG

Anil Kumar and Poonam Yadav

Abstract As the main characteristics of the centralized system, the world's energy supply system is a small unit, wide grid and high voltage system. While the world's power load is powered by such a single large power grid, demand for energy and power supply reliability in quality and security are increasing in today's society, and these requirements cannot be feed by a large power grid because of their own power shortcomings. The entire power grid is affected by the malfunctioning or the failure of any point in the large power grid, causing a large area of power outages or even the entire network to crash, having disastrous consequences, these accidents occur abroad and such a large power grid is very vulnerable to damage by war or terrorist powers. General military attacks have destroyed large power plants or power plants as one of the main objectives, once the large power grid has been disrupted, it will significantly jeopardize the security of the country. However, the centralized power grid cannot solve the problem of monitoring power charges, and for short peak loads, the construction of power plants becomes very costly and there are very small economic benefits.

Keywords Distribution generator · Power flow · Power system losses · Newton–Raphson · Wind energy

1 Introduction (Distribution System)

Generation system, transmission system and distribution system are the three main parts of the electric power system. Figure 1 shows the diagram of electric power system, which consists of these three subparts. The generating station generates electricity by transforming a primary energy source into electrical energy. Then by using step-up transformers, the output voltage of the generators stepped up according to the required appropriate transmission rates. Electrical power is transmitted from the end substation to the end substation through high-voltage transmission lines. When

A. Kumar (✉) · P. Yadav
Electrical Engineering Department, DCR University of Science & Technology, Murthal, Sonipat, Haryana, India

A. Tomar et al. (eds.), *Machine Learning, Advances in Computing, Renewable Energy and Communication*, Lecture Notes in Electrical Engineering 768,
https://doi.org/10.1007/978-981-16-2354-7_40

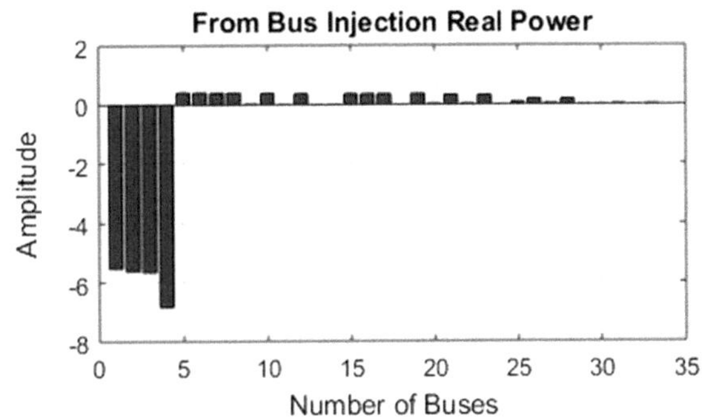

Fig. 1 From bus injection real power

the transmission line ended at substations, using the step-down transformers, voltage is then converted to lower value (say 11 or 33 or 66 kV). The secondary transmission system takes power from these substations at receiving end and transfers power to the secondary substation. At the secondary substation, there are switchgear, buses, voltage control facilities and two or more power transformers, stepping down to 11 kV at the secondary substation level. The circuits that take energy from transmission subsystem and provide energy to the distribution subsystem are designated by sub-transmission system. The sub-transmission system is typically supplied by the transmission substations, but still it is known as the sub-transmission. Before that, many sub-transmission networks were transmission lines. The customer derives electricity from the system of distribution. There are two branches of the distribution system: the primary distribution system and the secondary distribution system. The primary distribution system is a wire network of three phase 3 that usually operates at nominal voltages of 11/12.66 kV. Big customers such as factories and manufacturers may take power from the primary distribution networks directly. The secondary distribution network is a wire structure of three phase 4 that provides electricity to the end customers. The loads of the customer may be single or three phase. The primary feeder for distribution can be classified as: (1) Radial type and (2) Meshed/Looped or Ring main.

The radial distribution system is linear in nature. The substation feeds the power to the feeder lines. Loads are tapped onto the feeder. The distributors connected at various points along the length of the feeder then deliver the power to the service mains.

The major drawbacks associated with the radial distribution networks are: (1) In case of a total shutdown of the substation, the entire system is blacked out, (2) In case of any fault along the length of the feeder, the remaining loads connected after the fault point will not get any power and (3) Large voltage fluctuations are caused at the load points connected at the end of the feeder. The ring main distribution system is designed to compensate for the drawbacks of the radial feeder. The ring main system has two substations simultaneously delivering power to the mains so that even in case of one substation shut down, the other one will be able to continue the operation of the system. Moreover, several advanced AI and machine learning-based research analyses are represented in [1–7].

2 Calculation of Losses in Distribution System Using Wing (DG)

A. *Newton–Raphson method*

Since sparse matrix technology was applied to the Newton method in the 1960s, after decades of development, It has become the most commonly used approach for solving problems with power flow in power systems.

The power flow equation is a set of linear equations when node power is injected. Newton method is one of the most effective methods for solving nonlinear groups. The polar equation of Newton's method is

$$\begin{cases} \Delta P_i = P_i - U_i \sum_{i \ni j}^{\infty} U_j (G_{ij}\cos\theta_{ij} + B_{ij}\sin\theta_{ij}) \\ \Delta Q_i = Q_i - U_i \sum_{i \ni j}^{\infty} U_j (G_{ij}\cos\theta_{ij} + B_{ij}\sin\theta_{ij}) \end{cases}$$

The power flow algorithm of the Newton–Raphson method can be modified by taking one term from the Taylor series and can be written as follows:

$$\begin{bmatrix} \Delta P \\ \Delta Q \end{bmatrix} = -J \begin{bmatrix} \Delta\theta \\ \Delta U \end{bmatrix}$$

$$\mathrm{J} = \begin{bmatrix} \frac{\partial \Delta P}{\partial \theta} & \frac{\partial \Delta P}{\partial \Delta Q} \\ \frac{\partial \Delta Q}{\partial \theta} & \frac{\partial \Delta Q}{\partial U} \end{bmatrix}$$

In the formula: for the equations of power flow, the residual vectors are represented by ΔP, ΔQ, where $\Delta\theta$, ΔU are used for the bus voltage correction, J is the Jacobian matrix.

Fast Decomposition Method

Rapid decomposition is a product of computer practice. In 1974, Stott found that in various PQ decoupling methods, the coefficient matrix of the active phase angle correction equation is replaced by the coefficient matrix of the reactive voltage correction equation, and the active and reactive power deviation is removed by the voltage amplitude. The algorithm has the best convergence. It is the node susceptance matrix established with -1/x for the branch susceptance, and it is the imaginary part of the node admittance matrix. Stott called this method a fast decomposition method and the fast decomposition method for the current iteration formula can be written as

$$\Delta U^k = -B^{n-1} \Delta Q(\theta^k, U^k)$$

$$U^{k+1} = U^k + \Delta U^k$$

$$\Delta\theta^k = -B^{r-1}\Delta P\left(\theta^k, U^{-k+1}\right)$$

$$\theta^{k+1} = \theta^k + \Delta\theta^k$$

B. *Method for Calculating Power Flow of Bus Distribution Network*

Such algorithms include the Zbus method and the Ybus method. These two types of algorithms are essentially the same.

The Zbus algorithm is as follows:

(1) Calculate the voltage of bus j when the root node acts independently on the entire distribution network and all equivalent injections are off:
Where: is the root node voltage, is the network's corresponding impedance and is the frequency of the point to be solved.
(2) Calculate the equivalent injection current of busbar j
(3) Calculate the bus voltage when only the equivalent injection current is applied.

$$U^{''} = ZI^{''}$$

Apply Superposition

$$U_{new} = U^{'} + U^{''}$$

Formula

$$U^{'} = \left[U_{s,1}, U^{'}_{s,2}, \ldots\ldots\ldots\ldots\ldots U^{'}_{s,n}\right]^T$$

(4) Verify Iteration convergence conditions:

$$|U_{new} - U_{dd}| \leq \varepsilon$$

Know that the iteration conditions are met, stop the calculation, and do not satisfy the continuing iteration, which represents the voltage obtained from the previous iteration.

Calculation method for branch-type distribution network power flow.

Branch flow-based load flow calculation method:

In a radial distribution network, for branches, there are:

$$U_j = U_i - I_j\left(R_j + jX_j\right)$$

If the branch's endpoint is a distal point, then the current, that is, flows through these distal points is known as the branch current, which is equal to the distal point's current, that is, equal to the distal point's load current:

$$I_j = I_{L,j}$$

The consistent current can be expressed as:

$$I_{L,j} = \frac{P_{L,j} - jQ_{L,j}}{U_j}$$

Formula: $P_{i,j} - jQ_{i,j}$

where $P_{L,j} - jQ_{L,j}$ is the conjugate of the node's complex load power; it is the conjugate of the node voltage.

If the branch end is not tipped, the branch current shall be the sum of the current at the end of the branch and the currents of all its sub-branches:

$$I_j = I_{L,j} + \sum_{k \in d} I_k$$

where d is a collection ν_j of branches with nodes as parent nodes. Current calculations for each branch are done by the repetition from the power supply edge point. Each nodal voltage can be calculated by shifting from the power supply point to the endpoint.

3 Proposed Methodology

From the modeling analysis in the previous chapter, It is understood that four types of nodes can be grouped into specific distributed power sources: PQ(V) node, PI node, PV node and PQ node. For distributed power supplies of the PQ type, it is sufficient to simply handle them with a load whose load value is negative. This section mainly analyzes the processing of the other three types of distributed power in the program.

(1) P constant, V constant PV node

Newton method is used to replace the PV nodes directly. If n (n = 1, 2, 3) phase lines are used to connect the PV bus to the system, each node injects one half of the total injected power on the bus. The voltage step angle and the node's reactive power can be found after each iteration. If the estimated receptive power of the node exceeds the critical value, then this node is converted to the equivalent PQ node. The Q value is equal to the maximum reactive power that the distributed power source can send out. If in the subsequent iterations, the node voltage crosses the boundary, then this PQ node is transformed back as PV node.

(2) P is constant and I the current amplitude is also constant. The PI node substitution cannot be directly done into the Newton method. Hence, before applying each iteration, some processing must be done. If through n (n = 1, 2, 3) phase line system and PI bus are connected, then each node on the bus injects one-sixth of the total power that is injected into the bus. With the help of the voltages that are calculated in the previous iteration, the corresponding receptive power can be calculated at the given real power and the current magnitude:

$$Q_{k+1} = \sqrt{|I|^2\left(e_k^2 + f_k^2\right) - P^2}$$

For the distributed power supply, the receptive power value is, $\left(e_k^2 + f_k^2\right)Q_{k+1}$ for the k + 1 secondary iteration; f_k, e_k are imaginary and real subsequent of the voltage, respectively, K secondary iterations $\left(e_{k+j} f_k + V_k\right)$ For a constant distributed power supply, amplitude of the current phaser is I; the constant value of the real power is P.

Therefore, injected amount of the receptive power of the PI node can be calculated prior to the k + 1st iteration, in the calculations of power flow, and the PI node should be modified into active and reactive power outputs in the k + 1st iteration process, respectively. PQ node. Completion of iteration, then the calculations for receptive power and phase angle of voltage of each node should be completed. The distributed power supply of type PI also has the limitation of reactive output, but it can be seen from the standard $\left(e_k^2 + f_k^2\right)$ value of generally around 1.0, P and I are two values that must be maintained, so the final calculated value is affected. Only the given active power and current amplitude of the PI node, that is, if P and I are given reasonably, the calculated reactive power will not exceed the limit.

(3) P-Q (V) bus bars with Q limited by P and V, V indefinite, and P constant. Before each and every iteration, some processing is required because the substitution of P-Q (V) nodes into the Newton method is cannot be done directly. The output active power given by the P-Q(V) node is the output active power of the asynchronous motor. After each and every iteration, the nodal voltage U will be corrected. Receptive power Q that should be injected of the node is calculated as follows

$$s = \frac{r(U^2 - \sqrt{U^4 - 4\sigma x_\sigma^2 P^2}}{2P_e x_\sigma^2}$$

$$Q' = \frac{r^2 + x_\sigma(x_m + x_\sigma)s^2}{r x_m s}$$

$$Q_c = P_e * \left(\sqrt{\frac{1}{(\cos\varphi_1)^2} - 1} - \sqrt{\frac{1}{(\cos\varphi_1)^2} - 1}\right)$$

$$\lceil n \rceil = \frac{Q_c}{Q_n - unit}.$$

$$Q^{''} = n * Q_{N-Uit} * \frac{U^2}{U_N^2}$$

$$\mathrm{Q} = \mathrm{Q}' - \mathrm{Q}''''.$$

For asynchronous motor slip; Stator reactance and Rotor reactance; for excitation reactance; for the rotor resistance; For induction motor absorption reactive; Is the power factor of the asynchronous motor; is the power factor of the rear node of the shunt capacitor; general requirements in 0.9 above; Reactive power needed to compensate for shunt capacitor; The number of shunt capacitor groups for input; Reactive power compensation for each group of capacitors; The reactive power of the capacitor group is compensated effectively; inject reactive power to the node participating in the current iteration.

Q (V) bus and system pass N ($literal, 2, 3) are connected by the phase line, total injected power for the busbar is the power that is injected at each node. N, one of the points. In the flow calculation, the first k iteration can be the Q (V) node of the reactive absorption, in the k + 1, the q (V) node can be processed into active and reactive output as P and in the iteration process.

4 Results and Analysis

The transmission losses and the voltage security must be taken in consideration by the independent system operator during provide the service in the competitive environment of the management of the receptive power. A nondiscriminatory and the transparent procedure must be adopted by the system operator to provide the reactive power supply for optimal system deployment. Because the main sources of generation of the receptive power are generators, the marginal prices of the receptive and active power are affected by the cost of the receptive power generation. On the basis of marginal cost theory, in this paper, a methodology is proposed for the calculation of the locational marginal prices for the receptive power and active power. In this methodology for generators' receptive power support, different types of receptive power cost models are considered. As the FACTS controllers present in the system for the required flexible operation but due to these FACTS devices, it is not possible to ignore their impact on nodal prices for wheeling cost determination and consideration must also be given to their cost function. For the electric hybrid market model, these results were obtained, and also for the comparison pool model, results were calculated. With the use of GAMS and MATLAB interface, for the solving of the complex problems, an approach is formulated known as Mixed Integer Non-linear programming (MINLP). IEEE 33-bus Reliability Test System (RTS) is used to test this proposed approach (Tables 1, 2, 3).

Table 1 System configuration

Serial number	Configuration items	Quantity	
1	Buses	34	
2	Generator	2	
3	Committed Gens	2	
4	Loads	20	
5	Fixed	0	
6	Dispatchable	0	
7	Shunts	0	
8	Branches	33	
9	Transformer	0	
Serial number	Configuration items	Active power P(MW)	Reactive power Q (MVAr)
1	Total gen capacity	2.0	0.0–1.5
2	On-line capacity	2.0	0.0–1.5
3	Generation (actual)	9.8	4.1
4	Loads	0.4	0
5	Fixed	0.4	0
6	Dispatchable	−0.0 of −0.0	−0.0
7	Shunts(inj)	−0.0	0.0
8	Losses (I^2 * Z)	9.34	4.12
9	Branch charging (in	-	0.0
Parameter	Magnitude		
Voltage Magnitude	1.030 p.u		
Voltage Angle	−23.89 deg		
P Losses (I^2*R)	3.01 MW @ line 3–5		
Q Losses (I^2*X)	1.33 MVAr @ line 3–5		

Real power injected from the buses is shown in Fig. 1. From 1 to 5 buses, the power is received as negative power on the hand from 6 to onwards bus, the power is injected as positive power is shown in the figure.

The voltage phase angle graph is shown in Fig. 2. Bus 4 has maximum voltage phase angle as compared with other buses.

Reactive power injected from the buses is shown in Fig. 3. From 1 to 4 buses, the power is received as negative power on the hand from 6 to onwards bus, the power is injected as positive power and some are received as negative power is shown in the figure.

Figure 4 shows the injected receptive power to the bus. From 1 to 4 buses, the power is injected as positive power on the hand from 6 to onwards bus, the power

Table 2 Bus data

Bus	Voltage		Generation		Load	
	Mag (p.u)	Ang (deg)	P (MW)	Q (MVAr)	P (MW)	Q (MVAr)
1	1.042	0.099	–	–	0.02	0.01
2	1.050	0.165	–	–	–	–
3	1.204	1.218	–	–	0.01	0.00
4	1.233	1.411	7	2	–	–
5	1.194	1.018	–	–	–	–
6	1.186	0.857	–	–	-	–
7	1.186	0.851	–	–	–	–
8	1.186	0.850	–	–	0.00	0.00
9	1.186	0.851	–	–	0.01	0.01
10	1.183	0.788	–	–	0.01	0.01
11	1.185	0.879	–	–	0.01	0.02
12	1.183	0.783	13.00	14.00	0.00	0.00
13	1.183	0.788	–	–		
14	1.185	0.879	–	–	–	–
15	1.178	0.657	–	–	–	–
16	1.178	0.653	–	–	0.00	0.00
17	1.170	0.424	–	–	–	–
18	1.178	0.653	–	–	–	–
19	1.170	0.418	–	–	0.00	0.00
20	1.170	0.413	–	–	–	–
21	1.169	0.386	–	–	0.01	0.01
22	1.169	0.417	–	–	0.03	0.02
23	1.168	0.347	–	–	0.05	0.00
24	1.169	0.386	–	–	–	–
25	1.168	0.417	–	–	0.05	0.03
26	1.168	0.345	–	–	0.00	0.00
27	1.167	0.348	–	–	0.00	0.00
28	1.168	0.337	–	–	0.15	0.01
29	1.167	0.348	–	–	0.01	0.00
30	1.167	0.348	–	–	0.01	0.01
31	1.168	0.319	–	–	0.01	0.00
32	1.167	0.348	–	–	–	–
33	1.168	0.317	–	–	0.02	−0.13
34	1.030	0.000	−5.52	−1.52	–	–

Table 3 Losses after apply DG

Bus	Loss (I^2*Z) (Without DG)		Loss (I^2*Z) (With DG)	
	P(MW)	Q(MVAr)	P(MW)	Q(MVAr)
1	0.06272	0.02765	0.00532	0.02765
2	0.04236	0.01868	0.03256	0.01868
3	0.79999	0.3566	0.76999	0.34792
4	0.42511	0.4556	0.16651	0.06045
5	0.00328	0.00145	0.00228	0.00120
6	0.00260	0.00115	0.00130	0.00078
7	0.00009	0.00004	0.00002	0.00003
8	0.00003	0.00001	0.00001	0
9	0	0	0	0
10	0.00080	0.00035	0.00062	0.00028
11	0.00002	0.00001	0	0
12	0.00006	0.00003	0.00002	0.00001
13	0	0	0	0
14	0	0	0	0
15	0.00146	0.00064	0.00105	0.00045
16	0.00004	0.00002	0.00001	0
17	0.00261	0.00115	0.00261	0.00115
18	0	0	0	0
19	0.00029	0	0.00017	0
20	0.00001	0	0	0
21	0.00033	0	0.00014	0
22	0	0.00002	0	0
23	0.00001	0	0	0
24	0.00001	0	0	0
25	0	0.00002	0	0
26	0.00003	0	0	0
27	0	0	0	0
28	0.00003	0.00002	0	0
29	0	0	0	0
30	0	0	0	0
31	0.00004	0.00002	0	0
32	0	0	0	0
33	0.00001	0	0	0
34	0	0	0	0
Total Losses	**1.34193**	**0.86346**	**0.98261**	**0.4586**

Reduced loss	Active	Reactive
	0.15841	**0.01991**

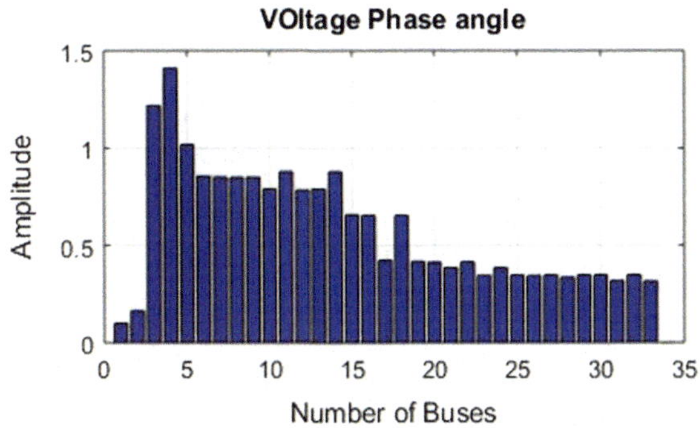

Fig. 2 Voltage phase angle

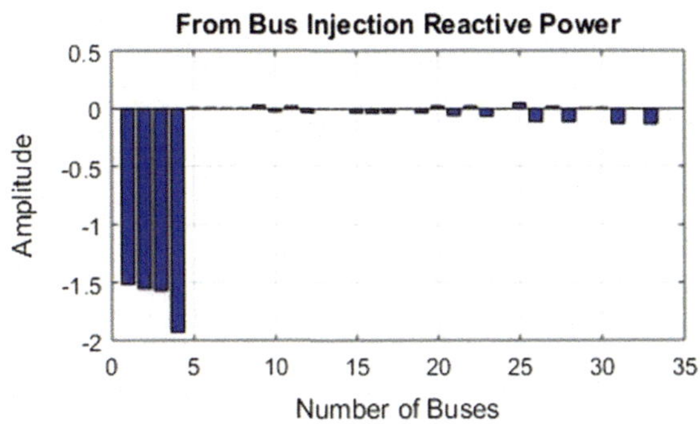

Fig. 3 From bus injection reactive power

is injected as positive power and some are received as negative power is shown in Fig. 4. It is just complement output of Fig. 3.

Real power injected into the buses is shown in Fig. 5. From 1 to 4 buses, the power is injected as positive power on the hand from 6 to onwards bus, the power is received as negative power is shown in Fig. 5.

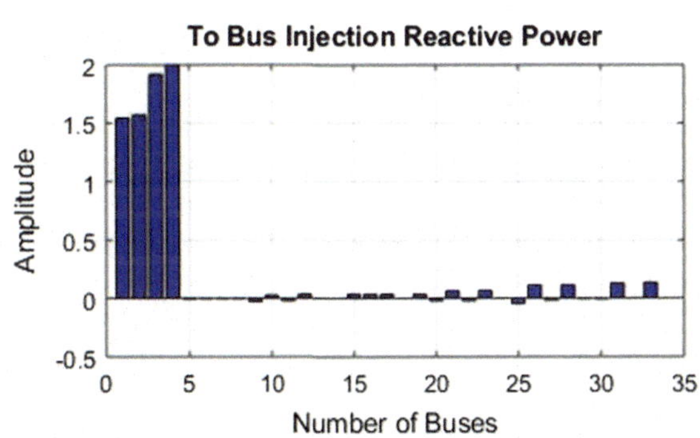

Fig. 4 To bus injection reactive power

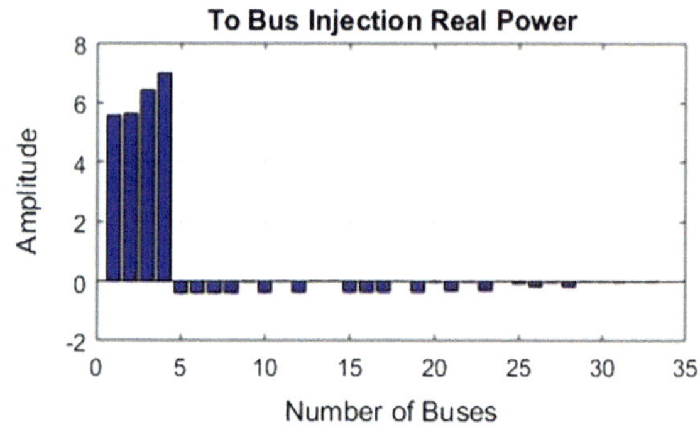

Fig. 5 To bus injection reactive power

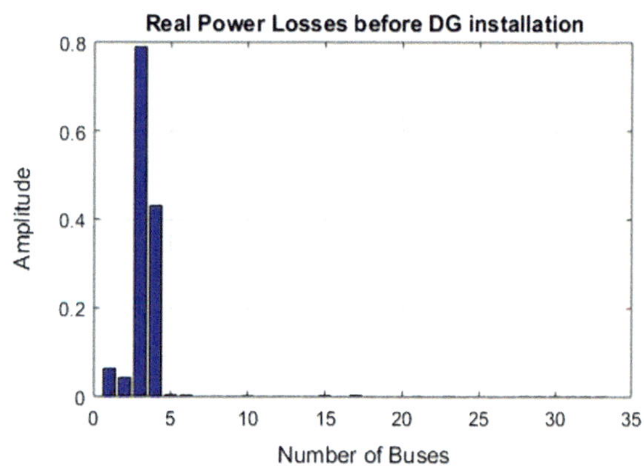

Fig. 6 Real power losses before DG installation

Real power losses before DG installation are shown in Fig. 6. The power losses at bus 4 are more as compared with other buses. Total active and reactive power loss at bus 4 is 0.42511 and 0.45, respectively.

Reactive power losses before DG installation are shown in Fig. 7. The power losses at bus 4 are more as compared with other buses. Total active and reactive power loss at bus 4 is 0.42511 and 0.4556, respectively.

Real power losses after DG installation are shown in Fig. 8. The power losses at bus 4 are more as compared with other buses. After apply DG, total active and reactive power loss at bus 4 is 0.16651 and 0.06045, respectively. Total loss decrease is 22.22%.

Reactive power losses after DG installation are shown in Fig. 9. The power losses at bus 4 are more as compared with other buses. After apply DG, total active and reactive power loss at bus 4 is 0.16651 and 0.06045, respectively. Total loss decrease is 22.22%.

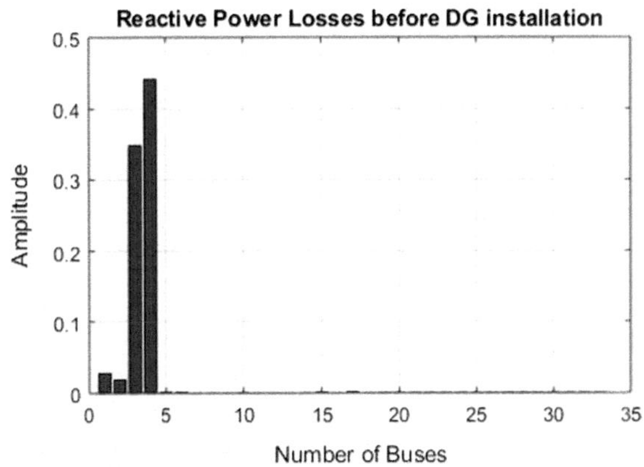

Fig. 7 Reactive power losses before DG installation

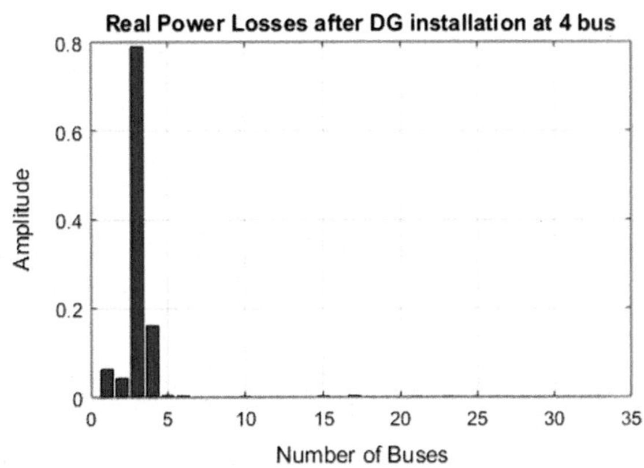

Fig. 8 Real power losses after DG installation

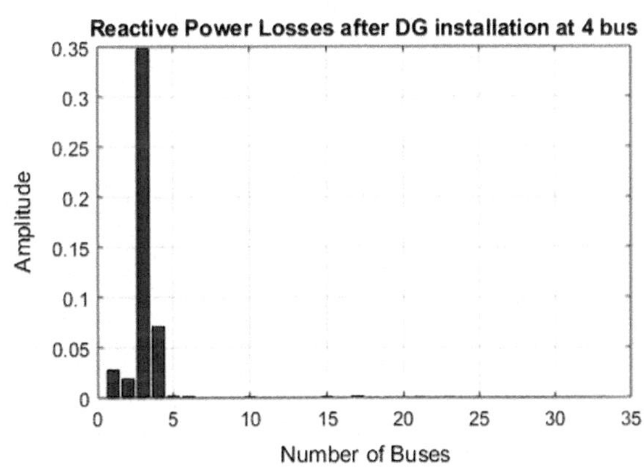

Fig. 9 Reactive power losses after DG installation

5 Conclusion

Newton–Raphson method is the best approach as compared with other techniques. The total benefits are achieved up to 1.9617e + 04. The reactive power loss bus 3 is more as compared with other busses. The relative percentage difference of the summer session is more as compared with the winter vacation. The offered retail price of summer is more than winter session. The reactive power of bus injection is up to five buses. The 34 Bus Systems is used. The following conclusions are made:

1. Total active and reactive power loss at bus 4 is 0.42511 and 0.4556, respectively.
2. After applying DG, total active and reactive power loss at bus 4 is 0.16651 and 0.06045, respectively. Total loss decrease is 22.22%.
3. Total active and reactive power loss at 1–34 buses is 0.15841 and 0.1991.
4. Total load benefit achieved is 1.9617.

References

1. SANDIA Report (2012) Project report: a survey of operating reserve markets in U.S. ISO/RTO-managed electric energy regions. http://prod.sandia.gov/techlib/accesscontrol
2. Chen Y, Gribik P, Gardner J (Mar. 2014) Incorporating post zonal reserve deployment transmission constraints into energy and ancillary services co-optimization. IEEE Trans Power Syst 29(2):537–549
3. MISO (2013) MISO resources not qualified for reserves in real-time market
4. Fatema N et al (2021) Intelligent data-analytics for condition monitoring: smart grid applications. Elsevier, 268 p ISBN: 978-0-323-85511-2. https://www.sciencedirect.com/book/9780323855105/intelligent-data-analytics-for-condition-monitoring
5. Smriti S et al (2018) Special issue on intelligent tools and techniques for signals, machines and automation. J Intell Fuzzy Syst 35(5):4895–4899. https://doi.org/10.3233/JIFS-169773
6. Jafar A et al (2021) AI and machine learning paradigms for health monitoring system: intelligent data analytics. Springer Nature, Berlin, 496 p https://doi.org/10.1007/978-981-33-4412-9. ISBN 978-981-33-4412-9
7. Sood YR et al (2019) Applications of artificial intelligence techniques in engineering, vol 1. Springer Nature, 643 p https://doi.org/10.1007/978-981-13-1819-1. ISBN 978-981-13-1819-1

Cause Analysis of Students' Dropout Rate Using PSPP

Sakshi, Chetan Sharma, Vinay Kukreja, and Divpreet Kaur

Abstract Student dropout is a crucial topic of universal essence, and evaluating the student dropout rate becomes a vital concern for any country as the nation spends a substantial amount of resources on the education system. It includes the money contribution of tax-payers and also a proportion of graduates and higher degree holders. Moreover, it becomes furthermore significant to realize and identify the leading causes of this dropout rate. This research article focuses on evaluating the causations for the student dropout and also endeavors to measure the relationships among these factors. It purely gravitates towards the development of a thought process that clarifies the rationalizations and underpinnings for the student dropout from universities and colleges. The research approach initiates at setting a hypothesis by consolidating the several considered factors that affect the dropout rate. After the stratification of influencing factors (sociodemographic, family background, emotional motivation, etc.), we performed statistical analysis using PSPP statistical tool. The data for the experimentation were mustered up from 328 university students, majorly belonging to the streams of science, engineering, and commerce. Furthermore, the collected data were processed as per the tool format, and frequency and descriptive analysis were accomplished to calculate the quality of the collected data. A chi-square test was employed to find the relationship between the considered factors and the dropout rate. The assumed hypothesis was verified, and the results reveal the list of factors that directly impacts the student dropout rate.

Keywords Student dropout · PSPP · Chi-square · Education · Statistical method

Sakshi (✉) · V. Kukreja · D. Kaur
Chitkara University Institute of Engineering and Technology, Chitkara University, Punjab, India
e-mail: sakshi@chitkara.edu.in

C. Sharma
Chitkara University, Himachal Pradesh, India

A. Tomar et al. (eds.), *Machine Learning, Advances in Computing, Renewable Energy and Communication*, Lecture Notes in Electrical Engineering 768,
https://doi.org/10.1007/978-981-16-2354-7_41

1 Introduction

The best investment is the investment done for education, i.e., seeking knowledge and skills for living. This term "education" not only defines your aptitude and attitude but certainly gives an altitude to your career. The academic centers for education are schools, colleges, and universities. A trend has been observed throughout that the number of students gets admitted to the number of students successfully passing out is significantly different. Over the years, there has been a significant concern of early student dropout issues, which has become one of the most hellacious problems in academic institutions spreading at a high rate over the entire world. Most of the freshman students are unable to make themselves prepare for the successful shift to college from high school and also unprepared to face a variety of challenges, which can be very stressful [1]. The definition of dropout differs among every researcher. Still, it usually is considered that if an institution loses a student by whatever means, the retention rate decreases of that particular institution [2]. So, it is essential to implement an effective program to decrease the dropout rate or to improve the retention rate. To retain the vulnerable students who are prone to drop their courses due to any reason, it is essential to understand the behavior of successful students.

The adverse outcomes of students dropping out of high school have a significant impact on society as well as on individuals. It also has a negative impact on university business as the admission decreases. Every institute is required to find out some solution to this issue, taking the feedback and suggestion from the students and must consider their feedback to build the correlation between student satisfaction and retention [3]. The ability to predict the dropouts and improve retention rate is challenging because it involves several intercorrelated and distinct elements. It is a challenging task to measure the academic performance of students as it hinges on diverse factors. The main factors that affect the low academic performance of students involve the university environment, the interaction between teacher and student, mental disability, health issues, etc. Students' characteristics like demographics, social, family background, and many more somehow influence institutes in their growth or failure [4]. As it is the most challenging task, every researcher undergoes alternative methods and techniques to improve the retention rate. In this paper, we are going to discuss the machine learning approach to find the dropout rate and to improve the retention rate.

2 Literature Review and Related Work

The author conducted their study on public universities located in the USA, wherein averages of around 23,000 students are enrolled. To maintain the student in universities, the author proposed an analytical model to predict and factors responsible for student retention. They exercise the 5-year university data with a total of 16,066 records using different data mining techniques. The methodology used in the study is

CRISP-DM, which is most prevalent in the data mining field. They applied the decision tree, SVM, ANN, and logistic regression with tenfold cross-validation. They found that SVM achieved the highest accuracy rate of 87.23%, and overall accuracy through SVM is 81.18%, which is higher in all data mining techniques used for the experiment. This model will help to find the last ratio of students who drop out the university before they graduate [5].

The author conducted their study on American universities by applying data mining techniques to solve the problem of student retention. Data are collected from the fresher's students in the aspect of different factors like academic, financial, and demographic information. They conduct their study through C4.5, One R, Byes classifier to predict the student dropout in the total number of classes the students enrolled. According to author, 10,000, students are enrolled in English classes, and students do not opt for residential services, which make the high risk of dropout ratio. The study concluded that 40% of retention is there in the third year out of total students enrolled in the course [6].

The author conducted a study to develop the model for student retention management. To achieve the present study, data are collected from the MCA department of the V.B.S Purvanchal University of Jaunpur of over 12 years, where the dataset is a collection of 432 records. From the dataset, it is noticed that out of 432 total students enrolled in 12 years, 34 students left the university during the first year. The author conducted their study using the WEKA software and implemented 16 classification algorithms on the considered dataset. They used tenfold cross-validation for their experiment on ID3, C4.5, and ADT and achieve an 82.8% accuracy rate from the ADT algorithm [7].

The author demonstrates the factors behind female students' low on-campus and off-campus academic performance and, consequently, high attrition. The author exercises both qualitative and quantitative research methods to gather the data. They circulated the questionnaire to the 600 students of the Bahir Dar University and analyzed the data using the SPSS 13.0 software. They conduct t-test, mean, standard deviations, and linear regression analysis on the selected dataset. Experiment results that 35% of the female student in the university goes through the sexual harassment per year, and teacher support and friends support have an effective relationship with their grades. The study also concludes that female students' personal, atmosphere in university, economic, and family problem factors are responsible for their dropout from the university [8].

The author study is to predict the dropout of students in an online course using data mining techniques. The study is conducted on 189 students who enrolled for an online IT certification course in the year 2007–2009. Data are collected through a questionnaire that included 10 variables and were circulated to the students through online mode. They used Matlab to conduct their study and applied decision tree, k-NN, NN, and Naive Bayes algorithms using tenfold cross-validation to train and test their dataset. The author achieved an 87% accuracy rate from the k-NN algorithm, which is the best performer among all the classifiers. They proposed to conduct their study on other courses in the future. This study helps to predict the students who will drop out of the course and reduce the dropout ratio [9].

The author conducted their study on students of flagship universities of the USA to investigate the characteristics related to their retention throughout their course. They surveyed 22,099 students who enrolled themselves for full time and part time in the year 1995–2005. They found that 33% of total enrolled student does not graduate and out of 33% of students, 39% of students have their GPA < 2.25 in the first semester. They used data mining techniques using SAS 9.2 software and conclude that students having a higher GPA in higher school have chances that they complete their graduation. The author focused on students' GPA in higher school and first semester to predict whether they complete their course [1].

They proposed the method to predict the student dropout after analyzing the data available after the completion of the course. They conducted a study on the 419 high school students of Mexico. They used classification techniques like SVM, NN, etc. to predict the student dropout in the early stage. They found that 419 students were at the age of 15 and enrolled in high school. After experimenting, they conclude that 57 students will drop out of 419 students [2].

The author depicts an exclusive study that is gravitated more on MOOC courses, which is an entirely new education approach, featuring in 2012. The study presents a depth observation scenario of MOOC data where statistical analysis is applied to students' behavioral information, and the result of the observation has been illustrated in the viewpoint of curriculum design on MOOC platforms. The authors implemented the experimentation using unsupervised learning methods [10].

Dropout of the students from the institutes was increasing rapidly and became the primary threat to the institutes. The dropout of students from the institutes is due to many factors. The author is given the methodology to predict the student dropout from the institute using the Naive Bayes classification algorithm in the R language. They develop the model to predict the reason related to student dropout in the early stage. According to the author demographical, psychological, behavior, and many more are the factors that are responsible for the student dropout in the first stage. The author achieved a 72% accuracy rate through the WEKA tool. This model helps institutes to retain the student in their academic program [11].

The author corroborates the improvement of the performance of a dropout early warning system. The author considered the dataset of NEIS, South Korea of the year 2014, which includes the data samples of 165,715 high school students. The experiment is conducted using the SMOTE function in R and employed 80% of data for training purposes and 20% of data for testing purposes. Tenfold cross-validation is used in the experiment, and random forest-boosted decision tree algorithms are used to train the model and predict the function that is used for testing. ROC and PR curves are generated from the results and found that ROC curves are not informative as compared to PR curves [12].

The author proposed the prediction model, which helps university in Peru in reducing their dropout ratio of students. The author considered the dataset of 500 undergraduate students who are enrolled in the private university in Lima. Bayesian network and decision tree, machine learning algorithms are used for the experiment purpose on the considered dataset. The author used an 80% dataset for training and a 20% dataset for testing purposes. The author achieved a 67.10% accuracy rate from the Bayesian algorithm and found it perform well than the decision tree on precision, accuracy, specificity, and error rate [13]. Moreover, numerous recent works have been published by using different AI and ML approaches [14–19].

3 Research Methodology

3.1 Data Collection

The data collection for experimentation is mustered through the questionnaire, which was created and effectuated with the help of Google forms. This form was circulated and, thus, filled by the respondents from the students of different streams. Data were collected from the students of varying backgrounds, i.e., engineering, commerce, computer sciences, application, etc. There are a total of 328 respondents who respond to the Google form. Likert Scale is used to collect the data from the different respondents. The purposive sampling method is applied to collect the data from our samples that are university/college students. Primary and secondary are the two types of data sources used to carry out the survey study; preliminary information included the data collection through the designed questionnaire and personal interviews of different respondents. The authors used the questionnaire only to carry out this research. Data for secondary sources are collected from sources like the internet, books, article, journals, etc.

3.2 Data Preparation

Data are collected through the Google form responses, and then the data are downloaded in the.csv file format. PSPP statistical tool is used to conduct the current study, so information is prepared in the.sav form, which is accepted by the PSPP tool used for the recent research. The total sampling size for the current study is 328, and the Non-Probability convenient sampling technique is used.

3.3 Data Analysis Techniques

A descriptive type of research is used for this study. We conduct descriptive analysis using the PSPP tool. Collected data are analyzed using the PSPP statistical tool. Frequencies and descriptive analyses are done on the collected data to calculate the quality of the data based on the mean, standard deviation, minimum, and maximum.

3.4 Scale Items

Independent and dependent variables are considered in this study to experiment. The questionnaire is formulated based on different scale items, which help the study to analyze the student dropout and factors responsible for the same, which are described in Table 1. Sociodemographic (age, gender, category, etc.) parents'

Table 1 Scale items

Family clime
Parents' highest education
Emotionally and financially support from the parents
Commitment towards the studies and responsibilities related to the family
I think expenses of education other than tuition fees are much beyond my capability to pay?
Individual attributes
Age, gender, category, exam passed
Initial commitments
Misfit in the current institute due to different factors of the dominant peer group (age group)
Graduation is essential to achieve career goals
External commitments
I prefer to spend time outside this university with my family and friends rather than being at the university
Relationships with other peer members have a positive influence on my personal and intellectual growth
Interaction with faculty
Interaction with a staff member after class hours had a positive influence on my personal/intellectual growth/career goals and aspirations
Educator relationship with students
Faculty members I have contacted are willing to discuss my issues other than my academics
The course curriculum and class hours are very stressful
Academic development
I am performing well in my academics
Institutes help me in attending cultural events, which I never heard before coming to this institute
Institutional choice and career goals
My decision is right in choosing this institute to achieve my educational and career goals
I go through different opportunities other than attending institutes and then will better than the institute I am currently attending
I will re-enroll myself at this institute in next semester
Problems with the residential/on-campus facilities make me rethink if I want to re-enroll at this institute in next semester
I feel comfortable around campus, in the department, and lectures

motivation (emotionally and financial support) and factors related to the institutes (facilities, faculty, mentoring, etc.) have been considered.

3.5 *Frequency Analysis*

Frequency analysis on the different independent variables is done through the considered tool, and a detail of frequency of each variable is as follows. First, we consider mother's highest qualification. Out of 328 respondents, 17 were primary school educated or less, 37 are high school or less, 59 are secondary school educated, 134 are graduates and 81 are postgraduate or more. This makes the percentage as 5, 18, 11.25, 17.99, 40.85, and 24.7, respectively. We find that most of the mothers are falling under the category of graduates. In the next step, we consider father's highest qualification. Most of the fathers fall under the same category of graduate as is the case with mothers. 146 (44.51%) of the respondents out of 328 are graduates, primary school or less are 9 (2.74%), high school or less are 43 (13.11%), secondary school are 46 (14.02%), and postgraduate are 84 (25.61%). Furthermore, we do a age analysis in the labels of 18–24, 25–30, 31–40 with frequencies of 326, 1, 1 out of the 328 responses, respectively. The respective percentages in age analysis are 99.39, 0.3, and 0.3%. The gender analysis of the respondents reveals that 60.67% are males and 39.33% are females out of the total respondents. As per the category (caste) analysis, it is found that 293 (89.33%) belong to general category, 26 (7.93%) belong to OBC category and 9 (2.74%) are SC/ST. 95.73% respondents have completed their 10 + 2 before joining their further studies, 3.35% had done diploma and 0.91% completed their metric. The analysis of medium of education for these 328 respondents reveals that 323 (98.48%) studied from English medium while 5 (1.52%) were from Hindi medium.

4 Hypotheses and Results

The author considers the following hypothesis regarding this study.

H0: No strong relationship between emotional motivation and student dropout.

H1: Strong relationship between emotional motivation and student dropout.

H2: No strong relationship between financial support and student dropout.

H3: Strong relationship between financial support and student dropout.

H4: No strong relationship between choosing university/college for career goal and student dropout.

H5: Strong relationship between choosing university/college for career goal and student dropout.

H6: No strong relationship between the facilities provided and student dropout.

H7: Strong relationship between facilities provided and student dropout.

Table 2 Test results of Hypotheses H0 and H1

O	E	(O–E)	(O–E)2	(O–E)2/E
3	82	−79	6241	76.11
7	82	−75	5625	68.60
60	82	−22	484	5.90
258	82	176	30,976	377.76
Total				528.37

Chi-square test is applied to the dataset collected to accept or reject the formulated hypothesis using the PSPP software. Results of Chi-square test on the considered dataset are given in Tables 2, 3, 4, and 5.

Chi-square test is used to find the relationship between the variable studied and there are some expected values so in the chi-square test values in below tables, O is representing the observed values and E is representing the expected values from the

Table 3 Test results of Hypotheses H2 and H3

O	E	(O–E)	(O–E)2	(O–E)2/E
3	82	−79	6241	76.11
1	82	−81	6561	80.01
100	82	18	324	3.95
224	82	142	20,164	245.90
Total				405.98

Table 4 Test results of Hypotheses H4 and H5

O	E	(O–E)	(O–E)2	(O–E)2/E
3	82	−79	6241	76.11
11	82	−71	5041	61.48
109	82	27	729	8.89
205	82	123	15,129	184.50
Total				330.98

Table 5 Test results of Hypotheses H6 and H7

Z	E	(O-E)	(O-E)2	(O-E)2/E
13	82	−69	4761	58.06
45	82	−37	1369	16.70
189	82	107	11,449	139.62
81	82	−1	1	0.01
Total				214.39

experiment. (O–E) is residual value, then square the "O–E" values and divide each by the relevant "expected" value to give (O–E)2/E. Add all the (O–E)2/E values and call the total "X2", which is further compared with the chi-square table to find the relationship between the formed hypothesis.

H0: No strong relationship between emotional motivation and student dropout.

H1: Strong relationship between emotional motivation and student dropout.

To conduct the experiment, we have considered the level of significance $\alpha = 0.05$, and degree of freedom is 3. Probability level for this is 7.81.

Chi-square value 528.37 > 7.81 probability level (Table 2: **Test results of Hypotheses H0 and H1**; Table 2).

Results depict that hypothesis H0 is rejected and H1 is accepted, so it concludes the occurrence of strong relationship between emotional motivation and student dropout.

H2: No strong relationship between financial support and student dropout.

H3: Strong relationship between financial support and student dropout.

To conduct the experiment, we have considered the level of significance $\alpha = 0.05$, and degree of freedom is 3. Probability level for this is 7.81.

Chi-square value 405.98 > 7.81 probability level (Table 3).

Results depict that hypothesis H2 is rejected and H3 is accepted, so it concludes the occurrence of strong relationship between financial support and student dropout.

H4: No strong relationship between choosing university/college for career goal and student dropout.

H5: Strong relationship between choosing university/college for career goal and student dropout.

To conduct the experiment, we have considered the level of significance $\alpha = 0.05$, and degree of freedom is 3. Probability level for this is 7.81.

Chi-square value 330.98 > 7.81 probability level (Table 4).

Results depict that hypothesis H4 is rejected and H5 is accepted, so it concludes the occurrence of strong relationship between choosing university/college for career goals and student dropout.

H6: No strong relationship between choosing facilities provided and student dropout.

H7: Strong relationship between facilities provided and student dropout.

To conduct the experiment, we have considered the level of significance $\alpha = 0.05$, and degree of freedom is 3. Probability level for this is 7.81.

Chi-square value 214.39 > 7.81 probability level (Table 5).

Results depict that hypothesis H6 is rejected and H7 is accepted, so it concludes the occurrence of strong relationship between the facilities provided and student dropout.

5 Conclusion and Future Scope

The chi-square test has been performed where the level of significance and the degree of freedom have values 0.05 and three, respectively, and, thus, the corresponding tabular value of chi-square is 7.81. The result analysis of the aforementioned experimentation process reveals the following conclusion.

The evaluated chi-square value of factors emotional motivation (258.37), financial support (405.98), the certainty of decision (choosing the university and career goals) (330.98), university reasons (214.39) are higher than the tabular value of chi-square, directs us to the conclusion that there exists strong relation between the dropout rate and the considered causation factors. The future investigations proposed would be implementing the advanced tool for training the model by gathering a larger dataset and employing machine learning and deep learning methods for predicting the student dropout to control the currently deteriorating scenario that helps to diminish the student dropout rate.

References

1. Raju D (2015) Exploring student characteristics of retention that lead to graduation in higher education using data mining models. J Coll Stud Retent Res Theory Pract 16(4):563–591
2. Márquez-Vera C, Cano A, Romero C, Noaman AYM, Mousa Fardoun H, Ventura S (2016) Early dropout prediction using data mining: a case study with high school students. Expert Syst 33(1):107–124
3. Ullah MA, Alam MM, Mahiuddin M, Rahman MM (2019) Predicting factors of students dissatisfaction for retention. In: Emerging technologies in data mining and information security. Springer, Singapore, pp 501–510
4. Tinto V (1975) Dropout from higher education: a theoretical synthesis of recent research. Rev Educ Res 45(1):89–125
5. Delen D (2010) A comparative analysis of machine learning techniques for student retention management. Decis Support Syst 49(4):498–506
6. Nandeshwar A (2011) Learning patterns of university student retention. Expert Syst Appl 38(12):14984–14996
7. Yadav SK, Bharadwaj B, Pal S (2012) Mining education data to predict student's retention: a comparative study. arXiv Prepr. arXiv1203.2987
8. Mersha Y (2013) Factors affecting female students' academic achievement at Bahir Dar University. J Int Coop Educ 15(3):135–148
9. Yukselturk E, Ozekes S, Türel YK (2014) Predicting dropout student: an application of data mining methods in an online education program. Eur J Open Distance e-learn 17(1):118–133
10. Chen Y, Zhang M (2017) Mooc student dropout: pattern and prevention. In: Proceedings of the ACM turning 50th celebration conference-China, pp 1–6
11. Hegde V, Prageeth PP (2018) Higher education student dropout prediction and analysis through educational data mining. In: 2nd international conference on inventive systems and control (ICISC), pp 694–699
12. Lee S (2019) The machine learning-based dropout early warning system for improving the performance of dropout prediction. Appl Sci 9(15):3093
13. Medina EC, Chunga CB, Armas-Aguirre J, Grandón EE (2020) Predictive model to reduce the dropout rate of university students in Perú: Bayesian networks versus decision trees. In: 15th Iberian conference on information systems and technologies (CISTI), pp 1–7

14. Aggarwal S, et al 2020 Meta heuristic and evolutionary computation: algorithms and applications 949. Springer Nature, Berlin. https://doi.org/10.1007/978-981-15-7571-6. ISBN 978-981-15-7571-6)
15. Yadav AK et al (2020) Soft computing in condition monitoring and diagnostics of electrical and mechanical systems 496. Springer Nature, Berlin. https://doi.org/10.1007/978-981-15-1532-3. ISBN 978-981-15-1532-3
16. Gopal et al (2021) Digital transformation through advances in artificial intelligence and machine learning. J Intell Fuzzy Syst, Pre-press, 1–8. https://doi.org/10.3233/JIFS-189787
17. Smriti S et al (2018) Special issue on intelligent tools and techniques for signals, machines and automation. J Intell Fuzzy Syst 35(5):48954899. https://doi.org/10.3233/JIFS-169773
18. Jafar A et al (2021) AI and machine learning paradigms for health monitoring system: intelligent data analytics 496. Springer Nature, Berlin. https://doi.org/10.1007/978-981-33-4412-9. ISBN 978-981-33-4412-9
19. Sood YR et al (2019)Applications of artificial intelligence techniques in engineering 1:643. Springer Nature. https://doi.org/10.1007/978-981-13-1819-1. ISBN 978-981-13-1819-1

Correlation-Based Short-Term Electric Demand Forecasting Using ANFIS Model

Seema Pal, Nitin Singh, and Niraj Kumar Chaudhary

Abstract Prediction of electrical demand in advance plays a key role in power system planning. More specifically for the utilities, market operators, and system aggregators, as it helps them to make critical decisions related to uncertainties in the generation side as well as the demand side. The accuracy of the demand forecasting model is influenced by multiple issues such as uncertainty in demand, weather factors, intermittency of renewable energy sources, etc. A small improvement in system-level and meter level forecasting can contribute to the proper utilization of renewable energy sources and can reduce the system cost. To achieve the better accuracy of demand forecasting, this paper shows an adaptive neuro-fuzzy inference system (ANFIS) for short-term demand forecasting. The ANFIS model can provide good results by considering correlated weather factors, hour of the day, and day type for demand forecasting. The mean absolute percentage error (MAPE), mean absolute error (MAE) and root mean square error (RMSE) of the ANFIS model are obtained for different day types. The results are compared with the ANN model and achieved better accuracy. Also, the impact of anomalous days on forecasting accuracy is analyzed.

Keywords Demand forecasting · Smart grid · Demand response · ANFIS · ANN · MAPE · Membership functions · Anomalous days · RMSE · MAE

1 Introduction

Electric demand forecasting aims at predicting the future demand of an electricity network. From the environmental concerns and problem of energy crisis point of view, nowadays more focus is on power generation from renewable energy sources (RES), but at the same time, due to the intermittent nature of renewable sources, many challenges are there. One direct challenge is in dealing with the uncertainties on the supply side as well as the demand side.

S. Pal (✉) · N. Singh · N. K. Chaudhary
Department of Electrical Engineering, MNNIT, Allahabad, Prayagraj, UP, India
e-mail: seema@mnnit.ac.in

A. Tomar et al. (eds.), *Machine Learning, Advances in Computing, Renewable Energy and Communication*, Lecture Notes in Electrical Engineering 768,
https://doi.org/10.1007/978-981-16-2354-7_42

Electric demand forecasting plays an important role in electric power management and coordinated energy dispatching issues [1]. Demand forecast comes under initial steps for power system planning and for providing continuous power supply to consumers in an economical way as well as in decision-making for generating companies. Therefore, demand forecasting plays a major role in the generating side as well as for the consumers to optimize the use of their electricity management systems. Also, an accurate electric demand forecast is required, so that the correct amount of electricity is purchased from the electricity market. Due to the distributed generation with renewable energy sources, there is a strong requirement for forecasting the demand for scheduling and dispatching the resources.

There are four different time horizons for load forecasting, long-term, medium-term, short-term, and very short-term load forecasting. Long-term forecasting ranges from 1 year to several years in advance give the basis for planning policies. Medium-term forecasting range from few weeks to few months in advance helps in scheduling for maintenance and fuel supplies. Prediction for the successive hours, days, or weeks comes under the category of short-term demand forecasting. It is a cornerstone in decision-making for residential demand response and also for utility. Forecasting of electrical demand from few minutes to several hours ahead comes under the category of very short-term demand forecasting, which is required for intelligent demand control and scheduling in real time [2]. The forecast's scale is defined as the size of the unit at which the load forecast is carried out. It can be divided into the meter-level load forecasting, integrated load forecasting (district, region, and country), and commercial building forecasting.

In the past decades, numerous studies have been proposed on STLF due to its big effect on the economy. In the field of demand forecasting, lots of work have been done on traditional methods such as regression methods, exponential smoothing, and time series methods. Several artificial intelligence methods like Artificial Neural Network, Pattern recognition, fuzzy neural network are paying more attention in the field of forecasting. Evolutionary optimization techniques such as particle swarm optimization, genetic algorithm have been used in the hybrid models for enhancing the accuracy as given in ref [3]. In ref [4], empirical mode decomposition(EMD) method is used as a preprocessing technique to decompose the time-series signals. A long–short-term memory (LSTM) method is based on the deep learning used with consumption sequences of appliances [5]. According to this, meter-level forecasting accuracy can be enhanced by using the LSTM method. In [6], author proposed a NILM (non-intrusive load monitoring) method for residential power prediction with the help of on and off-state duration of appliances. Demand prediction of anomalous days is proposed by using applying three predictors and trained using PSO (particle swarm optimization). So by combining multiple predictors and preprocessed inputs, prediction accuracy can be improved [7].

In this work, different factors and their impact on consumption of electric power in short-term load forecasting are analyzed by plotting various load curves. Also, the effect of anomalous days on prediction accuracy is analyzed. ANFIS is used to predict hourly electricity load and results are compared with ANN. ANFIS is an artificial intelligent method that is widely used in electric demand forecasting. It is an adaptive and hybrid network that combines the decision-making nature of fuzzy logic and the learning nature of neural network [8].

In Sect. 2, the data analysis and influencing factors for STLF are discussed. Section 3 gives an overview of ANFIS. Section 4 presents the result and discussion and section 5 describes the conclusion.

2 Data Analysis and Influencing Factors for Inputs Selection

2.1 Data Analysis

The hourly temperature and demand data have been taken from the Mendeley dataset that was used in ref. [9]. This dataset is from the power supply company of Johor city in Malaysia. The hourly load characteristics for the year 2009 are shown in Fig. 1, and the temperature variations for the same year are given in Fig. 2.

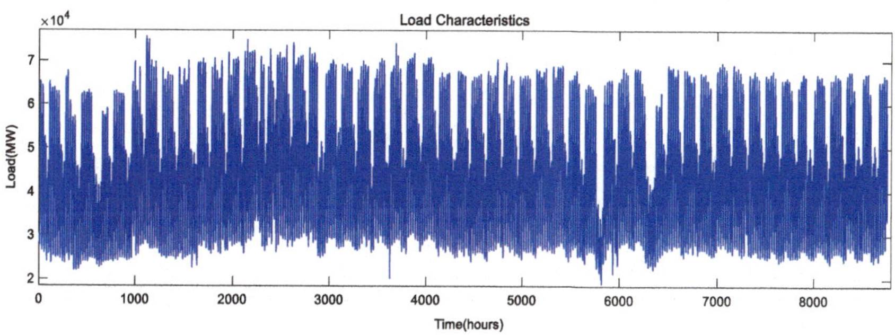

Fig. 1 Hourly electrical load consumptions for the year 2009

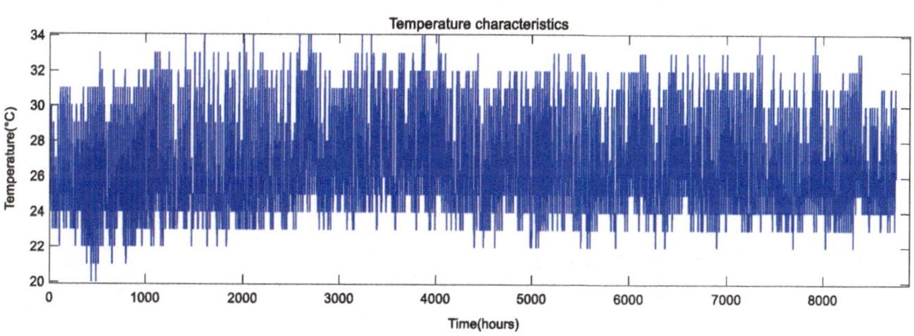

Fig. 2 Hourly temperature characteristics for the year 2009

2.2 Influencing Factors and Correlated Inputs

In general, the prediction accuracy of models depends on so many factors such as weather, time of the day, resident's behavior (appliance usage pattern), unexpected disturbances (recently a worldwide change in demand patterns due to lockdown as most of the industries, schools, colleges were closed and increased consumption of residential consumers during that period), cost of electricity, renewable energy sources, storage cells, economic factors, etc.

In this work, influencing factors having an impact on the demand (highly correlated variables) are selected as input to the ANFIS. The electricity consumption from 1 January to 31 January 2009 is used in this study. The load characteristic of the above period is shown in Fig. 3. The periodicity is very important in any load curve that is very useful for prediction.

Figure 3 represents periodicity for the day of the week and also shows the periodicity between times of the day. Another observation is that the weekend load curves are different from weekday load curves, i.e., Saturday and Sunday demand patterns are different from weekday demand patterns. This is due to the change in human activities. Also, load characteristics show a repetitive cycle. Figure 4 illustrates the variation of load with change in temperature, which is highly correlated. Among all of the influencing factors, weather factors have the most impact on electric demand because it decides the use of air conditioner, heating equipment, lighting devices, etc.

To examine the relationship between temperature and electrical demand, the Pearson correlation factor is used. The range of this factor lies between + 1 and −1 and indicates the strength of the relation between the variables. The correlation will be stronger when the absolute value is closer to 1 [10]. Table 1 shows the month-wise Pearson correlation coefficient between electricity demand and temperature for the year 2009. Table 1 shows that January month has the strongest relationship

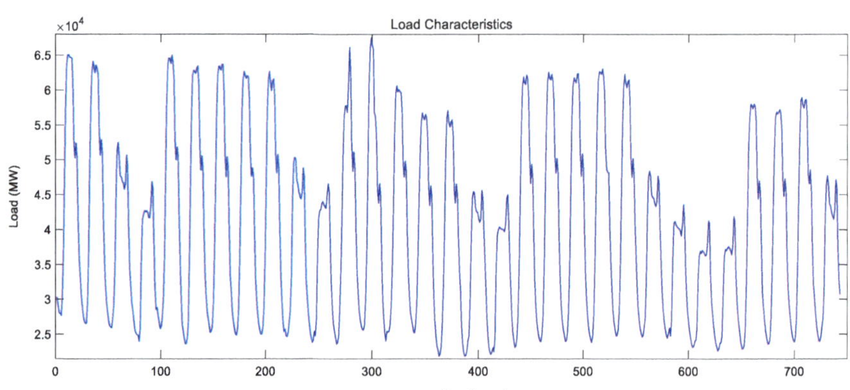

Fig. 3 Load characteristics for January 2009

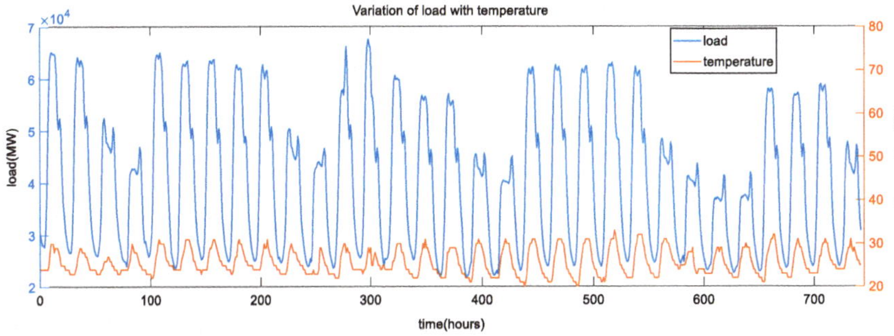

Fig. 4 Hourly variation of load with the change in temperature for January 2009

Table 1 Month-wise Pearson's correlation coefficient between temperature and electricity demand

S/no	Month	Pearson's correlation coefficient	S/no	Month	Pearson correlation coefficient
1	January	0.772526	7	July	0.68892
2	February	0.727475	8	August	0.645471
3	March	0.680858	9	September	0.702396
4	April	0.699624	10	October	0.666507
5	May	0.637736	11	November	0.658569
6	June	0.751382	12	December	0.740076

between temperature and electricity demand as compared with other months. In this paper, temperature, the hour of the day, and type of weekday are considered as a strong influencing factor for short-term electric demand forecasting.

3 Modeling of ANFIS

ANFIS is an approach for hybrid data learning that merges the properties of the fuzzy logic and neural network into a single method. The tuning of the fuzzy inference system parameters is performed by neural network learning methods. The ANFIS structure uses a first-order fuzzy Sugeno model with two if-then fuzzy rules [11].

$$\begin{aligned}&\text{rule}(1):\\&\text{IF x is } A_1 \text{ AND y is } B_1, \text{THEN } f_1 = p_1x + q_1y + r_1\end{aligned} \tag{1}$$

$$\begin{aligned}&\text{rule }(2):\\&\text{IF x is } A_2 \text{ AND y is } B_2, \text{THEN } f_2 = p_2x + q_2y + r_2\,.\end{aligned} \tag{2}$$

here, A_i and B_i are fuzzy sets, and inputs are denoted by x and y. f_i corresponds to output, and p_i, q_i, and r_i are the consequent design parameters.

The ANFIS structure is shown in Fig. 5. It is a five-layer architecture where the circle represents a fixed node that shows the fixed parameter sets in the system. The rectangle indicates an adaptive node that shows the adjustable parameter sets. The input is given to the layer 1, which gives output in terms of fuzzy membership grade. Due to the adaptive nature of the nodes in this layer, the parameters of the respective membership function are adjusted.

The next tier has fixed nodes. The firing strength of each rule is calculated with the multiplier represented by π. The strength of firing for a rule is illustrated by w_i. The third tier is for the normalization of firing strength. In this layer, fixed nodes are marked by N. Layer 4 computes the fuzzy rules output. As the nodes are adaptive in layer 4, the parameters are adjusted. Layer 5 receives the inputs from layer 4 and performs the summation to give the output of the network. The membership function parameters of a FIS are modified using the backpropagation algorithm or in combination with the least square method. So with the help of these five layers, FIS learns the relation of input and output.

The ANFIS-based forecasting model is designed in MATLAB (2013a). A matrix is created with 576 rows and 4 columns where row numbers show the time in hours from January 1st, 2009 to January 24, 2009. Out of four columns, the first column shows the time in hours of a particular day. The second and third columns represent the day of the week and temperature, respectively. The ANFIS model considers

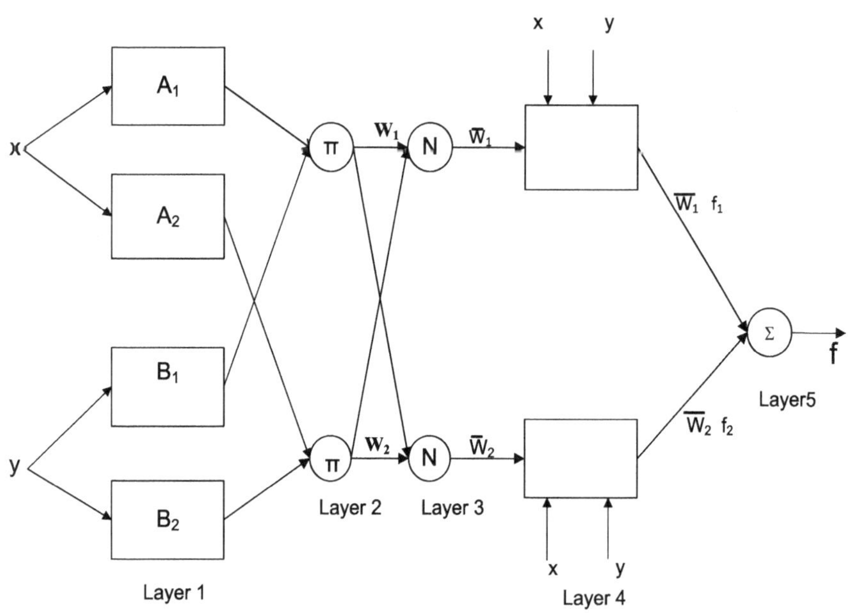

Fig. 5 ANFIS structure

Table 2 Inputs to the ANFIS model

Triangular membership function		
S/no	Inputs	Linguistic terms used
1	Hod (hour of the day)	Em (early morning), mng (morning), an (afternoon), evening, night, ln (late night)
2	Dow (days of the week)	Weekdays and weekends
3	Temperature	Low, medium, high

Table 3 Forecasting accuracy evaluation of different days using ANFIS

Date	Day	MAPE (%)	RMSE	MAE
25/01/2009	Sunday	10.99	4507.101	3851.625
26/01/2009	Monday	35.71	15,399.56	12,089.63
27/01/2009	Tuesday	35.75	15,420.05	12,029.38
28/01/2009	Wednesday	6.33	2911.013	2624
29/01/2009	Thursday	6.42	3378.343	2850.75
30/01/2009	Friday	4.72	2357.326	2046.333
31/01/2009	Saturday	3.679	1426.499	1280.708

the first three columns as input. The last column represents the target (demand) given to the ANFIS model. ANFIS structure is created by assigning numbers and types of membership functions as given in Table 2. The rules are generated by the grid partitioning method, which initializes the structure in FIS. During training, the parameters of the membership function improve themselves to acquire the relation of input and output.

After preparing inputs, the model is trained to minimize the mean square error. After that, the testing is performed on the ANFIS model. For the setting of parameters for the ANFIS model, the triangular membership function is used by giving linguistic terms according to Table 2.

4 Results and Discussion

The proposed model is trained from January 1st to January 24, 2009 and tested for 1 week from 25/01/2009 to 31/01/2009. Figure 6 represents the simulated results of actual and forecasted load for 7 days using ANFIS. The results are compared with the ANN method as shown in Fig. 7.

There are so many methods to evaluate the forecasting accuracy like absolute percentage error (APE), coefficient of determination (R^2), MAPE and RMSE [12, 13]. This paper includes various techniques like RMSE, MAE, and MAPE to calculate

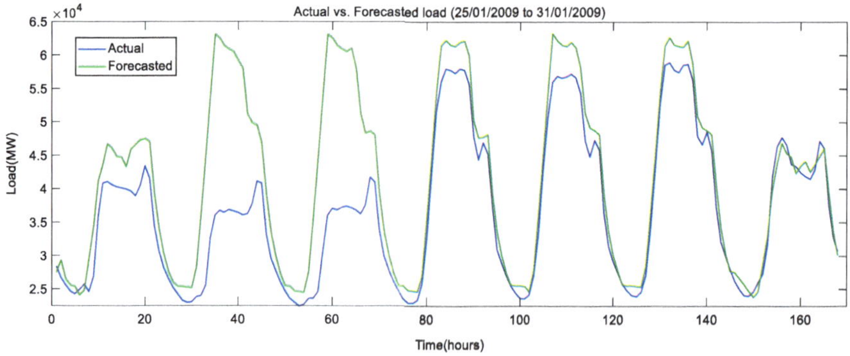

Fig. 6 Comparison of actual and forecasted load with ANFIS (25/01/2009–31/01/2009)

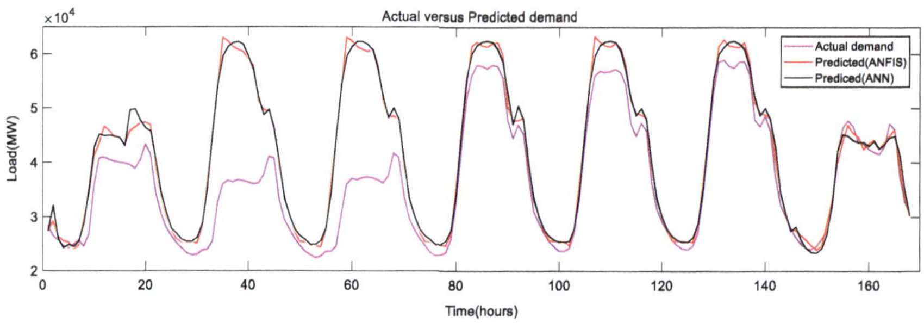

Fig. 7 Comparison of actual and forecasted load with ANN and ANFIS (25/01/2009–31/01/2009)

Table 4 Forecasting accuracy evaluation of different days using ANFIS and ANN

Comparison of MAPE using ANFIS and ANN							
Method/days	25-01–2009	26-01–2009	27-01–2009	28-01–2009	29-01–2009	30-01–2009	31-01–2009
ANFIS	10.99	35.71	35.75	6.33	6.42	4.72	3.61
ANN	11.41	35.26	35.49	6.88	6.41	5.08	3.63

the accuracy of the ANFIS model. MAPE is the mean of absolute percentage error as shown in Eq. (3) and assesses the size of the error. The smaller MAPE gives better forecasting results.

$$MAPE[\%] = \frac{1}{N}\sum_{i=1}^{N}\frac{|L_{A(i)} - L_{F(i)}|}{L_{A(i)}} \times 100. \tag{3}$$

The mean absolute error is given by (4).

$$MAE = \frac{1}{N}\sum_{i=1}^{N}|L_{A(i)} - L_{F(i)}|. \tag{4}$$

RMSE indicates the closeness of observed data points with the predicted values of the model.

$$RMSE = \sqrt{\frac{1}{N}\sum_{i=1}^{N}\left(L_{A(i)} - L_{F(i)}\right)^2} \tag{5}$$

where $L_{A(i)}$ is the real load, $L_{F\,(i)}$ is the forecasted load at the ith hour and N represents the number of samples. Figure 7 compares the forecasting result of the ANFIS model with the result of the ANN method. Table 3 shows the forecasting accuracy evaluation for 1 week.

The result of Table 3 and Fig. 8 shows that the MAPE is very large for 26 and 27 January 2009, i.e., 35.71 and 35.75, respectively, due to national holidays (Chinese New Year) [14]. This is because very less data used for these anomalous periods in the input. Figure 9 shows daywise comparison of forecasted demand. Table 4 shows that

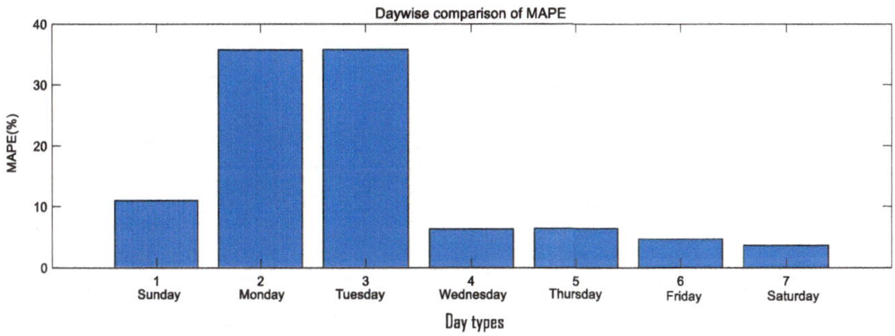

Fig. 8 MAPE values for different day types during 25/01/2009–31/01/2009

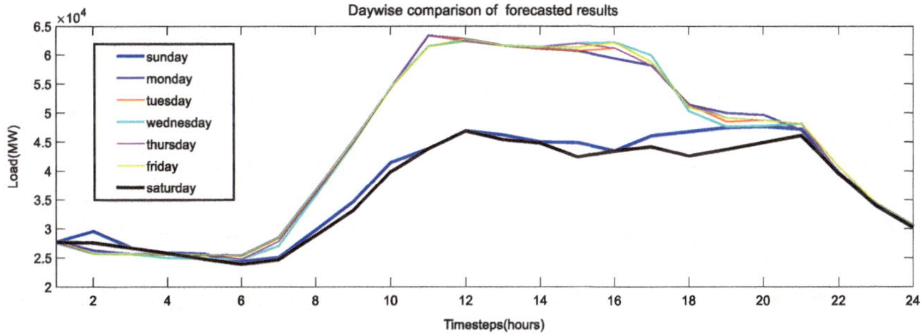

Fig. 9 Day-wise comparison of forecasted results

there is an increase in forecasting accuracy with the ANFIS model when compared with the ANN model except for the anomalous days. The impact of including anomalous days on forecasted results for the average hourly MAPE is observed in Fig. 10 and Table 5. It shows an increase in average hourly MAPE when included in forecasted results.

In conventional methods like ARMA, which shows good performance only for similar data patterns, without any uncertainty, but in the case of a new system, its performance decreases [15]. The ANFIS approach mixes the benefits of fuzzy logic in decision-making during uncertainty and neural network for learning the patterns.

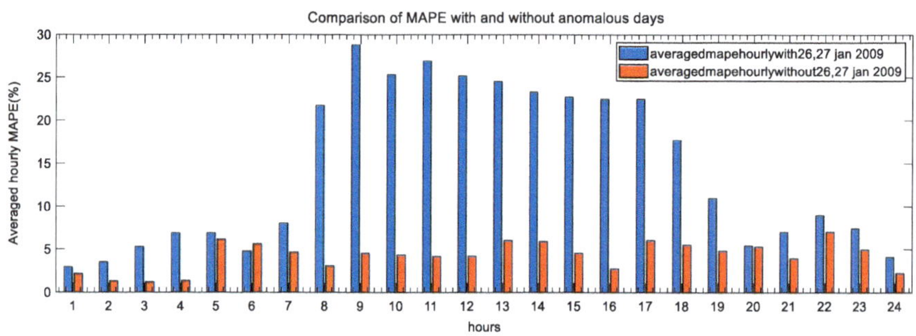

Fig. 10 Comparison of average hourly MAPE with and without anomalous days

Table 5 Averaged hourly MAPE with and without including anomalous days in forecasted results

Hour	Averaged hourly MAPE without including the anomalous day in forecasted results	Averaged hourly MAPE with including anomalous day in forecasted results	Hour	Averaged hourly MAPE without including the anomalous day in forecasted results	Averaged hourly MAPE with including anomalous day in forecasted results
1	2.170669232	2.923275239	13	6.068585829	24.48698498
2	1.330853225	3.531009501	14	5.936821662	23.26239405
3	1.202042675	5.33483362	15	4.595738912	22.70959226
4	1.346249733	6.908343509	16	2.744157152	22.47370244
5	6.199784337	6.926560535	17	6.072627934	22.45377558
6	5.659246092	4.811854564	18	5.557018046	17.70905159
7	4.700478993	8.047778931	19	4.852711411	10.97902037
8	3.075644223	21.72506097	20	5.340928049	5.452001936
9	4.575567586	28.86258337	21	3.938730529	7.025246604
10	4.372850865	25.30172246	22	7.039721758	9.037097076
11	4.157427644	26.92876291	23	5.028201621	7.440748265
12	4.196178424	25.14502613	24	2.232276608	4.083072518

So it is better because on any new system (having any uncertainties), it works based on the relations between input and target and decision capability.

5 Conclusion

To analyze the effect of anomalous days in forecasting, the ANFIS model is developed for hourly short-term electric demand forecasting. The periodicity of the load curve is observed for analysis. The prediction accuracy of the ANFIS is evaluated using various measures such as MAE, RMSE, and MAPE. The result from ANFIS model is compared with the result of ANN. The obtained result shows that ANFIS gives more accurate results than ANN. More influencing input parameters that are highly correlated with the demand can be used to reduce the error. The prediction errors due to anomalous days have been analyzed. The increase in prediction error is due to the very few data points of anomalous days. There should be a different model for different cases like for festivals and holidays for better accuracy.

In future work, anomalous days forecasting models can be designed by taking longer data length for appropriate training of the model. Further research can be done to explore the relationship between prediction error and electricity cost using hybrid machine learning methods.

References

1. Du Y, Wu J, Li S, Long C, Onori S (2019) Coordinated energy dispatch of autonomous microgrids with distributed MPC optimization. IEEE Trans Ind Inform 15(9):52895298. https://doi.org/10.1109/TII.2019.2899885
2. Laouafi A, Mordjaoui M, Laouafi F, Boukelia TE (2016) Daily peak electricity demand forecasting based on an adaptive hybrid two-stage methodology. Int J Electr Power Energy Syst 77:136144. https://doi.org/10.1016/j.ijepes.2015.11.046
3. Ma J, Ma X (2018) A review of forecasting algorithms and energy management strategies for microgrids. Syst Sci Control Eng 6(1):237248. https://doi.org/10.1080/21642583.2018.1480979
4. Semero YK, Zhang J, Zheng D (2020) EMD–PSO–ANFIS-based hybrid approach for short-term load forecasting in microgrids. Transm Distrib IET Gener 14(3):470475. https://doi.org/10.1049/iet-gtd.2019.0869
5. Kong W, Dong ZY, Hill DJ, Luo F, Xu Y (2018) Short-term residential load forecasting based on resident behaviour learning. IEEE Trans Power Syst 33(1):1087–1088
6. Dinesh C, Makonin S, Bajić IV (2019) ‘Residential power forecasting using load and graph spectral clustering. IEEE Trans Circuits Syst II Express Briefs 66(11):19001904. https://doi.org/10.1109/TCSII.2019.2891704
7. Raza MQ, Mithulananthan N, Li J, Lee KY (2020) Multivariate ensemble forecast framework for demand prediction of anomalous days. IEEE Trans Sustain Energy 11(1):2736. https://doi.org/10.1109/TSTE.2018.2883393
8. Zheng D, Eseye AT, Zhang J, Li H (2017) Short-term wind power forecasting using a double-stage hierarchical ANFIS approach for energy management in microgrids. Prot Control Mod Power Syst 2(1):13. https://doi.org/10.1186/s41601-017-0041-5

9. Sadaei HJ, de Lima e Silva PC, Guimarães FG, Lee MH (2019) Short-term load forecasting by using a combined method of convolutional neural networks and fuzzy time series. Energy 175, 365–377. https://doi.org/10.1016/j.energy.2019.03.081
10. Yadav HK, Pal Y, Tripathi MM (2018) Short-term pv power forecasting using adaptive neuro-fuzzy inference system. In: 2018 IEEE 8th power India international conference (PIICON), pp 1–6 https://doi.org/10.1109/POWERI.2018.8704445
11. Al-Hmouz A, Shen J, Al-Hmouz R, Yan J (2012) Modeling and simulation of an adaptive neuro-fuzzy inference system (ANFIS) for mobile learning. IEEE Trans Learn Technol 5(3):226–237. https://doi.org/10.1109/TLT.2011.36
12. Çevik HH, Çunkaş M (2015) Short-term load forecasting using fuzzy logic and ANFIS. Neural Comput Appl 26(6). https://doi.org/10.1007/s00521-0141809-4
13. Mollaiy-Berneti S (2016) Optimal design of adaptive neuro-fuzzy inference system using genetic algorithm for electricity demand forecasting in Iranian industry. Soft Comput 20(12):48974906. https://doi.org/10.1007/s00500015-1777-3
14. 'Malaysia Public Holidays & School Holidays (2009) One stop Malaysia. https://www.onestopmalaysia.com/holidays-2009. HTML 11 Mar 2020
15. Mahmud K, Ravishankar J, Hossain MJ, Dong ZY (2020) The impact of prediction errors in the domestic peak power demand management. IEEE Trans Ind Inform 16(7):45674579. https://doi.org/10.1109/TII.2019.2946292

Recent Trends of Fake News Detection: A Review

Anunay Gupta, Anjum Anjum, Shreyansh Gupta, and Rahul Katarya

Abstract In the age of the internet, the rate at which humans consume and produce information is so large that the lines between misinformation and Facts have blurred. A manual censor board is neither equipped nor capable of reviewing news on such a large scale. Humanity has been burdened with fake news. Given the seriousness of the problem, a lot of research is being carried out to understand the characteristics of fake news and thus build an automated fake news detection system. In this paper, we review, critique, and summarize the existing approaches in fake news detection.

Keywords Fake news · Graph neural networks · Explainable AI · Multimodel · Content-based

1 Introduction

With a total of 4.57 billion active users, the internet has truly transformed the world in all its breadth into a global village [1]. Though the internet has many advantages and has immensely helped advance the human race, there is a cost attached to such widespread connectivity. One such cost is the rising social evil of fake news. To understand the threat that fake news poses, one must understand the significance of the internet in humanity's way of learning new information. It has been estimated that 62% of the adults in the USA depend on social media sites for news, and about 75% of adults get their news through e-mail or social media site updates [2]. Similar trends can be noticed for other countries as well. The questions on which humanity seriously needs to ponder are—of the billions of social media posts sent every day. How many of these posts are verified? Who will be held accountable if a mishappening occurs due to such social media posts? Since there are no background checks on most social media platforms, how is the credibility of users ensured? The posed questions raise some severe concerns on social media platforms.

A. Gupta · A. Anjum (✉) · S. Gupta · R. Katarya
Delhi Technological University, Delhi 110042, India

A. Tomar et al. (eds.), *Machine Learning, Advances in Computing, Renewable Energy and Communication*, Lecture Notes in Electrical Engineering 768,
https://doi.org/10.1007/978-981-16-2354-7_43

Furthermore, humanity has already seen the disastrous effects of fake news propagated on social media can have. For example, in the recent Delhi riots, which led to the destruction of property worth 25,000 crore Rupees and cost India 42 lives [3], Fake news had a huge role in the Delhi riots. It helped in mobilizing people against people of other religions. It has been estimated that of the many news stories that went viral during the time, 150 were utterly baseless [4]. The election of USA 2016 is another glaring example of the effect fake news has on our lives.

It is evident that a proper protocol for verifying news on social media is the need of the hour. Nevertheless, this protocol cannot be similar to the protocols used for traditional media since the volume of news propagated per day is very large (500 million tweets per day, 500 million posts per day on Facebook) [5]. Protocols with manual components will not have the ability to handle such massive and velocity data.

Thus, automated detection of fake news is the only way of combating this social evil. However, before seeking information on the methods to detect fake news, it is of esteem importance that we understand what fake news is. Fake news is false information presented and propagated as if it is true; it is usually propagated to create unrest and disturb society's balance. Fake news can be classified based on its propagation agenda, namely—clickbait, propaganda, satire/parody.

Using various differentiating factors between fake news and factual information such as textual features, visual features, propagation characteristics, the credibility of people involved in the propagation, researchers have proposed many states of the art model on automated fake news detection. This paper explores all major approaches to detect fake news, observing its advantages and disadvantages.

2 Various Approaches to Detect Fake News

On reviewing almost 25 papers on automatic detection of fake news, we observed that three approaches could efficiently detect fake news.

1. Propagation-Based Fake News Detection.
2. Content-Based Fake News Detection.
3. Explainable AI (Artificial Intelligent)

2.1 Propagation-Based Fake News Detection

This approach is based on the relatively new and upcoming topic of graph machine learning. It is called so because instead of building a hypothesis based on traditional array type data here, a machine learning model learns on graphically structured data. In the coming subsection, we try to explain graph machine learning (graph neural networks to be specific) in the context of fake news detection [6] and the pros and cons of using GNN in fake news detection.

2.1.1 Graph Neural Networks

As shown in Fig. 1 initially, each node has some set of features represented by a distributed vector and visually represented alternate black and grey boxes. For fake news detection based on propagation, these features can be taken as characteristics that define a user. For example, the authors [7] take information about user profile (location, settings, word embedding of the profile description), the activity of a user (number of likes, statuses), user network (connections between the user, the number of followers and friends) and user's *polarization as initial features.*

Now we observe a set of features associated with each node in the final representation of the graph. One might ponder, then, what value does the application of GNN add to our understanding? The difference between the initial set of features and the final set of features is that the initial set of features only contain features relevant to one node. It does not take into account the relationships nodes have between each other while the final set of features, in addition to the information contained by initial features, also contain information about the neighborhood of the node. Nodes in GNN learn information about their neighborhood, bypassing their respective feature vectors to their neighbors at regular intervals, and integrating the features obtained from the neighborhood [7].

Once we have obtained the final features, various classification algorithms can be run to classify the node. However, this time, our classifying algorithm will also have information about the node's neighborhood in question.

2.2 *Content-Based Detection*

In content-based detection, fake news detection is based on the assumption that there is an inherent difference in the style of writing and presenting between fake and real news. Thus, we propose the hypothesis for fake news detection, which differentiates

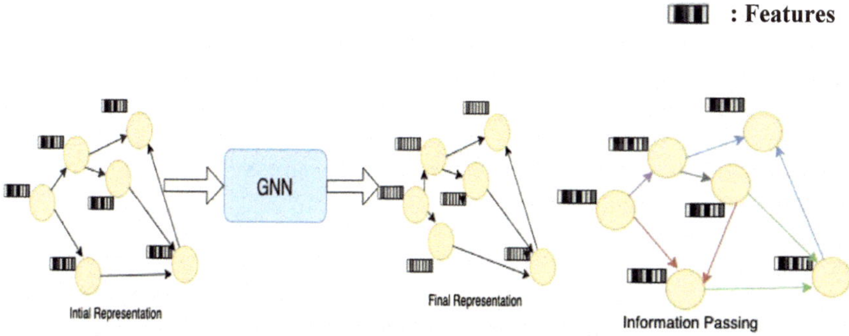

Fig. 1 GNN representation and training process

between fake news and real news based on features obtained from the content alone. Now there are two subapproaches in Content-Based Fake News Detection, namely Detection with textual features [8], Multimodal Detection of Fake news.

2.2.1 Multimodal Detection of Fake News

In Multimodal Detection of fake news, a hypothesis is built based on the features obtained from the text as well as the visual aids contained in it, such as images [9]. This approach proves to be particularly useful in detecting fake news on social media platforms and click baits. Multimodal feature detection in many ways is similar to ensemble learning. In order to understand the multimodal approach, we can bifurcate this approach into three smaller parts—(1) textual part, (2) visual part, and (3) concatenation part.

Textual Part—This part deals with the extraction of textual features. In all of the papers we reviewed, textual feature extraction was achieved by deep learning techniques such as GRU encoder, though this is not mandatory but is generally done to match the complexity of textual features with visual features extraction, as we will see in the next subsection.

Visual Part—This part deals with the extraction of visual features. This is done through a network of convolution layers to decrease the computation and only extract the essential and differentiating features.

Concatenation part—This is the part where a multimodal approach differs from the ensemble approach, and this is the part where various multimodal approaches differ. A direct concatenation of textual features and visual features may work but may not effectively detect fake news. Researchers employ various methods to concatenating the feature. For example, [9] used the similarity between the features obtained from textual and visual sources to classify. In contrast, [10] uses a latent vector and neural network to classify fake news.

We saw the workings of both approaches. Both attempts to solve the problem the fake news detection but widely differ in their methodologies. Both have their pros and cons. The advantage of the propagation-based feature detection approach is that this approach is a language-independent and a relatively objective approach compared with the content-based approach. While content-based fake news detection classifies based on content, history, and the source's location, it has no bearing on the decision since we do not want the news to be declared as fake just because it originated from a particular IP address.

2.3 *Explainable Artificial Intelligence*

As evident by its name, explainable artificial intelligence (XAI) is the artificial intelligence that humans can understand. With the advent of deep learning, feature engineering has virtually become extinct. However, deep learning techniques have had immense success and are being applied in a plethora of areas. These techniques lack interpretability; because of the complexity of deep learning techniques, it becomes a mammoth task to decode which factors are the model making a decision. Explainable artificial intelligence techniques try to find the factors based on which deep learning model makes its decision and presents them in a form that is understandable by humans. The advantages of adopting the principles of explainable machine learning are twofold (1) A comprehensive evaluation of machine learning models can be done. Parameters relevant to the specific problem can be used to improve models further and (2) deep learning models will become more reliable since the reason behind their decision will become understandable. The techniques of explainable machine learning in fake news detection [11] can have a considerable impact in the domain of natural language processing.

Explainable Artificial Intelligence (XAI) is relatively new in the field of NLP (Natural Language Processing). Earlier, the use of explainable artificial intelligence was only limited to computer vision, for example, using class activation maps from grad cam technique to show the regions where the deep learning model focuses on making the prediction [12]. Nowadays, XAI in NLP is generally used in:

1. **In Sentiment Analysis:** In this method, we determine the sentiment score of a sentence or a word in an article or a corpus. Various methods have been proposed, but the Auxiliary Prediction Task method [13] is one of the most commonly used methods. In this method, we first train a neural network to predict words in a sentence, and after training the neural network, we freeze its weights and transform it into a classifier and make classifications on it. Finally, this network can be used to predict the property or class of interest and we can finally assume that network has learned that property.
2. **Explaining Prediction in Sentence Translation:** In this, the translation of every word is explained. This can be implemented by making the model predict the prediction and explanations, but we would require manual annotations of the explanations for this. Nowadays, LIME [14] and SHAP [15] are being used to explain the prediction made by the machine learning models
3. **Detecting Fake News:** Using XAI, there has been some work done in this field to determine the fakeness of news. In [11], the author used coattention subnetworks to rank top k user comments and sentences in the article in terms of fakeness.

In the next subsection, we present our analysis of the recent and relevant research papers based on techniques discussed in the previous subsections.

2.4 Recent Approaches for Fake News Detection

In this subsection, various recent research papers in the domain of fake news detection have been discussed in detail. Table 1 succinctly discusses each paper's methodology, its contribution to the fake news detection domain, the advantage of the technique proposed, the dataset used, and challenges for the technique proposed.

Table 1 Recent paper analysis

Ref.	Methodology	Key contributions	Challenges	Dataset
[16]	Sentence comment coattention mechanism along with sentence encoding and word encoding is used on both news articles as well as user comments	Highlighted the significance of user comments for the classification of fake news. Achieved a 5% higher f1 score than benchmarked models	The credibility of users can be taken into account to improve the model's performance further	FakeNewsNet
[17]	Comparative analysis of supervised learning methods on lexical, semantic features along with features about publishers and sources are used	Compared the significance of various feature sets such as semantic and lexical features. Obtained a true positive rate of 100 and 40% misclassification of true news	The proposed model can be further improved by incorporating the features of explainable machine learning leading to a significant decrease in the misclassification rate	BuzzFace
[10]	Used Autoencoders to classify social media post containing both textual materials as well as images	Utilized both textual and visual(images) for the classification of social media posts. Outperforms state of the art models by 5% in terms of F1-score	Propagation features can be considered for better classification	Twitter Dataset Weibo Dataset
[7]	Utilized propagation-based features using geometric deep learning	A language-independent model for detecting fake news. Obtained 92.7% ROC AUC	Explainable machine learning methods can be adopted for better interpretability	Twitter self Collected
[18]	A novel Graph neural network approach is proposed for the detection of fake news using the content, and information about the publisher	Introduced a new hybrid feature learning unit (HFLU) for feature extraction and used deep diffusive neural network for classification of neural network	The principle of explainable machine learning can be incorporated for better understanding as well as performance	PolitiFact, Twitter

(continued)

Table 1 (continued)

Ref.	Methodology	Key contributions	Challenges	Dataset
[9]	A multimodal approach to detecting fake news is proposed, which exploits textual features, visual features as well as the relationship between using Text CNN	Proves the significance of using the relationship between textual features and visual features. Achieved an F1-score of 0.896, 0.895 on PolitiFact, GossipCop Datasets, respectively	Intramodal relationships can be explored for further improvement	PolitiFact, GossipCop
[8]	A combination of multiple stylometric features word vector models (like BoW, TFIDF, Skip-Gram.) using various ensemble techniques	Proved the importance of the relationship between stylometric features and prediction accuracy. Achieved a maximum accuracy of 95.48%	Future work can focus on reducing the time complexity of analysis to make the work ideal for real-time application	The FakeNewsNet & The McIntire Dataset
[19]	In this paper, the PSM (Propensity Score Matching) was used for feature selection, which helps in reducing the effect of confounding variables	They achieved more than a 10% increase in the AUROC score than the baseline models that used document frequency technique word to vector conversion techniques and basic classifier	Future work could zero in on alleviating the biases brought about by latent variables. Bayesian networks and structural equation can also be used with this for fake news detection	PolitiFact and GossiCop
[11]	Conducted an exploratory analysis to find the best set features, to be used in M.L models	Proposed the best combinations of features tailored for fake	Due to randomness, the validity of the feature combination can not be known	BuzzFace Dataset
[6]	A pattern-driven approach is proposed The approach uses the relationship between users, user's credibility	Patterns in text are explored and summarized An F1-Score of 0.93 and 0.84 is obtained on PolitiFact and BuzzFeed datasets	A component of content-based fake news detection can be added for detecting Fake news in the preliminary stages	PolitiFact and BuzzFeed
[20]	A multisource multiclass model, which measures the degree of fakeness using features extracted from different sources	Introduces the idea of measuring the degree of fakeness in the news articles using automated feature extraction	The proposed model can be enhanced by creating a difference between classes and quantitively classifying the degree of fake news	LIAR

(continued)

Table 1 (continued)

Ref.	Methodology	Key contributions	Challenges	Dataset
[21]	Utilized harmonic Boolean label crowd-sourcing algorithm	Achieved an accuracy of 99.4% using Harmonic algorithm	The use of textual as well as visual feature can control and detect fake news at preliminary stages	Facebook post-self-collected

Table 2 Datasets

Name	Aspects	Challenges
LIAR [22]	This dataset contains 12,836 short statements from politifact.com	No multimedia features exist
NELA GT 2018 [23]	This dataset contains 713,000 news articles from 194 news websites	Data does not have multimedia features
WEIBO [24]	Forty thousand posts with images and contextual information Suitable for textual analysis, image-based analysis, and propagation-based analysis	Data points are mostly in Chinese, and content is mostly related to China
BUZZFACE [11]	This dataset contains 2,282 articles related to the 2016 USA elections. Contains various features pertaining to user	This dataset only contains information only related to politics
MediaEval Datasets [25]	This dataset comprises 18,049 tweets and 438 images related to 17 events	An unbalanced dataset

3 Datasets

News can be accessed from anyplace on the internet utilizing sources like social media websites and search engines. Nevertheless, physically deciding the genuineness or validity of news can be tedious. Therefore, news stories with annotations are gathered from different dependable fact-checking websites, crowd-sourced workers, and so on. Even though there is no current benchmark dataset for fake news detection, some mainstream datasets for different methods presented are shown in Table 2 below.

4 Conclusion and Future Work

The paper analyzed, reviewed, and summarized the recent trends in automated fake news detection, which included a thorough qualitative and quantitative analysis of recent publications in the domain of fake news detection. We observed three central theories that form an integral part of fake news detection during the review, namely propagation-based, content-based, and explainable artificial intelligence.

We believe that the future of fake news detection will involve an amalgamation of all these approaches. Since the propagation-based approach is quick to detect any malicious activity, it can be used to tag malicious tweets quickly. Tagged tweets can then be classified using a content-based approach to confirm the detection. Finally, due to the use of explainable AI, false detection of fake news will be low, and the model can be continuously evaluated. Such a big data system will be relatively inexpensive in terms of computational power and will have low latency (a significant problem with machine learning solutions) and can, thus, be implemented in the near future.

References

1. Global Digital Overview—DataReportal—Global Digital Insights. In: Datareportal. https://datareportal.com/global-digital-overview
2. (2020) Social media as a news source—Wikipedia. In: Wikepedia. https://en.wikipedia.org/wiki/Social_media_as_a_news_source
3. TEAM DW (2020) Rs 25,000 crore loss estimated in Delhi riots. In: DNA NEWS INDIA. https://www.dnaindia.com/business/report-rs-25000-crore-loss-estimated-in-delhi-riots-2815581
4. (2020) Coronavirus: the human cost of fake news in India—BBC News. In: BBC. https://www.bbc.com/news/world-asia-india-53165436
5. Twitter Usage Statistics—Internet Live Stats. In: world wide web Found. https://www.internetlivestats.com/twitter-statistics/
6. Zhou X, Zafarani R Network-based fake news detection: a pattern-driven approach. 21:48–60.
7. Monti F, Frasca F, Eynard D, et al (2019) Fake news detection on social media using geometric deep learning. pp 1–15
8. Reddy H, Raj N, Gala M, Basava A (2020) Text-mining-based fake news detection using ensemble methods. Int J Autom Comput 17:210–221. https://doi.org/10.1007/s11633-019-1216-5
9. Zhou X, Wu J, Zafarani R (2020) SAFE: similarity-aware multi-modal fake news detection. Lect Notes Comput Sci (including Subser Lect Notes Artif Intell Lect Notes Bioinformatics) LNAI 12085:354–367. https://doi.org/10.1007/978-3-030-47436-2_27
10. Khattar D, Gupta M, Goud JS, Varma V (2019) MvaE: multimodal variational autoencoder for fake news detection. In: Proceedings of the world wide web conference WWW 2019. pp 2915–2921. https://doi.org/10.1145/3308558.3313552
11. Reis JCS, Correia A, Murai F, et al (2019) Explainable machine learning for fake news detection. In: WebSci 2019—Proceedings of the 11th ACM conference on web science. pp 17–26. https://doi.org/10.1145/3292522.3326027
12. Gupta A, Gupta S, Katarya R (2020) InstaCovNet-19: a deep learning classification model for the detection of COVID-19 patients using chest X-ray. Appl Soft Comput J. https://doi.org/10.1016/j.asoc.2020.106859
13. Radford A, Jozefowicz R, Sutskever I (2017) Learning to generate reviews and discovering sentiment
14. Ribeiro MT, Singh S, Guestrin C "Why should I trust you?" Explaining the predictions of any classifier. https://doi.org/10.1145/2939672.2939778
15. Lundberg SM, Erion G, Chen H et al (2020) From local explanations to global understanding with explainable AI for trees. Nat Mach Intell 2:56–67. https://doi.org/10.1038/s42256-019-0138-9

16. Cui L, Shu K, Wang S, et al (2019) dEFEND. pp 2961–2964. https://doi.org/10.1145/3357384.3357862
17. Reis JCS, Correia A, Murai F et al (2019) Supervised learning for fake news detection. IEEE Intell Syst 34:76–81. https://doi.org/10.1109/MIS.2019.2899143
18. Zhang J, Dong B, Yu PS (2020) FakeDetector: effective fake news detection with deep diffusive neural network. In: Proceedings of the international conference on data engineering 2020. pp 1826–1829. https://doi.org/10.1109/ICDE48307.2020.00180
19. Ni B, Guo Z, Li J, Jiang M (2020) Improving generalizability of fake news detection methods using propensity score matching
20. Karimi H, Roy PC, Saba-sadiya S, Tang J (2018) Multi-source multi-class fake news detection. pp 1546–1557
21. Tacchini E, Ballarin G, Della Vedova ML, et al (2017) Some like it hoax: automated fake news detection in social networks. In: CEUR workshop proceedings 1960. pp 1–12
22. Wang WY (2016) Liar, liar pants on fire
23. Nørregaard J, Horne BD, Adalı S (2019) NELA-GT-2018: a large multi-labelled news dataset for the study of misinformation in news articles
24. Jin Z, Cao J, Guo H (2017) Multimodal fusion with recurrent neural networks for rumor detection on microblogs. pp 795–803
25. MediaEval Datasets

An Unaccustomed AdaBoost Approach to Dig Out the Magnitude of Trash

Devliyal Swati, Jain Sourabh, and Sharma Ghai Anupriya

Abstract The waste generated by day to day use of domestic premises is called domestic refuse. This waste is not domestic if it is taken under a commercial arrangement. In this paper, a novel AdaBoost approach is used for feature extraction. AdaBoost is one of the most out-performing boosting algorithms. It has a strong hypothetical premise and has made incredible achievements in useful applications. AdaBoost is used to improve the performance of any machine learning algorithm. Due to the unavailability of the database, a new database is created to test the working of the foreseen model. This was tried on our dataset and furthermore contrasted and different calculations, for example, Support Vector Machine (SVM), K-Nearest Neighbor (KNN), Neural Network (NN), and Tree. Given a convergence result for the algorithm showing that prior knowledge can substantially improve classification performance. After testing, it is visualized that the highest accuracy (99.5%) was achieved by the novel AdaBoost approach.

Keywords AdaBoost · Garbage · NN · SVM · Stacking · Tree · KNN

1 Introduction

India is still facing challenges associated with garbage collection after things are getting automated day by day. Still, the situations are not good to handle the waste. Every year tons of garbage is generated from different sources—like industry waste, animal waste, farming waste, food waste, medical waste, etc. The government is putting so many efforts into collecting garbage by doing door-to-door collection by making things automated but still, there are some areas where things are not reaching.

Decline incorporates trash and garbage. Generally, decomposable food squander is trash. Trash is typically dry material like glass, paper, fabric, or wood. Trash is useless

D. Swati · J. Sourabh (✉) · S. G. Anupriya (✉)
Graphic Era Deemed to be University, Dehradun, Uttarakhand, India

S. G. Anupriya
Graphic Era Hill University, Dehradun, Uttarakhand, India

A. Tomar et al. (eds.), *Machine Learning, Advances in Computing, Renewable Energy and Communication*, Lecture Notes in Electrical Engineering 768,
https://doi.org/10.1007/978-981-16-2354-7_44

that includes massive items such as an old refrigerator. It requires a special collection and handling. Generally, solid waste is generated from residential, commercial, institutional, and industrial activities. Another type of solid waste is electronic waste including discarded computer equipment, television, telephones, and different types of other electronic devices [1].

Garbage is an opening or introductory stride toward contamination. Flaming or concealing scrap is an alternative worn by many territories. Blazing the rubbish is an indubitable root of air pollution together with fatal gases throughout the habitat. It will lead to global warming as well as torment reservoirs [2]. Rather than burning, concealing is also the origin of air and water pollution. As a consequence, it needs to be categorized that the domestic refuse moreover conducts them correspondingly.

Numerous published papers are constructed fundamentally on two algorithms of discerning debris, initial is a Reference counting algorithm, which states that this algorithm straps a counter in each entity, sustaining the value of counter identical to the number of entity or garbage references. This algorithm is tenacious in the distributed domain. This is the rationale for not contemplating this algorithm at a greater distance. The next algorithm is Vestal's tracing algorithm, which states that the globe is cleaved into areas within parallel throng might occur [3]. The algorithm tracer is used to notice the implementation of the graph beginning from the root and traverse up to the termination of the graph. If any entity remains distant is considered garbage.

AdaBoost, as well as Stacking, is also termed as "Ensemble learning". AdaBoost is the short form of Adaptive Boosting. It is utilized to improve the performance of any machine learning algorithm. This algorithm's most part utilized with weak learners. These are models that accomplish exactness simply above arbitrary possibility on an order issue [4]. A decision tree is the most suited and most commonly used algorithm with AdaBoost. It is versatile as in ensuing powerless students are changed for those cases misclassified by past classifiers. It is touchy to loud information and anomalies [5].

The difficulties that are confronted incorporate the inaccessibility of good quality pictures, a variety of pictures, unfocused pictures, so an effective dataset for best outcomes is required.

In this paper, an unaccustomed AdaBoost algorithm is used for portraying a diverse size of trash location is proposed. Several blends of info boundaries are utilized. It was seen that our calculation gives the best outcomes by utilizing qualities of steady boundaries. Further, by utilizing a few logical strategies, precision evaluation is finished. The most noteworthy order precision accomplished after a few mixes of boundaries is proposed by the AdaBoost approach is 95.04%.

This paper is divided into the following segments, respectively: Introduction (Sect. 1), Datasets (Sect. 2), Methodology (Sect. 3), Result and discussions (Sect. 4), and Conclusion (Sect. 5).

2 Datasets

The technique in AdaBoost approach requires a well-built dataset for training models. The better preparation of models will bring about additional exactness got subsequent to testing.

There are no datasets accessible for regions like trash discovery. Due to this reason, a new database is created with images, belonging to different classes as shown in Table 1.

In this database, good quality images have been selected, with proper dimension. The picture include vectors are produced utilizing measurable activities to process the actual appearance of the surface and a NN model is utilized to order the created surfaces with suitable names into classes [6]. It has been composed of the zone around Graphic Era University located in the clement town of Dehradun. The research area is primarily classified into five different categories specifically R1, R2, R3, R4, and R5. An entire 4.5 km is enfolding for all disparate classes. In R1, R2, R3, and R4, an integral of 600 divergent images are examined while in R5 100, images are clutching (Fig. 1).

3 Methodology

An ensemble of decision trees for the classification of garbage detection in order to learn its important features is used. The machine includes a RAM of 8 GB, an Intel i5 processor, and a 2 GB Graphic card on which work has been executed.

To find a strong rule, boosting is used, which combines weak and base leaner. To find the weak rule, the base learning calculation is applied with an alternate dispersion. Each time another feeble expectation rule is produced when the base learning algorithm is applied. This is an iterative cycle. The boosting algorithm at that point consolidates these shaped frail standards into a solitary solid forecast rule after much emphasis [7].

Steps for selecting the right distributions are:

Step 1: For each observation, the base learner takes all the distributions and assigns equal weight.

Step 2: Need to give higher consideration to recognitions if there is any expectation mistake brought about by a respectable starting point learning calculation. At that point, just the next base learning calculation taxi be applied.

Step 3: Finally continue emphasizing stage 2 till the most elevated precision is accomplished.

Finally, the yields from a delicate amateur are solidified and make a solid tenderfoot which enhances the projection intensity of the model in the end. A higher spotlight is paid on models that have higher blunders by going before powerless principles.

The misclassification rate is calculated as shown in Eq. 1 below

Fig. 1 Sample of garbage taken from dataset

$$\text{Error} = (\text{correct} - N)/N \tag{1}$$

where Error is the arrangement rate.

Right are the amount of getting ready event foreseen precisely in the model where N is the hard and fast number of planning cases as showed up in condition 1. For example, if the model foresaw 68 of 100 planning events precisely the goof of misconceive rate would be (68–100)/100 or 0.32.

The Classifier takes datasets of images as input and gives the output as the sum of all the models used as depicted in Fig. 2. Data are prepared by three steps in AdaBoost namely, "Quality data", "Outliers", and "Noisy data".

Quality data—Misclassifications in the training data are corrected by ensemble methods as well as the training data should be of high quality.

Outliers—This will force the ensemble methods for working hard to correct cases that are unrealistic. These types of unrealistic data could be removed from the training dataset.

Noisy data—One problem in the output variable can be noise. So it should be cleaned from the training dataset [7].

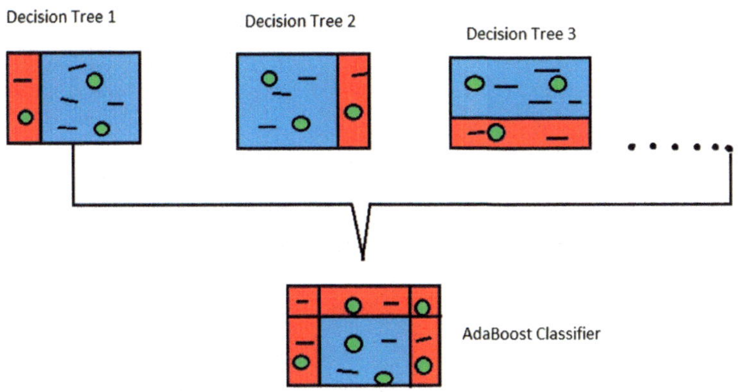

Fig. 2 AdaBoost classifier

4 Results and Discussions

This part shows the accompanying test situations into different scenarios by using AdaBoost, Neural Network, Tree, Stacking, kNN, and SVM. Reader may refer [8–14] for detailed information of the design and implementation of these six AI and machine learning techniques. The analyzed results are as follows:

Scenario 1 (AdaBoost)

In this section, the AdaBoost algorithm is applied. AdaBoost is used to enhance the performance of decision trees used in classification problems. AdaBoost is the best algorithm for effective learners. It can achieve accuracy above random chances of classification problems [7]. The most elevated characterization exactness accomplished after a few mixes of boundaries is 95.04% (Table 1).

Scenario 2 (Neural Network)

In this section, the Neural Network (NN) algorithm is applied. NN, in essence, matches the pattern based on your input and envision the output. NN is applied for complex mapping from the input to outer space [15]. Several combinations of input parameters are employed and the highest classification accuracy (94.10%) was achieved when employed five hidden layers (Fig. 3).

Scenario 3 (Tree)

In the concept of classification trees, this algorithm generates univariate trees with the help of constructive induction with linear function [16]. After using several combinations of parameters, the highest classification accuracy was achieved 93.63%.

Table 1 Cross-tabulation matrix of classified versus reference data for different garbage collection models

Predictive model	Classified data	Reference data					Producer's accuracy (%)	User's accuracy (%)	Overall accuracy (%)
		R1	R2	R3	R4	R5			
AdaBoost	R1	108	0	0	0	0	100	100	95.04
	R2	0	92	0	0	0	98.92	100	
	R3	0	1	71	0	20	100	77.17	
	R4	0	0	0	104	0	100	100	
	R5	0	0	0	0	28	58.33	100	
Neural network	R1	108	0	0	0	0	100	100	94.10
	R2	0	91	1	0	0	94.79	98.91	
	R3	0	2	83	0	7	86.45	90.21	
	R4	0	0	0	104	0	100	100	
	R5	0	3	12	0	13	65	46.42	
Tree	R1	108	0	0	0	0	100	100	93.63
	R2	0	87	4	0	1	98.86	94.56	
	R3	0	0	71	1	20	94.66	77.17	
	R4	0	0	0	104	0	99.04	100	
	R5	0	1	0	0	27	56.25	96.42	
Stack	R1	108	0	0	0	0	100	100	92.45
	R2	0	91	1	0	0	100	98.91	
	R3	0	0	88	0	4	75.86	95.65	
	R4	0	0	0	104	0	100	100	
	R5	0	0	27	0	1	20	3.57	

(continued)

Table 1 (continued)

Predictive model	Classified data	Reference data					Producer's accuracy (%)	User's accuracy (%)	Overall accuracy (%)
		R1	R2	R3	R4	R5			
KNN	R1	107	0	0	1	0	99.07	99.07	88.91
	R2	1	76	8	0	7	89.41	82.60	
	R3	0	5	72	0	15	83.72	78.26	
	R4	0	0	0	104	0	99.04	100	
	R5	0	4	6	0	18	45	64.28	
SVM	R1	108	0	0	0	0	100	100	87.97
	R2	0	78	14	0	0	76.47	84.78%	
	R3	0	10	82	0	0	75.22	89.13	
	R4	0	0	0	108	0	100	100	
	R5	0	14	13	0	1	100	50	

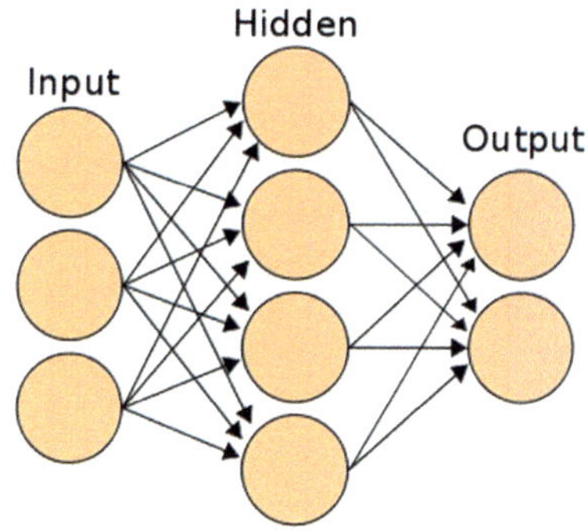

Fig. 3 Neural network

Scenario 4 (Stacking)

Stacking is the way to convene many different models together to get a prominent result. There are many ways to assemble the models but the one that is used above all is bagging. The main concept of bagging is to assemble different models group to give the result for the same problem [17]. The accuracy achieved using stacking is 92.45%.

Scenario 5 (K-Nearest Neighbor (KNN))

KNN is the best algorithm that is used for regression and classification. KNN uses data and then collocates the new data based on the similarity measures [18]. After using several combinations of parameters, the highest classification accuracy was achieved 88.91% (Fig. 4).

Scenario 6 (Support Vector Machine (SVM))

SVM algorithm is used to segregate two or more classes. SVM is often used for classification problems. It is used to plot each data in the N-dimensional plane [15].

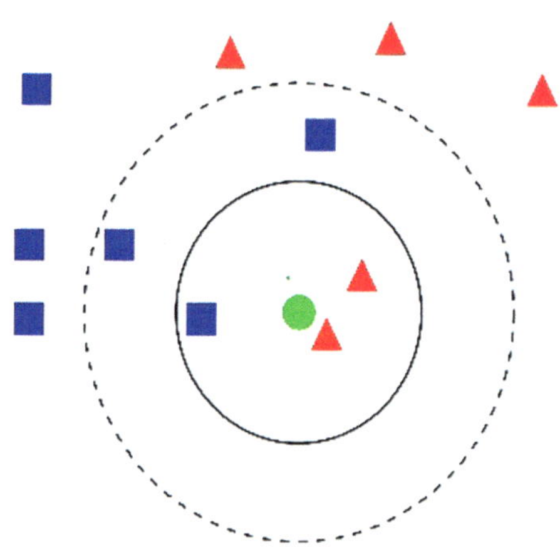

Fig. 4 KNN

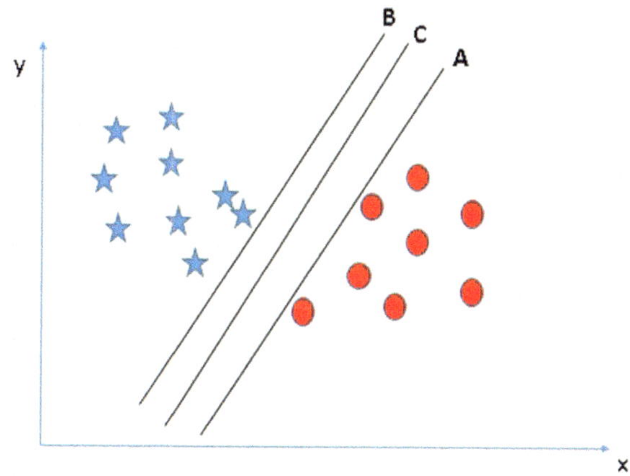

Fig. 5 Support vector machine

After using several combinations of parameters, the highest classification accuracy was achieved 87.97% (Fig. 5).

Out of the whole situation, AdaBoost is outflanking the other condition of-workmanship algorithm as appeared in Table 1. It is clearly visible that the AdaBoost algorithm is giving the best outcome because of the following two reasons: Tweaking of boundaries and diminishes predisposition for little fluctuation. Henceforth, it improves the exhibition of the grouping contrasted with other conditions of craftsmanship algorithm.

It can be visualized that by using Adaboost algorithm, the highest accuracy 99.5% with precision 100%, specificity 100%, AUC 99.5% and sensitivity 82.8% was achieved as shown in Table 2.

ROC bend portrayed in Fig. 6. The rating of trash has been determined. It is made by delineating "affectability" on y-pivot and "particularity" on x-hub. This bend is utilized to make TPR (True sure rate) against FPR (False certain rate). The TPR is otherwise called affectability [19].

The lift curve is used to pitch the performance of the target model for having an enhanced response, which is measured against a contingent targeting model [20]. By gain curve it is illustrated in Fig. 7. The rating of garbage has been illustrated.

5 Conclusion

In this paper, the AdaBoost approach is adapted to detect garbage. After using several combinations of parameters, the highest classification accuracy was achieved 95.04%. Good quality images were selected for this work, with proper dimension. The research area is primarily classified into five different categories specifically R1, R2, R3, R4, and R5. An entire 4.5 km is enfolding for all disparate classes. In R1, R2, R3, and R4, an integral of 600 divergent images are examining while in R5 100, images are clutch. It, very well, may be seen that AdaBoost gave the best outcome

Table 2 Measurement proportions of the proposed models for the training (T) and validation (v) stage for planning rating

Dataset	Model	Classification				
		Accuracy	Precision	Specificity	AUC	Sensitivity
	AdaBoost					
R3		0.995	1.000	1.000	0.995	0.828
	Neural network					
R3		0.993	0.929	0.978	0.993	0.866
	Tree					
R3		0.990	0.957	0.987	0.990	0.838
	Stack					
R3		0.989	0.814	0.924	0.989	1.000
	KNN					
R3		0.977	0.840	0.944	0.977	0.847
	SVM					
R3		0.977	0.835	0.956	0.977	0.819

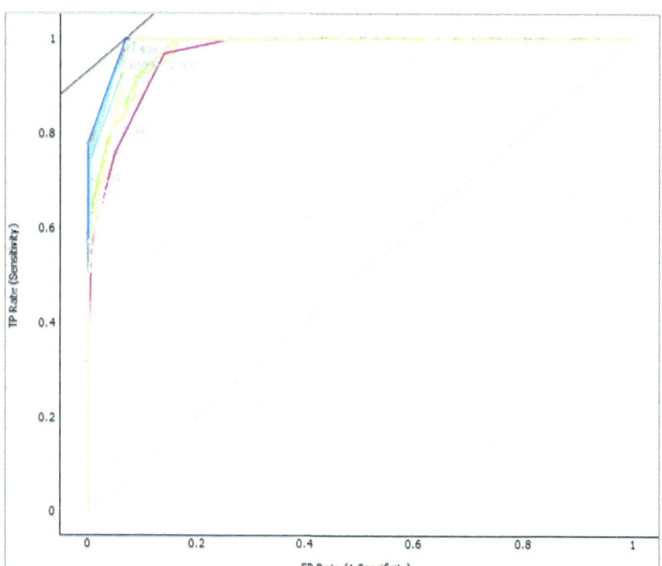

Fig. 6 ROC bend for explicitness and affectability of the prepared model

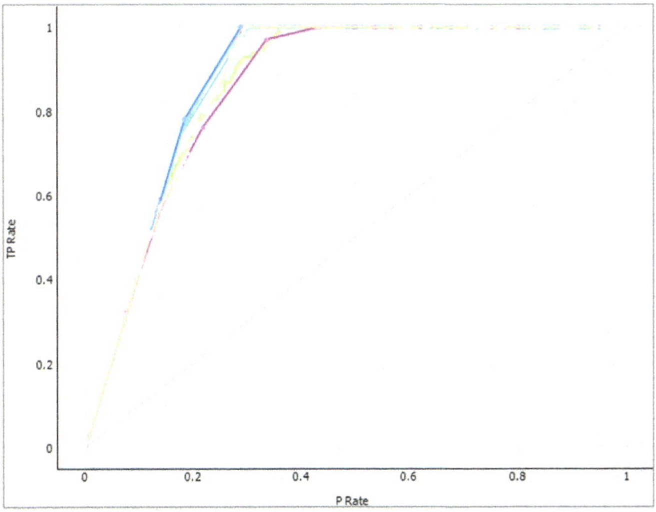

Fig. 7 Lift curve for the performance of trained model

when contrasted with other conditions of craftsmanship calculations on account of its tweaking highlight. This paper increased substantially more trouble because of the inaccessibility of the dataset of trash location. This model can, in like way, be applied for agnate datasets of trash. The exact assessment shows that different models are looked at utilizing changed mistake estimating measurements, for example, error variance, MSE, and RMSE. These boundaries approve the exhibition and vigor of the proposed models [21].

References

1. Shapiro M, Gruber O, Plainfossé D (1990) A garbage detection protocol for a realistic distributed object-support system. Diss. INRIA
2. Wang WJ, Varela CA (2006) Distributed garbage collection for mobile actor systems: the pseudo root approach. In: International conference on grid and pervasive computing. Springer, Berlin
3. Paul RW (1992) Uniprocessor garbage collection techniques. Memory management. Springer, Berlin, pp 1–42
4. Nijhawan R, Rana M (2016) Green computing and strategies for energy efficient cloud management. Int J Comput Appl 150(2)
5. Nijhawan R, Das J, Raman B (2018) A hybrid of deep learning and hand-crafted features based approach for snow cover mapping. Int J Remote Sens 1–15
6. Aggarwal A, Alshehri M, Kumar M, Alfarraj O, Sharma P, Pardasani RK (2020) Landslide data analysis using various time-series forecasting models. Comput Electr Eng 88:106858. ISSN 0045-7906

7. Kotsiantis SB, Zaharakis I, Pintelas P (2007) Supervised machine learning: a review of classification techniques. Emerging artificial intelligence applications in computer engineering, vol 160, pp 3–24
8. Aggarwal S et al (2020) Meta heuristic and evolutionary computation: algorithms and applications. Springer Nature, Berlin, p 949. https://doi.org/10.1007/978-981-15-7571-6. ISBN 978-981-15-7571-6
9. Fatema N et al (2021) Intelligent data-analytics for condition monitoring: smart grid applications. Elsevier, p 268. ISBN: 978-0-323-85511-2. https://www.sciencedirect.com/book/9780323855105/intelligent-data-analytics-for-condition-monitoring
10. Smriti S et al (2018) Special issue on intelligent tools and techniques for signals, machines and automation. J Intell Fuzzy Syst 35(5):4895–4899. https://doi.org/10.3233/JIFS-169773
11. Jafar A et al (2021) AI and machine learning paradigms for health monitoring system: intelligent data analytics Springer Nature, Berlin, p 496. https://doi.org/10.1007/978-981-33-4412-9. ISBN 978-981-33-4412-9
12. Sood YR et al (2019) Applications of artificial intelligence techniques in engineering, vol 1. Springer Nature, Berlin, p 643. https://doi.org/10.1007/978-981-13-1819-1. ISBN 978-981-13-1819-1
13. Gopal et al (2021) Digital transformation through advances in artificial intelligence and machine learning. J Intell Fuzzy Syst 1–8. https://doi.org/10.3233/JIFS-189787
14. Yadav AK et al (2020) Soft computing in condition monitoring and diagnostics of electrical and mechanical systems. Springer Nature, Berlin, p 496. https://doi.org/10.1007/978-981-15-1532-3. ISBN 978-981-15-1532-3
15. Waske B et al (2009) Machine learning techniques in remote sensing data analysis. Kernel methods for remote sensing data analysis, pp 3–24
16. Kühnlein M et al (2014) Improving the accuracy of rainfall rates from optical satellite sensors with machine learning—a random forests-based approach applied to MSG SEVIRI. Remote Sens Environ 141:129–143
17. Džeroski S, Ženko B (2004) Is combining classifiers with stacking better than selecting the best one? Mach Learn 54(3):255–273
18. Bijalwan V et al (2014) KNN based machine learning approach for text and document mining. Int J Database Theory Appl 7(1):61–70
19. Nijhawan R et al (2017) An integrated deep learning framework approach for nail disease identification. In: 2017 13th international conference on IEEE signal-image technology & internet-based systems (SITIS)
20. Vuk M, Curk T (2006) ROC curve, lift chart and calibration plot. Metodoloski zvezki 3(1):89
21. Aggarwal A, Kumar M (2020) Image surface texture analysis and classification using deep learning. Multimed Tools Appl. ISSN 1573-7721

Predicting Online Game-Addicted Behaviour with Sentiment Analysis Using Twitter Data

Ramesh Narwal and Himanshu Aggarwal

Abstract Social media give us the power to connect and share our views with millions of people; it does not matter how far away they are. Before social media, it is not an easy task to connect with millions of people. The Internet plays a vital role in this journey. Even a lot of development was done in the field of games. Like social media, people can communicate with each other; it does not matter how far away they are. People can play online games with others; it does not matter how much far away they are. The top 10 online games are Fortnite, PUBG, Overwatch, Apex Legends, Minecraft, Dota 2, Grand Theft Auto V, League of Legends, Sea of Thieves, Counter-Strike: Global Offensive (Top 10 Online Multiplayer Games | Game Development | Melior Games, https://meliorgames.com/game-development/top-10-online-multiplayer-games/. Accessed 20 October 2020, [1]). Sometimes, the government took some steps regarding these online games or apps. Like the Government of India (GoI) banned 118 mobile apps regarding privacy concerns of its people (India bans 118 more Chinese mobile apps including PUBG Mobile over privacy concerns—The Financial Express, https://www.financialexpress.com/industry/technology/india-bans-118-more-chinese-mobile-apps-including-pubg-mobile-over-privacy-concerns/2072830/. Accessed 20 October 2020). PUBG, a popular online game, is one of them that is forbidden by GoI. So, people expressed their views on Twitter and other social media. In this paper, we download tweets regarding PUBG; sentiment analysis was done using lexical analysis and machine learning techniques. We classify tweets in three factors like neutral, negative, and positive. Most of the persons who are posting negative comments mean they are using PUBG, and they are addicted. We are trying to find online gaming addictive behaviour of people.

Keywords Addictive · Online gaming · Sentiment analysis · Human behaviour

R. Narwal (✉) · H. Aggarwal
Computer Science & Engineering Department, Punjabi University, Patiala 147002, India

A. Tomar et al. (eds.), *Machine Learning, Advances in Computing, Renewable Energy and Communication*, Lecture Notes in Electrical Engineering 768,
https://doi.org/10.1007/978-981-16-2354-7_45

1 Introduction

Social media provide us with a platform where anyone can share their views with millions of other peoples. Now these days, the population of social media is increasing day by day. In 2020, around 3.81 billion users are active on social media in all over the world. Around 29% population of India are using social media [3]. Internet along with the latest technologies transforming our lives. For example, people are moving from offline games to online games. These online games help many people develop leadership skills, complex problem-solving skills, and deal with unexpected situations. The researcher found that playing online games excessively can increase anxiety, depression, and physical damages in players. Even it can increase blood pressure and increase heartbeats due to too much stress and excitement [4]. Online game addiction (OGA) is a genuine problem for many people. There are two types of online video games, single-player and multi-player. OGA disorder has severe consequences for the people who are suffering from it. Sometimes its symptoms and signs are tough to recognize. Single-player games are less prone to addiction because there is a clear mission or goal related to beating a high score or completing the task. But multi-player online video games users are more prone to addiction because it is played with many players, and generally, they have no end.

Sometimes some actions are taken by the government and other bodies by using that we can predict people' addictive behaviour. For example, the Government of India (GoI) banned 118 mobile apps concerned with users' privacy issues [2]. PUBG is one of the apps that is forbidden by GoI. When these things happen in our society, people post their views on the Internet using social media and blogs. In this paper, we use people these views, which are posted on Twitter. Twitter is among the most trendy social media apps and has the largest psychological database. So, we collected people's tweets regarding the PUBG ban. We did their sentiment analysis. Then we will apply various classification algorithms.

Using sentiment analysis, we extracted the people's perception towards a specific brand, scheme, issue, etc. from text data. It can be used in product review analysis, brand monitoring, and policy framing [5]. In this research article, sentiment exploration of tweets related to PUBG ban is performed. Emotions, actions, and cognition are essential factors of human behaviour. In general, language how anybody interacts and acts also define his behaviour.

Figure 1 depicts sentiment analysis data, techniques and its application. It can be done on various types of textual data like blogs, social media, text in various languages. Sentiment analysis can be done by using NLP (natural language processing) and machine learning algorithms. It has various types of application domain, which we classify as business-oriented and human behaviour oriented [5]. Business-oriented application includes various sectors like finance, entertainment, security, society, travel, medical, and others. Human behaviour-oriented application includes analysis of interaction, influence, truth, language, behaviour, and emotions. So, in this paper, twitter data are collected and we work on behaviour-oriented application. We tried to find out online game-addicted behaviour of people [6].

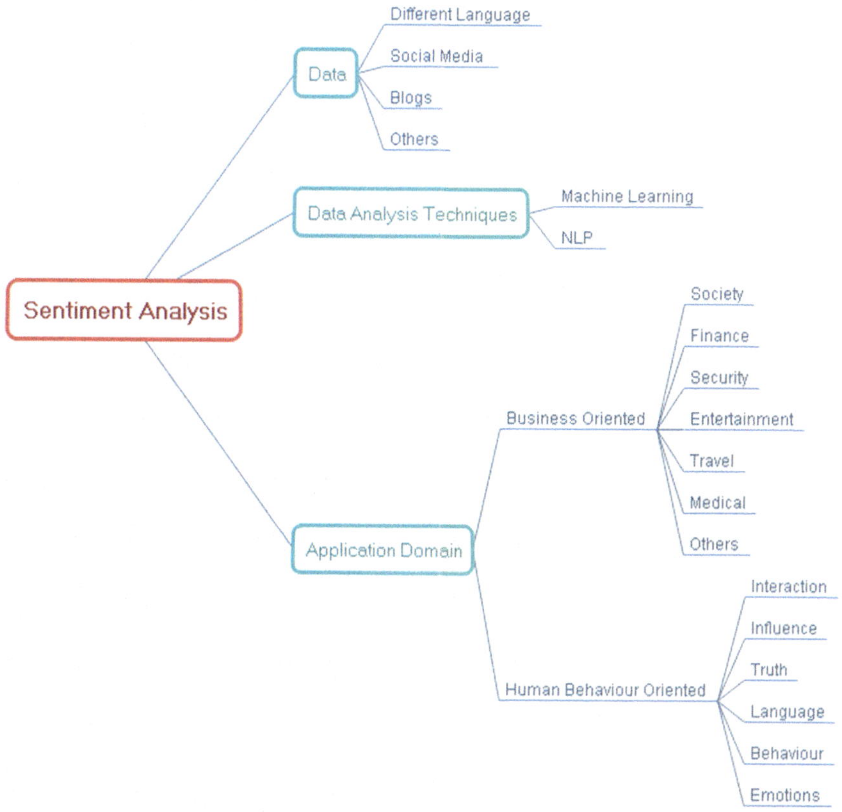

Fig. 1 Sentiment analysis data, analysis techniques, and application domain

2 Related Studies

Social media is like double-edge sward because there are many facilities like video, audio calls, sending and receiving multimedia files, messaging, etc. which are provided by social media help a lot to connect with our relatives or loved ones. On the other hand, it causes addiction that affects our life in the negative side, hate speeches on social media also create social disturbances [7]. Authors found nonlinear associations among social media addiction and neuroticism using multilayer perceptron [8]. With the advent of online multiplayer games, online gaming has become one of the utmost popular.

Online Gaming is the Leisure activity among online users worldwide. Prior studies have shown that extreme video gaming is linked with negative effects and can also lead to different problems [9, 10]. Various factors like individual factors, social interactions, psychological factors, etc. were related with online gaming addiction [11]. Researchers found out that the visual design, look of online gaming applications and their consequence will be enlarged as the domination increases. It is also noticed that game players want to run off from actuality and events that make them unhappy or stress in daily life [12].

3 Methodology and Results

Figure 2 depicts the workflow of the proposed system. Using tweepy, library tweets are collected and then useful features were selected. After that data scrubbing is done like retweet symbols, removing URLs and hashtags. Tweets sentiments are labelled using SentiWordNet (SWN) lexicon. After that various operations like Stop words identification and removal, stemming and lemmatizing are carried out on collected data. In the next step, data splitting and its mathematical representation are done using Count Vectorizer (CV) and Tfidf Vectorizer (TfV). Then vectors extracted using CV and TfV are passed through various Machine Learning (ML) classification algorithms. In the end, models are trained using DBOW, DMM, DMC doc2vec models, and the resultant vectors acquired using these models are used for classification. From all of the classifiers, logistic regression gives the best result, which is later evaluated using the test data and we got 78.92% accuracy.

3.1 Sentiment Analysis Approaches

Sentiment analysis can be carried out with machine learning (ML) and lexicon-based approach (LBA). Lexicon-based method basically calculates the polarity of tweets using dictionary words with their respective semantic scores allocated to them. Each word sentiment score is calculated and then all scores of sentence are added. If we got a positive score that means the tweet is positive and if we got negative score it infers as negative. Like that all tweets are labelled as positive, negative, and neutral if we get sentence sentiment score zero. We use SentiWordNet (SWN) lexicon approach to label tweets. After pre-processing, the data machine learning approach can be applied to tackles text classification problem. In this paper, for pre-processing, we use SentiWordNet then we apply machine learning classification models.

3.2 Collecting Data

Now there is a need of textual data, which convey people's views or sentiment related to PUBG ban. Textual data can be downloaded using various social media APIs like Facebook, Reddit, Twitter, etc. and from web pages. We use Twitter Tweepy API to gather tweets related to PUBG ban, as Twitter is considered as a large psychological database [13]. Tweepy Library used for Tweets collection for various keywords and hashes like #PUBG, #PUBGBAN, #PUBG_BAN,

#PUBGBANNED, #PUBG_BANNED, PUBG, PUBG_BAN, PUBGBAN, PUBG-BANNED, PUBG_BANNED.

3.3 Data Cleaning

Various insignificant objects like—indices, metadata, screen_name exist in collected data. We keep objects like—full_text(tweet full text), created_at(tweet time and date), retweet_count, follower_count, favorite_count, coordinates, geo(geographic location), created_at(tweet time and date), place, id_str(tweet id), screen_name. We keep the above objects in our data and remove the rest. To tackle with emojis, we encode the data file into unicode_escape. As data are collected using different hashtags and keywords so there are duplicate tweets are there in the data. So, duplicate tweets and unnecessary columns were removed.

In the tweet text, there are unnecessary symbols like #, @, URL, punctuations and numbers which is not important for our analysis. So, all of these symbols are replaced with whitespace. Only # symbol is removed not complete hashtag, because it may contain some sentiment information. After cleaning, there are total 4455 tweets we have for further processing.

3.4 POS Tagging and Sentiment Labelling

In ML algorithms, tweet 'full_text' acts as a predictor variable. Sentiment score is target variable for our data, which is calculated using SentiWordNet (SWN). SWN is huge corpus of Part-of-Speech (POS)-tagged English words with their sentiment score, which is explicitly used for opinion mining and sentiment classification applications. It gives positive and negative scores for each word. Then these scores of tweets are added, if its value is greater than 0 then it is labelled as positive tweet and if it is less than 0, tweet is labelled as negative otherwise it is considered as neutral tweet. Before using SWN, POS tagging of tweets was done using NLTK library.

Then data encoding is done for negative tweets -1 is used, for positive tweets 1 is used and for neutral tweets 0 is used. There are 1839 positive tweets, 1029 neutral tweets, and 1587 negative tweets as depicted in Fig. 3. TextBlob and Afinn were also used for sentiment allocation to tweets. In the next step, sentiment labelled textual data are prepared for machine learning classifiers.

Fig. 2 Workflow of the proposed system

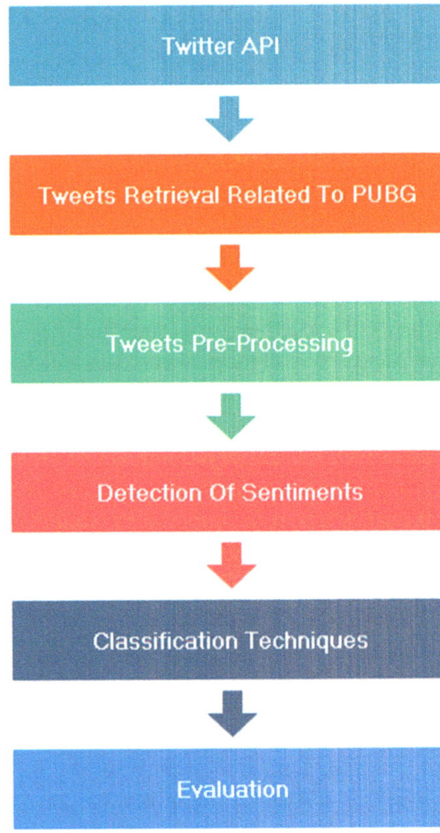

3.5 Removal of Stop Words

Stop words include frequently used words like 'for', 'the', of', etc., these words were removed using NLTK (Natural Language ToolKit) library [14]. Stop words raise the dimensionality of collected data. So, these stop words are removed.

Before applying ML algorithms on data, it is converted into vectors and vectors with similar meaning confined or reduced to their root word. For example, words with similar meaning like—'play', 'playing', 'played' convey same meaning reduced to word 'play'. This is achieved by using lemmatizing and stemming.

3.6 Lemmatizing and Stemming

Lemmatizer and stemming algorithms did same job of diminishing the words to their stems but the major difference is that lemmatizer keeps words linguistics in context.

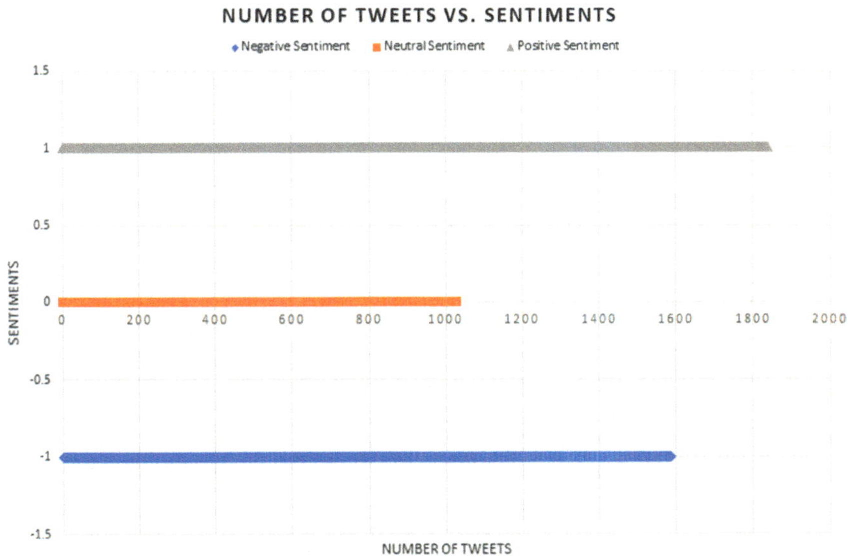

Fig. 3 Number of tweets versus sentiment score

For example, all of the above 'play'-related words will be reduced to 'play'. These processed texts are substituted in place of old ones. Stop word removal, stemming and lemmatizing are not applied before sentiment labelling because they can cause deformity in sentiment detection. All tweets are also converted to lowercase letters.

3.7 Splitting the Data

In this step, data were split into training, validation and test sets. The data are divided as follows—90% for training data, 5% for validation data, and 5% for test data.

3.8 Vectorizing Text of Tweets

Before implementing various ML classifiers, text data are converted into vectors. It is important because ML text classifiers expect data in mathematical form rather than textual form.

For vectorization, two vectorizers—Count and Tfidf (term frequency-inverse document frequency) Vectorizers are used. They are part of Bag-of-Words (BOW) models.

Count Vectorizer (CV) computes how frequently a word present in tweet and uses it as its weight. In TF-IDF, assigned weight of a token is calculated as product of word frequency and the inverse document frequency of word. Thus, as number of times a word appears in tweets, less IDF weight is assigned to that word in TD-IDF vectorizer. For frequency, TF-IDF gives fractional vales and CV gives integer values.

3.9 N-grams

N-grams is defined as combinations of adjacent words that we can discover in tweets. There is ngram_range parameter is used in CV and Tfidf for the calculation of n-grams. In ngram_range, uni-gram is denoted as (1, 1), bigram as (1, 2), trigram as (1, 3), and so on. CV and Tfidf are implemented using uni, bi, and trigrams and their respective accuracy score is noted. After vectorizing of tweets, classification algorithms were implemented.

3.10 Machine Learning Classifiers Implementation

The following classifiers are implemented—linear models (Perceptron, RidgeClassifier, LogisticRegression, PassiveAggressiveClassifier), Naive Bayes models (MultinomialNB, BernoulliNB), Ensemble models (AdaBoostClassifier, RandomForest classifier) and SVM model (LinearSVC). Accuracy score is used for the performance measurement of these models (Table 1).

In the above table, 'cv_1' means CountVectorizer with unigrams and 'tf_1' mean Term Frequency-Inverse Document Frequency Vectorizer with unigrams. Both LogisticRegression (Tfidf-unigrams) and Perceptron (Tfidf-bigrams) give the maximum accuracy of 85.6502 and 84.3043%. But LogisticRegression takes only 13.5 s for training and perceptron takes 6.97 s. It is very important to note that the used dataset is disproportion—as positive (+ve) and negative (−ve) tweets are more than neutral tweets. That is why; there are chances that accuracy scores can take us wrong conclusion. So, to avoid that, Random Forest Classifier is used with balanced class weights. It performs better with 74.51% average accuracy score and 77.58% with maximum accuracy.

3.11 Gensim Model Implementation

CV and Tfidf bag-of-words models used earlier are simple and not able to preserve the relationship between words. So, to preserve the relationship among words, sophisticated models are used because they give better representation of words mathematically. Doc2Vec that is an extension of Word2Vec models that can do it.

Doc2Vec algorithms, DBOW (Distributed Bag of Words), DMM (Distributed Memory Mean), DMC (Distributed Memory Concatenated), and their combinations

Table 1 All classifiers results using CV, TF-IDF and with Uni, Bi, and Tri-grams

Sr. No	Vec_Gram	Classifier	Accuracy	Time (in Seconds)
1	cv_1	MultinomialNB()	0.627802691	3.167022943
2	cv_1	BernoulliNB()	0.704035874	3.204078436
3	cv_1	LogisticRegression()	0.753363229	10.97595263
4	cv_1	LinearSVC()	0.820627803	3.942568541
5	cv_1	AdaBoostClassifier()	0.64573991	13.95540357
6	cv_1	RidgeClassifier()	0.784753363	3.68473196
7	cv_1	PassiveAggressiveClassifier()	0.802690583	3.558809042
8	cv_1	Perceptron()	0.811659193	3.21301651
9	cv_1	RandomForest Classifier	0.748878924	48.5113616
10	cv_2	MultinomialNB()	0.69955157	6.343557596
11	cv_2	BernoulliNB()	0.708520179	6.532910824
12	cv_2	LogisticRegression()	0.757847534	28.85697794
13	cv_2	LinearSVC()	0.838565022	7.767507315
14	cv_2	AdaBoostClassifier()	0.65470852	25.09738135
15	cv_2	RidgeClassifier()	0.811659193	7.50293088
16	cv_2	PassiveAggressiveClassifier()	0.834080717	6.816093206
17	cv_2	Perceptron()	0.834080717	6.902798653
18	cv_2	RandomForest Classifier	0.757847534	91.86163807
19	cv_3	MultinomialNB()	0.708520179	10.00783253
20	cv_3	BernoulliNB()	0.64573991	10.43630242
21	cv_3	LogisticRegression()	0.748878924	55.03128695
22	cv_3	LinearSVC()	0.811659193	12.32500815
23	cv_3	AdaBoostClassifier()	0.641255605	36.22467995
24	cv_3	RidgeClassifier()	0.807174888	10.97327328
25	cv_3	PassiveAggressiveClassifier()	0.820627803	10.03281617
26	cv_3	Perceptron()	0.834080717	9.599086523
27	cv_3	RandomForest Classifier	0.69058296	128.3715734
28	tf_1	MultinomialNB()	0.677130045	2.928195238
29	tf_1	BernoulliNB()	0.704035874	3.070109606
30	tf_1	LogisticRegression()	0.856502242	13.49957085
31	tf_1	LinearSVC()	0.829596413	11.94485855
32	tf_1	AdaBoostClassifier()	0.659192825	11.20278502
33	tf_1	RidgeClassifier()	0.775784753	5.468579054
34	tf_1	PassiveAggressiveClassifier()	0.834080717	3.859255791
35	tf_1	Perceptron()	0.820627803	3.515517473
36	tf_1	RandomForest Classifier	0.775784753	55.1449635

(continued)

Table 1 (continued)

Sr. No	Vec_Gram	Classifier	Accuracy	Time (in Seconds)
37	tf_2	MultinomialNB()	0.766816143	6.122223377
38	tf_2	BernoulliNB()	0.708520179	6.300122499
39	tf_2	LogisticRegression()	0.834080717	37.7722218
40	tf_2	LinearSVC()	0.829596413	18.90667868
41	tf_2	AdaBoostClassifier()	0.641255605	20.92410684
42	tf_2	RidgeClassifier()	0.798206278	10.88392544
43	tf_2	PassiveAggressiveClassifier()	0.829596413	7.236640453
44	tf_2	Perceptron()	0.843049327	6.969199657
45	tf_2	RandomForest Classifier	0.762331839	104.0579317
46	tf_3	MultinomialNB()	0.766816143	9.508183956
47	tf_3	BernoulliNB()	0.64573991	9.382218838
48	tf_3	LogisticRegression()	0.834080717	75.00797677
49	tf_3	LinearSVC()	0.807174888	30.11644745
50	tf_3	AdaBoostClassifier()	0.64573991	31.33968806
51	tf_3	RidgeClassifier()	0.780269058	16.65845656
52	tf_3	PassiveAggressiveClassifier()	0.825112108	10.66542888
53	tf_3	Perceptron()	0.825112108	9.845933914
54	tf_3	RandomForest Classifier	0.735426009	143.4576552

are used from Gensim library of python. Gensim is used to automatically dig out semantic topics effectively from documents. These models are used to train all classifier, which we did earlier. The below table depicts results of all Gensim models with classifiers (Table 2).

LinearSVC (DBOW + DMM) gives the maximum accuracy of 63.67%. If training time is considered then logistic regression (DBOW + DMM) is the optimum one among all (with 0.76 s). Logistic regression (DBOW + DMM) also depicts fine performance with 63.22% accuracy after 12.05 s for training. DBOW + DMM model surpasses the previous models with a mean accuracy of 55.47%. The Gensim models demonstrate slightly lesser performance with respect to CV and Tfidf models. Among all the models, LogisticRegression (Tfidf-unigrams) best performs with an accuracy of 85.6502%. So it is evaluated using test data and it gives an accuracy of 78.92%.

4 Conclusion and Future Work

Now these days, digital world is an important part of our life. People posts their views on social media, which is part of digital world about various events happen around us. People normally play online games to get relaxed from stress and entertainment.

Table 2 Results using Gensim models

Sr. No	Model	Classifier	Accuracy	Time
1	DBOW	BernoulliNB()	0.479820628	0.190983295
2	DBOW	LogisticRegression()	0.596412556	0.510683775
3	DBOW	LinearSVC()	0.600896861	3.904580593
4	DBOW	AdaBoostClassifier()	0.470852018	4.326505184
5	DBOW	RidgeClassifier()	0.596412556	0.299863577
6	DBOW	PassiveAggressiveClassifier()	0.421524664	0.107691288
7	DBOW	Perceptron()	0.448430493	0.053966045
8	DMC	BernoulliNB()	0.322869955	0.017988443
9	DMC	LogisticRegression()	0.372197309	0.237054586
10	DMC	LinearSVC()	0.421524664	0.311802387
11	DMC	AdaBoostClassifier()	0.439461883	4.646183014
12	DMC	RidgeClassifier()	0.403587444	0.024474621
13	DMC	PassiveAggressiveClassifier()	0.376681614	0.087949038
14	DMC	Perceptron()	0.399103139	0.064957619
15	DMM	BernoulliNB()	0.439461883	0.019988775
16	DMM	LogisticRegression()	0.488789238	0.510682583
17	DMM	LinearSVC()	0.506726457	6.2501719
18	DMM	AdaBoostClassifier()	0.457399103	4.606146336
19	DMM	RidgeClassifier()	0.475336323	0.033673763
20	DMM	PassiveAggressiveClassifier()	0.349775785	0.133915424
21	DMM	Perceptron()	0.439461883	0.066961765
22	DBOW + DMM	BernoulliNB()	0.466367713	0.036981106
23	DBOW + DMM	LogisticRegression()	0.632286996	0.76252079
24	DBOW + DMM	LinearSVC()	0.6367713	12.05359554
25	DBOW + DMM	AdaBoostClassifier()	0.533632287	10.45952153
26	DBOW + DMM	RidgeClassifier()	0.605381166	0.045982838
27	DBOW + DMM	PassiveAggressiveClassifier()	0.470852018	0.312793732
28	DBOW + DMM	Perceptron()	0.538116592	0.105933428
29	DBOW + DMC	BernoulliNB()	0.408071749	0.037977695
30	DBOW + DMC	LogisticRegression()	0.587443946	0.828485966
31	DBOW + DMC	LinearSVC()	0.600896861	6.187167406
32	DBOW + DMC	AdaBoostClassifier()	0.547085202	9.44914484
33	DBOW + DMC	RidgeClassifier()	0.596412556	0.058964014
34	DBOW + DMC	PassiveAggressiveClassifier()	0.470852018	0.193883419
35	DBOW + DMC	Perceptron()	0.479820628	0.110931635

According to some research, people get addicted to these online games and sometimes even they do not know it. When some events stop them to play these online games, they post their views on social media platforms. In this paper, we collected tweets of one social event or incident of banning PUBG gaming application. Then we did sentiment labelling, we classify tweets into positive, negative, and neutral tweets. The people who posted negative tweets on PUBG ban. That means most of the peoples whose sentiments are negative, they are addicted to that online gaming application. After banning PUBG game that people are not able to use that game, so they are posting negative comments about it. In future, we can collect more textual data from other apps like Reddit, Facebook, and Blogs regarding events like PUBG ban, and we can improve our classification algorithm accuracy by applying deep learning and other models.

References

1. Top 10 Online Multiplayer Games | Game Development | Melior Games. https://meliorgames.com/game-development/top-10-online-multiplayer-games/. Accessed 20 October 2020
2. India bans 118 more Chinese mobile apps including PUBG Mobile over privacy concerns—The Financial Express. https://www.financialexpress.com/industry/technology/india-bans-118-more-chinese-mobile-apps-including-pubg-mobile-over-privacy-concerns/2072830/. Accessed 20 October 2020
3. How Many People Use Social Media in 2020? (65+ Statistics). https://backlinko.com/social-media-users. Accessed 20 October 2020
4. Zamani E, Chashmi M, Hedayati N (2009) Effect of addiction to computer games on physical and mental health of female and male students of guidance school in city of isfahan. Addict Heal 1:98–104
5. Mäntylä MV, Graziotin D, Kuutila M (2018) The evolution of sentiment analysis—a review of research topics, venues, and top cited papers. Comput Sci Rev 27:16–32. https://doi.org/10.1016/j.cosrev.2017.10.002
6. Dwyer R, Fraser S (2016) Addicting via hashtags: how is twitter making addiction? Contemp Drug Probl 43:79–97. https://doi.org/10.1177/0091450916637468
7. Moqbel M, Kock N (2018) Unveiling the dark side of social networking sites: personal and work-related consequences of social networking site addiction. Inf Manag 55:109–119. https://doi.org/10.1016/j.im.2017.05.001
8. Leong LY, Hew TS, Ooi KB, Lee VH, Hew JJ (2019) A hybrid SEM-neural network analysis of social media addiction. Expert Syst Appl 133:296–316. https://doi.org/10.1016/j.eswa.2019.05.024
9. Kuss DJ, Griffiths MD (2012) Internet gaming addiction: a systematic review of empirical research. Int J Ment Health Addict 10:278–296. https://doi.org/10.1007/s11469-011-9318-5
10. King DL, Haagsma MC, Delfabbro PH, Gradisar M, Griffiths MD (2013) Toward a consensus definition of pathological video-gaming: a systematic review of psychometric assessment tools. Clin Psychol Rev 33:331–342. https://doi.org/10.1016/j.cpr.2013.01.002
11. Hyun GJ, Han DH, Lee YS, Kang KD, Yoo SK, Chung US, Renshaw PF (2015) Risk factors associated with online game addiction: A hierarchical model. Comput Hum Behav 48:706–713. https://doi.org/10.1016/j.chb.2015.02.008
12. Arı E, Yılmaz V, Elmastas Dikec B (2020) An extensive structural model proposal to explain online gaming behaviors. Entertain Comput 34:100340. https://doi.org/10.1016/j.entcom.2020.100340

13. Antonakaki D, Fragopoulou P, Ioannidis S (2021) A survey of twitter research: data model, graph structure, sentiment analysis and attacks. Expert Syst Appl 164:114006. https://doi.org/10.1016/j.eswa.2020.114006
14. Natural Language Toolkit—NLTK 3.5 documentation. https://www.nltk.org/. Accessed 20 October 2020

Exploring the Strengths of Neural Codes for Video Retrieval

Vidit Kumar, Vikas Tripathi, and Bhaskar Pant

Abstract Websites like YouTube, Facebook, Twitter, etc. encounter large amounts of videos every day, mostly uploaded from mobile devices, digital cameras, etc. These videos rarely have metadata (semantic tags) attached, without which it is very difficult to retrieve similar videos without using content-based search techniques. More recently, two-dimensional convolutional networks (2d-CNN) have shown breakthrough performance over hand-engineered methods on image-related tasks in all aspects of computer vision field. The video is also composed of 2D frames arranged along time dimension, which can also be processed by 2d-CNN. In this paper, we investigate the significance of activations of CNN layers for video representation and analyzed its performance on the basis of nearest the neighbor search task, i.e. video retrieval. Three well-known CNN networks (AlexNet, GoogleNet and ResNet18) are exploited for feature extraction, and UCF101 dataset is chosen to conduct the experiment. The results showed that feature fusion of multiple CNN layers can strengthen the video representation.

Keywords CNN · Content-based search · Deep learning · Video feature extraction · Video retrieval

1 Introduction

With the availability of cheap devices such as digital cameras, smartphones, etc., video has become an essential part of the multimedia communication environment. As a result of these advances in technology, we are seeing a sudden increase in videos with or without semantic tags on social networking sites. According to YouTube statistics, approximately 200 hours of video content is uploaded to YouTube every minute and approximately 11 million videos are posted to Twitter every day without bad text or tags. As online videos without semantic tags are on the rise in popularity, robust content-based video analysis techniques are demanding. With content-based

V. Kumar (✉) · V. Tripathi · B. Pant
Graphic Era Deemed to be University, Dehradun, India

A. Tomar et al. (eds.), *Machine Learning, Advances in Computing, Renewable Energy and Communication*, Lecture Notes in Electrical Engineering 768,
https://doi.org/10.1007/978-981-16-2354-7_46

video retrieval (CBVR) as a technology, it opens and provides solutions to applications like in-video advertising, content filtering, video navigation, video indexing and video surveillance. In-video advertising, the goal is to retrieve target videos that are similar and suitable to include advertisement between it. In content filtering, unappropriated activity is excluded, which can also be solved by retrieving unappropriated videos to an unappropriated query video.

Significant research progress has been made over the last decades in image retrieval [1] including fine-grained search [2], but CBVR has received insufficient attention in the multimedia community compared with image retrieval domain. Traditional search techniques are difficult to process large-scale database videos due to the high cost of computing. Lots of efforts have been applied in this field. In [3], a video content indexing by objects is presented. In their approach, moving object is detected in wavelet domain by a combination of morphological color segmentation at a lower scale with global motion estimation. Then histograms of wavelet coefficients of objects at multi-scale are computed and matched with database for retrieval of similar videos. The limitation is that the system is dependent on how much object segmentation is accurate, and technique is needed to exploit temporal dynamics even if the objects are roughly segmented.

In recent, CNN shows tremendous success in the field of computer vision, especially for the tasks like image classification, object detection, segmentation and image retrieval. This progress also led to video retrieval problem. Like, Lou et al. [4] propose compact and discriminative CNNs descriptor for video retrieval. Limitation is that they do not consider the relationship between feature maps of CNN, which can be incorporated to compute temporal features. Podlesnaya et al. [5] use CNN features for video clip representation. Its limitation is that the size of feature vectors causes cost complexity while matching videos. The feature vector dimensionality can be reduced in order to search in log time scale. Kumar et al. [6] deal the problem of movie scene retrieval with CNN and LSTM. There are also works done in the scope of hashing-based video search like [7–9]. All these works learn a new subspace in binary (hash) domain where similar videos are closer and dissimilar videos are far away. For instance, [8] proposes a deep auto encoder–decoder framework utilizing two-layered hierarchical LSTM to learn binary codes. Kumar et al. [9] also exploit CNN with lstm for video retrieval problem. Moreover, several recent research studies are performed by researchers using AI/ML approaches [10–12].

In this paper, we investigate the significance of the 2d-CNN's middle and higher layer's features for video representation. First, we conduct systematic assessment of the performance of features from different layers of CNN in video retrieval tasks. Then, we find which features fusion combination can boost the performance.

2 Materials and Method

2.1 CNN Architecture

CNN can be considered as an extension of the multi-layer perceptron (MLP) that exploits the rich 2D spatial structure of image that MLP fails to do, where initial layers (convolutional) are responsible for sensing spatial relationship within nearby pixels and the final layers are responsible for generating lower dimensional representation with higher level abstraction of image. The general CNN network looks as in Fig. 1. Once the network is trained with a sufficiently large dataset (proportional to the number of network's parameters), then each layer extracts the rich information present in the image in a hierarchical manner. The early layers extract the low-level image properties like edges, objects contour. Middle layers extract the shape, color, texture, and higher layers extract features responsible for global level abstraction like face, month, nose, etc.

Three types of CNN architectures are used in this paper: AlexNet [13], GoogleNet [14] and ResNet18 [15]. Tables 1, 2 and 3 show the respective CNN's layers name and its output's sizes. Moreover, reader may refer [13–15] for detailed information on the implementation of CNN.

AlexNet: This CNN consists of five convolutional layers and three dense layers. It achieves first position in ILSVRC 2012. It takes 227 × 227 × 3 RGB image as an input and passes it through all intermediate layers to output final class score. Due to the dense layers at the end, the network makes over 61 M parameters. *GoogleNet*: This CNN is deeper compared with AlexNet and introduces a concept of inception block and achieves first position in ILSVRC 2014. Each inception module consists of multiple convolutions of kernel sizes 1 × 1, 3 × 3 and 5 × 5. The 1 × 1 convolutional layers in the middle are for the dimensionality reduction of the feature space. In total, nine inception modules are connected sequentially. More info can be found in [14].

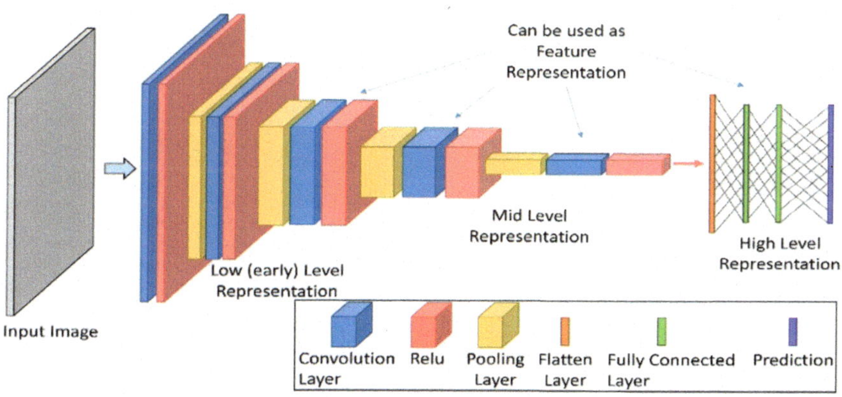

Fig. 1 General CNN architecture

Table 1 AlexNet architecture

Layer	Output size
Conv1	$55 \times 55 \times 96$
Pool1 (max)	$27 \times 27 \times 96$
Conv2	$27 \times 27 \times 256$
Pool2 (max)	$13 \times 13 \times 256$
Conv3	$13 \times 13 \times 384$
Conv4	$13 \times 13 \times 384$
Conv5	$13 \times 13 \times 256$
Pool5 (max)	$6 \times 6 \times 256$
F6	4096
F7	4096
F8	1000

Table 2 ResNet18 architecture

Layer	Output size
Conv1	$112 \times 112 \times 64$
Conv2_x	$56 \times 56 \times 64$
Conv3_x	$28 \times 28 \times 128$
Conv4_x	$14 \times 14 \times 256$
Conv5_x	$7 \times 7 \times 512$
Pool5 (Avg)	$1 \times 1 \times 512$
Fc	1000

ResNet18: This is another CNN that introduces the residual connection by which it solves the problem of vanishing gradient in training deeper CNN. With the inclusion of residual connection in CNN, it provides the shortcut to the gradients so that it can easily reach the input without vanishing that much as in without residual case. It is the winner of ILSVRC 2015. In this model, there are four residual blocks of length {2, 2, 2, 2}. For more info, refer [15].

2.2 Feature Extraction

To represent the video frame, activations from the particular layer of CNN can be extracted. Following the findings of [16] and [17], we choose the last two convolutional and fully connected layers as feature representation (see in Tables 1, 2 and 3, bolded font ones used as descriptors). Let f_{CNN} be the feature transformation function that maps the $R^{m \times n}$video frame ($m \times n$ is resolution of video) to $R^{u \times v}$($u \times v$ is size of feature map) feature space. Given a set of T consecutive frames of ith clip sampled from the nth video, the feature vector for ith clip for a particular Lth

Table 3 GoogleNet architecture

Layer	Output size
Conv1	112 × 112 × 64
Pool1 (max)	56 × 56 × 64
Conv2	56 × 56 × 192
Pool2 (max)	28 × 28 × 192
Inception 3a	28 × 28 × 256
Inception 3b	28 × 28 × 480
Pool3 (max)	14 × 14 × 480
Inception 4a	14 × 14 × 512
Inception 4b	14 × 14 × 512
Inception 4c	14 × 14 × 512
Inception 4d	14 × 14 × 528
Inception 4e	14 × 14 × 832
Pool4 (max)	7 × 7 × 832
Inception 5a	7 × 7 × 832
Inception 5b	7 × 7 × 1024
Pool5 (avg)	1 × 1 × 1024
Fc	1000

layer is denoted as $CFmax_{ni}^{L}$and $CFmean_{ni}^{L}$, which is computed as:

$$CFmax_{ni}^{L} = \max\left(f_{CNN}^{L}\left(V_{ni}^{(1:T)}\right)\right) \tag{1}$$

$$CFmean_{ni}^{L} = \sum_{t} f_{CNN}^{L}\left(V_{ni}^{(t)}\right) \Big/ T \tag{2}$$

where, $CFmax$ and $CFmean$represent the features associated with max and mean pooling over temporal dimension.

All clip level features are averaged to generate the descriptor at video level.

3 Experimental Settings

3.1 Dataset and Setting

We conduct the experiments on UCF-101 dataset [18], which consists of 13 k videos from 101 categories. The standard train/test split 1 of the dataset is used. Retrieval is done by assuming the videos of the testing set as queries and training

Table 4 Spatial pooling strategy in different layers

Network (layer)	Pooling kernel/(stride)	Output feature dimension
AlexNet (Conv4 and Conv5)	3 × 3/(2, 2)	13,824, 9216
GoogleNet (Inception 4e)	5 × 5/(3, 3)	13,312
GoogleNet (Inception 5a)	3 × 3/(2, 2)	7488
ResNet18 (Conv4b)	5 × 5/(3, 3)	4096
ResNet18 (Conv5b)	7 × 7/(1, 1)	512

videos as retrieval set. We adopted the standard mean average precision (mAP@k) for evaluation purposes. Matlab 2019b and tesla k40 GPU are employed for all experiments.

3.2 *Implementation*

First, 10 clips per video evenly sampled from each video, then following [13] each clip undergoes through spatial center cropping of network's input size. All the networks are pretrained on imagenet and are not trained on video dataset, which confirms the experiments are conducted under unsupervised settings. Features are extracted as discussed in the Sect. 2.2 and we choose $T = 16$ frames per clip. Convolutional features costs in higher dimensionality, so we applying spatial average pooling (see Table 4 for filter size) to extract lower dimensional features from it. For matching the video clips, the cosine distance is adopted.

4 Results

In this section, we first explore the effectiveness of individual features, then we see the usefulness of fusion of these features.

4.1 *Effectiveness of Different Layer's Features*

In the following, we inspect each network's performance.

Table 5 Neural codes of AlexNet and its performance analysis on basis of mAP@k

Descriptor	Spatio-temporal pooling									
	Max pooling					Average pooling				
	k = 1	k = 5	k = 10	k = 20	k = 50	k = 1	k = 5	k = 10	k = 20	k = 50
Conv4	**36.33**	**23.02**	**19.25**	**12.17**	**9.81**	32.13	20.42	17.12	11.03	06.76
Conv5	**43.59**	**34.92**	28.21	**22.29**	14.37	43.33	34.02	**28.25**	22.17	**14.81**
Fc6	**53.95**	**46.04**	**39.92**	**32.54**	**22.92**	50.04	41.52	35.69	28.58	19.70
Fc7	**53.30**	**45.55**	**39.66**	**32.59**	**22.67**	52.13	43.61	37.65	30.84	21.53

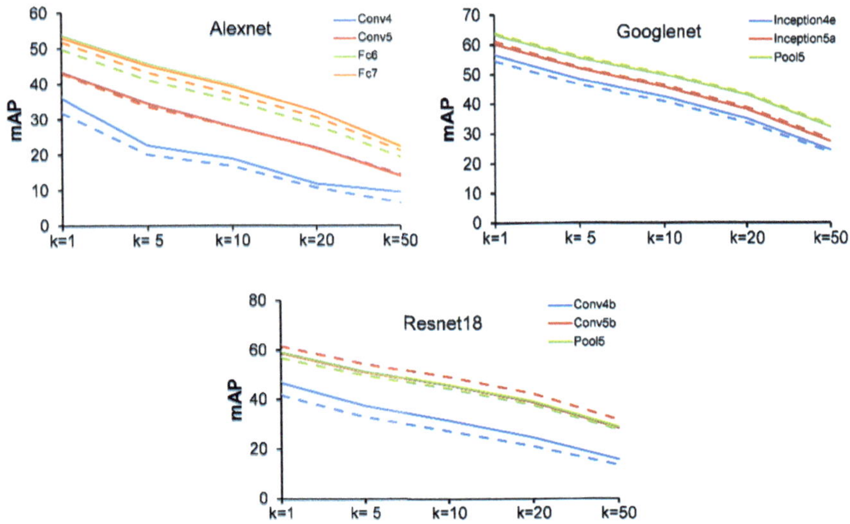

Fig. 2 mAP of different layers of three different networks; dotted line denotes performance under temporal mean pooling otherwise max pooling

Experiment using AlexNet

In Table 5, we can see that higher layers (fc6 and fc7) outperform the middle layers (conv4 and conv5), the reason being that the higher layer captures rich global level distinctive features with high level abstraction compared with middle layers. Using temporal max pooling, fc6 (*53.95 mAP@1*) performs slightly better than fc7 (*53.30 mAP@1*). The reason seems to that the last fully connected layer has class-specific features that are generalized to only seen classes, but fc6 is better to generalize to unseen classes. Also, we can observe temporal max pooling performs better than temporal mean pooling (see Fig. 2).

Experiment using GoogLenet

In case of the last two layers of GoogleNet, temporal mean pooling performs better than max pooling as reported in Table 6. But for the second last inception block Inception4e, temporal max pooling performs better. Contrary to AlexNet, last layer (pool5) outperforms others.

Experiment using ResNet18

With similar findings in the above two networks, max pooling performs better in the second last residual block conv4b a.k.a res4b, and also in the case of pool5 but not true for the con5b. Using mean pooling, conv5b outperforms others (see Table 7 and Fig. 2).

Table 6 Neural codes of GoogleNet and its performance analysis on basis of mAP@k

Descriptor	Spatio-temporal pooling									
	Max pooling					Average pooling				
	k = 1	k = 5	k = 10	k = 20	k = 50	k = 1	k = 5	k = 10	k = 20	k = 50
Inception4e	**56.75**	**48.63**	**42.73**	**35.34**	**24.84**	54.64	46.82	41.09	33.94	24.02
Inception5a	60.24	52.19	45.90	38.33	27.52	**61.12**	**52.47**	**46.48**	**38.97**	**28.26**
Pool5	63.44	55.68	50.11	43.29	32.49	**63.92**	**56.29**	**50.49**	**43.80**	**33.18**

Table 7 Neural codes of ResNet18 and its performance analysis on basis of mAP@k

Descriptor	Spatio-temporal pooling									
	Max pooling					Average pooling				
	k = 1	k = 5	k = 10	k = 20	k = 50	k = 1	k = 5	k = 10	k = 20	k = 50
Conv4b	**47.11**	**37.82**	**31.71**	**24.83**	**16.06**	42.08	33.31	27.45	21.27	13.79
Conv5b	59.32	51.52	45.79	38.93	28.55	**62.01**	**54.72**	**49.13**	**42.35**	**31.87**
Pool5	**59.58**	**51.78**	**45.94**	**39.24**	**29.10**	57.28	50.07	44.43	37.79	27.87

4.2 *Influence of Fusion of Multiple layer's Features*

With the above finding, we wish to investigate the significance of fusion of different features in context of video search. In this experiment, we use the CNN layers with best performed temporal pooling for fusion (denoted as subscript in Table 8). We also use two handcrafted features: LBP [19] and HOG [20] for sake of comparison. Both LBP and HOG features are computed for each clip's frame (grayscale frame) and then averaged across all clips of the video to generate a video descriptor. For fusion of CNN's activations, first, we apply L2 norm on individual features and then fusion (concatenate) of features.

Table 8 reports the mAP@k on a different combination of features. We can observe that deep leaning features easily outperform the handcrafted ones with large margin. We can also see the fusion of either combination of layers does not improve performance as much as compared to the best standalone layer feature. For example, in case of GoogleNet, $Pool5_{mean}$'s performance is higher than its fusion with other lower layers feature. The reason seems to be that the higher layer captures the essential compact information from preceding layer, by fusing the lower layer with higher

Table 8 Comparison of mAP@k of different layers fusion strategies

Descriptor		k = 1	k = 5	k = 10	k = 20	k = 50
LBP		21.97	15.27	10.45	08.28	2.97
HOG		30.12	23.29	18.59	13.51	8.59
AlexNet	$Conv5_{max}$	43.59	34.92	28.21	22.29	14.37
	$\mathbf{Fc6_{max}}$	**53.95**	**46.04**	**39.92**	**32.54**	**22.92**
	$Fc7_{max}$	53.30	45.55	39.66	32.59	22.67
	$Conv5_{max} + Fc6_{max}$	49.14	44.88	35.11	28.14	18.15
	$Fc6_{max} + Fc7_{max}$	53.69	45.72	39.78	32.58	22.78
	$Conv5_{max} + Fc6_{max} + Fc7_{max}$	53.25	45.24	39.13	31.55	22.32
GoogleNet	$Inception4e_{max}$	56.75	48.63	42.73	35.34	24.84
	$Inception5a_{mean}$	61.12	52.47	46.48	38.97	28.26
	$\mathbf{Pool5_{mean}}$	**63.92**	**56.29**	**50.49**	**43.80**	**33.18**
	$Incep.4e_{max} + Incep.5a_{mean}$	57.15	49.77	43.49	36.67	25.47
	$Incep.5a_{mean} + Pool5_{mean}$	62.21	54.18	48.24	40.19	29.17
	$Incep.4e_{max} + Incep.5a_{mean} + Pool5_{mean}$	59.80	50.25	46.68	37.24	26.96
ResNet18	$Conv4b_{max}$	47.11	37.82	31.71	24.83	16.06
	$\mathbf{Conv5b_{mean}}$	**62.01**	**54.72**	**49.13**	**42.35**	**31.87**
	$Pool5_{max}$	59.58	51.78	45.94	39.24	29.10
	$Conv4b_{max} + Conv5b_{mean}$	54.62	46.50	38.71	35.18	23.04
	$Conv5b_{mean} + Pool5_{max}$	61.21	53.03	47.12	40.15	30.12
	$Conv4b_{max} + Conv5b_{mean} + Pool5_{max}$	57.02	48.14	44.13	37.80	25.84

Table 9 Comparison of mAP@k of different networks fusion strategies

Fusion feature	k = 1	k = 5	k = 10	k = 20	k = 50
$Fc6_{max}$(AlexNet)	53.95	46.04	39.92	32.54	22.92
$Conv5b_{mean}$(ResNet18)	62.01	54.72	49.13	42.35	31.87
$Pool5_{mean}$(GoogleNet)	63.92	56.29	50.49	43.80	33.18
$Fc6_{max}$(AlexNet) + $Conv5b_{mean}$(ResNet18)	64.00	56.55	50.80	43.70	32.68
$Fc6_{max}$(AlexNet) + $Pool5_{mean}$(GoogleNet)	64.13	56.95	51.15	44.13	33.46
$Conv5b_{mean}$(ResNet18) + $Pool5_{mean}$(GoogleNet)	66.06	59.04	53.55	46.81	35.96
$Fc6_{max}$(AlexNet) + $Pool5_{mean}$(GoogleNet) + $Conv5b_{mean}$(ResNet18)	**66.69**	**59.19**	**53.72**	**46.85**	**35.99**

layer makes redundant feature (logically). Hence, the direct fusion of layers within same network is not feasible.

4.3 Effectiveness of Fusion of Different Network Features

Next, we wish to explore the influence of fusion of different network's features on nearest neighbor search. The results are reported in Table 9, where we can see that any combination of fusion performs better than standalone performing layer. This suggests that multi-model fusion works superior.

5 Conclusion

This paper analyzes and discusses the significance of different layer's features of network under the nearest neighbor search task. In particular, AlexNet, GoogleNet and ResNet18 are deployed to extract features to represent videos. We explored the effectiveness of each layer features and their fusion on the performance on video retrieval. Results suggest that direct fusion of middle level features with higher layer features of the same network architecture does not seem to boost the performance than standalone features. In the future, we will investigate how to tackle this issue. Results also suggest that on fusion of different networks features can boost the performance, but this also increases the memory requirement, time complexity etc. In addition, learning video representations require a large dataset that is labor-intensive. Future work will include to explore self-supervised learning approach as it is a promising direction to tackle the need for large-scale video datasets.

References

1. Zhou W, Li H, Tian Q (2017) Recent advance in content-based image retrieval: a literature survey. arXiv:1706.06064
2. Kumar V, Tripathi V, Pant B (2020) Content based fine-grained image retrieval using convolutional neural network. In:7th international conference on signal processing and integrated networks (SPIN), pp 1120–1125
3. Morand Cl, Benois-Pineau J, Domenger J-Ph, Zepeda J, Kijak E, Guillemot Ch (2010) Scalable object-based video retrieval in hd video databases. Sig Process Image Commun 25(6):450–465
4. Lou Y, Bai Y, Lin J, Wang S, Chen J, Chandrasekhar V, Duan L-Y, Huang T, Kot AC, Gao W (2017) Compact deep invariant descriptors for video retrieval. In: Data compression conference (DCC), pp 420–429
5. Podlesnaya A, Podlesnyy S (2016) Deep learning based semantic video indexing and retrieval. In: Proceedings of SAI intelligent systems conference, pp 359–372
6. Kumar V, Tripathi V, Pant B (2019) Content based movie scene retrieval using spatio-temporal features. Int J Eng Adv Technol 9(2):1492–1496
7. Hao Y, Mu T, Hong R, Wang M, An N, Goulermas JY (2016) Stochastic multiview hashing for large-scale near-duplicate video retrieval. IEEE Trans Multimed 19(1):1–14
8. Song J, Zhang H, Li X, Gao L, Wang M, Hong R (2018) Self-supervised video hashing with hierarchical binary auto-encoder. IEEE Trans Image Process 27(7):3210–3221
9. Kumar V, Tripathi V, Pant B (2019) Learning compact spatio-temporal features for fast content based video retrieval. Int J Innov Technol Explor Eng 9(2):2402–2409
10. Ozkan S, Akar GB (2021) Exploiting Local Indexing and Deep Feature Confidence Scores for Fast Image-to-Video Search. In: 2020 25th international conference on pattern recognition (ICPR). IEEE
11. Wu J, Ngo C-W (2020) Interpretable Embedding for Ad-Hoc Video Search. In: Proceedings of the 28th ACM international conference on multimedia. ACM. https://doi.org/10.1145/3394171.3413916
12. Nguyen P-A, Ngo C-W (2021) Interactive Search vs. Automatic Search. ACM Trans Multimedia Comput Commun Appl 17:1–24. https://doi.org/10.1145/3429457
13. Krizhevsky A, Sutskever I, Hinton GE (2012) Imagenet classification with deep convolutional neural networks. In: Advances in neural information processing systems, pp 1097–1105
14. Szegedy C, Liu W, Jia Y, Sermanet P, Reed S, Anguelov D, Erhan D, Vanhoucke V, Rabinovich A (2015) Going deeper with convolutions. In: Proceedings of the IEEE conference on computer vision and pattern recognition, pp 1–9
15. He K, Zhang X, Ren S, Sun J (2016) Deep residual learning for image recognition. In: Proceedings of the IEEE conference on computer vision and pattern recognition, pp 770–778
16. Babenko A, Slesarev A, Chigorin A, Lempitsky V (2014) Neural codes for image retrieval. In: European conference on computer vision, pp 584–599. Springer, Cham
17. Yu W, Yang K, Yao H, Sun X, Xu P (2017) Exploiting the complementary strengths of multi-layer CNN features for image retrieval. Neurocomputing 237:235–241
18. Soomro K, Zamir AR, Shah M (2012) UCF101: a dataset of 101 human actions classes from videos in the wild. arXiv:1212.0402
19. Ojala T, Pietikainen M, Maenpaa T (2002) Multiresolution gray-scale and rotation invariant texture classification with local binary patterns. IEEE Trans Pattern Anal Mach Intell 24(7):971–987
20. Dalal N, Triggs B (2005) Histograms of oriented gradients for human detection. In: IEEE computer society conference on computer vision and pattern recognition (CVPR '05), San Diego, CA, USA, pp 886–893

Security Issues and Application of Blockchain

Sandeep Saxena, Umesh Kumar Gupta, Renu, and Vimal Dwivedi

Abstract Blockchain is the most important IT invention in current technology. Blockchain is much secured, easily shared and works with various distributed ledgers. Blockchain is changing the working environment from centralized to decentralize. Blockchain has removed the problem of single point of failure (SPOF). In the blockchain, you can easily track all previous transactions and can validate future transactions. Due to various benefits in blockchain, the number of organizations works on blockchain technology for providing solutions and many others are working on security issues in blockchain like 51% vulnerability, private key security, double spending, criminal activities, transaction leakage, smart contracts and many others.

Keywords Blockchain · Blockchain security · DDoS · Blockchain application

1 Introduction

Blockchain is one of the propitious and emerging technologies in cybersecurity. Cybersecurity involves various aspects of social life and it is an interest of the people. In its germinal state, blockchain technology is successfully replaced traditional transaction systems in many organizations. It moves from a centralized system to a distributed system. Most of the industries move toward blockchain technology because it promises a secure distributed framework to smooth the progress of sharing, exchanging and assimilation of information across all users and the third party. Blockchain technology will play a key role in the identification and defending of various attacks like Distributed Denial of Service attack (DDoS), data integrity, data

S. Saxena (✉) · Renu
Galgotias College of Engineering and Technology, Greater Noida, India

U. K. Gupta
School of Computing Science and Engineering, Galgotias University, Greater Noida, India

V. Dwivedi
Department of Software Science, Tallinn University of Technology, Tallinn, Estonia

A. Tomar et al. (eds.), *Machine Learning, Advances in Computing, Renewable Energy and Communication*, Lecture Notes in Electrical Engineering 768,
https://doi.org/10.1007/978-981-16-2354-7_47

confidentiality, etc. Blockchain is the next revolution technology, as it redeveloped the way by which we will work and live.

Blockchain is a chain of blocks; each block is stored the fixed size of records. Blockchain will grow as the number of transaction performs. Blocks are linked together using cryptography and cryptographically secure from tampering and modification. Each block generates a cryptographic hash value using a timestamp and transaction data (generally represented as a Merkle tree) stored in the block. This cryptographic hash value will use to connect the new block in the blockchain. All these blocks are open access by all clients in the blockchain network and each new transaction will verifiable by distributed ledgers then only append into the blockchain.

History of Blockchain

Blockchain technology comes from a past distributed system problem: the byzantine failures, it is a basic issue in peer-to-peer network communication proposed by Lesley Lambert [1]. In byzantine failure, it is impossible to achieve consistency on an unreliable channel of lost information. In most cases, we assume that the channel is reliable so there is no such problem. On the internet, when information exchange with strangers then the central node never guarantees complete trust. In this situation, nodes on the network never reach a consensus to identify malicious nodes. Blockchain technology provides a solution to this biggest problem. Blockchain technology can achieve consensus without relying on a single node to which are a network of consistency.

Blockchain comes into perspective to replace current/traditional banking system. If one technology replaces the other then there must be some problem in the current scenario. The issues in current banking system are high third-party transaction fees, double spending (i.e. money is spent twice), financial crises and crashes. These all issues can be solved by blockchain that makes rise to use blockchain technology. Blockchain solves issue of high transaction fees by its decentralized nature and no third-party involvement, blockchain has a distributed ledger and can be publicly accessible. Double spending is not allowed because of the basic structure of block transactions.

Blockchain technology is the spine of the Bitcoin system. Blockchain is basically public ledger database, which holds the encrypted ledger. According to IBM, blockchain is a decentralized ledger that stores or provides the process for record-keeping and tracking in a network. It keeps the details of all properties or things and their transactions over the network. Each transaction is encrypted using cryptographic functions and then all these transactions are stored in a block. Hence, with the help of cryptography, these blocks are linked together so that no modification can be done. The whole process created an immutable and unforfeited record of the transaction.

Blockchain technology is not a new technology, it is a combination of various existing technologies. These technologies are jointly configured so that we achieve a decentralized and trustful network. In blockchain technology, any system can work as a data center to store data and any system can work as miners. Blockchain technology involves cryptocurrency, which is earned when any node performs mining. In cryptocurrencies had a computational problem is related to a double-spending problem, which related to how to ensure that some amount of digital case was not already spent without the validation of a trusted third party (e.g. Bank), which keeps all transaction and user balances. Although blockchain has been developed as a tool for cryptocurrency, it is not necessary to use blockchain for cryptocurrency. Blockchain is used to develop decentralized applications.

Similarly, IoT is a popular technology and plays an important role in our society, in civilian and military contexts. It developed the Internet of Drones, Internet of Military Things and Internet of Battlefields. IoT security is a big research topic and blockchain plays a big role in IoT security. In the IoT network, various third parties and devices are used, which are not completely reliable. In such a network, blockchain is a good solution to made secure and reliable communication on an unreliable channel.

Like a blockchain, IoT also has a similar problem, since, in the IoT system, there are various entities (nodes, gateways, routers, and users) that are not essentially trusted on each other whenever performing any transaction. So blockchain can bring into IoT despite its current limitations, and blockchain-based IoT (BIoT) architecture has been proposed.

2 Overview of Blockchain Technology

In this section, we first explain the basic trust mechanism (i.e. consensus mechanism) used in blockchain and then discuss the synchronization process between nodes.

Consensus mechanism

In blockchain system, third-party trusted authority does not require. For maintaining the reliability and consistency of the data and transaction, blockchain used decentralized consensus mechanism. In blockchain, four consensus mechanisms exist: Proof of work (PoW), Proof of Stack (PoS), Delegated Proof of Stack (DPoS) and Practical Byzantine Fault Tolerance (PBFT). Other than these consensus mechanisms, there exists Proof of Bandwidth (POB), Proof of Authority (PoA), Proof of Elapsed Time (PoET), etc.

PoW mechanism is used to find the solution of puzzles to verify the credibility of data. Generally, puzzles are computationally hard and easily verifiable problem. When any node creates a block, it will add in blockchain only after resolving a PoW puzzle (Fig. 1).

Proof of stack (PoS) is used for the verification of ownership of cryptocurrency and also credibility of the data. Blockchain used PoS at the time of block creation and transaction, for this process, users have to pay some amount of cryptocurrency. If the block is validated eventually then cryptocurrency is refunded to the original node/user. Otherwise, it will be charged.

Fig. 1 PoW consensus mechanism

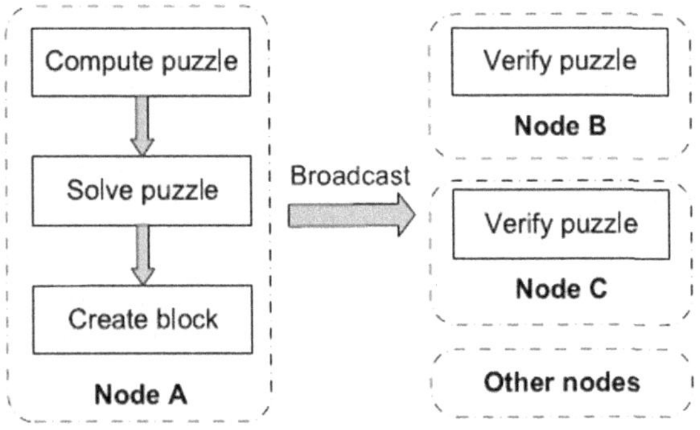

PoW in blockchain required a lot of calculations, its result is computing resource wastage. On the other side, PoS can reduce computation power drastically, it will increase the throughput of the entire blockchain system.

Terminologies

1. *BitCoin*: BitCoin is the first decentralized digital currency, which is introduced by Satoshi Nakamoto in 2009, at the same time, he proposed the idea of blockchain as well.
2. *Block*: Block stores the verified transaction and previous block hash function in the block header and one block consists of only one parent block.
3. *Transaction*: A transaction is a sequence of read and write operations performed atomically.
4. *Ledger*: A ledger is a set or database of transaction records.
5. *Mining*: The process of verifying the transaction is called mining and the nodes or participants who perform mining are called miners. In this process, they have to solve a computationally high mathematical problem and who solves first gets incentive accordingly.
6. *Consensus Algorithm*: Consensus algorithm is a technique through which all the miners need a common agreement about the present state of ledger. They decide whether to add a particular block to the blockchain or not. Consensus algorithms are of two type's proof-based consensus algorithms and voting-based consensus algorithms. Proof of Work (PoW) and Proof of Stake (PoS) are the most commonly used proof-based consensus algorithms.
7. *Smart Contract*: Smart Contract was founded by Nick Szabo in 1994. It was made to execute contracts digitally between participating parties (Michael Crosby 2016).
8. *Hyperledger fabric*: Hyperledger fabric is a distributed ledger platform that runs smart contracts and modular architecture permits for pluggable implementation.
9. *Hash*: Encrypted data value in the block.

Characteristics

Blockchain has the following characteristics that made its application apart from cryptocurrency (Karim Sultan 2018).

1. *Decentralization*: For almost a decade we are using client–server architecture in any computing system or other. The server has all authority and manages client. This client–server architecture has some disadvantages like authority is given to one it can come to unfair chance but when the decision taken by group of people or power is distributed then the chance of biasing becomes less.
2. *Immutability*: This is the biggest advantage or feature of blockchain, i.e. the records that are stored once cannot be deleted. As the copies of blocks or records are distributed among all the nodes or participants then if someone wants to delete it becomes impossible as it leads to inconsistency.
3. *Transparency*: Nothing can be hiden, i.e. what we call transparency. What records are added, what transaction is taken place, previous records and so on they all are transparent to everyone this leads to no corruption or fraud.
4. *Consensus driven*: Every block or new transaction is added and verified by the group of people or nodes to gain trust this procedure is known as mining and those nodes are called miners.

Working Principles behind blockchain and tools used

Blockchain is a decentralized and distributed ledger for the peer-to-peer network. The working mechanism is followed as: Suppose a new transaction is issued, it will broadcast to all P2P network then network of nodes or miners verifies the transaction by solving computationally high hash function or follow some other consensus algorithm based on the approach used. Once the transaction is corroborated, the transaction becomes a chunk of new block for the ledger then this current block is added to the existing blockchain and then the transaction succeeds.

There are many platforms used for blockchain development like Ethereum, Hyperledger Fabric, IBM blockchain, Ripple and so on.

The most used platform is Ethereum; it is a blockchain platform that is open source and highly vital that forms the station for the other applications to be developed. The programming language supported by this is Solidity. It uses Proof of Work (PoW) consensus algorithm and approachable to all Hyperledger fabric is an open-source permission or private blockchain platform used for running smart contracts.

3 Security Issues in Blockchain Technology

Protect the Infrastructure: DDoS Attack

The DDoS attack is performing by combining multiple computers as attacker that will launch DDoS attack against single or multiple targets. The denial of service attack exactly targets the CIA (confidentiality, integrity and availability). There are various ways to perform DDoS attack, one is flooding of packets [2].

Another is to send a large amount of requests to the server to perform DDOS.

Data Security

For data security, developers use cryptography and digital signatures to maintain the security and integrity of data values.

Data integrity is also checked by the hash value of datastore and the previous calculated hash value of data from hash algorithms.

Common Risks to Blockchain

51% vulnerability

In blockchain technology, consensus mechanism has 51% vulnerability and if this is exploited by attacker then attacker controls the entire blockchain.

Private Key security

In blockchain, if the user's private key is lost, it will never be recovered. So if the attacker has stolen the user's private keys then user blockchain account faces the major risk of being account modification by others and it is very difficult to track attacker behaviors to know the modification in blockchain information.

4 Application of Blockchain Technology

In the research domain, there are numerous applications and implementation of blockchain technology such as: education [3–5], smart cities [6, 7], solid waste management [8, 9], digital voting [10], digital identity [11], supply chain management [12, 13], music industry [14], health care [15], real estate [16] and others applications [17–23].

Education: From starting till the present, only paper-based certification process is running constantly. There is a limitation in paper-based certification, i.e. the validity. There is no system to verify or gain the trust of certificate issues. Blockchain provides a way to gain trust in certification by transparency that makes its application in Education. [3–5].

Smart cities: A city is a smart city if it provides an easy and luxurious lifestyle and solves all urban problems like water sanitation, pollution, etc. Many technologies are working to upgrade working or implementing smart city concept, blockchain is one of them. Blockchain can be applied to smart cities due to its features like transparency, decentralization, etc. [6, 7].

Solid Waste Management: Solid waste is a global issue; the amount of solid waste is increasing day by day that has both social and economic impacts. Solid waste management is a way how to manage or procedure to recycle or reuse solid waste in a systematic manner. Solid waste management can be implemented through blockchain. Various tools or platforms that are already working on blockchain that helps in solid waste management are Plastic bank, Swacch Coin, Recereum [8, 9].

Digital Voting: As the fraud and corruption in voting increases, it leads to digital voting. Bringing blockchain technology to digital voting comes with benefits like transparency [10].

Digital Identity: The digital identity means information about people, organization or individual in a digital or online world. There associated various hacks or stealing of data associated with the information. As the data are shared on client–server architecture, the data can be mishandled; there is no transparency that leads to fake identities. So, to overcome these limitations, blockchain comes to rescue due to its features like decentralization, Transparency [11].

Supply Chain Management: Supply chain Management is defined as the flow of goods or products at each stage of production to supply to the consumer. There can be any skip or fraud that occurs at any stage between the production and supply due to lack of transparency. Blockchain brings transparency to supply chain management procedure. Each step is added as a block in the network [12, 13].

Music Industry: The piracy and copyright issues are the issues that every industry has to deal with, and music industry is one of it. Another issue that the music industry to face is royalty payment, the originator or creator of music does not get enough amounts because of the intermediate parties. So, to overcome the above described the limitations blockchain technology is used [14].

Health Care: Blockchain has usage in Healthcare by storing and safe delivery of health care data that, in turn, leads to effective diagnosis and treatment. The application of blockchain in healthcare is not limited to this only, it can apply in drug supply chain management, so that proper records of medicines or drugs can be maintained and so that they cannot be used in a wrong manner. The various fields in healthcare that blockchain can be used are Clinical Research, Electronic Health Records, Medical Fraud Detection, Pharmaceutical Industry and Research, Neuroscience Research [15].

Real Estate: Real estate can be land or property or building. The problem with the current real estate system is transparency, not accessible to all, high fees, lack of liquidity, pricing commitments, etc. These issues can be handled by blockchain by providing transparency, not any intermediate, less costly and accessible to all. Blockchain came up with the concept of smart contracts and tokens [16].

5 Conclusion

This paper describes the basic concept of blockchain and security issues in blockchain technology. We discussed basic principles and characteristics of blockchain after which I did discuss the existing security issue in blockchain technology, which will use by future researchers for their research. In this paper, we discussed the application areas of blockchain technology in which blockchain will use in the future for technology enhancement. Blockchain applications range from financial to non-financial sectors. Non-financial sectors include various areas such as Healthcare, Education, Business and Industry, etc.

References

1. Mei H, Liu J (2016) Industry present situation, existing problems and strategy suggestion of blockchain. J Telecommun Sci 32(11):134–138
2. Li Y, Xin Y, Han Y, Li W, Xu Z (2017) A survey of DoS attack in content centric networking. J Cyber Secur 2(1):91–108
3. Han M, Li Z, He J, Wu D, Xie Y, Baba A (2018) A novel blockchain-based education records verification solution. In: 19th annual conference on information technology education (SIGITE18). New York, NY, USA, October 3–6
4. Liu Q, Zhu H, Green G, Guan Q, Yang X, Yin S (2018) Education-industry cooperative system based on blockchain. In: IEEE international conference on hot information-centric networking (HotICN)
5. Turkanovic M et al. (2018) EduCTX: a blockchain-based higher education credit platform. IEEE Access
6. Ibba S, Pinna A, Seu M, Pani FE (2017) CitySense: blockchain oriented smart cities. In: Proceedings of XP '17 Workshops. Cologne, Germany, May 22–26
7. Michelin RA, Dorri A, Steger M, Lunardi RC, Kanhere SS, Jurdak R (2018) Speedy Chain: a framework for decoupling data from blockchain for smart cities
8. Shah PJ, Anagnostopoulos T, Zaslavsky A, Behdad S (2018) A stochastic optimization framework for planning of waste collection and value recovery operations in smart and sustainable cities. Elsevier Waste Management, pp 104–114
9. Zhang D (2018) Application of blockchain technology in incentivizing efficient assessing the feasibility of using the heat demand-out. In: 10th international conference on applied energy (ICAE2018). Hong Kong, China, 22–25 August 2018
10. Ayed AB (2017) A conceptual secure blockchain based electronic voting system. Int J Netw Secur Appl (IJNSA)
11. Daniel Augot YZ, Herv'e Chabanne X, Olivier Cl'emot X, George W Transforming face-to-face identity proofing into anonymous digital identity using the Bitcoin blockchain.
12. Chen S, Shi R, Ren Z, Yan J, Shi Y, Zhang J (2017) A blockchain based supply chain quality management framework. In: The fourteenth IEEE international conference on e-business engineering. China
13. Wu H, Cao J, Yang Y, Tung CL, Jiang S, Tang B, Liu Y, Wang X, Deng Y Data management in supply chain using blockchain: challenges and a case study. Alibaba Innovative Research
14. Nucciarelli CS, Alberto (2018) The impact of blockchain on the music industry. In: 29th European regional conference of the international telecommunications society. Milan, Italy
15. Siyal AA, Junejo AZ, Zawish M, Ahmed K, Khalil A, Soursou G (2019) Applications of blockchain technology in medicine and healthcare: challenges and future perspectives
16. Karamitsos I, Papadaki M, Al Barghuthi NB (2018) Design of the blockchain smart contract: a use case for real estate. J Inform Secur 177–190
17. Aggarwal A et al. (2020) Meta heuristic and evolutionary computation: algorithms and applications. Springer Nature, Berlin, p 949.doi: https://doi.org/10.1007/978-981-15-7571-6. ISBN 978-981-15-7571-6
18. Yadav AK et al. (2020) Soft computing in condition monitoring and diagnostics of electrical and mechanical systems. Springer Nature, Berlin, p 496. https://doi.org/10.1007/978-981-15-1532-3. ISBN 978-981-15-1532-3
19. Gopal et al. (2021) Digital transformation through advances in artificial intelligence and machine learning. J Intell Fuzzy Syst 1–8. doi: https://doi.org/10.3233/JIFS-189787

20. Fatema N et al. (2021) Intelligent data-analytics for condition monitoring: smart grid applications. Elsevier, p 268. ISBN: 978-0-323-85511-2. https://www.sciencedirect.com/book/9780323855105/intelligent-data-analytics-for-condition-monitoring
21. Smriti S et al (2018) Special issue on intelligent tools and techniques for signals, machines and automation. J Intell Fuzzy Syst 35(5):48954899. https://doi.org/10.3233/JIFS-169773
22. Jafar A et al. (2021) AI and machine learning paradigms for health monitoring system: intelligent data analytics. Springer Nature, Berlin, p 496. https://doi.org/10.1007/978-981-33-4412-9. ISBN 978-981-33-4412-9
23. Sood YR et al. (2019) Applications of artificial intelligence techniques in engineering, vol 1. Springer Nature, p 643. https://doi.org/10.1007/978-981-13-1819-1. ISBN 978-981-13-1819-1

Comprehensive Study on Heterojunction Solar Cell

Pranava Sai Aravinda Pakala, Amruta Pattnaik, Shivangi, and Anuradha Tomar

Abstract Renewable energy is gaining momentum worldwide due to the increasing concern over traditional fuels' sustainability and environmental impact. Solar photovoltaics is a serious contender, owing to its many advantages: no harmful greenhouse gas emissions; free and abundant; minimal maintenance costs. PV panels can provide an effective solution for peak demand needs. They are also easy to install and can be easily integrated with the existing systems. This paper focuses on the Heterojunction Intrinsic Thin Film (HIT) technology. This technology combines both crystalline and amorphous solar cells' best qualities, thus maximizing its overall efficiency. This paper analyses the HIT cells with advantages, disadvantages, and comparisons. A case study also verifies the performance. In this case study, the effect of crystalline Si and amorphous Si-based heterojunction photovoltaic cell with ZnS nanoparticle layer has been studied. ZnS nanoparticle layer was deposited on crystalline Si and amorphous Si-based HJ photovoltaic cell via the spin coating process. This study described the effect of zinc sulfide nanoparticles (ZnS NP) embedded in polymethyl methacrylate (PMMA) as a top layer of a crystalline Si and amorphous Si-based HJ photovoltaic cell. The cell characterizations have been carried out by quantum efficiency (QE). ZnS NP/PMMA film acts as an antireflective layer to utilize high-energy photons.

Keywords Renewable energy · Solar cell · HIT technology · ZnS nanoparticles

P. S. A. Pakala (✉) · A. Pattnaik
ADGITM, FC-26, Shastri Park, New Delhi 110053, India

Shivangi
BHEL-ASSCP, Gwal Pahari 122001, Haryana, India

A. Tomar
Netaji Subhas University of Technology, Delhi 110078, India

A. Tomar et al. (eds.), *Machine Learning, Advances in Computing, Renewable Energy and Communication*, Lecture Notes in Electrical Engineering 768,
https://doi.org/10.1007/978-981-16-2354-7_48

1 Introduction

Solar energy, as a technology, has attracted academic as well as industrial attention for a long time. This attention is not unwarranted because of several compelling reasons. Conventional forms of energy rely heavily on fossil fuels. Not only do they have adverse effects on the environment but they are also limited in terms of resources. This leads to a volatile market. One can easily protect himself/herself from fluctuating utility prices by investing in solar. It also boosts grid security, and transmission losses can be significantly minimized.

The study that constitutes the conversion of sunlight to electric current is referred to as solar photovoltaics. This process happens inside a solar cell. A solar cell is fundamentally a p–n junction diode. This diode is made of silicon. It is preferred for various reasons: first, it is economical; it has stability and favorable physical, chemical, and electronic properties. The generation of electric current happens inside the depletion region of the diode [1].

Heterojunction photovoltaic cells are known to possess superior Voc, increased efficiencies, and lower temperature coefficients [2–4], making them better than the conventional c-Si solar cells for many applications. The use makes such types of solar cells of both the crystalline-silicon and the amorphous-silicon layers. Such solar cells provide stable junction relatively better than other varieties owing to the presence of the amorphous-Si film in the SHJ solar cell that can passivate the dangling bonds on the crystalline Si surface, which obviates the recombination at the junction of amorphous Si and crystalline Si [5, 6]. The configuration of a SHJ photovoltaic cell was first introduced in 1983, whose cell efficiency was not more than 12% [7].

Heterojunction technology consists of passivated contact technology, comprising of HJs. Greater cell efficiency can also be achieved through the Interdigitated Back Contact Technology (IBC) [8]. The latter first came into existence in 1977, and a full-fledged IBC-based cell was developed in 1984 by Swanson et al. [2]. In 2014, Masuko et al. reported achieving an efficiency of 25.6% by a HIT back contact cell [9]. In 2016, Smith et al. reported a total area efficiency of 25.2% on an IBC cell with passivated contact [3, 10]. Sanyo (now Panasonic) first developed the heterojunction technology that used an amorphous silicon layer to passivate the crystalline silicon surface in 1990. The work went on to be published in 2000 [4]. When first introduced, HIT cells had an efficiency of 14.4% and produced 170 W. Panasonic's latest offering is a 60 cell model whose efficiency is around 20%, and it produces 330 W [5]. This technology can achieve greater PCEs through thin film processes (to reduce production costs) than c-Si cells. In 2019, Haque et al. reported that the capabilities of HIT photovoltaic cells could be ameliorated by augmenting the work function of the TCO layer. Indium tin oxide was used as the TCO layer [11]. Delavaran and Shahhoseini reported in 2019 that deploying mc-Si O: F: H in the emitter of a heterojunction photovoltaic cell lead to a reduction in the effect of recombination and, consequently, an increase in the Isc in the cell, this was owing to the utilization of SiF4 in the dropping process of mc-Si O:F: H. Additionally, enhancing the lifetime of carriers located in the emitter layer resulted in the accumulation of a greater number of minority carriers in the front layer [12]. This improved the short circuit current and resulted in higher efficiency [12]. Huang et al. reported a design for an amorphous silicon/crystalline silicon HJ photovoltaic cells with a localized structure (HACL) cell to derive greater PCEs and mitigate production cost. It was reported that the PCE and the Isc density of the HACL cell could potentially be almost 28.18% and 43.06 mA/cm^2, respectively. The main reasons for this significant improvement were (i) diminished optical absorption loss of a-Si: H and (ii) minimized photocarrier recombination for the HACL cell [13]. Bifacial solar cells, therefore, benefitted from the double-side local junction [13]. Kato et al. fabricated Al_2O_3 by adopting the process of atomic layer deposition for preventing the corrosion, also known as passivation, of loosely arranged bonds. Nevertheless, their research revealed that the electron carriers were unable to reach the external circuit because the Si nanowires were completely covered by Al_2O_3. Their research adopted the procedure of chemical–mechanical polishing to purge the remnants of the oxide layer from the top of the Si nanowire. They then constructed SHJ photovoltaic cells possessing an efficiency of 1.6% making use of amorphous silicon. They reported that within 340 nm wavelength, the IQE of the silicon nanowire photovoltaic cell exceeds in comparison to the crystalline silicon variety, thus enhancing the absorption of the Si nanowire cells, proposing that Si nanowire could hold great potential for crystalline-silicon thinning [14]. In 2020, Khokhar et al. observed that by overturning the p–n junction from the anterior to the posterior, augmented flexibility for anterior design was obtained, with no adverse impact on the efficacy of the device [15]. In the same year, Srisantirut et al. reported about the prospect of employing DLC films as protective ARCs for SHJ photovoltaic cells. They noted that the heterojunction solar cell's I–V characteristic response showed an improvement after the DLC antireflection coating deposition. The efficiency of the photovoltaic cells is improved by approximately 1%. The anti-reflection coating did not have much effect on the Voc [16].

There are several advantages of this technology over the conventional c-Si cells:

(1) **Greater efficiency**: HIT cells combine the benefits of c-Si cells with good absorption and better passivation characteristics and the widespread availability of thin films. The superior surface passivation capabilities of c-Si result in high V_{oc} and high overall efficiencies. Carrier lifetime, specifically the minority charge carrier lifetime, helps to measure the extent of surface passivation. The efficiency of HJT panels currently available in the market ranges between 19.9 and 21.7%.

(2) **Lower production cost**: HJT panels cost cheaper as compared with the industry-standard PERC technology. This is because of the employment of thin-film technology. They require lesser manufacturing steps, in comparison to other technologies.

(3) **Lower temperature coefficient**: It is an established fact that the performance capabilities of solar cells greatly reduce in high temperatures. HJT panels can produce these record efficiencies even in high temperatures. A low-temperature coefficient not only reduces manufacturing steps but also prevents bulk degradation.

(4) They generate **higher power for the same system size**.
(5) **Maximum utilization of available roof space** [7].

2 Types of Silicon Heterojunction Solar Cells

SHJ cells can be grouped according to two categories, based on the location of the emitter, as shown in Fig. 1a, b [15].

The transparent conductive oxide layer on the front side is responsible for charge collection and acts as an ARC for the SHJ photovoltaic cell. A disadvantage of employing a front transparent conductive oxide is its parasitic absorption at higher energy because of inter-band absorption in proximity to the bandgap energy and in the near-infrared region as to permit carrier immersion [18, 19]. In the rear emitter SHJ photovoltaic cells, the bulk carrier carriage to the metal grating is achieved together using the forward-facing transparent conductive oxide and the silicon wafer, thus enabling the utilization of transparent conductive oxides with reduced conductivity than those required for front emitter configuration [20]. In such cells, the sheet resistance (Rsh) of the front transparent conductive oxide does not affect the cell performance much, to rephrase it [21]. Hence, similar, if not better, results than front emitter cells can be achieved.

In 2016, Kobayashi et al. fabricated heterojunction photovoltaic cells on epitaxial developed silicon sheet. They fabricated three categories of epitaxial grown silicon HJ cells: (i) A cell with three busbars containing stacking fault (SF) lumps on the n-type a-Silicon film side; (ii) a cell containing three busbars with SF lumps on the pa-Silicon film side; and (iii) a busbar-less cell with SF bulges on the pa-Silicon film side. The third cell was fabricated to evaluate their best cell's J–V characteristics, as shown in figure [22] (Figs. 2 and 3).

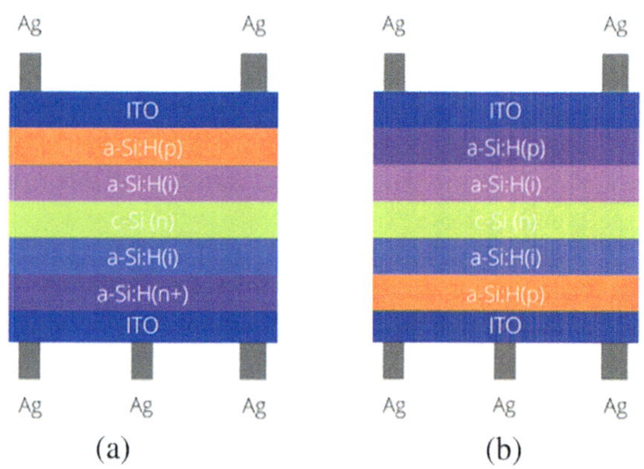

Fig. 1 Front emitter SHJ solar cell (**a**) and rear emitter SHJ solar cell (**b**) [17]

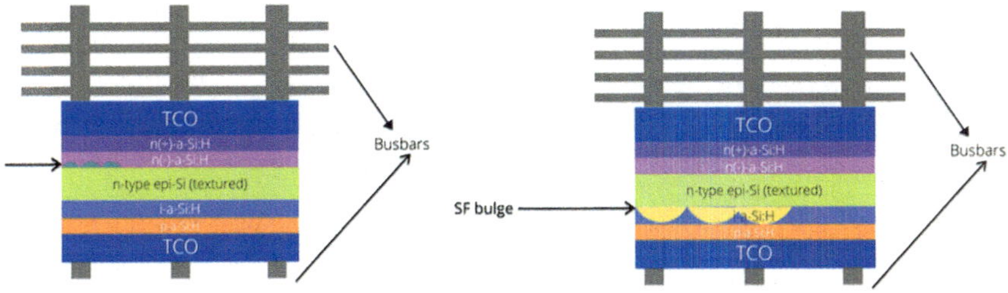

Fig. 2 The schematic model 3-busbar cell with stacking fault (SF) bulges on the n-a-Si layer side (**a**) and 3-busbars cell with SF bulges on the pa-Si layer side (**b**) [22]

Fig. 3 Bus-bar less cell with SF bulges on the pa-Si layer side

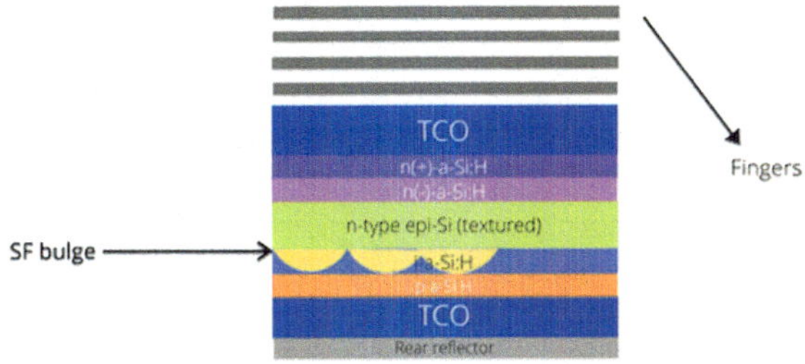

3 Case Study

In this paper, two types of structures of HIT solar cells have been discussed. Heterojunction solar cells possess greater open-circuit voltages, increased efficiencies, and low-temperature coefficients [23–26], which makes them superior to c-Si solar cells. ZnS is an encouraging material for optical studies such as phosphor material, flat panel displays, electro-luminescent, and IR devices [26–29]. Nanomaterials have attracted significant traction owing to their unique optical field properties, as opposed to the bulk materials [30]. In this research work, an attempt was made to study the performance of heterojunction-based solar cell design along with and without ZnS nanomaterial.

The synthesis, characterizations, and deposition of ZnS nanomaterial were already discussed in our previous publication [31].

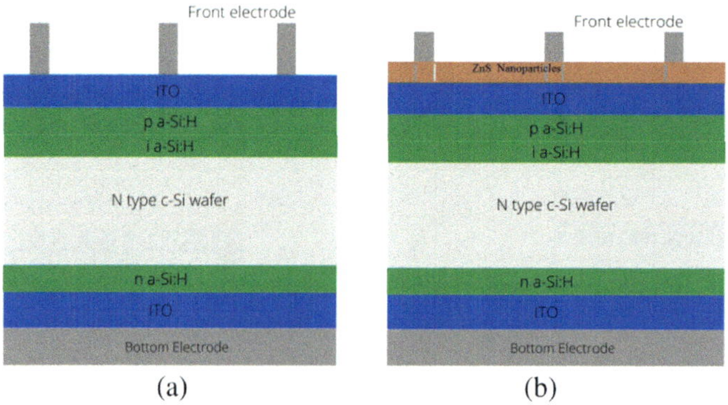

Fig. 4 The schematic model of amorphous silicon/crystalline silicon HIT photovoltaic cell not using (**a**) and using (**b**) ZnS nanoparticle (ZnS NP) layer

An illustration of the amorphous silicon/crystalline silicon heterojunction photovoltaic cell deployed with and without ZnS nanoparticle structure is shown in Fig. 4. A p-type amorphous silicon was placed on the forward-facing to the sun, and an n-type amorphous silicon layer was situated on the backside of the photovoltaic cell [6]. ITO layers were placed on the entire surface of the rear and topsides to collect the photons efficiently [32]. The silver layer was used as a back electrode. The external quantum efficiency (EQE) is a suitable tool for computing photons' total accumulation at each wavelength [6]. The external and internal quantum efficiency (EQE)/(IQE), as well as reflectance, was measured using a Bentham PVE300 system. The parameters of reflection and parasitic absorption in the indium tin oxide film are responsible for the current loss in a typical amorphous silicon/crystalline silicon SHJ cells [33]. High front layer recombination can also decrease the EQE in the UV region, as the UV light is absorbed near the surface [6, 30]. We recall here our previous research, wherein we reported that ZnS NP/PMMA was found to be an appropriate material to ameliorate the miniscule wavelength of ultraviolet response [31].

Figure 5a describes the EQE spectra of heterojunction photovoltaic cells not using the ZnS nanoparticle/PMMA coating. In EQE spectra of photovoltaic cell with ZnS nanoparticle/PMMA layer, it is noticed that the short wavelength is enhanced from 300 to 500 nm, and the EQE is reduced from 450 to 850 nm as matched with heterojunction solar cell without ZnS nanoparticle/PMMA layer [15, 31]. It can be speculated that the ITO absorbs the light at 600 nm, but the coating of ZnS nanoparticle/PMMA has increased the level of reflectivity from 450 to 620 nm, as shown in Fig. 5b. It can be the result of either thickness or scattering of light due to the ZnS nanoparticle.

Figure 5c describes the internal quantum efficiency spectra of heterojunction photovoltaic cells using and not using ZnS nanoparticle/PMMA film. It is noted from the IQE spectra of the solar cell with ZnS nanoparticle/PMMA layer that the short wavelength response is partially enhanced (1–2%) from 300 to 450 nm [31]. In

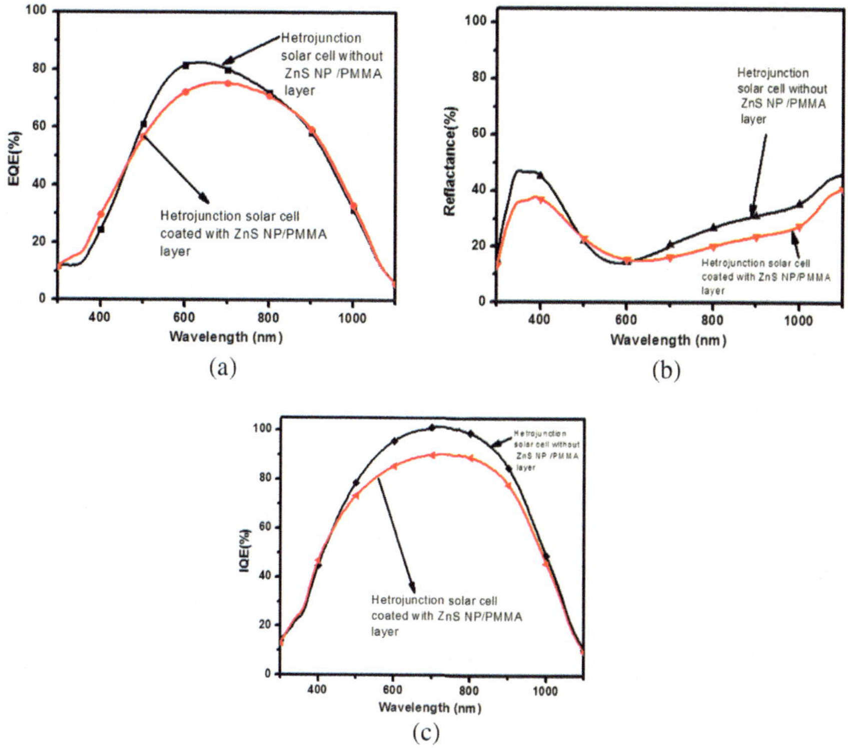

Fig.5 EQE (**a**), Reflectance (**b**) and IQE (**c**) spectra of Heterojunction solar cell using and not using a coating of ZnS NP/PMMA on HIT solar cell [31]

other words, the IQE response is wholly degraded in between the wavelength of 450–1000 nm as juxtaposed to the photovoltaic cell without the ZnS nanoparticle/PMMA layer. Figure 5b represents the solar cell's reflectance spectra with and without the ZnS nanoparticle/PMMA layer. It is noticed that the reflectance is reduced due to the layer of ZnS nanoparticle/PMMA layer. The reflectance is high from 500 to 620 nm because of the scattering of light.

4 Conclusions and Future Scope

The performance of SHJ solar cells and the ZnS nanoparticle layer in PMMA have been studied. The upgradation in the ultraviolet response of the heterojunction PV cell was confirmed. It can be concluded that the overall performance of the heterojunction PV cell with ZnS nanoparticle/PMMA layer has deteriorated as opposed to the heterojunction PV cell without ZnS nanoparticle.

Several improvements can be made to this technology to increase its performance. First, optical losses need to be reduced to improve Jsc. The surface texturing of wafers can result in better light trapping, optimization of TCO, and a-Si: H layers to facilitate their absorption, and the aspect ratio of the grid electrodes should be improved to diminish the covered area. Second, recombination losses need to be reduced to increase Voc [34]. It can be ensured by the elimination of any metallic contamination and particles from the surface through cleaning the wafer surfaces before a-Si: H deposition. Third, resistance losses affecting the fill factor (FF) need to be reduced. This is accomplished by reducing the series resistance of the device by deploying highly conductive TCO and superior ohmic contacts at the contact interfaces to minimize resistance [31, 34].

Acknowledgements The authors would like to recognize Dr. Anil Kumar Saxena, Dr. B. Pant, and Mr. Vinayan Bhardwaj [BHEL-ASSCP] for immense technical insights.

References

1. Zeman M, Zhang D (2012) Heterojunction silicon based solar cells. In: van Sark WGJHM, Korte L, Roca F (eds) Physics and technology of amorphous-crystalline heterostructure silicon solar cells. Engineering materials. Springer, Berlin. https://doi.org/10.1007/978-3-642-22275-7_2
2. Swanson RM et al (1984) Point-contact silicon solar cells. IEEE Trans Electron Devices 31:661–664. https://doi.org/10.1109/T-ED.1984.21586
3. Smith DD, Reich G, Baldrias M, Reich M, Boitnott N, Bunea G (2016) Silicon solar cells with total area efficiency above 25%. In: Proceedings of the IEEE 43rd photovoltaic specialists conference (PVSC), pp 3351–3355. https://doi.org/10.1109/PVSC.2016.7750287
4. Taguchi M, Kawamoto K, Tsuge S, Baba T, Sakata H, Morizane M, Uchihashi K, Nakamura N, Kiyama S, Oota O (2000) HIT™ cells—high-efficiency crystalline Si cells with novel structure. Prog Photovolt Res Appl 8:503–513. https://doi.org/10.1002/1099 159X(200009/10)8:5%3c503::AID-PIP347%3e3.0.CO;2-G
5. Panasonic (2019) Photovoltaic module HIT, VBHN330SJ47/VBHN325SJ47 datasheet
6. Kim N, Um HD, Choi I, Kim KH, Seo K (2016) 18.4%-efficient heterojunction Si solar cells using optimized ITO/Top electrode. ACS Appl Mater Interfaces 8:11412–11417. https://doi.org/10.1021/acsami.6b00981
7. Olson JM (1994) Heterojunction solar cell. US Patent 5,342,453
8. Yoshikawa K et al (2017) Exceeding conversion efficiency of 26% by heterojunction interdigitated back contact solar cell with thin film Si technology. Sol Energy Mater Sol Cells 173:37–42. https://doi.org/10.1016/j.solmat.2017.06.024
9. Masuko K et al (2014) Achievement of more than 25% conversion efficiency with crystalline silicon heterojunction solar cell. IEEE J Photovolt 4(6):1433–1435. https://doi.org/10.1109/JPHOTOV.2014.2352151
10. Ru X, Qu M, Wang J, Ruan T, Yang M, Peng F, Long W, Zheng K, Yan H, Xu X (2020) 25.11% efficiency silicon heterojunction solar cell with low deposition rate intrinsic amorphous silicon buffer layers. Sol Energy Mater Sol Cells 215:110643. https://doi.org/10.1016/j.solmat.2020.110643
11. Haque MM, Dickens Tusha Falia C, Hasan M (2019) Investigating the performance of nanocrystalline silicon HIT solar cell by Silvaco ATLAS. In: Proceedings of the 22nd international conference on computer and information technology (ICCIT). Dhaka, Bangladesh, pp 1–6. https://doi.org/10.1109/ICCIT48885.2019.9038595

12. Delavaran H, Shahhoseini A (2019) Hydrogenated fluorinated microcrystalline silicon oxide (μc-SiO:F:H): a new material for the emitter layer of HIT solar cells. In: Proceedings of the 27th Iranian conference on electrical engineering (ICEE), Yazd, pp 36–40. https://doi.org/10.1109/IranianCEE.2019.8786742
13. Huang H, Zhou L, Yuan J, Quan Z (2019) Simulation of a-Si:H/c-Si heterojunction solar cells: from planar junction to local junction. Chin Phys B 28:128503. https://doi.org/10.1088/1674-1056/ab5212
14. Kato S, Kurokawa Y, Gotoh K et al (2019) Silicon nanowire heterojunction solar cells with an Al_2O_3 passivation film fabricated by atomic layer deposition. Nanoscale Res Lett 14:99. https://doi.org/10.1186/s11671-019-2930-1
15. Khokhar MQ, Hussain SQ, Kim S et al (2020) Review of rear emitter silicon heterojunction solar cells. Trans Electr Electron Mater 21:138–143. https://doi.org/10.1007/s42341-020-00172-5
16. Srisantirut T, Pengchan W, Phetchakul T (2020) Diamond-like carbon thin-film coating for application on heterojunction solar cells by ECR-CVD system. IOP Conf Ser Mater Sci Eng 855:012009. https://doi.org/10.1088/1757-899X/855/1/012009
17. Li S, Tang Z, Xue J, Jian-jun G, Shi Z, Li X (2018) Comparative study on front emitter and rear emitter n-type silicon heterojunction solar cells: the role of folded electrical fields. Vacuum 149:313–318. https://doi.org/10.1016/J.VACUUM.2017.12.041
18. Green MA, Hishikawa Y, Dunlop ED, Levi DH, Hohl-Ebinger J, Ho-Baillie AWY (2018) Solar cell efficiency tables (version 51). Prog Photovolt Res Appl 26:3–12. https://doi.org/10.1002/pip.2978
19. Meza D, Cruz A, Morales-Vilches AB, Korte L, Stannowski B (2019) Aluminum-doped zinc oxide as front electrode for rear emitter silicon heterojunction solar cells with high efficiency. Appl Sci 9(862). https://doi.org/10.3390/app9050862
20. Pankove JI, Tarng ML (1979) Amorphous silicon as a passivant for crystalline silicon. Appl Phys Lett 34:156–157. https://doi.org/10.1063/1.90711
21. Wang J, Meng C, Zhao L, Wang W, Xu X, Zhang Y, Yan H (2020) Effect of residual water vapor on the performance of indium tin oxide film and silicon heterojunction solar cell. Sol Energy 204:720–725. https://doi.org/10.1016/j.solener.2020.04.086
22. Kobayashi E, Watabe Y, Hao R, Ravi TS (2016) Heterojunction solar cells with 23% efficiency on *n*-type epitaxial kerfless silicon wafers. Prog Photovolt Res Appl 24:1295–1303. https://doi.org/10.1002/pip.2813
23. Taira S, Yoshimine Y, Terakawa A, Maruyama E, Tanaka M (2006) Temperature properties of high-Voc HIT solar cells. Renew Energy 115–118
24. Mulligan WP, Rose DH, Cudzinovic MJ et al (2004) Manufacture of solar cells with 21% efficiency. In: Proceedings of the 19th EPVSEC, p 387
25. Green MA (1987) High efficiency silicon solar cells. In: Proceedings of the seventh EC photovoltaic solar energy conference, pp 681–687. https://doi.org/10.1007/978-94-009-3817-5
26. Navaneethan M, Archana J, Nisha KD, Ponnusamy S, Arivanandhan M, Hayakawa Y, Muthamizhchelvan C (2012) Synthesis of Wurtzite ZnS nanorods by microwave assisted chemical route. Mater Lett 66:276–279. https://doi.org/10.1016/j.matlet.2011.08.082
27. Omurzak E, Mashimo T, Sulaimankulova S, Takebe S, Chen L, Abdullaeva Z, Iwamoto C, Oishi Y, Ihara H, Okudera H, Yoshiasa A (2011) Wurtzite-type ZnS nanoparticles by pulsed electric discharge. Nanotechnology 22:365602. https://doi.org/10.1088/0957-4484/22/36/365602
28. She Y, Yang J, Qiu K (2010) Synthesis of ZnS nanoparticles by solid-liquid chemical reaction with ZnO and Na_2S under ultrasonic. Trans Nonferrous Metals Soc China 20:211–215. https://doi.org/10.1016/S1003-6326(10)60041-6
29. Onwudiwe DC, Ajibade PA (2011) ZnS, CdS and HgS nanoparticles via alkyl-phenyl dithiocarbamate complexes as single source precursors. Int J Mol Sci 12:5538–5551. https://doi.org/10.3390/ijms12095538
30. Farfan W, Mosquera Vargas E, Marin C (2011) Synthesis and blue photoluminescence from naturally dispersed antimony selenide (Sb2Se3) 0-D nanoparticles. Adv Sci Lett 4:85–88. https://doi.org/10.1166/asl.2011.1185

31. Pattnaik A, Jha S, Tomar M, Gupta V, Prasad B, Mondal S (2018) Improving the quantum efficiency of the monocrystalline silicon solar cell using erbium-doped zinc sulphide nanophosphor in downshift layer. Mater Res Express 5:095014. https://doi.org/10.1088/2053-1591/aad602
32. Dai X, Chen T, Cai H, Wen H, Sun Y (2016) Improving performance of organic-silicon heterojunction solar cells based on textured surface via acid processing. ACS Appl Mater Interfaces 8:14572–14577. https://doi.org/10.1021/acsami.6b03164
33. Lei C, Peng C, Zhong J, Li H, Yang M, Zheng K, Qu X, Wu L, Yu C, Li Y, Xu X (2020) Phosphorus treatment to promote crystallinity of the microcrystalline silicon front contact layers for highly efficient heterojunction solar cells. Sol Energy Mater Sol Cells 209:110439. https://doi.org/10.1016/j.solmat.2020.110439
34. Zeman M, Schropp REI (2012) Thin-film silicon PV technology. In: Sayigh A (ed) Comprehensive renewable energy, vol 1. Elsevier BV, pp 389–398. https://doi.org/10.1016/B978-0-08-087872-0.00119-0

A New Non-isolated High Gain DC–DC Converter for Microgrid Applications

Abbas Syed Nooruddin, Arshad Mahmood, Mohammad Zaid, Zeeshan Sarwer, and Adil Sarwar

Abstract In this paper, a new high gain chopper with substantially reduced voltage stress has been documented. Its application can be in DC microgrids, which have become increasingly popular due to the increased use of photovoltaic systems and fuel cells to tackle problems such as climate change, increasing pollution, etc. However, the output voltage of these systems is quite low. The conventional non-isolated converter is popular but it only boosts the voltage levels at high duty ratio values resulting in high stress on switches. The proposed topology can convert low voltage levels to high ones (DC) and uses only two inductors and hence is cost-effective. The switches are operated simultaneously at same duty ratios. A comparison of this topology with other existing topologies has been carried out in detail.

Keywords Microgrids · Voltage stress · Voltage gain · Non-isolated

1 Introduction

A conventional DC to DC converter is a type of power electronic converter, which can either operate in buck or boost mode and change DC voltage levels according to the requirement. From being used as power optimizers in renewable energy generation systems to portable electronic devices such as phones and laptop computers, these power electronic converters have a wide range of uses. DC–DC high voltage gain converters are utilized for several applications such as automobile lighting systems (HID lamp ballasts), battery backup systems for UPS [1]. The high gain DC–DC converter can have isolated as well as non-isolated structures. The main challenge is to develop new converters with reduced stress on switching components along with high gain. Immense research has been carried out and various research papers have been written on the high gain converters. In traditional DC–DC boost converter, higher voltage gain is obtained at an extreme value of duty ratios resulting in high stress and poor efficiency [2].

A. S. Nooruddin (✉) · A. Mahmood · M. Zaid · Z. Sarwer · A. Sarwar
Department of Electrical Engineering, Zakir Husain College of Engineering and Technology, Aligarh Muslim University, Aligarh 202002, India

A. Tomar et al. (eds.), *Machine Learning, Advances in Computing, Renewable Energy and Communication*, Lecture Notes in Electrical Engineering 768,
https://doi.org/10.1007/978-981-16-2354-7_49

To overcome challenges like high stresses on switches and get high gains, high efficiency increased range of duty cycle with avoiding operation of converters at the utmost duty ratio and low voltage stresses on switches at particular voltage gain, several DC to DC converters have been proposed. In [3], a coupled inductor-based converter has been proposed. Another new boost converter for microgrid applications with reduced stress is proposed in [4]. In [5], a new converter with twice the gain as conventional boost converter is proposed. Switched inductors and capacitors are also employed to increase the gain of the converter [6, 7]. In [8], a new novel chopper topology is proposed for wind energy farm applications. Some other new structures of converters with different boosting techniques are proposed by authors in [9–11]. New non-isolated converters with high gain are proposed by authors in [12–14].

A high gain DC–DC converter is proposed with low voltage stresses across the switches and other power components. However, the gain will decrease with non-ideal components after the duty ratio of 0.9. This converter has more components but a very light, expensive, and simple structure. A schematic diagram of the proposed converter topology has been shown in Fig. In sect. 2, circuit description, working, and mode of operation have been detailed. This section has the calculation of voltage gain, inductor current, voltages across capacitors, and voltage stresses of devices also. Further, in Sect. 3, design parameters such as the value of inductances and capacitances have been calculated. Comparison with other existing topologies is shown in Sect. 4. Simulation results are shown in Sect. 5. Conclusions are written in Sect. 6.

2 Proposed Converter Steady-State Analysis in CCM

2.1 *Circuit Details*

The presented converter in Fig. 1 includes single switch S_1, S_2 two inductors L_1 and L_2, four capacitors indicated as C_1, C_2, C_3, and C_4, and the four diodes specified as D_1, D_2, D_3, D_4 and load resistance R. All capacitances and inductances are taken as large as that their voltages and currents respectively are with a very low ripple or constant. This converter has two operating states within one switching Period T_s. All analysis is done in CCM mode. All components are assumed without parasitic resistances.

2.2 *Modes of Operation*

The circuit has been analyzed in two modes: the first one is analyzed during switch ON and the second one is during switch OFF. Both modes of the circuit have been

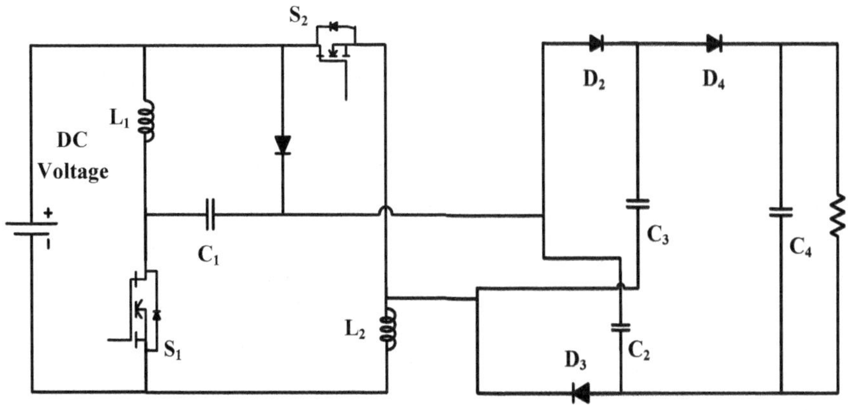

Fig. 1 Proposed converter

shown in Fig. 3, and some graphs related to converters such as diode voltage V_{D1}, inductor current I_{L1}, I_{L2}, and some more are in Fig. 2.

I. *When both Switches are ON* ($t_0 < t < t_1$)

When both switches are turned ON, in Fig. 3a, D1 and D4 conducts, while D_2, D_3 are operating in reverse biased conditions. The voltage across L_1 and input voltage V_{in} both are equal to each other, and inductor currents I_{L1} and I_{L2} rise also at the same time. Using Kirchhoff's voltage and current laws, the following voltages and currents are derived from Fig. 3a:

$$\begin{cases} L_1 \frac{dI_{L1}}{dt} = V_{L1} = V_{in} \\ L_2 \frac{dI_{L2}}{dt} = V_{L2} = -V_{C1} \\ C_1 \frac{dV_{C1}}{dt} = I_{C1} \\ C_2 \frac{dV_{C2}}{dt} = I_{C2} \\ C_3 \frac{dV_{C3}}{dt} = I_{C3} \\ C_4 \frac{dV_{C4}}{dt} = I_{C4} = I_{C2} - \frac{V_0}{R} = I_{C3} - I_0 \end{cases} \quad (1)$$

II. *When switches are OFF* ($t_1 < t < t_2$)

When switches are OFF, as shown in Fig. 3b D_2 and D_3 conduct while D_1 and D_4 are reverse biased. During the OFF period, both inductor currents decrease at the same time and the load is fed by capacitor C_4. The following are the derived voltage and current relations from Fig. 3b:

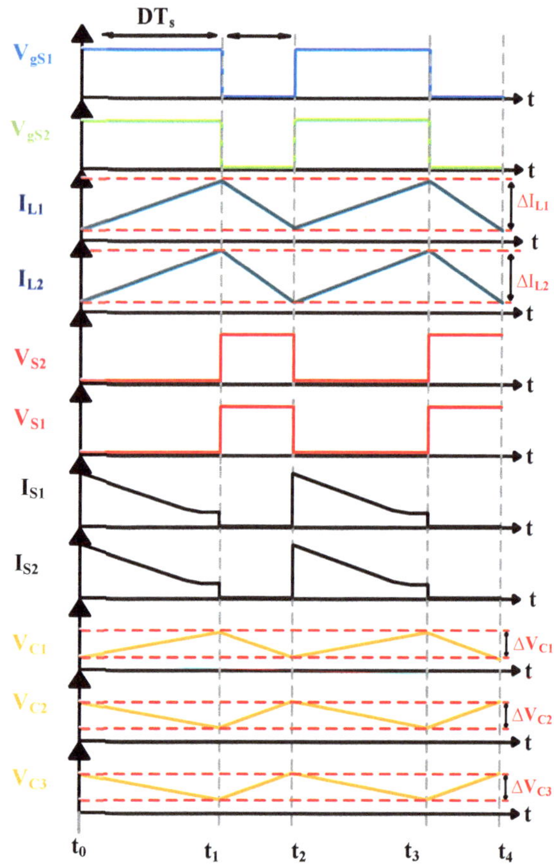

Fig. 2 Some key graphs of converters in CCM

$$\begin{cases} L_1 \frac{dI_{L1}}{dt} = V_{L1} = \frac{V_{in} - V_{C1} - V_{C2}}{2} \\ L_2 \frac{dI_{L2}}{dt} = V_{L2} = \frac{V_{in} - V_{C1} - V_{C2}}{2} \\ C_1 \frac{dV_{C1}}{dt} = I_{L1} \\ C_2 \frac{dV_{C2}}{dt} + C_3 \frac{dV_{C3}}{dt} = I_{L1} \\ C_4 \frac{dV_{C4}}{dt} = -\frac{V_0}{R} = -I_0 \\ V_{C2} = V_{C3} \end{cases} \tag{2}$$

Other results that can be interpreted from the circuit are as follows:

$$V_0 = V_{C2} + V_{C3}, I_{C2} = I_{C3}$$

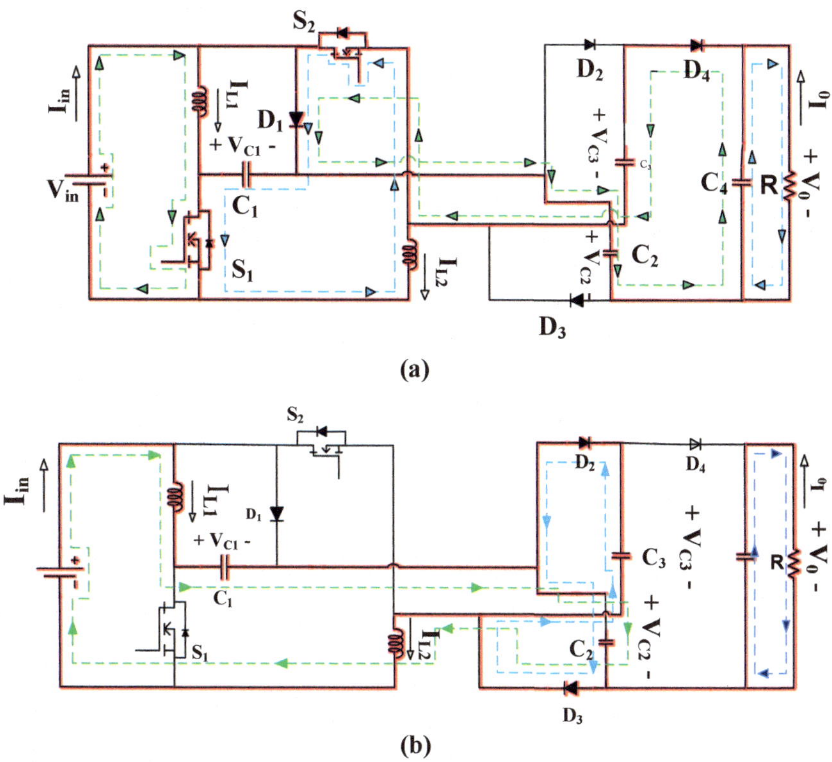

Fig. 3 Modes of operation in CCM **a** When the switches are ON **b** When switches are OFF

2.3 Voltage Gain (M) Calculation

Applying the volt-second balance principle on L_1 and using results from (1) and (2), the following equations have been derived:

$$DV_{in} + (1 - D)\left(V_{in} - \frac{V_0}{4}\right) = 0 \quad (3)$$

Using Equs. (1), (2), (3), the following results can be interpreted:

$$M = \frac{V_0}{V_{in}} = \frac{4}{1 - D} \quad (4)$$

2.4 Stress Calculations of Power Devices

(a) **Voltage Stress:**

The voltage stresses across switch S_1 and all diodes have been mentioned below:

$$\begin{cases} V_{S1} = \frac{V_0}{4}, V_{S2} = \frac{V_0}{4} \\ V_{D1} = \frac{V_0}{4}, V_{D2} = \frac{V_0}{2} \\ V_{D3} = \frac{V_0}{2}, V_{D4} = \frac{V_0}{2} \end{cases} \tag{5}$$

From (4) and (5), it can be observed that converter has a high gain with low voltage stress across diodes and switches, which will ultimately lead to the selection of low voltage rating devives and increases the efficiency of this proposed converter.

Current Stress:

Considering zero losses in the circuit (ideal conditions), input power is transferred to the load completely:

$$V_{in}I_{in} = V_0 I_0 \quad \mathrm{P_{in}} = \mathrm{P_{out}} \tag{6}$$

$$\frac{I_{in}}{I_0} = \frac{V_0}{V_{in}} = M = \frac{4}{1 - D}$$

From Fig 3.

$$I_{in} = \frac{4}{1 \quad D} I_0 \tag{7}$$

Applying ampere second balance on C_1 and C_2, we can derive the average inductor currents as:

$$I_{L1} = I_{L2} = \frac{2}{1 - D} I_0 \tag{8}$$

Therefore, current stress across switches and diodes are as follows:

$$\begin{cases} I_{S1} = \frac{2}{1-D} I_0 \\ I_{S2} = \frac{1+D}{1-D} I_0 \\ I_{D1} = I_{D2} = I_{D3} = I_{D4} = I_0 \end{cases} \tag{9}$$

3 Design Parameters

A. *Duty Cycle Calculation*

To achieve the required output voltage at a given input voltage, duty cycle D of the converter can be calculated using (4):

$$D = \frac{V_0 - 4V_{in}}{V_0} \tag{10}$$

B. *Passive Component Design:*

For a given suitable value of ripple of inductor currents ΔI_{L1} and ΔI_{L2} at a fixed value of switching frequency f_s for this converter, inductances can be extracted from (1) as:

$$\begin{cases} L_1 = \frac{V_{in}D}{\Delta I_{L1} f_s} \\ L_2 = \frac{V_{in}D}{\Delta I_{L2} f_s} \end{cases} \tag{11}$$

Moreover, within a valid range of voltage ripple capacitances can be taken from (2) as:

$$\begin{cases} C_1 = \frac{2V_0}{R\Delta V_{C1} fs} \\ C_2 = \frac{V_0}{R\Delta V_{C2} fs} \\ C_3 = \frac{V_0}{R\Delta V_{C3} fs} \\ C_4 = \frac{(1-D)V_0}{R\Delta V_{C4} fs} \end{cases} \tag{12}$$

By knowing the value of D, f_s, V_0, V_{in}, R and suitable current and voltage ripples, convenient inductors and capacitors can be chosen.

4 Comparison Among Other High Gain Converters

Comparison among different topologies with relatively high gains in terms of some components and other expressions are listed in Table 1. Comparison has been shown based on the following: the number of components, voltage stresses on switch, and voltage gain. Figure 4 presents the graph of voltage gain versus duty ratio (D) for the converters listed in Table 1 from where it can be inferred that this particular proposed converter has the highest gain among all the topologies. The converter proposed in [14] has utilized a total of 16 components but its gain is essentially less than the proposed converter. Other converters have less components than the proposed converter but their gain is also much lower. In Fig. 5, the graph of voltage

Table 1 Comparison between various high gain converters

Number of elements/converters	Conventional boost converter	Converter in [11]	Converter in [12]	Converter in [13]	Converter in [14]	Proposed topology
Switches	1	2	1	2	2	2
Inductors	1	2	1	2	4	2
Diodes	1	2	3	2	9	4
Capacitors	1	3	3	2	1	4
Voltage Gain (V_o/V_i)	$\frac{1}{1-D}$	$\frac{2}{1-D}$	$\frac{3D}{1-D}$	$\left(\frac{D}{1-D}\right)^2$	$1+(\frac{4D}{1-D})$	$\frac{4}{1-D}$
Voltage Stress on Switches(V_s)	V_o	$S_1 \rightarrow \frac{V_o}{4}$. $S_2 \rightarrow \frac{V_o}{4}$	$\frac{2V_o}{3}$	$S_1 \rightarrow \left(\frac{1-D}{D^2}\right)V_0$ $S_2 \rightarrow \frac{V_o}{D}$	$S_1 \rightarrow \left(\frac{1+D}{1+3D}\right)V_0$ $S_2 \rightarrow \left(\frac{2D}{1+3D}\right)V_o$	$S_1 \rightarrow \frac{V_o}{4}$, $S_2 \rightarrow \frac{V_o}{4}$

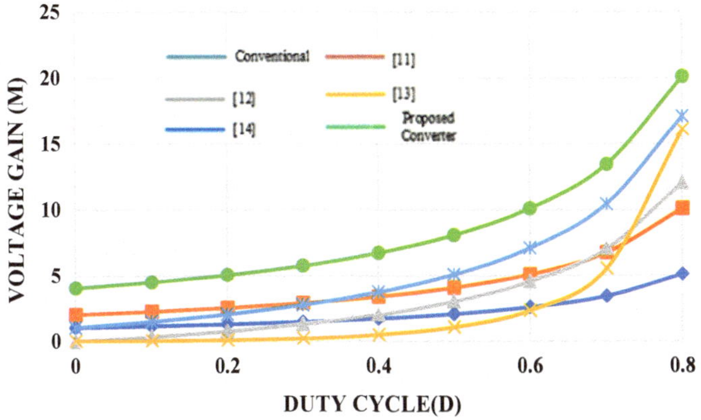

Fig. 4 Voltage gain

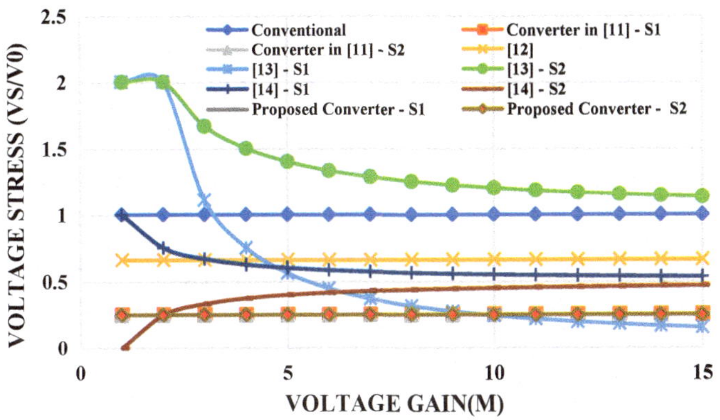

Fig. 5 Voltage stress

stress on switches versus is shown. The power switches of the proposed converter have one of the lowest voltage stresses among other listed converters at a particular gain.

5 Simulation Results

Simulation results have been drawn to verify and conclude that this circuit works with the agreement of the above theoretical analysis and to testify the steady-state analysis of the proposed converter. Detailed simulations of the circuit have been carried out using PLECS software. Table 2 has all values of the circuit parameters, which have been taken for simulation purposes. Diodes have been taken as ideal

Table 2 Specifications of the proposed converter

Parameters	Values
Switching frequency (f_s)	50 kHz
Inductor (L_1)	0.5 mH
Inductor (L_2)	0.5 mH
Capacitor (C_1)	80 μF
Capacitor (C_2)	80 μF
Capacitor (C_3)	80 μF
Capacitor (C_4)	80 μF
R_{on} of switch	0.1 $m\Omega$
Load resistance (R)	150 Ω
Duty Cycle (D)	0.4
Input Voltage (V_{in})	12 V

with zero forward voltage drop and zero R_{on}. From (1) and (2), it has been drawn out that $V_{C3} = V_0/2$. Theoretically, at given 12 V input voltage with a duty cycle of 0.4 output voltage comes out of 66 V and V_{C3} is 33 V. After the simulation, these values have been found closely near to theoretical values (Fig. 6).

6 Conclusions

The proposed converter is derived from the H bridge structure of the inverter. The converter gain is considerably higher than the conventional boost converter. All calculations have been enumerated under ideal conditions and in CCM mode. The converter proposed has relatively high gain and a low voltage stress across its components such as switches and diodes. Operating principles and theoretical analysis with simulation have been documented. Although the input current is not continuous, a small LC filter can be used to make the current continuous at the input. The proposed boost converter can be used in solar PV application automobile lamps and DC microgrids applications.

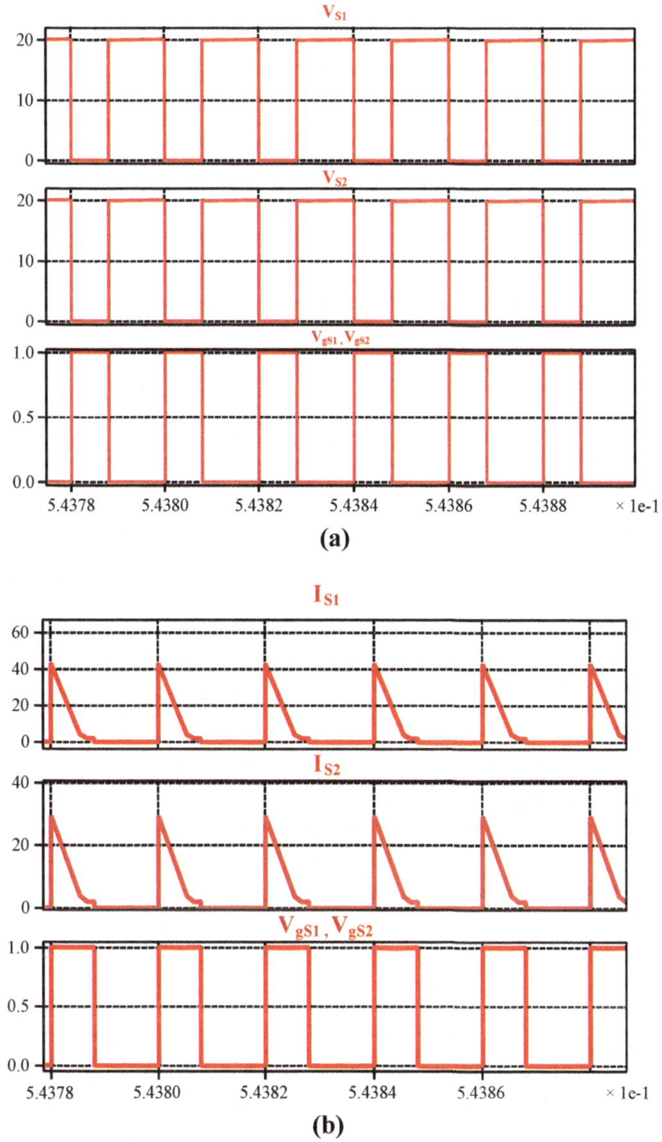

Fig. 6 Simulation results: **a** V_{S1}, V_{S2} and V_{gS1}, V_{gS2} **b** I_{S1}, I_{S2} and V_{gS1}, V_{gS2} **c** V_{C1}, V_{C2}, V_{C3} and V_{gS1}, V_{gS2} **d** V_{D1}, V_{D2}, V_{D4} and V_{gS1}, V_{gS2}

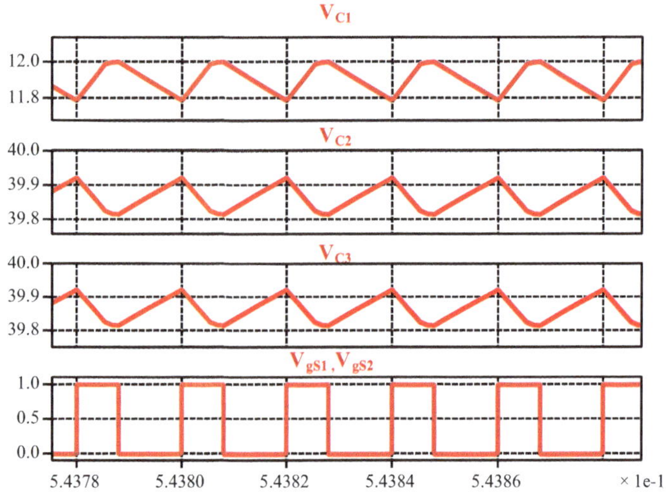

(c)

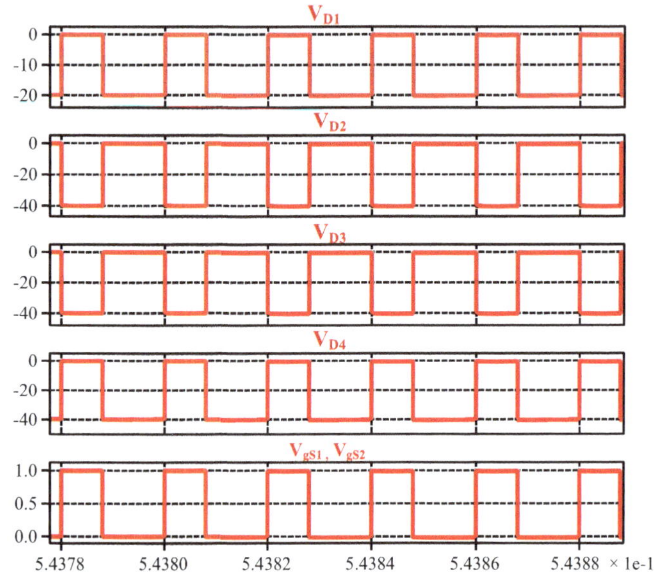

(d)

Fig. 6 (continued)

References

1. Sadaf S, Mahajan SB, Meraj M, Iqbal A, Alemadi N Transformer-less boost converter with reduced voltage stress for high voltage step-up applications.IEEE Trans Ind Electron https://doi.org/10.1109/TIE.2021.3055166
2. Sarikhani A, Allahverdinejad B, Hamzeh MA non-isolated buck-boost DC-DC converter with continuous input current for photovoltaic applications. IEEE J Emerg Sel Top Power Electron
3. Lee S, Do H (2019) Quadratic boost DC–DC converter with high voltage gain and reduced voltage stresses. IEEE Trans Power Electron 34(3):23972404. https://doi.org/10.1109/TPEL.2018.2842051
4. Lakshmi M, Hemamalini S (2018) Nonisolated high Gain DC–DC converter for DC microgrids. IEEE Trans Industr Electron 65(2):12051212. https://doi.org/10.1109/TIE.2017.2733463
5. Zeng Y, Li H, Wang W, Zhang B, Zheng TQ (2020) Cost-effective clamping capacitor boost converter with high voltage gain. IET Power Electron 13(9):1775–1786
6. Dobakhshari SS, Fathi SH, Milimonfared J, Tazehkand MZ (2021) A dual active clamp DC–DC converter with high voltage gain. IEEE Trans Power Electron 36(1):597606. https://doi.org/10.1109/TPEL.2020.3000460
7. Kumar A et al (2020)Switched-LC based high gain converter with lower component count.IEEE Trans Ind Appl 56(3):2816–2827. https://doi.org/10.1109/TIA.2020.2980215
8. Xu B, Gao C, Zhang J, Yang J, Xia B, He Z (2021) A novel DC chopper topology for VSC-based offshore wind farm connection. IEEE Trans Power Electron 36(3):30173027. https://doi.org/10.1109/TPEL.2020.3015979
9. Zhao J, Chen D, Jiang J (2021) Transformerless high step-up DC-DC converter with low voltage stress for fuel cells. IEEE Access 9:1022810238. https://doi.org/10.1109/ACCESS.2021.3050546
10. Afzal R, Tang Y, Tong H, Guo Y (2021) A high step-up integrated coupled inductor-capacitor DC-DC converter. IEEE Access 9:1108011090. https://doi.org/10.1109/ACCESS.2020.3048354
11. Nguyen AD, Jason Lai J, Chiu H (2019)Analysis and implementation of a new non-isolated high-voltage-gain boost converter. In: 2019 IEEE energy conversion congress and exposition (ECCE), Baltimore, MD, USA, pp 1251–1255
12. Mohammadzadeh Shahir F, Babaei E (2017)A new structure for non-isolated boost dc- dc converter based on voltage-lift technique. In: 2017 8th power electronics, drive systems & technologies conference (PEDSTC). Mashhad, pp 25-30
13. Miao S, Wang F, Ma X (2016) A new transformerless buck-boost converter with positive output voltage. IEEE Trans Industr Electron 63(5):29652975. https://doi.org/10.1109/TIE.2016.2518118
14. Gupta N et al (2020)Novel non-isolated quad-switched inductor double-switch converter for DC microgrid application. In:2020 IEEE (EEEIC/I&CPS Europe). Madrid, Spain, 2020, pp 1–6.https://doi.org/10.1109/EEEIC/ICPSEurope49358.2020.9160839

Investigation into the Correlation of SFRA Numerical Indices and Short Circuit Reactance Measurements of Transformers

V. Sreeram, S. Sudhakara Reddy, T. Gurudev, M. Maroti, and M. Rajkumar

Abstract Sweep frequency response analysis (SFRA) has been in use for condition monitoring of transformers for two decades. However, there have been concerted efforts to derive an objective evaluation criterion for SFRA measurements. The short circuit reactance/impedance measurement has been used as a direct indication of winding movement in transformers, mostly in the context of evaluation of short circuit performance as prescribed by IEC and IEEE standards. Derived numerical indices have formed one method of interpretation of SFRA data. This paper aims to investigate the correlation between various numerical indices derived from SFRA and short circuit reactance measurements performed as part of an evaluation of the short circuit test.

Keywords Frequency response · Power transformers · Statistical analysis

1 Introduction

The sweep frequency response analysis technique has been a foremost tool in condition monitoring of transformers. The method which measures the transfer function of the transformer and expresses it in the frequency domain theoretically captures data that is reflective of the characteristics of all its components. However, the interpretation of the data to infer the characteristics of its components has been proven to be vexing. Direct reconstruction of the transformer internal structure with sufficient high-level resolution is likely to remain a theoretical possibility at least for the near future since the transformer is a very complex electromagnetic structure. Hence SFRA has been used to predict a defect in the internal structure of the transformer by comparing the traces before and after events or a significant maintenance window. The evaluation of differences between the measurements has been done using subjective expert knowledge or by using statistical techniques applied to data sets. Statistical

V. Sreeram (✉) · S. S. Reddy · T. Gurudev · M. Maroti · M. Rajkumar
High Power Laboratory, Central Power Research Institute, Bengaluru, India
e-mail: sreeram@cpri.in

A. Tomar et al. (eds.), *Machine Learning, Advances in Computing, Renewable Energy and Communication*, Lecture Notes in Electrical Engineering 768,
https://doi.org/10.1007/978-981-16-2354-7_50

techniques have produced numerical indices which give a single number indication of the differences between the before and after SFRA traces.

CIGRE working group A2.53 had been constituted to formulate an evaluation plan for SFRA measurements which can be used to produce objective judgments that can lead to an evaluation criterion for the short circuit tests in the future [1]. The main objective evaluation criterion for short circuit tests as per the present IEC [2] and IEEE [3] standards is the short circuit reactance or leakage impedance measurement. This has been the gold standard for short circuit performance for many decades and has served well as a measure of mechanical damage inside the transformer. However, the SFRA measurement has the potential to serve as a better indicator of mechanical damage than the short circuit reactance measurement which is essentially a sample of a simplified transfer function at one frequency.

SFRA interpretation has been carried out through several techniques. The most prominent ones are:

(1) Comparison of traces using circuit modelling
(2) Comparison of traces using artificial intelligence techniques
(3) Comparison of traces using numerical indices
(4) Comparison of traces by estimation of a rational function.

Abundant literature is available for comparison using numerical indices because of its objective nature and simplicity in computations. The Chinese standard DL/T 911-2004 [4] even proposed an evaluation criterion based on a numerical index computed for different frequency ranges.

This paper uses the SFRA measurements and short circuit reactance measurements before and after short circuit tests on a few transformers, conducted at High Power Laboratory, Central Power Research Institute, Bengaluru, India to investigate the correlation between the two. The main aim is to identify the presence of an SFRA numerical index that has the potential to directly replace the short circuit reactance measurement by eliciting a strong correlation with it.

2 SFRA and Evaluation by Numerical Indices

The transfer function of any system, generally expressed in the frequency domain, describes the system behaviour exhaustively. The transfer function when experimentally determined has to be in the time domain. The transfer function when expressed in the time domain is the impulse response of the system. Hence two variations have developed in the transfer function method applied to transformers—impulse frequency response analysis (IFRA) and sweep frequency response analysis (SFRA). IFRA involves applying an impulse to the transformer terminal and measuring the time-domain response which can deliver the frequency-domain response through Fourier transform. SFRA involves the direct measurement of transformer response for a sufficient number of frequency points in a given range of frequency. The SFRA has since then emerged as the preferred of the two methods of frequency response analysis. IEC 60076-18 was published in 2012 giving guidelines for measurements and recommended practices [5]. The corresponding IEEE standard C57.149 was also published in 2012 elaborating measurement methods and recommended practices [6]. Both standards do not, however, specify a criterion for interpretation of the measurements. The Chinese standard mentioned earlier is the only standard outlining an interpretation criterion [4]. Samimi and Tenbohlen [7] and Samimi et al. [8] present various numerical indices which can be used in the interpretation of SFRA results and their relative advantages. A plethora of numerical indices are available in the literature. The following indices are chosen for this study [7, 8]:

1. Correlation Coefficient:

$$CC = \frac{\sum_{i=1}^{N} Y(i) \cdot X(i)}{\sqrt{\sum_{i=1}^{N} [X(i)]^2 \sum_{i=1}^{N} [Y(i)]^2}} \tag{1}$$

2. Sum Squared Ratio Error

$$SSRE = \frac{\sum_{i=1}^{N} (\frac{Y(i)}{X(i)} - 1)^2}{N} \tag{2}$$

3. Root Mean Square Error

$$RMSE = \sqrt{\frac{1}{N} \sum_{i=1}^{N} (\frac{(|Y(i)| - |X(i)|}{\frac{1}{N} \sum_{i=1}^{N} |X(i)|})^2} \tag{3}$$

4. Absolute Sum of Logarithmic Error

$$ASLE = \frac{\sum_{i=1}^{N} |20 \log Y(i) - 20 \log X(i)|}{N} \tag{4}$$

5. Lin's Concordance Coefficient

$$LCC = \frac{2S_{XY}}{(\overline{Y} - \overline{X})^2 + S_X^2 + S_Y^2} \tag{5}$$

$$S_{XY} = \frac{\sum_{i=1}^{N} (Y(i) - \overline{Y}) \cdot (X(i) - \overline{X})}{N} \tag{6}$$

$$S_X^2 = \frac{\sum_{i=1}^{N} (X(i) - \overline{X})^2}{N}; \; S_Y^2 = \frac{\sum_{i=1}^{N} (Y(i) - \overline{Y})^2}{N} \tag{7}$$

$$\overline{Y} = \frac{\sum_{i=1}^{N} Y(i)}{N}; \; \overline{X} = \frac{\sum_{i=1}^{N} X(i)}{N} \tag{8}$$

In the above equations $Y(i)$ and $X(i)$ represent the SFRA traces after and before events or maintenance periods. The above-mentioned five indices are those that treat all data points equally. The SFRA measurements will contain crests and troughs corresponding to resonance and anti-resonance frequencies. The values of resonance and anti-resonance frequencies are representative of the internal structure of the transformer and hence some indices are calculated based on the change in the resonance frequencies. Hence two indices based on the resonance and anti-resonance frequencies are also used in this paper [7, 8]:

1. Index of Frequency Deviation

$$\mathrm{IFD} = \sum_{i=1}^{M} \frac{|f_y(i) - f_x(i)|}{|f_x(i)|} \tag{9}$$

2. Index of Amplitude Deviation

$$\mathrm{IAD} = \sum_{i=1}^{M} \frac{|A_y(i) - A_x(i)|}{|A_x(i)|} \tag{10}$$

In the above equations, f_y and f_x represent the resonance and anti-resonance frequencies of the SFRA traces after and before events, respectively, and A_y and A_x represent the corresponding amplitudes.

3 Short Circuit Testing of Transformers—SFRA and Short Circuit Reactance Measurements

The short circuit test on transformers is a special test which is conducted to verify the ability of transformers to withstand the dynamic effects of short circuit. The test is listed as a special test in IEC 60076-1 and the test is conducted as per IEC 60076-5 [2]. The corresponding IEEE standard is C57.12.90 which describes many tests including the short circuit test [3]. The transformer is evaluated based on its behaviour during the short circuit test and its ability to withstand the listed routine tests afterwards. The change in the short circuit reactance is measured as per the IEC standard and compared to thresholds. As per IEEE standard, the short circuit leakage impedance is measured. The final step is the out-of-tank inspection of the core-coil-assembly of the transformer to check for mechanical damages or signs of electrical discharge. The short circuit test is one of the most strenuous tests for the transformer and provides a unique opportunity for a total investigation of the transformer. High Power Laboratory, Central Power Research Institute, Bengaluru has been conducting short circuit testing on transformers for more than two decades. This paper uses the data on eight oil-immersed transformers that were short circuit tested at High Power Laboratory. The SFRA measurements are not presently part of the evaluation of

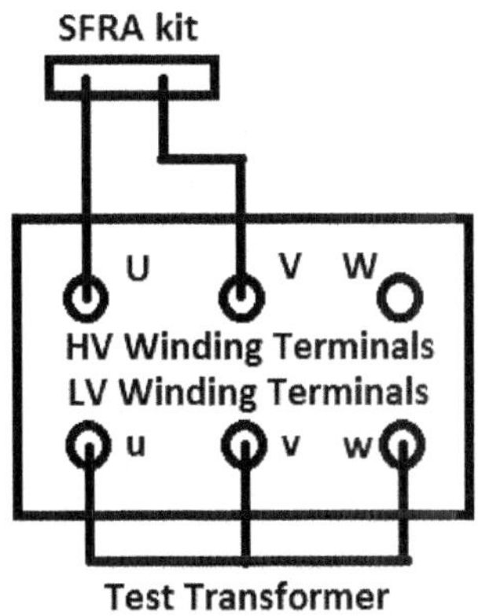

Fig. 1 Measurement configuration for SFRA and short circuit reactance

performance in short circuit tests as per any standard. They are, however, conducted in many cases for additional diagnostic information. The transformers listed in the study had SFRA measurements conducted before and after the short circuit test. The short circuit reactance measurements were also carried out as per the routine procedure. The short circuit reactance was indeed the main criterion of evaluation. The measurements were carried out on the high-voltage windings of the transformers while the low-voltage windings were short-circuited, which is the configuration for short circuit tests as well. The measurements configuration are depicted in Fig. 1.

The measurements were carried out across two phases or across one phase and neutral, depending on the vector groups and ratings of the transformers. For large power transformers with higher voltage ratings, short circuit tests are carried out on individual phases separately. When the high-voltage windings are star-connected and neutral is brought out as is the case with most large power transformers, the SFRA measurements are carried out across individual phases and neutral. The short circuit reactance is measured across two phases and hence the individual phase values are derived through calculation. When the SFRA measurements are carried out across phases, no computation is required on the short circuit reactance values and they can be used as such. The SFRA measurements were carried out using MEGGER FRAX 101 SFRA test kit and the short circuit reactance measurements were carried out using an inductance–capacitance meter at a frequency of 1 kHz using an inductance–resistance series equivalent circuit. The SFRA measurements were carried out in the IEC specified range of 20 Hz to 2 MHz. The transformers were selected from across the spectrum of voltage and power ratings. The details are listed in Table 1.

4 Methodology and Results

After the SFRA and short circuit reactance measurements were done, the seven numerical indices were calculated in each case. Measurements were carried out on three-phase combinations for each transformer in Table 1, while nine-phase combinations were carried out for transformers in S. no. 6. The transformer in S. no. 8 is

Table 1 Details of transformers included in the study

S. no	Power rating (MVA)	Voltage rating (kV)	Vector group
1	31.5	132/33	YNyn0
2	50	132/33	YNyn0
3	2	11/0.433	Dyn1
4	16	11/11.5	YNd11
5	31.5	132/33	YNyn0
6	160	220/132/33	YNa0d11
7	63.5	220/33	YNyn0
8	5	$(132/\sqrt{3})/33$	Single phase

a single-phase transformer and hence only one set of measurements is available. A total of 28 measurement combinations were used in the study as listed in Table 2.

The first five numerical indices were calculated in the full range of frequencies but the last two were calculated up to 1 MHz only. This was done since the portion after 1 MHz is especially prone to measurement errors due to high-frequency noise. The resonance and anti-resonance frequencies were identified by checking the slope of the magnitude curve. The measurement uncertainties also cause the curve to have local maxima and minima which are insignificant. Hence those significant resonant and anti-resonant points which were present in both the curves were chosen out of those detected. None of the SFRA curves after the short circuit test had the appearance or disappearance of resonance and anti-resonance points compared to curves before the short circuit test—there were only changes to the magnitude of response and location. All the indices can be calculated using absolute values or dB values. The dB values were used for the calculation of all indices, whereas for ASLE it is the only option. The first numerical index is the correlation coefficient (CC). CC and LCC, however, have a value of 1 for a perfectly matched case and 0 for perfect mismatch while it is inverse for the others. Hence 1-CC and 1-LCC are considered for normalisation.

Table 2 Measurement combinations

S. no. of transformer	Tap no	Measurement combination
1	1	U-N; V-N; W-N
2	1	U-N; V-N; W-N
3	1	U-W; V-U; W-V
4	1	U-N; V-N; W-N
5	1	U-N; V-N; W-N
6	1	U-N; V-N; W-N
6	13	U-N; V-N; W-N
6	17	U-N; V-N; W-N
7	1	U-N; V-N; W-N
8	1	1.1–1.2

Table 3 Correlation coefficient

S. no	Numerical index	Correlation coefficient
1	1-CC	0.284
2	SSRE	−0.230
3	RMSE	0.203
4	ASLE	0.187
5	1-LCC	0.318
6	IAD	0.173
7	IFD	0.145

The correlation coefficient of the seven numerical indices with percentage variation in short circuit reactance was calculated and are presented in Table 3.

From Table 3, it can be inferred that the numerical indices are not showing a significant correlation with the short circuit reactance measurements. Of all the indices, it is 1-LCC that shows the greatest correlation. The scatter plot of 1-CC against percentage variation in short circuit reactance is shown in Fig. 2.

The first five indices can be improved by adding the phase data to the magnitude data [9]. When the SFRA measurements are carried out, the transfer function data is recorded as two separate plots—magnitude plot and phase plot. In order to add the phase data, the transfer function data was converted to a complex form and the real part was used for calculating the indices. The correlation coefficient was calculated and is presented in Table 4.

Table 4 shows that the addition of phase data has not significantly improved the correlation. However, 1-CC still shows the strongest correlation and it has improved

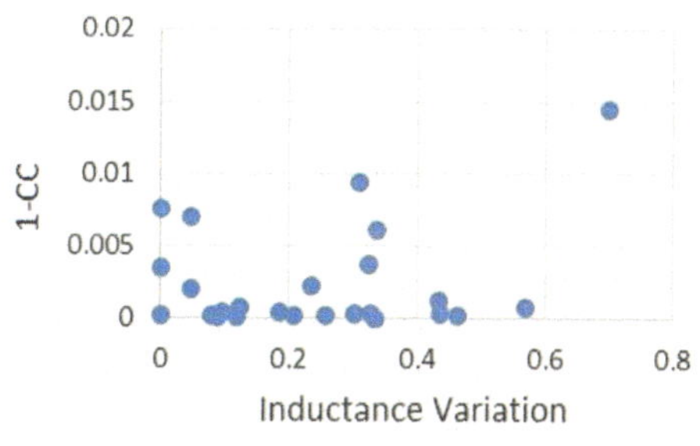

Fig. 2 Scatter plot of 1-CC versus percentage variation in short circuit reactance

Table 4 Correlation coefficient with real components

S. no	Numerical index	Correlation coefficient
1	1-CC	0.332
2	SSRE	−0.129
3	RMSE	0.162
4	ASLE	0.213
5	1-LCC	0.312

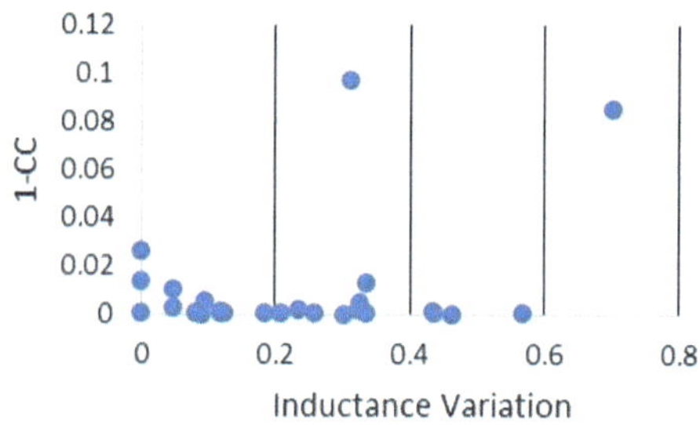

Fig. 3 Scatter plot of 1-CC with real components versus percentage variation in short circuit reactance

by an insignificant measure. The scatter plot of 1-CC versus percentage variation in short circuit reactance is shown in Fig. 3.

The measurements of SFRA carried out corresponding to the short circuit measurements prescribed in IEC 60076-18. In this configuration, the properties of the transformer core do not have an effect on the traces. However, [1] suggests that changes in the winding structure and inter-winding interaction is mostly evident in the frequency region from 2 kHz to 1 MHz. Below 2 kHz, the properties of the core dominate and beyond 1 MHz, the terminal connections influence the traces. Hence the indices were re-calculated for this particular region and the correlation coefficient was determined. The correlation coefficient with complete and real components are presented in Tables 5 and 6, respectively.

The removal of error-prone low- and high-frequency regions do not result in a higher correlation with short circuit reactance measurements. In fact, the highest correlation coefficient achieved is 0.343 for 1-CC in the range. The scatter plots of 1-CC with percentage variation in short circuit reactance in the two cases are shown in Figs. 4 and 5.

Table 5 Correlation coefficient from 2 kHz to 1 MHz

S. no	Numerical index	Correlation coefficient
1	1-CC	0.343
2	SSRE	0.107
3	RMSE	0.174
4	ASLE	0.187
5	1-LCC	0.244

Table 6 Correlation coefficient with real components from 2 kHz to 1 MHz

S. no	Numerical index	Correlation coefficient
1	1-CC	0.291
2	SSRE	0.084
3	RMSE	0.146
4	ASLE	0.188
5	1-LCC	0.255

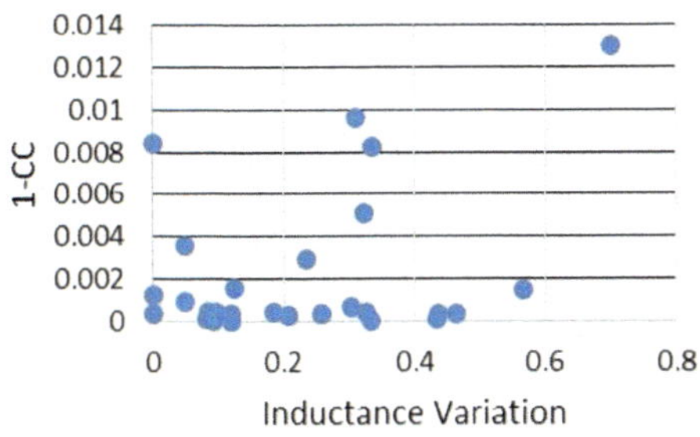

Fig. 4 Scatter plot of 1-CC from 2 kHz to 1 MHz versus percentage variation in short circuit reactance

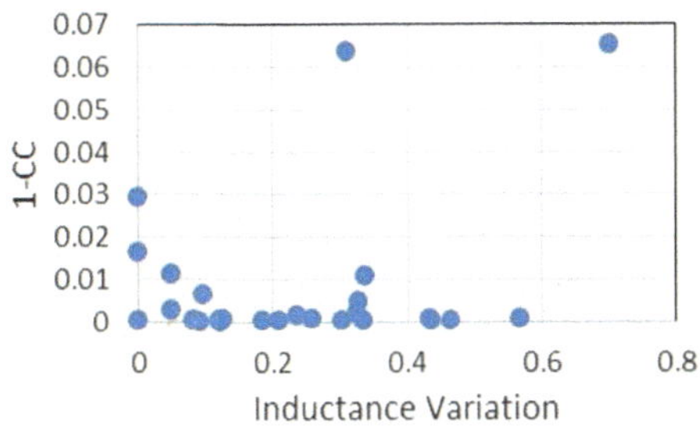

Fig. 5 Scatter plot of 1-CC from 2 kHz to 1 MHz with real components versus percentage variation in short circuit reactance

Hence it may be possible that the numerical indices presently under investigation do not actually serve to directly replace short circuit reactance measurements by strong correlation. However, it may be possible to derive a numerical index that is more sensitive than short circuit reactance changes, to structural changes in the transformer which may replace short circuit reactance as the main evaluation criterion for short circuit tests.

5 Conclusion

The SFRA technique is currently the mainstay of transformer diagnostics. The interpretation of SFRA data is largely done through the comparison and computation of numerical indices. Efforts are on to establish SFRA as a major evaluation criterion of short circuit tests on transformers. Furthering this effort, this paper has investigated the correlation between a few commonly used SFRA numerical indices and short circuit reactance measurements, which is presently a major criterion of evaluation of short circuit tests. However, the indices investigated did not show a strong linear correlation with short circuit reactance measurements. The indices may share a nonlinear relationship or may turn out to be more sensitive to winding structural changes than short circuit reactance measurements. Moreover, testing data across voltage and power ratings of transformers are relatively scarce and do not provide for a statistical study with large sample size. Further investigation is required before the indices can be accepted as an evaluation criterion.

Acknowledgements The authors would like to express their profound gratitude to the management of Central Power Research Institute, Bengaluru, India for the facilities provided for carrying out this work.

References

1. Pitcher P, Tenbohlen S, Lachman M, Scardazzi A, Patel P (2017) Current state of transformer FRA interpretation. Procedia Eng 202:3–12
2. IEC 60076–5: Ability to withstand short circuit (2006)
3. C57.12.90: IEEE standard test code for liquid-immersed distribution, power, and regulating transformers (2015)
4. DL/T 911, 2016: frequency response analysis on winding deformation of power transformers
5. IEC 60076–18: Measurement of frequency response (2012)
6. C57.149: IEEE guide for the application and interpretation of frequency response analysis for oil-immersed transformers (2012)
7. Samimi MH, Tenbohlen S (2017) FRA interpretation using numerical indices: state-of-the-art. Electri Power Energy Syst 89:115–125
8. Samimi MH, Tenbohlen S, Shayegani AA, Mohseni H (2017) Evaluation of numerical indices for the assessment of transformer frequency response. IET Generat Trans Distribut 11(1):218–227
9. Samimi MH, Tenbohlen S, Shayegani AA, Mohseni H (2016) Improving the numerical indices proposed for the FRA interpretation by including the phase response. Electri Power Energy Syst 83:585–593

Mineral Oil-Filled Transformer DGA from Detective Correction to Strategic Prevention

G. T. Naidu, U. Mohan Rao, and Suresh Kumar Sudabattula

Abstract Power transformers, being one of the most vital among the major assets in an electric power system network, require proper care during service. The electrical health of a power transformer is determined primarily by the quality of insulating oil which needs proper attention. The present paper discusses offline dissolved gas analysis with in-service transformer case studies. Additionally, an expert system-based procedure is proposed to avoid the violations in the existing diagnostic methods. The discussions span to highlight important developments in this area and cover the latest trends in the field of DGA that move it from detective corrective mode to strategic preventive mode, which would increase the accuracy of DGA and prove beneficial for utility engineers and technologists.

Keywords DGA · Expert system · Transformer

1 Introduction

Transformers are one of the important equipments in the electrical network whose unexpected or unscheduled failure leads to a lot of economic damage accompanied with reduced reliability and stability of power system network [1]. This may also lead to interruption to the power supply and possible failures. The cost of these failures includes equipment replacement cost, cleanup, and may also include penalties for the unplanned outages. Considering all these undesirable factors gives the importance of condition monitoring of a transformer. Thus, *"better is the condition monitoring better is the risk assessment of a power transformer"*. There exist many tools for the condition monitoring of a power transformer in which a dissolved gas analyzer (DGA) is one of the most effective tools to diagnose a transformer and estimate its

G. T. Naidu · S. K. Sudabattula
School of Electronics and Electrical Engineering, Lovely Professional University, Phagwara, India

U. M. Rao (✉)
Department of Applied Sciences, Université du Québec À Chicoutimi, Saguenay, QC G77J3, Canada
e-mail: mohan.ungarala1@uqac.ca

A. Tomar et al. (eds.), *Machine Learning, Advances in Computing, Renewable Energy and Communication*, Lecture Notes in Electrical Engineering 768,
https://doi.org/10.1007/978-981-16-2354-7_51

condition. Here transformer insulating liquid is analyzed for dissolved gases and the results are interpreted using analytical and interpretive methods, including key gas method [2], Dornerburg [3], Rogers ratio method [2], and Duval methods [4]. Therefore, this comes under detective corrective mode because the oil analysis should detect the problem that needs to be corrected manually to maintain the good health of a transformer. This can be overcome if the transformer is kept under continuous monitoring using online methods at regular intervals and store the analyzed results data in a database. This can be much developed if we introduce an expert system to the existing system functioning with a rule base. The rule base is to be made from the database stored from the test history. Interpreting this data gives a picture that helps to understand the condition of the transformers. Thus an appropriate action may be taken to premature aging and catastrophic failures. One may refer to this method of avoiding failures progress as "strategic preventative". DGA of the transformer insulating fluid is a widely used diagnostic tool to access the overall condition of the insulation system [5]. With improvements in scientific research and due to technological advancements offline DGA (collecting oil sample and bringing it to the laboratory for testing) is shifted to the online monitoring of fault gasses (online DGA). This enables monitoring of the transformer with enhanced reliability and an effective way of monitoring.

2 Offline DGA and Its Practical Approach

2.1 Case Study

A transformer of 16/20MVA, 132/11 kV, Mfg.Year-2001, Comm.Date-07-04-2004 has been tested for dissolved gases on 13-04-2018 and the DGA reports are shown in Table 1.

Table 1 Sample report of DGA for an in-service transformer

S. no	Dissolved gases and test date	Gases (13.04.2018) (ppm)	Gases (25.05.2018) (ppm)
1	Hydrogen (H_2)	<0.5	0.5
2	Carbon dioxide (CO_2)	12850	12441
3	Carbon monoxide (CO)	283	258
4	Ethylene (C_2H_4)	07	05
5	Ethane(C_2H_6)	08	06
6	Methane(CH_4)	10	08
7	Acetylene (C_2H_2)	<0.5	<0.5

Therefore, the total dissolved combustible gas (TDCG) = 0.5 + 283 + 7 + 8 + 10 + 0.5 = 309. On observing the gas composition according to *IS-10593/1992 standards*, it has been found that CO_2 gas is in the warning stage. This indicates that excess CO_2 is being generated in the transformer which may be a result of pyrolysis of cellulose or simply due to excess temperature. So the cooling equipment of the unit is enhanced in order to maintain temperature balance and proper loading of the transformer is assured. But the frequency of fault gas monitoring may be increased. Thus accordingly, the DGA test is scheduled for a period of 40 days. It is to be mentioned that temperature and loading are properly maintained after a period of 40 days and performed DGA on 24-5-2018 and the gas composition shown in Table 1 is noticed. Total dissolved combustible gas (TDCG) is noticed as (0.5 + 258 + 5 + 6 + 8 + 0.5) = 278 (Fig. 1).

It is seen that TDGC and all the individual gas composition is reduced as depicted in Fig. 2. However, as per IS-10593, CO_2 is still in the warning stage. Therefore, it is assumed that high CO_2 may be due to thermal degradation of insulation paper which may be confirmed by furan analysis or methanol measurement. To conclude with the diagnosis of this unit based on fault gas analysis, DGA directed the path to diagnostic action but DGA alone failed to give a complete solution for diagnosis. In general, condition monitoring of transformers is mostly based on scheduled activity. For instance, if a maintenance schedule period is planned for six months, where fault gas analysis results are normal for a test date, the next test date is scheduled after six months from the date of the previous test.

Thus, any fault occurrence in the duration between the first test and second test may not be detectable immediately. If the fault intensity is low, it may be less detrimental to the insulation system. In case the fault is of high energy discharge or of high intensity, this may lead to total flashover of the system. In addition, manual sampling in the offline DGA may also include scope for the misleading of fault gas analysis. For instance, poor sampling technique may allow liquid to interact with environmental

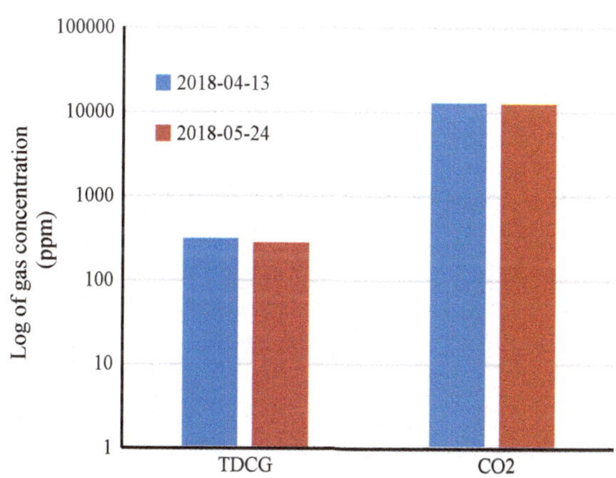

Fig. 1 Comparison of concentration of CO_2 and TDGC on test dates

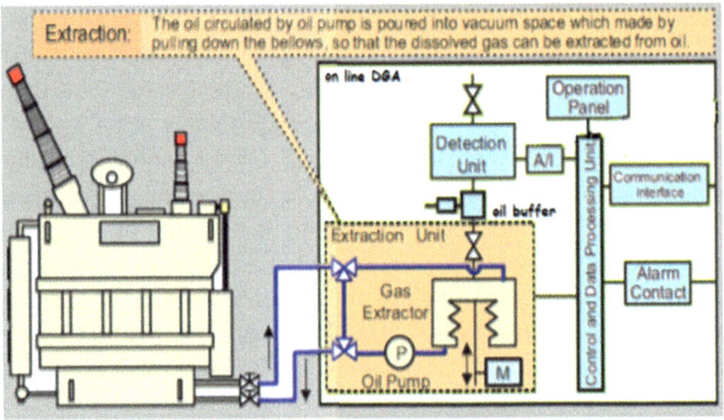

Fig. 2 Representation of online DGA

air and external moisture. This will manipulate the actual oxidation and hydrolysis byproducts in the oil. In other words, there is also scope for gases to escape from the liquid phase to the solid phase. Therefore, there is a high scope for the DGA results of the same sample to vary from a series of tests or one laboratory to the other laboratory. These are the potential limitations of the detective corrective mode of maintenance. To overcome these limitations, there is a need for a strategic preventive mode of maintenance.

3 Strategic Prevention Mode

Detective corrective mode maintenance is a method of reacting after consequences are in the picture and situations are beyond the hands of an engineer [6]. Nowadays, utility is stepping to online monitoring modules while accumulating huge monitoring data. However, the demand for high reliability of electrical energy has been set on the high side. This being a strong commitment of major utilities led the condition monitoring engineers to develop sensors and auto detective actions in addition to online monitoring modules to prevent failures. Thus the data captured while online monitoring is critical for data analytics in developing a possible expert system and intelligent modules.

With online DGA monitoring and moving ahead for automatic analysis of the fault gases using an expert system, a crisp understanding of the situations is possible. This method may be referred to as strategic preventative mode. With automated monitoring, condition monitoring engineers may identify the fault easily at an appropriate time.

In addition, this is applicable for mineral oil-filled transformers only. The reason for the limited scope of online monitoring for ester-filled transformers is the lack of data. Thus, there is also a need to understand the traditional oil DGA with ester-filled transformers DGA, provided the loading profile is maintained the same for both the transformers.

3.1 Online DGA

Online DGA monitoring helps in avoiding unplanned outages, reduces maintenance costs, and improves transformer service life. This is generally carried out by installing the pertinent sensors and online modules to the transformer as shown in Fig. 2. Online DGA allows recording the load profile data and gas concentrations of the transformer fleet [7]. These parameters, including loading, gases, and temperature, may be used for developing expert systems and specified input ranges to detect elements.

Once the gas composition is obtained from the detection unit, necessary diagnostic action is done. From communicating interface connected to a data processing unit, one can have an access to the data. These monitoring devices may carry out continuous monitoring for the transformers, with alarms that are dependent on gas concentrations and the rate of the change of the gas ratios [7]. Sometimes, sampling the liquid after a long time of an event (fault) results in non-tracing of the fault gas which has been actually generated. This is because the gases are no longer available in the insulating liquid for measurement by traditional means. This is due to the aliasing effect, and this effect finds no place in online DGA analyzer as opposed to the traditional DGA laboratory method.

3.2 Correlating Online DGA Results to Transformer Load

When only one or two tests per year are done then it is very difficult, and even a big deal, to relate the gassing events to transformer events like load and temperature. Continuous online monitoring of dissolved gases provides an idea required to understand the relationship between the fault gases and other causing parameters [6], thus avoiding the problems associated with the offline DGA. With such a detailed understanding of various parameters, it is always possible to lay efforts on correlating the DGA information with the supervisory control and data acquisition (SCADA) monitoring of the transformer. But for doing the same, continuous monitoring of the transformer is required. This is explained by a review case study reported in [7] where a 40-year-old 500 kV transformer was selected for understanding the relationship of online DGA with the load. Now one could find the similarity between gassing and load guide on a linear scale. From the said case study, it is understood that online DGA monitoring reveals the combustible gases generated and dissolved in the transformer insulating liquid if the load profile exceeds half of the rated load [3]. Therefore, it is essential to monitor the load and combustible gases for the loading condition that is close to 50% of the specified load and balance the load accordingly.

Therefore, it is of high concern that the temperature is maintained within the specified level in accordance with the load. This demands the insulating liquid to exhibit high thermal conductivity and has high fire and flashpoints. Alternative insulating liquids synthetic ester and natural ester pose a 30% high thermal profile (flash and fire points) as compared to that of the mineral insulating liquids.

3.3 Development of DGA Using an Expert System

To diagnose the nature of incipient deteriorations and advanced intelligent methodologies for diagnosis are reported [8, 9]. Each of these has advantages and limitations. These techniques may or may not necessarily arrive at a similar conclusion.

The accuracy depends upon the expertise of the person handling the analysis. But it has been found that some ratios fall beyond the fault ranges mentioned in standards for identifying the faults. This limitation initiated to a system that has a reliable and wide range of fault determination characteristics and much accuracy. If the accuracy of the person handling is more, then the accuracy of the test will be better, so if this analyzing job is given to an expert system then its accuracy will be at its best. An expert system will have knowledge about the specific problem and exhibits the ability for applying this knowledge. Ideally, an expert system may self-learn from the mistakes. The procedure for expert diagnosis is as shown in Fig. 3. Here if the standard method fails to characterize the fault then the job is to be handled by the expert system for diagnosis, and corresponding maintenance is to be done.

4 Review of Case Study: Mineral Oil-Filled Unit

For the practical approach of an expert system in the field of transformer diagnosis, DGA test records of the past 10 years from different utilities report a new diagnosing method [5]. A rule base is reported in [5] for knowledge representation in order to develop or overcome the limitations of the ratio method in the case if the ratio is out of range or unable to detect the fault. This attempt has been made to increase the effectiveness of transformer diagnostics in DGA. To overcome this limitation, additionally 18 new combinations to the existing nine combinations, so a total of 28 combinations [6] are developed with separate characterizations of the faults. A transformer of 100 MVA, 132/11 kV, date of installation: 06/07/97 has been taken and subjected to DGA on Test date-1 and found that all the gases are within the permissible limits in ppm, as depicted in Table 2. In addition, the same transformer has been tested again for three months and the results are as shown in the same. Then it has been found that according to IEC standards [9], i.e., $C_2H_2/C_2H_4 = \mathbf{0.2}$, $CH_4/H_2 = \mathbf{1.2}$, $C_2H_4/C_2H_6 = \mathbf{2}$.

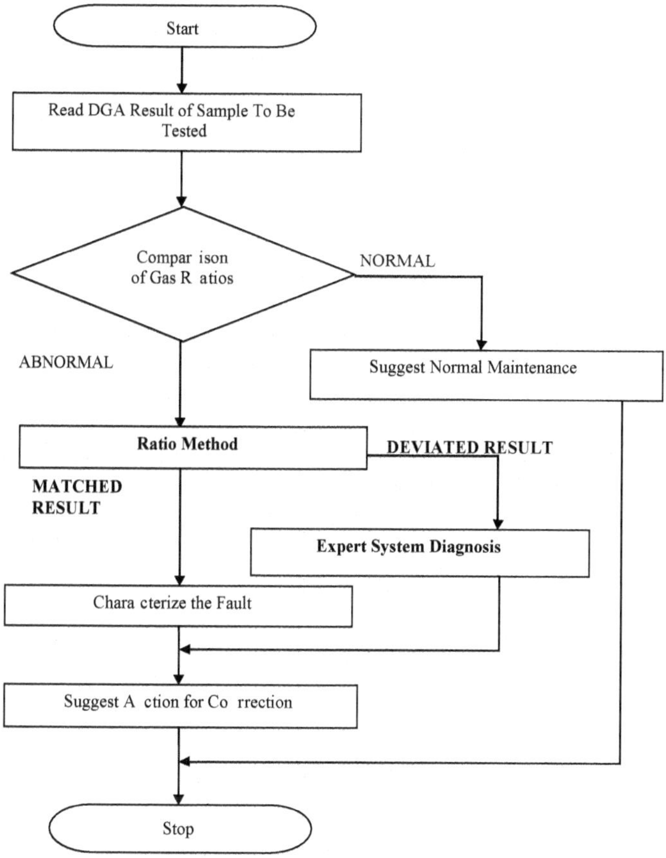

Fig. 3 Flowchart representing an expert system approach for diagnosis

Table 2 Gas concentrations on test dates in ppm

Gases*	Test date-1	After 3 months in service
C_2H_2	02	32
C_2H_4	43	109
CH_4	38	487
H_2	65	383
C_2H_6	40	53
CO	150	173
CO_2	843	932

Thus, a fault is unidentifiable but according to the new codes reported in [6] the characteristic fault is "Discharge of high energy thermal fault 300–700 °c". Hence if there is a high energy thermal fault, the oil is to be degassed immediately and should be reset within a month. Finally, a diagnostic approach is done by using the expert system diagnosis.

5 Conclusion

Precise condition monitoring of power transformer fleet is still a challenge in the area of power transformer diagnosis. The aim of this article is to promote online monitoring of DGA and facilitate asset management in addition to enhancing reliability. This is possible by using load monitoring, temperature monitoring, and finding possible relationships. Thus, several case studies have been reported and reviewed to justify the importance of online DGA and alternative insulating liquids with high thermal performance. Dissolved gas analysis, being the most vital and effective tool for diagnosing transformers, needs automation to survive with its flexibility and accuracy. This paper highlights the trends of DGA from "detective-corrective" mode to "strategic-prevention" mode. A conceptual procedure with an expert system for transformer diagnostics based on DGA data is proposed in this paper.

References

1. Mohan Rao U, H. Pulluri, Kumar NG (2018) Performance analysis of transformer oil/paper insulation with ester and mixed dielectric fluids. IEEE Trans Dielectr Electr Insul 25(5):1853–1862
2. Rogers RR (1978) IEEE and IEC codes to interpret incipient faults in transformers, using gas in oil analysis. IEEE Trans EI–13(5):349–354
3. Dornerburg E, Strittmatter W (1974) Monitoring oil cooling transformers by gas analysis. Brown Boveri Rev 61:238–247
4. Duval M (1989) It can save your transformer. IEEE Electr Insul Mag 5(6)
5. Lindgren SR, Fischer P Correlating DGA results to transformer load on-line DGA. Electric Energy Publications
6. Golarz J Automating the thinking and reducing guesswork about the health of electrical transformers. Electric Energy Publications
7. Wenlong Z, Yufeng Y, Chunyu G, Hengzhen L, Gang L (2011) Study on intelligent development of power transformer on-line monitoring based on the data of DGA. In: IEEE conference (APPEEC)
8. Lewand L Techniques for interpretation of data for DGA from transformers. http://www.transformerscommittee.org/info/F06/F06-DGA.pdf
9. Mohan Rao U, Sood YR, Jarial RK (2015) Subtractive clustering fuzzy expert system for engineering applications. Procedia Comput Sci 48, 77–83

Calculation of Health Index for Power Transformer Solid Insulation Using Fuzzy Logic

Teruvai Manoj and Chilaka Ranga

Abstract A new fuzzy logic-based technique to determine the health index of solid insulation of a power transformer is proposed in the present paper. The health condition of a power transformer mostly depends upon the health of its solid insulation. Continuous health index assessment of transformers and their solid dielectrics is possible with its continuous monitoring by performing various significant diagnostic tests, like dissolved gas analysis (DGA) and furan analysis (FA). The proposed fuzzy model in the present work is designed based on DGA and FA. The samples used to test the validity, reliability and efficiency of the proposed model are collected from the different 25 transformers owned by Indian Railways traction substations. It is observed from the output results that the proposed model with modified fuzzy rules is very accurate and yields the best results. It can overcome the shortcomings of previous conventional methods which were generally designed with complex mathematical theories. The proposed method is very convenient to not only experienced engineers but also inexperienced engineers. It shall also help in providing a way to rectify the fault that occurs in the power transformer.

Keywords Transformer · Insulation · Solid insulation · Fuzzy logic · Health index

1 Introduction

In the modern and competitive world, everything is interlinked and the maximum percentage of people depends upon electricity either directly or indirectly. So, it is important to maintain power quality and ensure valuable service. A transformer is one of the important and useful equipment in electrical power transmission and distribution [1]. The failure of transformers may lead to a burden for industries as well as commercial and also residential sectors in either financial or technical aspects [2]. To avoid these issues, condition monitoring of major equipments is very

T. Manoj · C. Ranga (✉)
Department of Electrical Engineering, National Institute of Technology Srinagar, Jammu and Kashmir 190006, India
e-mail: chilakaranga@nitsri.net

A. Tomar et al. (eds.), *Machine Learning, Advances in Computing, Renewable Energy and Communication*, Lecture Notes in Electrical Engineering 768,
https://doi.org/10.1007/978-981-16-2354-7_52

important and required [3]. Condition monitoring helps in achieving more reliable and efficient operation from the transformers. The lifetime of a transformer is decided by various factors. Among them, one of the major factors is insulation, i.e., solid or liquid insulation. The condition of the insulation decides the health status of the transformer.

The lifetime of a transformer depends upon the mathematical or experimental evaluation in view of the electrical, mechanical, physical and chemical characteristics of the insulation. Deterioration of insulation is caused due to electrical and thermal stresses which occur during the normal operation of transformers [4]. The power transformer consists of insulation with both cellulose paper and insulating mineral oil. Cellulose paper consists of cellulose, hemi-cellulose and lignin. Also, the life estimation of cellulose paper is up to 40 years [5–7]. The gases like carbon dioxide (CO_2) and carbon monoxide (CO) are dissolved into the oil. And these concentrations are the indicators for the faults generated in the transformer [6, 8]. Due to the decomposition of cellulose, these gases are generally generated. These gases are used to calculate the paper health index which is also termed as solid insulation index. In addition to this, one more concentration that reduces the lifetime of the transformer is furan content.

Various researchers across the world performed several diagnostic tests on the transformers and used different methods to estimate the status of transformers. As per a wide literature survey, for calculating health indices artificial neural network (ANN) and probabilistic approaches are used. However, the outputs are not satisfied and did not achieve the permanent solution in obtaining the accurate health status of transformers. In general, the estimation of remnant life of transformers can be done by using mathematical methods and traditional schemes such as ranking methods using weights. Although the existing methods have individual speciality and strengths in determining the health index of transformers, however, accuracy and reliability are not up to considerable and promised limits. The conventional methods have a large set of equations which increases the complexity of the problem. Also, such models are time-consuming. The electrical and thermal criticalities of paper dielectrics are explained in [9]. These criticalities cannot decide the overall decision-making of transformers [10, 11].

To overcome these shortcomings, a new fuzzy logic model is proposed to determine the paper health index (PHI) in the present study. Also, the comparison between trapezoidal and Gauss2 membership functions is discussed. The inputs having more than three MFs require a large number of possible fuzzy rules. And such reduction of rules results in the accuracy of the output of the fuzzy models [1]. In this present work, the fuzzy model consists of three inputs that decide the paper health index of a transformer. These three inputs have four membership functions each. Also, 64 fuzzy rules were formed to determine PHI. The proposed model helps in calculating health status with less time. Also, it can access accurate and reliable results as compared to conventional methods.

2 Fuzzy Logic Approach

The world is heavily dependent on artificial intelligence (AI) in various sectors. Soft computing techniques come under this category and it focuses mainly on the evaluation of complex problems by using various computing models. Fuzzy logic which comes under these techniques proved that it is a very useful tool to evaluate the health indices of the transformers [12]. It also provides very reasonable and accurate decision-making for transformers, assuring reliability and proper maintenance [13]. The steps included in fuzzy logic are explained in the following subsections.

2.1 Initialisation of Membership Functions and Fuzzification

Fuzzification converts values from precise sets to fuzzy sets. A curve called membership functions (MF) is defined as given input mapped with the degree of membership (DOM), between 0 and 1 of fuzzy sets (imprecise sets) [14, 15]. There are various types of membership functions such as triangular, sigmoidal, trapezoidal, Gaussian and Gauss2 shapes used in the fuzzy logic method [16, 17]. Among these membership functions, trapezoidal and Gauss2 are the most used MF shapes and also accurate as compared to other MF shapes. The figure Gauss2-shaped MFs are shown in Fig. 1. The range and limits of CO, CO_2 are considered from [18]. Similarly, the range and limits of 2-FAL are considered from [19]. The information related to limits is specified in Table 1.

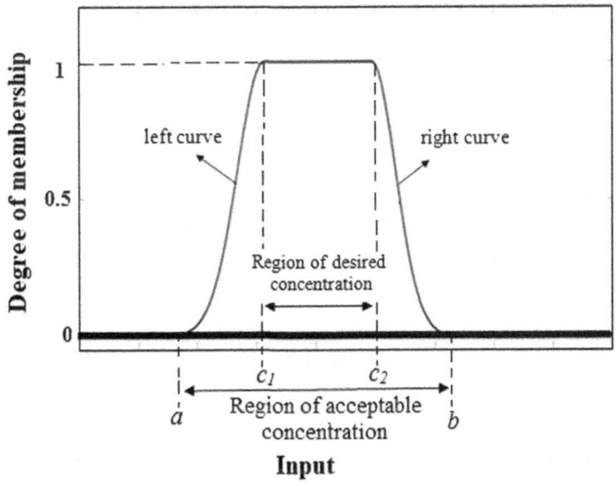

Fig. 1 Gauss2-shaped membership function

Table 1 Lower and upper limits of each gas, namely 2-FAL, CO and CO_2 denoted with a and b, respectively, for each four membership functions

Input	Low		Medium		High		VHigh	
	A	B	A	b	a	b	a	b
2-FAL	0	2	3	4	5	6	7	15
CO	0	300	400	520	630	1350	1450	2000
CO_2	0	2300	2700	3800	4200	9800	10,200	15,000

2.2 Gauss2-Shaped Membership Function

The shape of Gauss2 is shown in Fig. 1. The product of two Gaussian functions results in MF given in Eq. (1).

$$MF = e\left[\frac{-(x-c_1)^2}{2\lambda_1^2}\right] \times e\left[\frac{-(x-c_2)^2}{2\lambda_2^2}\right] \quad (1)$$

where MF is membership function, input value denoted as x, and centres of the two exponential functions are c_1 and c_2. The standard deviations of centres are λ_1 and λ_2 [16]. The area between (c_1, c_2), (a, c_1), (c_2, b) varies according to the input MFs. The input between c_1 and c_2 achieves unity DOM which is maximum and the area between (a, c_1) and (c_2, b) have less than 1. The furan content with four MFs is shown in Fig. 2.

Also, one more term λ_1 and λ_2 are set equal values for each MF. It is because the unequal values of λ_1 and λ_2 result in asymmetrical shapes of MFs which leads to inaccurate outputs [17]. Hence, λ_1 and λ_2 were fixed as 0.75 ppm for furan content, 100 ppm for CO and 675 ppm for CO_2 for all MFs. The overlap of 50% has been set for MFs to get accurate results, i.e., paper health index.

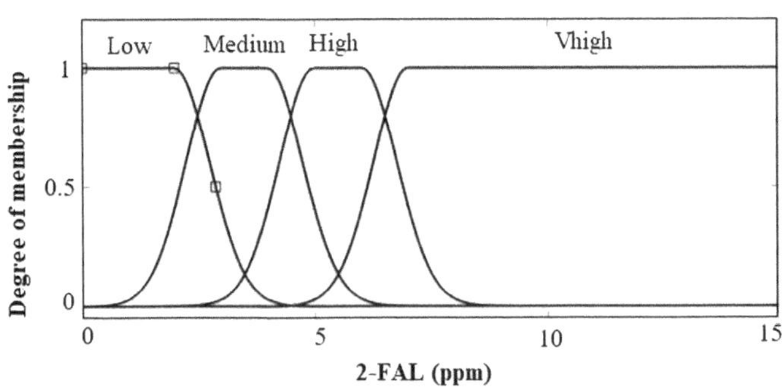

Fig. 2 Membership functions and DOM designed for 2-FAL

2.3 Fuzzy Inference

Fuzzy inference is the mapping of input values with specially designed rules to obtain the desired and exact output. It acts as an intermediate between fuzzification and defuzzification. Mamdani maximum–minimum method, a very popular and also widely used method, is implemented as a fuzzy inference method in the present work [16]. The output depends on the set of fuzzified inputs and also designed rules in this method. Further, the method truncates the output MF at the minimum DOM value.

2.4 Defuzzification

A precise value from the truncated output MF after fuzzy inference is considered in the defuzzification. In the present work, a popular method, namely the centre of gravity method is used to perform defuzzification. The area covered by the truncated output MF evaluates the centroid or centre (Z_o) of gravity [16, 17]. It is obtained by

$$Z_o = \frac{\int z.\mu(z)dz}{\int \mu(z)dz} \tag{2}$$

where z is the output variable and $\mu(z)$ is the DOM of truncated output MF.

3 Furan Analysis

Core and winding are the two most important and major parts of the transformer where both have solid dielectric. Since dielectric is made up of cellulose which consists of a long-chain molecular structure [20], due to ageing, the long chains in cellulose are broken into small particles which are large in number [21, 22]. When solid dielectric comes in contact with the transformer oil, it will be damaged due to heat. Further, it dissolves into the oil along with the gases such as carbon monoxide (CO) and carbon dioxide (CO_2) [23, 24]. These gases are derived from the fur-furaldehyde group. The most predominant component in the group is 2-furfural [25]. The life of paper insulation directly depends on the rate of rising of furfural products in oil with respect to time. Furfural compounds are used to assess the condition of paper insulation. A fur-furaldehyde analysis is very sensitive [20].

4 Paper Health Index of Transformer Based on Proposed Fuzzy Logic Method

The paper health index of the transformer is determined by using a fuzzy logic model. The fuzzy model for the health assessment of solid insulation of transformers is shown in Fig. 3.

Solid insulation is so-called the heart of transformers and its health status heavily influences the overall health of the power transformers. Therefore, it is a prerequisite to evaluate the condition of transformer solid insulation separately [22, 26]. CO and CO_2 are dangerous gases accumulated due to the deterioration of cellulose present in solid paper insulation and also release the furan compounds into the transformer oil [5, 19]. Paper decomposition leads to generate a most prominent component called 2-furaldehyde (2-FAL) [27]. The degree of polymerisation varies according to this component. CO, CO_2 and 2-FAL have been assigned as inputs for the current proposed FL model because of their influence on solid insulation. Also, the paper health index (PHI) has been considered as output. The concentrations of CO, CO_2 and 2-FAL are given in Table 2. These inputs individually fuzzified into four membership functions, namely low, medium, high and very high (Vhigh).

The maximum and minimum limits of output, i.e., PHI of transformers are considered from [18]. The output of the present FL model ranges from 0 to 1 and divided into four Gauss2 MF ranges, namely 'Excellent', 'Good', 'Poor' and 'Worst'. These terms are familiar in the ranking of systems in the competitive world shown in Fig. 4. In accordance with [18] the limits have been taken which suit the model.

Worst MF has maximum and minimum limits as 0 and 0.15, poor has 0.25 and 0.45. Similarly, good MF has 0.55 and 0.65, excellent has 0.75 and 1. An overlap of 50% is maintained for input MFs and the same has been applied to the output. The overlap between output MFs is 0.1 [27]. The necessity of overlap is to get accurate and reliable output, i.e., PHI. More information about the formation of MFs and DOM is given in [16, 17].

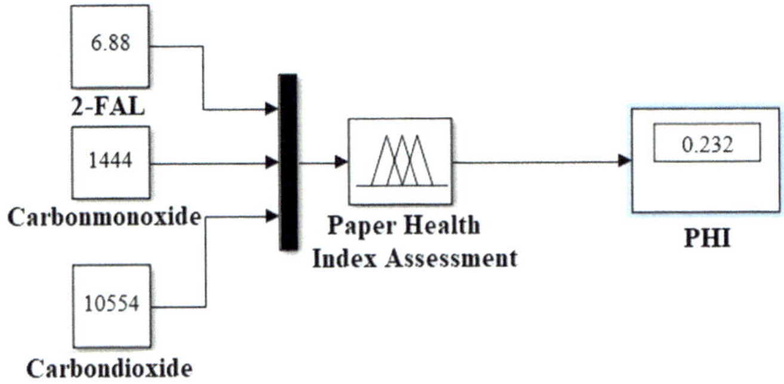

Fig. 3 Fuzzy logic model designed for health assessment of solid insulation

Table 2 The concentration of 2-FAL, CO and CO_2 (in ppm) obtained from furan analysis of 24 transformers, rating 5–50 MVA, and voltage of 6–200 kV

Paper sample number	2-FAL	CO	CO_2	Paper sample number	2-FAL	CO	CO_2
1	0.831	84.6	674	13	3.85	566	4155
2	3.96	853	8520	14	2.363	532	3776
3	0.82	374	3247	15	2.86	569	4474
4	6.699	1485	10,240	16	4.83	535	4155
5	4.56	679	8639	17	4.39	640	3756
6	1.27	363	1679	18	1.89	377	2830
7	5.55	1708	10,359	19	0.673	251	1932
8	2.571	629	7917	20	1.99	277	2441
9	1.63	250	2684	21	3.51	1132	4341
10	2.75	385	3930	22	1.89	259	2551
11	2.43	357	2349	23	6.91	1437	6183
12	4.418	385	4185	24	2.94	397	1475

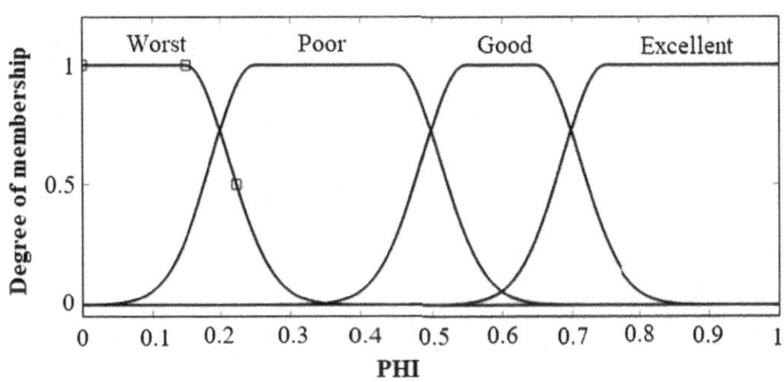

Fig. 4 Membership function and DOM of output paper health index

5 Results and Discussion

In the present work, the proposed FL method is evaluated by using Gauss2 shaped MF. Also, it is compared with trapezoidal-shaped MF. These two MFs are having their own importance in health assessment but will be differentiated in the output values. As discussed in Sect. 2, the proposed FL model is carried out by using three stages, namely fuzzification, fuzzy inference and defuzzification. The inputs have been converted from precise values to fuzzy values.

The DOM of all inputs can be obtained by using Eq. (1) for 2-FAL, CO and CO_2. Consider the formula m^n to get the number of rules, where m is the number of member functions for each input and n is the number of inputs. The fuzzy rules are designed in the form of IF-AND-THEN fuzzy rules. To get accurate results from the inputs, a total of 64 fuzzy rules are designed and more priority has been given to 2-FAL in the proposed model.

Therefore, furan content concentration is the deciding factor to determine PHI. It is because the influence of 2-FAL is more on the health of solid insulation as compared to CO and CO_2. The designed fuzzy rules are shown in Fig. 5. The slanted black and yellow-filled shapes represent the membership functions of all inputs considered to

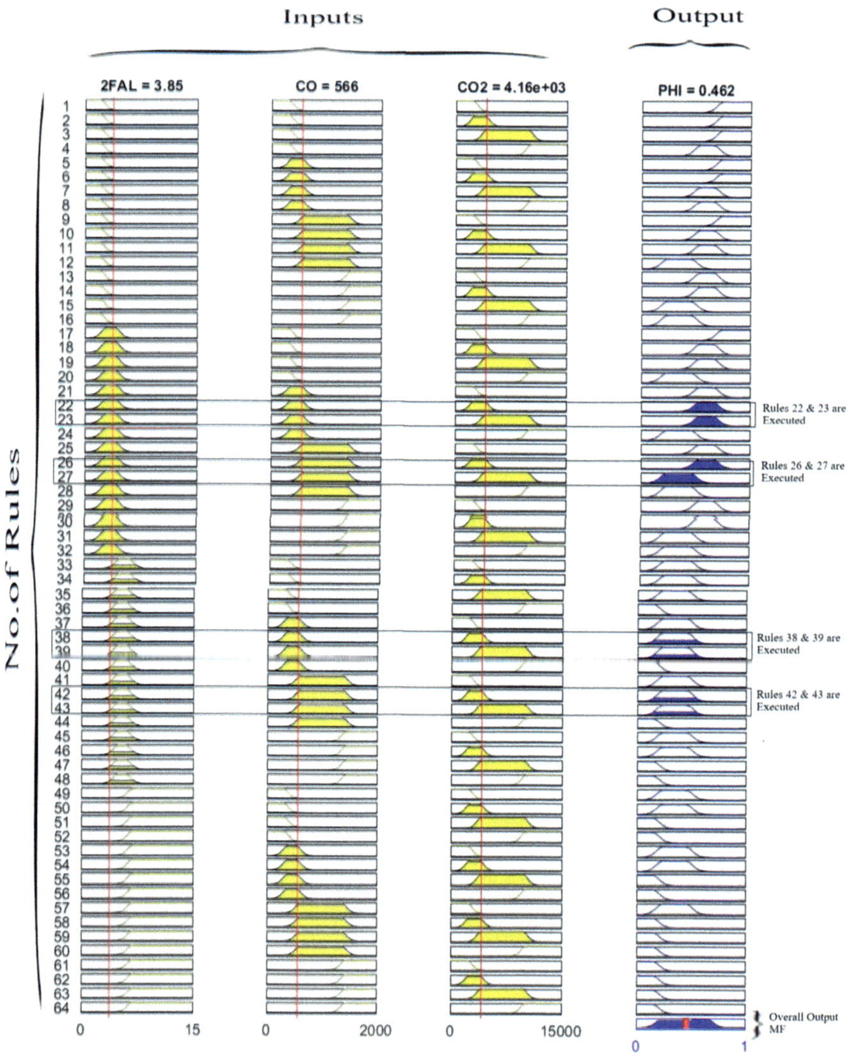

Fig. 5 Graphical representation of the 64 fuzzy rules developed to determine the effect of 2-FAL, CO, CO_2 on solid insulation of the transformer

design fuzzy rules. Similarly, output MFs are represented by slanted blue and blue-filled shapes. Columns 1 to 3 are having yellow-filled shapes that represent the MFs for 2-FAL, CO and CO_2. Similarly, Column 4 with blue-filled shapes represents the output MFs, i.e., health index of solid insulation. To analyse the effect of 2-FAL and remaining gases, sample number 13 has been considered, which is shown in Fig. 5.

All slanted and colour-filled limits are framed according to the input and output MFs in the designing of fuzzy rules.

Consider transformer 13 (Table 1). The input values of 2-FAL, CO and CO_2 are 3.85, 566 and 4155, respectively. The input values have been imposed on 64 fuzzy rules designed where 2-FAL is the deciding factor. The output results in the case of the FL method using trapezoidal is 'Good', whereas the output of the proposed Gauss2-shaped method is 'Poor'. It is because the value of 2-FAL is 3.85 which is between MFs 'Medium' and 'High'. Since the health status depends upon furan content, CO and CO_2 also having 'High' result in 'Poor' output condition. This did not actually happen in the trapezoidal-shaped FL method which resulted as 'Good' for the same. Therefore, the Gauss2-shaped method monitors better than the trapezoidal method and the comparative results are shown in Table 3.

According to the classification from Table 4, the number of transformers which attained the status of 'Excellent and Good' is 17 by using conventional methods. Similarly, it attained 15 by using trapezoidal-shaped MFs and 13 by using Gauss2-shaped MFs. The number of transformers which attained the status of both 'Poor and Worst' is 8 by using conventional methods; similarly, 10 by using trapezoidal-shaped MFs and 12 by using Gauss2-shaped MFs. In the proposed Gauss2-shaped MFs, consider the data given in Table 4, the percentage of transformers status with 'Excellent and Good' is $13/25 = 52\%$. Similarly, the percentage of 'Poor and Worst' is $12/25 = 48\%$.

The percentage of transformers status with 'Excellent and Good' is 68% and 60% in conventional methods and trapezoidal-shaped MFs, respectively. Similarly, the percentage of transformers status with 'Poor and Worst' is 32% and 40% and in conventional methods and trapezoidal-shaped MFs, respectively. Therefore, the percentage of transformers with 'Excellent and Good' is more. And, the percentage of transformers status with 'Poor and Worst' is less in the case of Gauss2-shaped MFs. It means that the correct status of the transformers has been strongly executed in Gauss2 as compared to trapezoidal and conventional methods used by diagnostic experts. This proves the reliability of Gauss2-shaped MFs to find PHI of transformers is more as compared to different methods.

To prove the reliability of the proposed model, consider the example of transformer 17, where the values of 2-FAL, CO and CO_2 are 4.39, 640 and 3756, respectively. The 2-FAL value lies in the range of 'High'. Similarly, the value of CO lies under 'High' and the value of CO_2 lies between 'Medium' and 'High'. According to rules mentioned in the proposed fuzzy logic method, the result should be 'Poor' but the results attained in both conventional methods and trapezoidal-shaped MFs represent 'Good' which proves that these methods did not determine the correct results. Similarly, sample numbers 13 and 14 also come under this category.

Table 3 Health indices for 25 test case transformers

PHI using trapezoidal shaped MF	Status	PHI using Gauss2-shaped MF	Status	PHI using conventional methods	PHI using trapezoidal-shaped MF	Status	PHI using Gauss2-shaped MF	Status	PHI using Conventional methods
0.851	E	0.833	E	E	0.468	G	0.462	P	G
0.35	P	0.354	P	P	0.771	E	0.612	E	E
0.845	E	0.772	E	G	0.462	P	0.541	G	G
0.106	W	0.224	W	W	0.383	P	0.426	P	P
0.35	P	0.394	P	G	0.475	G	0.454	P	G
0.843	E	0.818	E	E	0.846	E	0.765	E	G
0.105	W	0.16	W	W	0.851	E	0.833	E	E
0.435	P	0.444	P	P	0.843	E	0.774	E	G
0.85	E	0.802	E	E	0.35	P	0.437	P	P
0.681	E	0.688	G	G	0.843	E	0.782	E	E
0.756	E	0.73	E	E	0.14	W	0.229	W	P
0.468	G	0.46	P	G	0.621	G	0.697	G	G

where E = Excellent, G = Good, P = Poor, W = Worst

Table 4 Classification of transformers according to the health status obtained in the proposed method and conventional methods

The PHI of trapezoidal MFs		The PHI of Gauss2 MFs (proposed method)		The PHI using conventional methods by diagnostic experts	
E and G	P and W	E and G	P and W	E and G	P and W
15	10	13	12	17	8

where E = Excellent, G = Good, P = Poor, W = Worst

Therefore, it is proved that the results obtained using the proposed Gauss2 method finds out the correct status of the transformer. Among the conventional, trapezoidal-shaped MFs and Gauss2-shaped MFs methods, the proposed one, i.e., the Gauss2-shaped MFs method is a more reliable method.

6 Conclusion

In the present paper, a new fuzzy logic model is proposed to determine the health index of solid insulation in the transformer. It is designed with a specific set of fuzzy rules which helped in obtaining accurate results. A new innovative method has been implemented on 25 test case oil samples collected from traction substations owned by Indian Railways. Significant diagnostic tests such as DGA and FA are performed to obtain the values of furan content and gases CO and CO_2. The proposed model demonstrates that the Gauss2-shaped model is good in terms of accuracy, efficiency and reliability as compared to the trapezoidal-shaped model. It has been proved by the results obtained in the present paper. This proposed method overcomes the shortcomings such as complexity, inaccuracy and non-reliability in conventional methods. The proposed method is implemented and explained in a perfect manner that is convenient not only to experienced engineers but also to inexperienced engineers. It is the way useful to improve the health status of a transformer. The proposed method can be applied anywhere in the world and it will be very easy to apply on a large number of transformers. Fuzzy logic has future scope because the modern world heavily depends upon artificial intelligence.

References

1. Bakar NA, Abu-Siada A (2016) Fuzzy logic approach for transformer remnant life prediction and asset management decision. IEEE Trans Dielectr Electr Insul 23(5):3199–3208
2. Guo C et al. (2020) Transformer failure diagnosis using fuzzy association rule mining combined with case-based reasoning. IET Generation Transm Distrib 14(11):2202–2208
3. Li S, Li J (2017) Condition monitoring and diagnosis of power equipment: review and prospective. High Voltage 2(2):82–91

4. Wang K et al. (2019) Thermal aging characteristics of newly synthesized triester insulation oil. IEEE Access 7:175576–175583
5. Vasovic V et al. (2019) Aging of transformer insulation - experimental transformers and laboratory models with different moisture contents: part I DP and furans aging profiles. IEEE Trans Dielectr Electr Insul 26(6):1840–1846
6. Madavan R, Saroja S (2020) Decision making on the state of transformers based on insulation condition using AHP and TOPSIS methods. IET Sci Measure Technol 14(2):137–145
7. Mishra D, Baral A, Haque N, Chakravorti S (2020) Condition assessment of power transformer insulation using short-duration time-domain dielectric spectroscopy measurement data. IEEE Trans Instrum Measure 69(7):4404–4411
8. Ranga C, Chandel AK, Chandel R (2017) Condition assessment of power transformers based on multi-attributes using fuzzy logic. IET Sci Measure Technol 11(8):983–990
9. Arshad M, Islam SM, Khaliq A (2014) Fuzzy logic approach in power transformers management and decision making. IEEE Trans Dielectr Electr Insul 21(5):2343–2354
10. Khan SA, Equbal MD, Islam T (2015) A comprehensive comparative study of DGA based transformer fault diagnosis using fuzzy logic and ANFIS models. IEEE Trans Dielectr Electr Insul 22(1):590–596
11. Ranga C, Chandel AK (2017) Expert system for health index assessment of power transformers. Int J Electr Eng Inform 9(4)
12. Liao R, Liu J, Yang L, Zhang Y, Gao J, Ma Z, Hao J (2015) Extraction of frequency domain dielectric characteristic parameter of oil-paper insulation for transformer condition assessment. Electric Power Compon Syst 43(5):578–587
13. Jahromi A, Piercy R, Cress S, Service J, Fan W (2009) An approach to power transformer asset management using health index. IEEE Electr Insul Mag 25(2):20–34
14. Taha IBM, Hoballah A, Ghoneim SSM (2020) Optimal ratio limits of rogers' four-ratios and IEC 60599 code methods using particle swarm optimization fuzzy-logic approach. IEEE Trans Dielectr Electr Insul 27(1):222–230
15. Lin Y, Wei C, Tao F, Li J (2019) Aging assessment of oil-paper insulation of power equipment with furfural analysis based on furfural generation and partitioning. IEEE Trans Power Delivery 34(4):1626–1633
16. Karray F, de Silva CW (2006) Soft computing and intelligent systems design, theory, tools and applications. IEEE Trans Neural Netw 17(3):825–825
17. Nedjah N, Mourelle LM (2005) Fuzzy systems engineering theory and practice. Springer
18. (1992) IEEE Guide for the Interpretation of Gases Generated in Oil-Immersed Transformers. IEEE Std C57.104-1991, pp 0–1
19. Abu-Elanien AEB, Salama MMA, Ibrahim M (2012) Calculation of a health index for oil-immersed transformers rated under 69 kV using fuzzy logic. IEEE Trans Power Delivery 27(4):2029–2036
20. Ashkezari AD, Ma H, Saha TK, Ekanayake C (2013) Application of fuzzy support vector machine for determining the health index of the insulation system of in-service power transformers. IEEE Trans Dielectr Electr Insul 20(3):965–973
21. Thiviyanathan VA, Ker PJ, Leong YS, Jamaluddin MZB, Mun LH (2020) Detection of 2FAL furanic compound in transformer oil using optical spectroscopy method and verification using Morse oscillation theory. IEEE Access 8:76773–76779
22. Abd El-Aal RA, Helal K, Hassan AMM, Dessouky SS (2019) Prediction of transformers conditions and lifetime using furan compounds analysis. IEEE Access 7:102264–102273
23. Jiang J, Chen R, Chen M, Wang W, Zhang C (2019) Dynamic fault prediction of power transformers based on hidden Markov model of dissolved gases analysis. IEEE Trans Power Delivery 34(4):1393–1400
24. Li, Wu G, Dong H, Yang L, Zhen X (2020) Probabilistic health index-based apparent age estimation for power transformers. IEEE Access 8:9692–9701
25. Feng D, Yang L, Zhou L, Liao R, Chen X (2019) Effect of oil–paper–pressboard mass ratio on furfural content in transformer oil. IEEE Trans Dielectr Electr Insul 26(4):1308–1315

26. Suna HC, Huanga YC, Huang CM (2002) A review of dissolved gas analysis in power transformers. Energy Procedia 14:1220–1225
27. Ranga C, Chandel A, Chandel R (2017) Expert system for condition monitoring of power transformer using fuzzy logic. J Renew Sustain Energy 9(4)

Millimeter-Wave Wideband Koch Fractal Antennas

S. B. T. Abhyuday, R. Ramana Reddy, and N. K. Darimireddy

Abstract This work is conducted to develop a compact, wideband patch fractal antenna suitable to operate in an untapped potential region of the millimeter-wave spectrum. The antenna is developed in various configurations with Koch fractal geometry forming a snowflake patch structure on substrate Rogers RT5880. The antenna is of dimensions 6.66 $\lambda_0 \times 6.66\ \lambda_0 \times 0.26\ \lambda_0$ at 100 GHz. The proposed iteration 4 design provides the highest gain of 11.6 dB and iteration 5 developed operates in the range of 85.20–330.62 GHz, giving the best operational bandwidth performance. The antenna developed is suitable for millimeter-wave applications of mobile communications and security systems. The simulations are done using Ansys HFSS software.

Keywords Fractal · Millimeter-wave antenna · Koch geometry

1 Introduction

With the discovery of untapped potential in the millimeter region (30–300 GHz) of the electromagnetic spectrum, there is an enormous demand for low-profile antennas to operate in this region as they are suitable to be embedded into various devices because of their miniature size. The efforts toward millimeter wireless network realization are to attain benefits like miniaturization with system-on-chip antennas, high data rates, short-range communications, automotive radar, radio astronomy, and security screening devices operating at high-frequency bands [1]. Antennas for high-frequency systems at millimeter-wave spectrum are attained through fractal geometry which forms a mathematical representation of irregular sets to describe many naturally occurring phenomena. Fractals offer maximization of the effective

S. B. T. Abhyuday (✉) · R. R. Reddy
JNTUACEP, Pulivendula, AP, India

N. K. Darimireddy
Lendi IET (A), Vizianagaram, AP, India

A. Tomar et al. (eds.), *Machine Learning, Advances in Computing, Renewable Energy and Communication*, Lecture Notes in Electrical Engineering 768,
https://doi.org/10.1007/978-981-16-2354-7_53

length of electromagnetic material on the same given surface area which allows them to be fitted into small space forming high-frequency radios.

Fractals show the redundancy property at different scales by going through infinite iterations to achieve multi-band characteristics with resonant frequencies. They follow that if an antenna after scaling by a factor λ got identical form, then the boundary conditions for Maxwell's equations also remain identical, so the radiation properties are expected to behave similarly when the frequency is scaled by $1/\lambda$ and wavelength by λ. Many research works are reported in the open literature on the development of antennas to cover multiple frequencies and wideband.

Triangular fractal antenna with hexagonal patch operating from 3 to 25.2 GHz frequency offering wide bandwidth of 22.2 GHz with gain varying from 3 to 9.8 dBi is reported [2]. Asymmetric and symmetric bow-tie slots in the circular patch for circular polarization offer 350 MHz bandwidth, 5 dBic gain for the asymmetric bow-tie, and 100 MHz bandwidth. The 5.1 dBic gain for symmetric bow-tie is presented [3]. Scaled-up symmetric bow-tie slot offers 340 MHz bandwidth and 5 dBic gain, whereas scaled-down version gives 710 MHz bandwidth and 5.25 dBic gain. Koch fractal patch operates at Ka-band [4] offering a gain of 9.7 dBi covering a bandwidth of 26–40 GHz. Feeding of the antenna is with a coplanar waveguide and contains DGS. Antenna with Koch fractal offering circular polarization [5] contains defected ground plane, multi-substrate. The antenna developed offers a bandwidth of 91 MHz with a gain of 5.45 dBi and a return loss of −39 dB.

A 4×4 millimeter-wave antenna [6] operates at 45 GHz and contains "L" branches with truncated corners offering 17 dBic gain and 3 dB axial ratio from 41.9 to 49.1 GHz. Reconfigurable microstrip antenna with the cut ring [7] contains diodes to switch between linear polarization, left and right-hand circular polarizations offering 10 dB gain at each of them and operates from 3.86 to 3.98 GHz. Circular polarization in [8] is attained with the fractal structure in the ground plane forming defected ground surface offering circular polarization from 1.572 to 1.578 GHz with a gain of 1.7 and 2.2 dBic, respectively. Reconfigurable microstrip antenna with cavity-backed proximity-coupled feed [9] containing PIN diodes to switch between different linear polarizations at 0°, 45°, and 90°. The gain is ranging from 7.2 to 8.1 dBi. Wearable antenna with Koch fractal, meandering slits, and defected ground [10] operates at 2.45 GHz offering 2.06 dBi gain. Hexagonal fractal antenna [11] with DGS operates at 3.79 GHz and 5.5 GHz with a gain of 6.2 dBi and 6.8 dBi, respectively.

Fractal antenna with wheel structure on circular patch [12] operating in the range of 2.93–9.53 GHz with a gain of 5.17 dBi is reported. The double branch line fractal patch antenna [13] operating at dual bands from 1.1 to 2.2 GHz and 4.8 to 7 GHz with a gain of 2.9 dBi and 3 dBi, respectively, is presented. Sierpinski Carpet Fractal antenna with six resonant frequencies between 4.825 and 9.145 GHz [14] covering C and X bands useful for satellite and space applications is reported. Giuseppe Peano fractal antenna developed with the resonant frequency of 4 GHz and for different iterations antenna works for WLAN, Bluetooth, X band [15]. This glance at literature shows that there is a need for the development of antennas to operate at higher frequencies of the millimeter-wave spectrum. A compact, wideband patch fractal antenna operating in the millimeter-wave spectrum is reported in this paper. The antenna designed is of dimensions 6.66 $\lambda_0 \times 6.66\ \lambda_0 \times 0.26\ \lambda_0$ operating in the frequency range of 85.2–330.62 GHz.

2 Antenna Modeling

2.1 Basic Antenna Iterations

The antenna in iteration 1 is designed as a traditional triangular patch antenna as shown in Fig. 1a. The triangular radiating element is mounted upon the Rogers RT Duriod 5880 substrate. In iteration 2 conventional star-shaped patch structure is formed with an equilateral triangle and is given in Fig. 1b. The depicted Koch

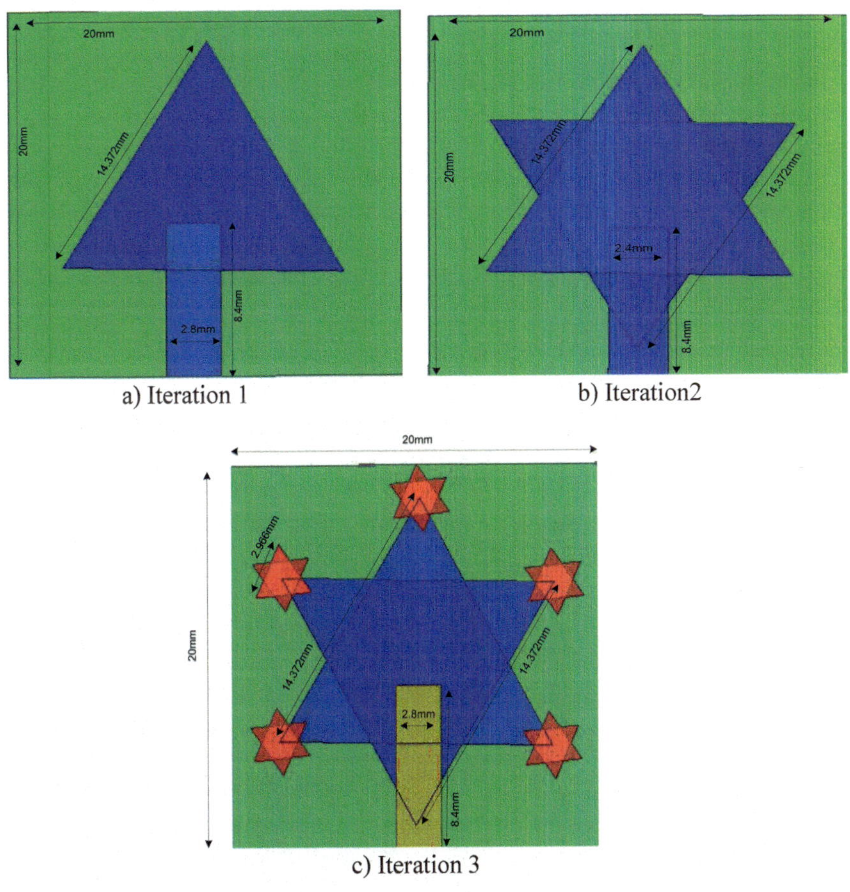

Fig. 1 Fundamental antenna iterations

fractal structure in iteration 3 is obtained forming further iterations in iteration 2 by repeating star structures on the five corners of iteration 2 with smaller equilateral triangles as shown in Fig. 1c. Iteration 3 is developed in [16] to operate over a region of 17.22–180 GHz covering a bandwidth of 162.78 GHz using Koch fractal structure.

2.2 *Proposed Antenna*

The work developed in Fig. 2 is an effort to further enhance the performance of millimeter-wave antennas by relying on the design presented in [16] and many more. The dimensions of the proposed antenna are given in Table 1.

2.2.1 Iteration 4

The outer structure of the patch in Fig. 3 is based on Koch geometry. The Koch geometry is formed by mirroring two equilateral triangles forming a star shape which enhances the radiating area within the same surface area. Mathematical modeling is initialized with the equilateral triangle in the first iteration, which will be mirrored to form a star structure in the second iteration; in the third iteration, each triangle side is segmented into three, where the middle segment further forms a small equilateral triangle resulting in five-arm star. This method optimizes the equilateral triangle angles increasing the radiating area occupied by the triangle within the same surface area.

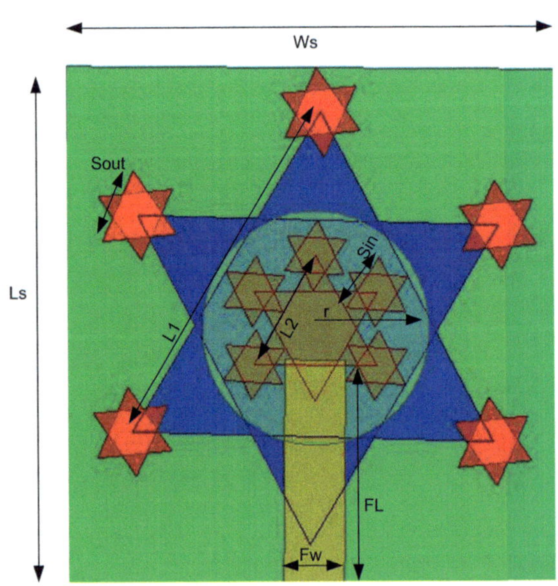

Fig. 2 Proposed antenna dimensions

Table 1 Proposed antenna dimensions in "mm"

Symbol	Parameter	Value (mm)
L1	Side of an outer equilateral triangle	14.372
L2	Side of an inner equilateral triangle	4.849
S_{out}	Side of outer arm equilateral triangle	2.966
S_{in}	Side of inner arm equilateral triangle	2.339
r	The radius of the circular disc	4.47
F_L	Length of feedline	8.8
F_W	Width of feedline	2.38
L_s	Length of substrate	20
W_s	Width of substrate	20
T	Thickness of substrate	0.787

Fig. 3 Proposed iteration 4

The effective side length of the triangle and a_{eq} is given by Eq. (1); "S" forms the original area of the equivalent triangle.

$$S = \pi a_{eq} \tag{1}$$

Equation (2) gives in each iteration a count of sides and Eq. (3) after each iteration gives the length of each side [4].

$$N_n = (3)(4)^n \tag{2}$$

$$L_n = (S)\left(\frac{1}{3}\right)^n \quad (3)$$

The inner structure of the patch is designed with an equilateral triangle of side length 1/3 unit of the outer equilateral triangle. The inner triangle got inversed to form a star-shaped structure. Further star-shaped structures are then repeated on the five corners of the inner star with smaller equilateral triangles of 1/2 unit of inner equilateral forming inner Koch fractal structure. This inner structure is subtracted from the outer structure to create a slot in the patch which will enhance the performance of the antenna.

2.2.2 Iteration 5

(a) Without DGS

The formation of iteration 5 in Fig. 4a is similar to that of iteration 4 added with a circular slot of 4.47 mm radius made in the outer structure to enclose the inner structure intact. The ground plane is of the same dimensions 20 × 20 mm^2 with no defects.

Mathematical modeling of the equilateral triangle is done by including the effect of the fringing field around the equilateral triangle which can be considered as an equivalent circular disk that has an equivalent area given as a triangular patch. The effective radius a_e of the circular disc is given in Eq. (4) [17].

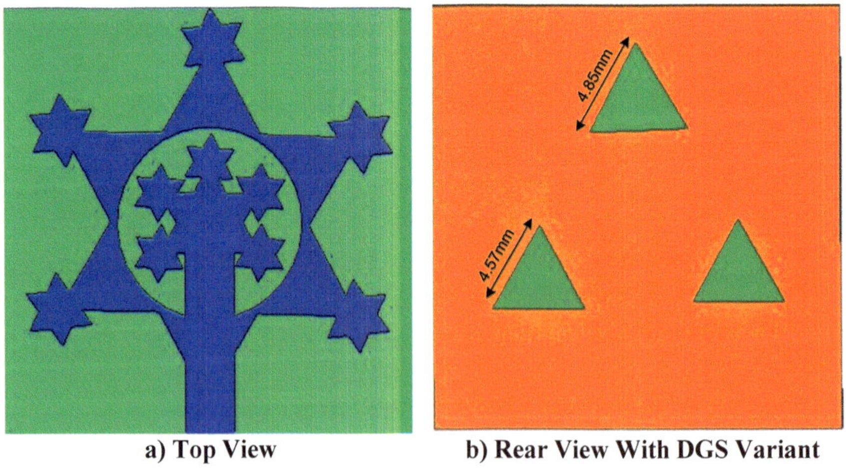

Fig. 4 Proposed iteration 5

$$a_e = a\sqrt{1 + \frac{2h}{\pi \in_r a}\left(ln\frac{\pi a}{2h} + 1.7726\right)} \tag{4}$$

The feedline is extended to meet the inner structure; therefore, the supplied power will radiate both the inner and outer fractals.

(b) With DGS

The ground plane of iteration 5 shown in Fig. 4b is with DGS which is for bandwidth enhancement offering resonances. The circular polarization design for the microstrip antenna is shown in iteration 5 rearview which contains three equilateral triangular slots in the ground plane [17]. The top equilateral is of side length 4.845 mm and the bottom two equilateral triangles are of side length 4.57 mm. The ground plane is of dimension $20 \times 20\ mm^2$.

3 Results

3.1 Return Loss

Return loss plots for the first three iterations are given in Fig. 5. For Iteration 1, which is a triangular patch, the minimum reflection coefficient observed is −37.7 dB

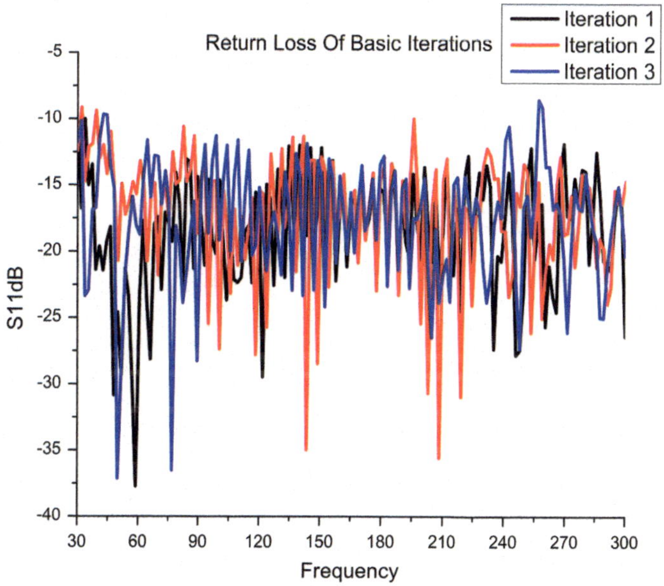

Fig. 5 Return loss performance of basic iterations

at 58.8 GHz. Iteration 2, which is a star-shaped patch, exhibits a shift in performance toward the higher regions of the spectrum in the middle frequencies. The antenna performance begins around 95 GHz and the highest performance is observed at 208.2 GHz with a return loss of −35.56 dB. Iteration 3, which a Koch fractal curve that extends the radiating area in the same footprint, exhibits a minimum reflection coefficient of −37.1 dB at 49.8 GHz and shows better performance at multi-frequencies millimeter-wave region of 50–100 GHz.

The proposed design in iteration 4, which comes with a slot in the Koch fractal curve, further extends the performance with a minimum reflection coefficient of −40.73 dB at 115.05 GHz and covers a wide bandwidth of 180.55 GHz showing better performance at multiple frequencies starting from 93.79 to 274.34 GHz as in Fig. 6.

Iteration 5, which brings the inner fractal intact without DGS, results in an antenna with wider bandwidth of 230.10 GHz starting from 67.46 to 297.56 GHz, exhibiting the best performance of −39.96 dB at 162.4 GHz as in Fig. 7. Iteration 5 with bringing the inner fractal intact and DGS introduced results in a wider bandwidth of 245.42 GHz ranging from 85.20 to 330.62 GHz, exhibiting a return loss of −37 dB at 281.2 GHz as shown in Fig. 7.

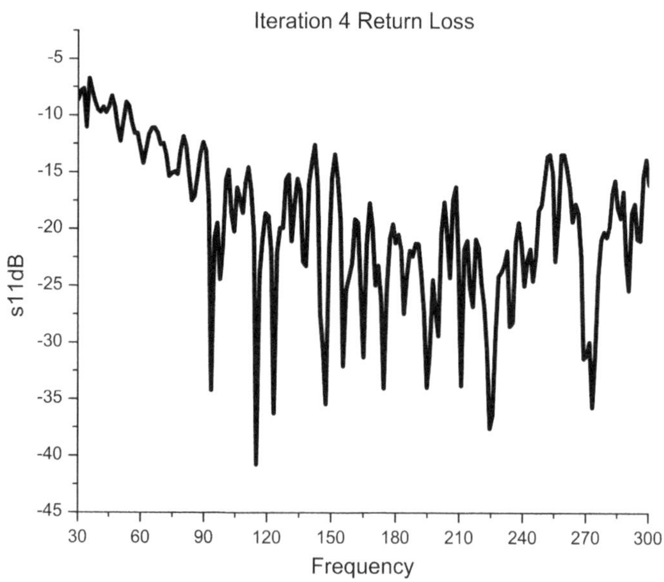

Fig. 6 Proposed iteration 4 return loss

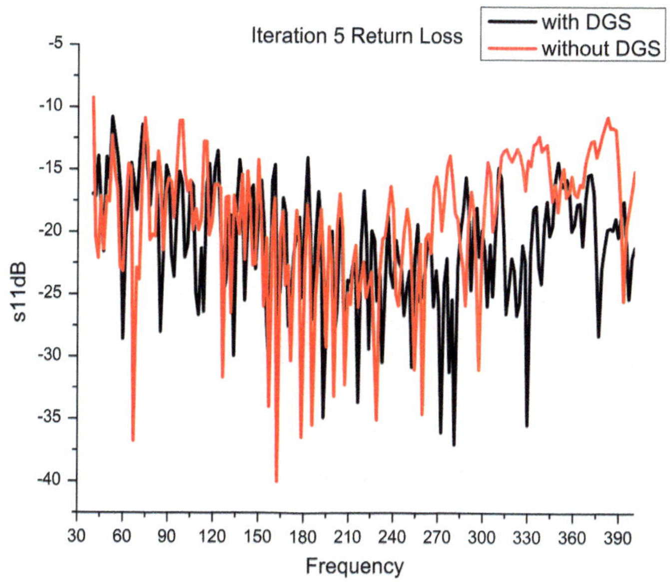

Fig. 7 Return loss for iteration 5 with and without DGS

3.2 *Gain*

Of all the designs rendered iteration 4 shows the highest gain of 11.6 dB at 111.05 GHz as given in Fig. 8. Iteration 5 without DGS yields a gain of 11.0 dB and 10.6 dB with

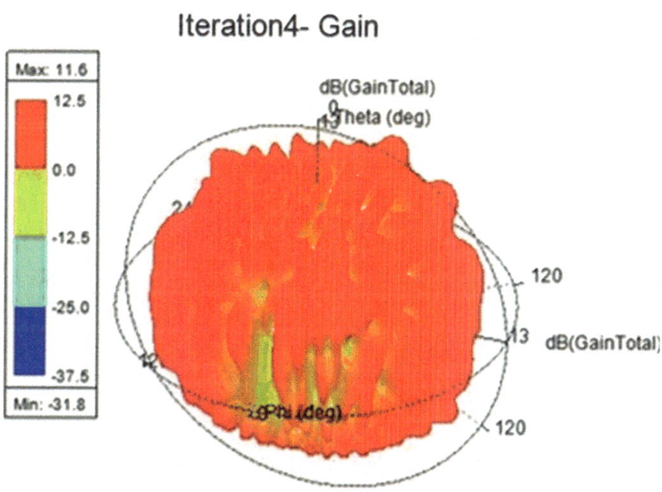

Fig. 8 Iteration 4 gain

DGS, as shown in Fig. 9 and Fig. 10, respectively. The conventional designs of 1, 2, and 3 result in 7.5 dB, 8.9 dB, and 11.1 dB, respectively.

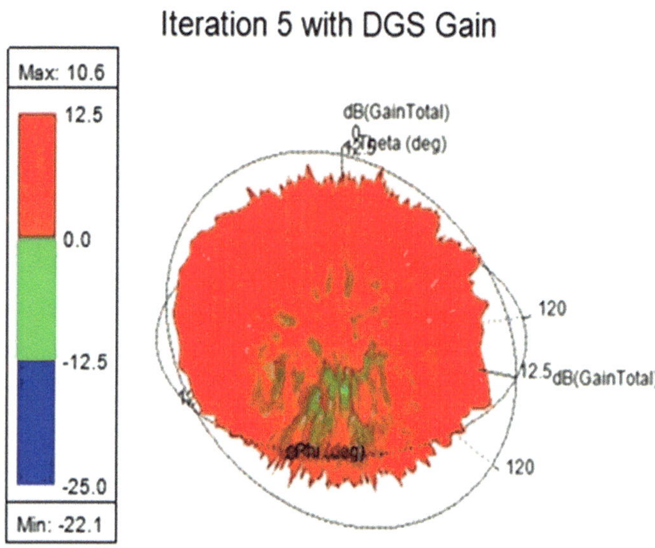

Fig. 9 Iteration 5 without DGS gain

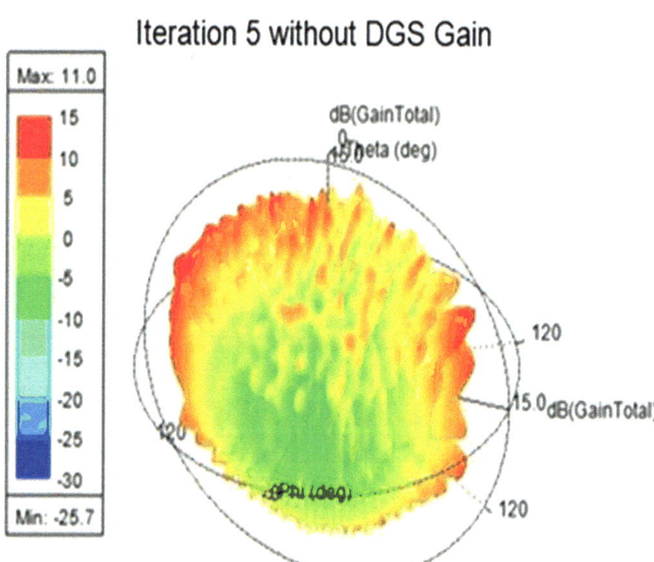

Fig. 10 Iteration 5 with DGS gain

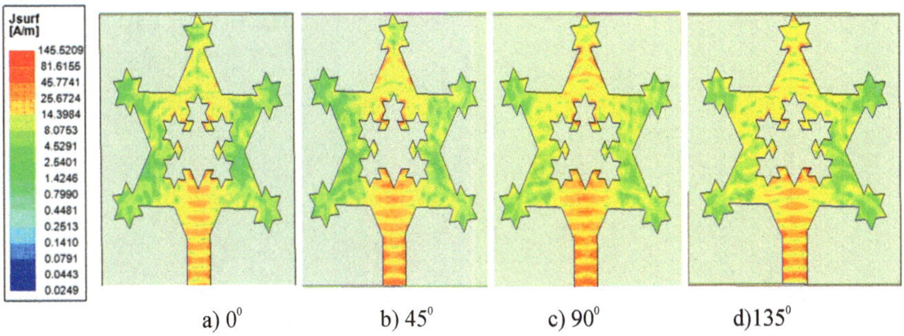

Fig. 11 Iteration 4 surface current at various phases **a** 0°, **b** 45°, **c** 90° and **d** 135°

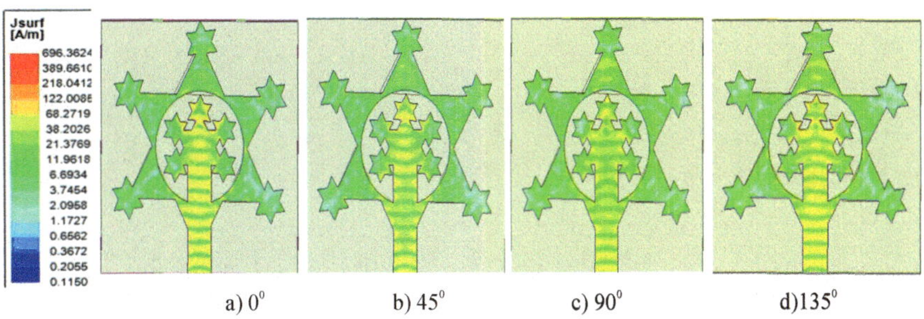

Fig. 12 Surface current distribution at various phases for iteration 5 with DGS **a** 0°, **b** 45°, **c** 90° and **d** 135°

3.3 Current Distribution

The current distribution of the antenna along the surface is observed at 100 GHz for iterations 4 and 5. The wideband behavior is due to current distribution along the edges of the inner triangle as observed in Fig. 11 adding more frequencies resulting in improved gain, with an electrical length increase of the fractal in iteration 4.

The broader bandwidth in iteration 5 with DGS is because of residual charge on edge of triangular along boundaries of the inner triangle and outer triangle, which supports the antenna wideband behavior of the fractal as presented in Fig. 12.

3.4 Radiation Pattern

Radiation pattern (RP) obtained at peak gain frequencies for iteration 4 at 115.05 GHz, iteration 5 without DGS at 162.4 GHz and with DGS at 281.2 GHz is shown in Figs. 13, 14, and 15, respectively.

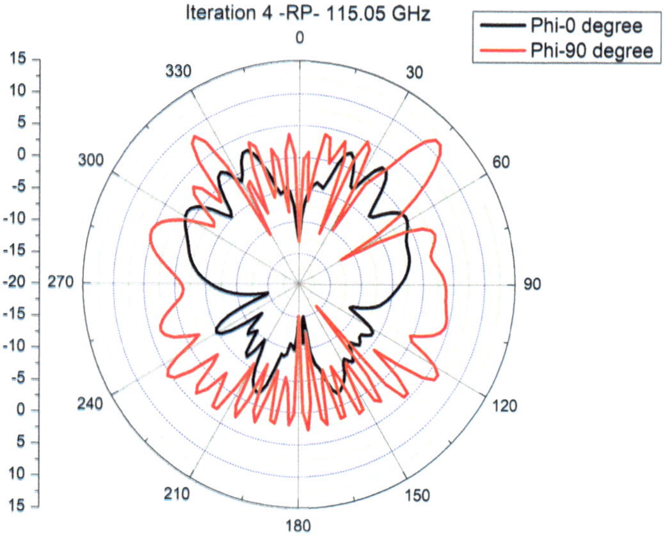

Fig. 13 Iteration 4 RP

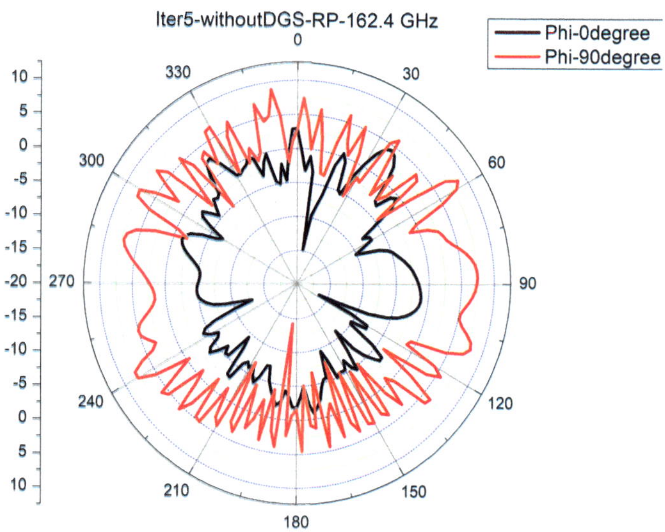

Fig. 14 Iteration 5 without DGS RP

4 Conclusion

Millimeter-wave technology has the potential for products in communication and security domains. The antenna is developed with substrate RogersRT5880 following Koch fractal geometry and is of dimensions 6.66 $\lambda_0 \times 6.66\ \lambda_0 \times 0.26\ \lambda_0$ at 100 GHz.

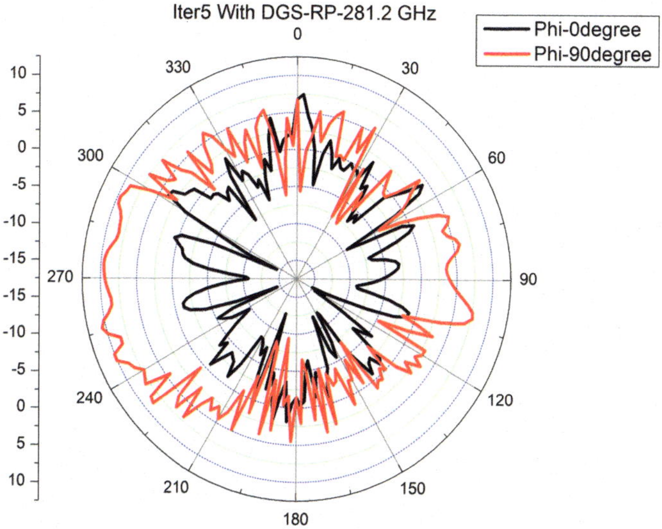

Fig. 15 Iteration 5 with DGS RP

The proposed iteration 4 design provides the highest gain of 11.6 dB and iteration 5 developed operates in the range of 85.20–330.62 GHz. Further scope involves developing antenna arrays to enhance gain, directivity, and beam-steering.

References

1. Du Preez J, Sinha S Millimeter-wave antennas: configurations and applications.Springer, Germany. https://doi.org/10.1007/978-3-319-35068-4
2. Darimireddy NK, Reddy RR, Prasad AM (2018) A miniaturized hexagonal-triangular fractal antenna for wide-band applications [Antenna applications corner]. IEEE Antennas Propag Mag 60(2):104–110. https://doi.org/10.1109/MAP.2018.2796441
3. Darimireddy NK, Reddy RR, Prasad AM (2018) Asymmetric and symmetric modified bow-tie slotted circular patch antennas for circular polarization. ETRI J 40:561–569. https://doi.org/10.4218/metric.2018-0008
4. Jilani SF, Aziz AK, Abbasi QH, Alomainy A (2018) Ka-band flexible koch fractal antenna with defected ground structure for 5g wearable and conformal applications. In: Proceedings of the 2018 IEEE 29th annual international symposium on personal, indoor and mobile radio communications (PIMRC). Bologna,Italy, pp 361–364. https://doi.org/10.1109/PIMRC.2018.8580692
5. Prajapati PR, Murthy GGK, Patnaik A, Kartikeyan MV (2014) Design of compact circular disc circularly polarized antenna with Koch curve fractal defected ground structure. In: Proceedings of the 2014 XXXIth URSI general assembly and scientific symposium (URSI GASS), Beijing, pp 1–4. https://doi.org/10.1109/URSIGASS.2014.6929151
6. Gan Z, Tu Z, Xie Z, Chu Q, Yao Y (2018) Compact wideband circularly polarized microstrip antenna array for 45 GHz application. IEEE Trans Antennas Propag 66(11):6388–6392. https://doi.org/10.1109/TAP.2018.2863243

7. Chen Q, Li J, Yang G, Cao B, Zhang Z (2019) A polarization-reconfigurable high-gain microstrip antenna. IEEE Trans Antennas Propag 67(5):3461–3466. https://doi.org/10.1109/TAP.2019.2902750
8. Wei K, Li JY, Wang L, Xu R, Xing ZJ (2017) A new technique to design circularly polarized microstrip antenna by fractal defected ground structure. IEEE Trans Antennas Propag 65(7):3721–3725. https://doi.org/10.1109/TAP.2017.2700226
9. Chen S, Qin P, Ding C, Guo YJ (2017) Cavity-backed proximity-coupled reconfigurable microstrip antenna with agile polarizations and steerable beams. IEEE Trans Antennas Propag 65(10):5553–5558. https://doi.org/10.1109/TAP.2017.2735484
10. Arif A, Zubair M, Ali M, Khan MU, Mehmood MQ (2019) A compact, low-profile fractal antenna for wearable on-body WBAN applications. IEEE Antennas Wirel Propag Lett 18(5):981–985. https://doi.org/10.1109/LAWP.2019.2906829
11. Desai A, Upadhyaya TK, Patel RH, Bhatt S, Mankodi P (2018) Wideband high gain fractal antenna for wireless applications. Prog Electromagn Res Lett 74:125−130
12. Gupta M, Mathur V (2017) Wheel shaped modified fractal antenna realization for wireless communications. Int J Electron Commun. https://doi.org/10.1016/j.aeue.2017.06.017
13. Sedghi T (2018) Compact fractal antenna for WiMAX 1.4 GHz and IEEE 802.11a using double branch line. J Instrum 13(9):P09021. https://doi.org/10.1088/1748-0221/13/09/P09021
14. Sivia JS, Kaur G, Sarao AK (2017) A modified sierpinski carpet fractal antenna for multiband applications. Wirel Pers Commun 95:4269–4279. https://doi.org/10.1007/s11277-017-4079-5
15. Sharma N, Singh G P, Sharma V (2016) Miniaturization of the fractal antenna using novel Giuseppe Peano geometry for wireless applications. In: Proceedings of the 2016 IEEE 1st international conference on power electronics, intelligent control and energy systems (ICPEICES), Delhi, pp 1−4. https://doi.org/10.1109/ICPEICES.2016.7853633
16. Malik R, Singh P, Ali H, Goel T (2018) A star shaped superwide band fractal antenna for 5g applications. In: Proceedings of the 2018 3rd international conference for convergence in technology (I2CT) Pune, pp 1–6. https://doi.org/10.1109/I2CT.2018.8529404
17. Garg R, Long SA (1988) An improved formula for the resonant frequencies of the triangular microstrip patch antenna. IEEE Trans Antennas Propag 36(4):570. https://doi.org/10.1109/8.1148

MB-ZZLBP: Multiscale Block ZigZag Local Binary Pattern for Face Recognition

Shekhar Karanwal and Manoj Diwakar

Abstract This work presents the LBP variant so-called MB-ZZLBP under illumination and expression variations. In MB-ZZLBP, initially mean is computed for all the square sub-blocks (size 2×2) of the 6×6 pixel window. After the mean computation, 3×3 window (pixel) is produced. Then ZigZag pixels are compared with each other, or in other words, the higher-order pixels are subtracted from the lower-order pixels. Differences of those which produce value ≥ 0 are allocated the label 1, else 0. After encoding each pixel position (binary pattern) eventually results in MB-ZZLBP transformed image. The MB-ZZLBP image is further divided into 3×3 sub-regions for extraction of histograms. The fused histogram is the feature size of MB-ZZLBP. The proposed FR approach attains remarkable results on EYB and Faces94 databases.

Keywords Multiscale Block ZigZag LBP · SVM · NN

1 Introduction

In recent years numerous feature extraction algorithms are developed for face recognition (FR). These algorithms achieve good results under the controlled conditions. But there are many challenges that arise in front of the robust FR under uncontrolled conditions. These uncontrolled conditions include illumination, pose, expression and occlusion variations. Extraction of features and classification are the two major steps of FR. If the extracted features are not discriminative then even the best classifiers fail to produce the desired results. The latter part discusses the work performed by the researchers under controlled and uncontrolled conditions.

Zhao et al. [1] discovered the Sobel-LBP for FR. Initially, the Sobel edge detector is utilized from which the gradient magnitude is produced. Then LBP feature extraction is performed from gradient magnitude. In addition, Sobel-LBP is also employed on imaginary and real parts of Gabor-filtered images. Experiments on FERET clearly

S. Karanwal (✉) · M. Diwakar
CSE Department, Graphic Era University (Deemed), Dehradun, Uttarakhand, India

A. Tomar et al. (eds.), *Machine Learning, Advances in Computing, Renewable Energy and Communication*, Lecture Notes in Electrical Engineering 768,
https://doi.org/10.1007/978-981-16-2354-7_54

demonstrate the capability of Sobel-LBP over LBP. Dong et al. [2] first extracted the local features by employing the rotation-invariant uniform LBP. To reduce the dimension, PCA is applied further. Then NN is adopted for classification. On Yale and JAFFE, the proposed method attains very good results. Wang et al. [3] presented the pyramid-based multiscale LBP for different unconstrained conditions; first by employing the multiscaling the face pyramid is created. Then LBP features are extracted from each level of the pyramid under different scales. Finally, feature fusion takes place for whole feature representation. Experiments conducted on FERET and ORL show the efficacy. Lei et al. [4] proposed an effective facial representation feature for FR called LBP eigenfaces. Initially, the original image is transformed to the LBP face space by employing the LBP. Then PCA is employed further which transforms the LBP image to the LBP eigenfaces. LBP eigenfaces capture local and global structures of the images (face). The LBP eigenfaces carry out superior outcomes than eigenfaces and the LBP histogram on EYB and AR databases. Yang et al. [5] introduced the WTP and WBP. The WTP and WBP descriptors engage the intensity difference (relative) between the neighborhoods and the center pixel. Both WBP and WTP are more robust under illumination variations than LBP and LTP.

Karanwal and Diwakar [6] invented the CZZBP and CMBZZBP descriptors. Both the proposed descriptors attain extraordinary outcomes (as compared to other operators) on two color databases. Among all, CMBZZBP is the most effective. Various techniques are outclassed by CMBZZBP. Nguyen and Caplier [7] presented ELBP for FR in uncontrolled environments. In ELBP, the elliptical pixels in vertical and horizontal directions are compared similarly as compared in LBP (i.e. with center pixel). Therefore, after comparison, VELBP and HELBP images (transformed) are built. Then from both the images (transformed) the sub-regional histograms (uniform pattern-based) are extracted. The integrated histograms are the size of ELBP descriptor. Further whitened PCA (WPCA) is applied next. The ELBP WPCA achieves astonishing results on three databases. Li et al. [8] give the novel approach in which histogram equalization (HE) and GLP filtering are employed first for illumination normalization. Then GWT is used for feature extraction. The feature length is compacted by PCA and matching is conducted by SVM. The proposed method attains better results than several approaches from the literature. Dabagh et al. [9] introduced a novel FR approach for single sample training problems by utilizing the LBP, Fisher's linear discriminant (FLD) and SVM. LBP is employed for preprocessing and FLD is employed for feature extraction. Finally, SVM is adopted for classification. The proposed method reached excellent outcomes at Yale.

Zhou et al. [10] discovered the improved CS-LBP for FR by incorporating the information of center pixel into CS-LBP called CS-LBP/Center. Specifically, the image is split into different sub-regions, and then the CS-LBP/Center histograms are extracted which are weighted by entropy image. Then all the features are combined serially for producing whole feature representation. Results on different datasets justify the efficiency of the proposed method. Li et al. [11] give the new method for FR based on CS-LBP and 2D-DWT. In the proposed method, CS-LBP is combined with the fusion of horizontal and vertical component images produced after employing 2D-DWT. The proposed method carries out encouraging results on EYB. Jun et al.

[12] invented the compact LBP (CLBP) for FR and FER. Specifically, an efficient code selection method is proposed to produce the CLBP by utilizing the MMI. The proposed descriptor achieves astonishing results.

In this paper, the novel LBP variant is presented for FR, the so-called MB-ZZLBP under illumination and expression variations. The main contributions are:

1. In LBP, the neighbors are used for comparison with the center pixel. However, in MB-ZZLBP initially mean is computed and then features are extracted in ZigZag style. Specifically, the mean of square blocks (of size 2×2) is computed by utilizing the 6×6 pixel window. After the mean computation 3×3 pixel window is produced. Then ZigZag pixels are compared with each other, or in other words, the higher-order pixel is subtracted from the lower-order pixels. Differences of those which have value $\geq$ to 0 are allocated the label 1, else 0.
2. After encoding every pixel position (binary pattern) gives the MB-ZZLBP image. The image is further decomposed into 3×3 sub-regions for extraction of histograms. The fused histogram is the feature size of the MB-ZZLBP descriptor. The huge dimension produced is compacted by FLDA [13] and classification is done by SVM [14, 15] and NN [16, 17].
3. The models (multiclass) are constructed by ECOCs [18]. The proposed approach achieves encouraging results on EYB [19, 20] and Faces94 [21].

The remainder of the paper is shaped as follows: The proposed FR descriptor is reported in Sect. 2, results are outlined in Sect. 3, discussions are clarified in Sect. 4 and Sect. 5 gives the conclusion.

2 Multiscale Block ZigZag Local Binary Pattern (MB-ZZLBP)

In MB-ZZLBP, initially, the mean of the square blocks is computed by utilizing the 6×6 pixel window. In the 6×6 pixel window, each square block size is 2×2. For this research work 6×6 pixel window is used, the other higher-order pixel windows can also be utilized for the mean computation. After computation of mean 3×3 pixel window is produced. Then pixels that lie according to ZigZag are compared with each other, or in other words, the higher-order pixels are subtracted from the lower-order pixels. Differences of those which produces value ≥ 0 are allocated a label 1, else 0. By assigning weights the pattern (8-bit) is converted into decimals. After encoding each pixel position's binary pattern, MB-ZZLBP transformed image is produced. MB-ZZLBP image is further divided into 3×3 sub-regions for extraction of histograms. The fused histogram is of the size of the MB-ZZLBP. Therefore, MB-ZZLBP feature size is 2304. Figure 1 shows the MB-ZZLBP demonstration and Fig. 2 shows the MB-ZZLBP transformed image with sub-regional histograms. The sample image used in Fig. 2 is taken from EYB. MB-ZZLBP equation-wise is expressed as

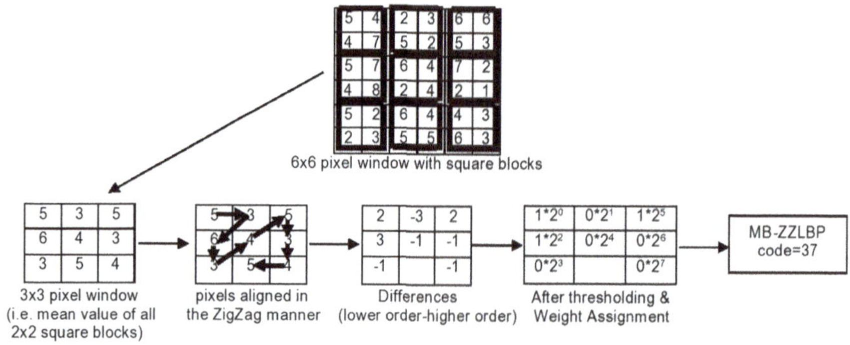

Fig. 1 Demonstration of MB-ZZLBP descriptor

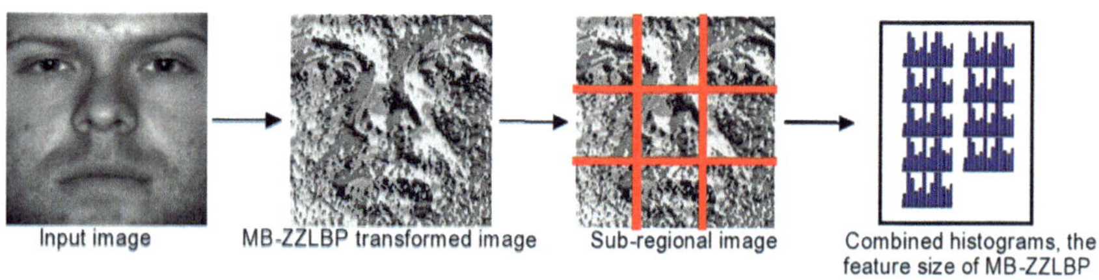

Fig. 2 The MB-ZZLBP transformed image with sub-regional histograms

$$V_{i,j} = \text{mean}(L_{i,j}) \tag{1}$$

$$\text{MB} - \text{ZZLBP}_{P,R}(x_c) = \sum_{p=0}^{P-1} k\left(V_{R,p} - V_{R,p+1}\right)2^{p} \tag{2}$$

where $k(y) = \begin{pmatrix} 1 y \geq 0 \\ 0 y < 0 \end{pmatrix}$ for $p = (0, \ldots P-1)$.

$V_{i,j}$ calculates the square blocks mean (specified as $L_{i,j}$). P and R interpret the neighborhood size and radius of 3×3 window (after mean computation). $V_{R,p}$ and $V_{R,p+1}$ denote the ZigZag pixels or the lower-order and the higher-order pixels. After employing FLDA, the classification is conducted by SVM and NN. From SVM, the RBF and POLY are utilized. From NN, the ESA is adopted with cosine distance. Therefore, the results are finally achieved from three classifiers.

Table 1 Illustration of two databases used for the evaluation

Databases	Total	Class images	Size	Total images	Challenges
EYB	38	64	192 × 168	2432	Variations in illuminations
Faces94	152	20	200 × 180	3040	Expression variations and slight head movement

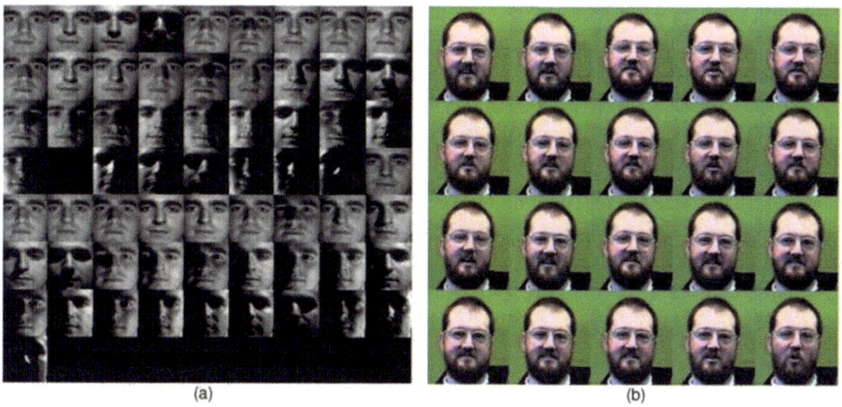

Fig. 3 The sample images of EYB (**a**) and Faces94 (**b**) databases

3 Experiments

3.1 Description of Databases

The two databases utilized are EYB and Faces94. Table 1 provides the illustration of both the databases and Fig. 3 shows some images.

3.2 Technical Details of the Descriptor

On EYB, the images are resized to 47 × 44 before doing feature extraction. On Faces94 the color images are changed to gray and then images are resized. Then feature extraction is done by the proposed descriptor MB-ZZLBP. The feature size constructed by MB-ZZLBP is 2304. Then FLDA is employed to bring off the discriminative feature for classification. The PCA feature sizes are 142 and 513, and the LDA feature size is 27 and 192 on EYB and Faces94. MATLAB R2018a is used for all simulations.

Table 2 Performance analysis by using three classifiers

Proposed FR approach	Training images			
	Tra = 5	Tra = 10	Tra = 20	Tra = 30
	False counts/accuracy			
MB-ZZLBP + FLDA + SVM (RBF)	42/98.12	27/98.68	17/98.98	08/99.38
MB-ZZLBP + FLDA + SVM (POLY)	60/97.32	41/98.00	26/98.44	12/99.07
MB-ZZLBP + FLDA + NN (ESA cosine distance)	60/97.32	42/97.95	26/98.44	17/98.68

3.3 *Accuracy Observed on Subsets*

For observing the accuracy several subsets are considered. The demonstration is given below.

3.3.1 EYB Face Database

On EYB (5, 10, 20 and 30) training samples are considered and the remaining ones are for testing. The partition ratio of the training and test images is set by using the holdout method. On each subset, false counts are measured (on test set). The formula is given as $\left[\text{Accuracy} = \frac{\text{Testsize} - \text{Falsecounts}}{\text{Testsize}} * 100\right]$. The maximum accuracy is observed after 37 runs. The best results of MB-ZZLBP descriptor are exploited by the RBF classifier after applying FLDA. Table 2 shows the presentation analysis.

3.3.2 Faces94 Database

On Faces94 (3, 6 and 10) training samples are utilized. The maximum accuracy is measured after 12 runs. On Train = 3 and Train = 6, the performance (of the MB-ZZLBP descriptor) exploited by the POLY and (ESA cosine distance) are better than the RBF classifier, after applying FLDA. On Train = 10, all classifiers achieve similar results. Table 3 gives the analysis.

3.4 *Results Comparison*

In the EYB dataset, MB-ZZLBP + FLDA + SVM (RBF) is utilized for comparison because it produces the best results. In the Faces94 database, the MB-ZZLBP + FLDA + SVM (POLY) is used for comparison.

Table 3 Performance analysis by using three classifiers

Proposed FR approach	Training images		
	Tra = 3	Tra = 6	Tra = 10
	False counts/accuracy		
MB-ZZLBP + FLDA + SVM (RBF)	04/99.84	02/99.90	00/100
MB-ZZLBP + FLDA + SVM (POLY)	00/100	00/100	00/100
MB-ZZLBP + FLDA + NN (ESA cosine distance)	00/100	00/100	00/100

3.4.1 EYB Database

In EYB database 13 approaches are considered for comparison. The descriptions of accuracy attained by them are: RSLDA [22], DLA [22], LRDLSR [23] and CRC [23] reach the rate of [87.46% 93.26%], [87.78% 93.09%], [91.18% 96.84%] and [91.85% 96.39%] when 10 and 20 samples per individual are utilized for training. LRR 8 × 8 overlapping [24], LRR 8 × 8 non-overlapping [24], LSDF [25], LSRP [28], MVP [28] and MR-ELM [29] achieve accuracy of [89.76% 96.15% 97.97% 98.58%], [88.04% 95.15% 97.37% 98.22%], [50.60% 61.70% 70.20% 72.80%], [71.69% 83.64% 91.61% 94.51%], [67.91% 81.93% 90.98% 93.35%] and [52.68% 75.25% 85.61% 92.78%] when 5, 10, 20 and 30 per individual samples are consumed for training. GRSDA [26] and DSNPE [26] reached the rate of [82.70% 89.70% 93.40%] and [81.60% 87.60% 91.70%] when 10, 20 and 30 samples are consumed for training. GODRSC [27] reaches recognition accuracy of 85.16% on training size of 10. The *proposed method* blasted all approaches. Table 4 gives the results comparison.

3.4.2 Faces94 Database

In the Faces94 database, six approaches are considered for comparison. The accuracy achieved by them are: (FDCT + QSVD)-RWN [30], PCA + LDA [30], 2DPCA [32], MP2DPCA [32] and P2DPCA [32] pulls off the recognition rate of 99.42%, 99.29%, 88.90%, 92.90% and 89.60% on training sample size of 10. Ridgelet transforms [31] attain an accuracy of 97.14% on a training sample size of 3. The *proposed method* outclassed the other approaches. Table 5 shows a comparison of the results.

Table 4 Results comparison on EYB face database

All FR approaches	Training images			
	Tra = 5	Tra = 10	Tra = 20	Tra = 30
	Accuracy in %			
RSLDA [22]	N/A	87.46	93.26	N/A
DLA [22]	N/A	87.78	93.09	N/A
LRDLSR [23]	N/A	91.18	96.84	N/A
CRC [23]	N/A	91.85	96.39	N/A
LRR 8 × 8, overlapping [24]	89.76	96.15	97.97	98.58
LRR 8 × 8, non-overlapping [24]	88.04	95.15	97.37	98.22
LSDF [25]	50.60	61.70	70.20	72.80
GRSDA [26]	N/A	82.70	89.70	93.40
DSNPE [26]	N/A	81.60	87.60	91.70
GODRSC [27]	N/A	85.16	N/A	N/A
LSRP [28]	71.69	83.64	91.61	94.51
MVP [28]	67.91	81.93	90.98	93.35
MR-ELM [29]	52.68	75.25	85.61	92.78
MB-ZZLBP + FLDA + SVM (RBF)	98.12	98.68	98.98	99.38

N/A—Not available

Table 5 Results comparison on Faces94 database

All FR approaches	Training images		
	Tra = 3	Tra = 6	Tra = 10
	Accuracy in %		
(FDCT + QSVD)-RWN [30]	N/A	N/A	99.42
PCA + LDA [30]	N/A	N/A	99.29
Ridgelet transforms [31]	97.14	N/A	N/A
2DPCA [32]	N/A	N/A	88.90
MP2DPCA [32]	N/A	N/A	92.90
P2DPCA [32]	N/A	N/A	89.60
MB-ZZLBP + FLDA + SVM (POLY)	100	100	100

N/A—Not available

4 Discussions

This work launched the novel descriptor in expression and illumination variations of the so-called MB-ZZLBP. After computing square blocks mean, the ZigZag pixels are compared, or in other words, the higher-order pixels are subtracted from the lower-order pixels. The MB-ZZLBP transformed image produced after is then divided into

3×3 sub-regions for extraction of histograms. The fused histograms are of the size (feature) of MB-ZZLBP. The reduction in dimension is done by FLDA. The proposed method carries out remarkable outcomes on EYB and Faces94 databases. In the EYB database, 13 approaches are considered for comparison and in the Faces94 database, 6 approaches are considered for the comparison. The proposed method outclassed all the approaches.

5 Conclusion

This work presented a novel descriptor in illumination and expression variations of the so-called MB-ZZLBP. After computing the mean of square blocks, the ZigZag pixels are compared with each other, or in other words, the higher-order pixels are subtracted from the lower-order pixels. The MB-ZZLBP transformed image produced after is then divided into 3×3 sub-regions for extraction of histograms. The fused histograms are of the size (feature) of MB-ZZLBP. The reduction in dimension is done by FLDA. The proposed method carries out remarkable outcomes on EYB and Faces94 databases. In the EYB database 13 approaches are considered for comparison and in the Faces94 database six approaches are considered for the comparison. The proposed method outclassed all the approaches.

References

1. Zhao S, Gao Y, Zhang B (2008) SOBEL-LBP. In: Proceedings of the ICIP. pp 2144–2147
2. Dong EZ, Fu YH, Tong JG (2015) Face recognition by PCA and improved LBP fusion algorithm. Appl Mech Mater 734:562–567
3. Wang W, Chen W, Xu D (2011) Pyramid-based multi-scale LBP features for face recognition. In: Proceedings of the ICMSP
4. Lei L, Kim DH, Park WJ, Ko SJ (2014) Face recognition using LBP eigenfaces. IEICE Trans Inf Syst E97-D(7):1930–1932
5. Yang Z, Jiang Y, Wu Y, Lu Z, Li W, Liao Q (2015) WBP and WTP for illumination-robust face recognition. In: Proceedings of APSAIPAASC. pp 1050–1053
6. Karanwal S, Diwakar M (2020) Two novel color local descriptors for face recognition. Optik 1–15
7. Nguyen HT, Caplier A (2012) Elliptical local binary patterns for face recognition. In: Proceedings of the ACCV. pp 85–96
8. Li M, Yu X, Ryu KH, Lee S, Umpon NT (2018) Face recognition technology development with Gabor, PCA and SVM methodology under illumination normalization condition. Clust Comput 21(1):1117–1126
9. Dabagh MZNA (2014) Face recognition using LBP, FLD and SVM with single training sample per person. IJSER. 5(5):180–183
10. Zhou N, Constantinides AG, Huang G, Zhang S (2018) Face recognition based on an improved CS-LBP. Neural Comput Appl 30(12):3791–3797
11. Li C, Zhao S, Xiao K, Wang Y (2017) Face recognition based on enhanced CSLBP. In: international conference on FIT. pp 539–544

12. Jun B, Kim T, Kim D (2011) A compact local binary pattern using MMI for face analysis. Pattern Recogn 44(3):532–543
13. Belhumeur PN, Hespanha JP, Kriegman DJ (1997) Eigenfaces vs. Fisherfaces: recognition using class specific LP. IEEE Trans PAAMI 19(7):711–720
14. Vapnik V (1998) Statistical learning theory. Wiley, New York
15. Kotsia I, Pitas I (2007) Facial expression recognition in image sequences using geometric deformation features and SVM. IEEE Trans Image Process 16(1):172–187
16. Wang S, Liu Z (2010) Infrared face recognition based on histogram and k-nearest neighbor classification. In: Proceedings of the ISNN. pp 104–111
17. Sohail ASM, Bhattacharya P (2007) Classification of facial expressions using k-nearest neighbor classifier. In: Proceedings of the ICCV/CGCT. pp 555–566
18. Kittler J, Ghaderi R, Windeatt T, Matas J (2003) Face verification via ECOC. Image Vis Comput 21(13–14):1163–1169
19. Georghiades AS, Belhumeur PN, Kriegman DJ (2001) From few to many: illumination cone models for face recognition under variable lighting and pose. IEEE Trans PAAMI 23(6):643–660
20. Lee KC, Ho J, Kriegman DJ (2005) Acquiring linear subspaces for face recognition under variable lighting. IEEE Trans PAAMI 27(5):684–698
21. http://cswww.essex.ac.uk/mv/allfaces/faces94.html
22. Wen J, Fang X, Cui J, Fei L, Yan K, Chen Y, Xu Y (2019) Robust sparse linear discriminant analysis. IEEE Trans Circuits Syst Video Technol 29(2):390–403
23. Chen Z, Wu XJ, Kittler J (2020) Low-rank discriminative least squares regression for image classification. Signal Process 173
24. Xue H, Zhu Y, Chen S (2009) LRR for face recognition. Neurocomputing 72(4–6):1342–1346
25. Zhao J, Lu K, He X (2008) Locality sensitive semi-supervised feature selection. Neurocomputing 71:1842–1849
26. Lou S, Zhao X, Chuang Y, Yu H, Zhang S (2016) Graph regularized sparsity discriminant analysis for face recognition. Neurocomputing 173:290–297
27. Zhang L, Chen S, Qiao L (2012) Graph optimization for dimensionality reduction with sparsity constraints. Pattern Recogn 45:1205–1210
28. Lai Z, Li Y, Wan M, Jin Z (2013) Local sparse representation projections for face recognition. Neural Comput Appl 23(7–8):2231–2239
29. Liu B, Xia SX, Meng FR, Zhou Y (2016) Manifold regularized ELM. Neural Comput Appl 27(2):255–269
30. Wan W, Zhou Z, Zhao J, Cao F (2015) A novel face recognition method: Using RWN and quasi-SVD. Neurocomputing 151:1180–1186
31. Kautkar S, Koche R, Keskar T, Pande A, Rane M, Atkinson GA (2010) Face recognition based on ridgelet transforms. In: Proceedings of the ICEBT, vol 2. pp 35–43
32. Wang H, Chen S, Hu Z, Luo B (2008) Probabilistic two-dimensional PCA and its mixture model for face recognition. Neural Comput Appl 17(5–6):541–547

A Critical Review on Secure Authentication in Wireless Network

Manoj Diwakar, Prabhishek Singh, Pramod Kumar, Kartikay Tiwari, and Shashi Bhushan

Abstract Wireless network connectivity is capable of addressing various mobility issues and helps users of smartphones to navigate around and remain connected to the network without taking control of their location. The 802.11 architecture is similar to cell architecture. This paper provides a short overview of wireless networks, their benefits over wired networks and urgent exposure to security concerns. The 802.11 architecture and the different facilities it delivers are pursued and then the motivation for doing the study is followed.

Keywords Handshake protocol · Network security · Denial of service · WLAN

1 Introduction

Wireless network communication is able to address various mobility issues and provides freedom to mobile users to roam around and still remaining connected to the network, without worrying about their location [1–6]. The 802.11 architecture is similar to the cellular architecture. The whole system is divided into different cells called basic service set (BSS) where each cell is controlled by its respective stations (access points). Now, in order to support mobility issues, AP of the respective cells are connected by some backbone system, generally a distributed system, which is a wired network [7–12]. This whole system of interconnected cells which includes

M. Diwakar
Graphic Era Deemed to be University, Dehradun, India

P. Singh (✉)
Amity School of Engineering and Technology, Amity University, Noida, Uttar Pradesh, India

P. Kumar
Krishna Engineering College, Ghaziabad, Uttar Pradesh, India

K. Tiwari
Thapar University, Patiala, Punjab, India

S. Bhushan
UPES, Dehradun, India

A. Tomar et al. (eds.), *Machine Learning, Advances in Computing, Renewable Energy and Communication*, Lecture Notes in Electrical Engineering 768,
https://doi.org/10.1007/978-981-16-2354-7_55

their respective APs and the distribution system is called extended service set (ESS) [13]. Various components of the architecture are[1]:

- **Stations**

Any entity that can be connected to a wireless network is termed as stations. These stations are generally battery-driven and include laptops, palmtops and notebooks [14–18]. All of them have a network interface card (NIC) which has a unique MAC address and helps in identifying the system over the network. Stations can be classified into two sub-categories [3]: access points and clients. Access points are normal wireless clients with have higher computational power and other resources. They are connected to a distributed system which in turn is connected to other wired networks and thus enables wireless clients to transmit and receive radio frequencies. Wireless clients include mobile and portable devices like palmtops, notebooks having wireless network interface card.

- **Basic Service Set**

It is the atomic unit of IEEE 802.11 WLAN comprising some stations which run the copy of similar MAC protocol and compete with one other for getting access to the wireless medium shared between them. The BSS resembles the cell as present in cellular architecture. Every BSS has its id known as BSSID that serves the wireless clients within that BSS. BSS exists in two modes [19–23]: independent BSS and infrastructure BSS. IBSS is generally like ad hoc networks in which stations communicate with one another in a direct manner and is set up for a very short period or interval; when the communication ends, it gets dissolved while in infrastructure BSSs if two nodes wish to communicate, then they are able to perform this by means of AP, i.e., first they send data to AP which then sends it to other communicating nodes [24–28].

- **Extended Service Set**

BSS makes the communication over a small range, i.e., within the coverage range of AP. Therefore, in order to enhance and lengthen or expand the range of the AP, i.e., the coverage area, BSSs are linked to each other by having some backbone network (distributed system) in the back of the network to form a region known as extended service set (ESS) [29]. All the APs within the ESS have the same service SET identifier (SSID).

- **Distributed System**

The main role of DS is to connect several BSSs to the wired network to result in an ESS. Several BSSs are connected via their respective APs which are connected to a

[1] Please note that the LNEE Editorial assumes that all authors have used the Western naming convention, with given names preceding surnames. This determines the structure of the names in the running heads and the author index.

distributed system which in turn gets connected to different 802.1x wired networks [4]. When a frame is received by the distributed system, it checks the MAC address and relays it to the appropriate AP, which in turn relays the frame to the destination client.

- **Distribution System Services**

The major role of these services is to interconnect various BSSs with one another with the help of connecting their respective APs to the distributed system so that services of the wired networks can be extended to WLANs by connecting DS to integrated IEEE802.1x LANs. These services [11–15] can be implemented within the respective APs of the BSSs or can be provided by using some special-purpose devices which are attached to the DS.

- **Station Services**

Providing station services is a basic feature of any IEEE 802.11 complying station which also includes access points [16–20]. These services are essential in order to deliver messages to the intended recipients. They provide confidentiality and privacy services in order to protect the messages being communicated between the stations. Also included are the authentication services in order to confirm the identity of the client so that they can avail access to other services.

2 IEEE 802.1X Framework

It provides a port-based access control mechanism to devices connected through various 802 LANs for authorization and authentication services [20]. It also serves the purpose of distributing the secure keys by use of various encryption techniques between different compatible clients, supplicants and access points, thus optimizing the public key authentication.

It has been proved that earlier methods of authentication, namely open system authentication and shared key authentication are not secure, therefore in order to counter the attacks, IEEE802.11i defined RSNA as a mechanism to provide strong mutual authentication and generate fresh temporal keys in order to provide strong confidentiality services. In network discovery, a wireless client always searches the available channels for these Beacon frames and responds with Beacon response frames to the access points depending on the available signal strength. In authentication and association, once the supplicant is authenticated, it sends the association request frame to the AP and indicates its security capabilities. AP replies with the association response frame indicating the association result. After this stage, the client/supplicant is said to be authenticated and associated.

3 A Comparative Study

However, the authentication achieved is not very strong, therefore subsequent phases are followed in order to make it more secure. Here, the RADIUS server and the client execute a mutual authentication protocol, i.e., EAP-TLS between them and AP just acting as a relay to forward messages. At the end of this stage, a shared key called PMK is generated between the two which is used for the derivation of subsequent keys. The authenticator only permits the 802.1X messages to allow it through port (off) before the client is being authenticated. The EAP messages or frames from the client are then relayed to the authentication server by means of an authenticator port access entity (PAE) [20].

3.1 Temporal Key Integrity Protocol

With many inherent weaknesses found in the use of WEP, a new scheme was introduced which can provide far better security. An attacker can easily get the secret key being used in the WEP technique within few minutes and in some situations even can decrypt the packets without having any apprehension about the secret key, thus is prone to very serious attacks. TKIP [18, 22] was used on top of an already used scheme, i.e., WEP in order to make it more secure and hide its weaknesses.

TKIP made many modifications in WEP which can limit many of the earlier attacks on WEP:

- Use of MIC as a means to protect the integrity of the generated message by making use of a new algorithm called Michael.
- Involving the use of a per-packet sequence counter in order to protect the entities from replay attacks.
- Use of per-packet key-mixing technique (function) in order to make it secure against weak-key attacks of the attacker on WEP secret key.
- Use of some countermeasures to handle attacks against MIC since due to some design constraints it is not deemed to be very secure.

3.2 Vulnerabilities of IEEE802.11i Standard

- Prone to denial of service (DoS) and DoS flooding attacks like RF jamming, session hijacking.
- Unprotected management frames lead to pinpoint the location of devices, thus making them vulnerable to DoS attacks and to guess the network topology.
- Control frames are also unprotected and send in plain text over a network.
- Possibility of de-authentication and disassociation attacks is very high.
- Vulnerable to offline guessing attacks.

- No protection for EAPOL frames.

The temporal key (TK) is generated by means of the EAPOL handshake procedure. The very first step of this technique is to get the per-packet key which is done in two phases. The first phase key mixing procedure takes a temporal key (TK), transmitting station address (TA) and 32 MSBs of TKIP sequence counter (TSC) as its input and outputs TTAK which is of 80 bits. The second phase key mixing procedure takes TK, TTAK and 16 LSBs of TSC as its input which results in the generation of WEP seed represented as 128-bit key (104-bit RC4 secret key and 24-bit IV for WEP).

TKIP also introduces a mechanism for checking the integrity of the message called MIC which is generated by means of the Michael algorithm which takes three inputs. Then the computed MPDU plus generated MIC is fragmented based on network packet size if required which is then send for WEP encapsulation as plain text.

3.3 *Flaws in WPA*

- Use of pre-shared keys as an alternate mechanism for providing authentication is a serious drawback.
- Dictionary or brute-force attacks are still possible.
- Vulnerable to DoS and DoS flooding attacks.

3.4 *Wi-Fi Protected Access (WPA)*

In 2002, Wi-Fi Alliance (WFA) presented a new mechanism called WPA [10] as a temporary or provisional solution to counter the attacks which were prevalent in WEP. Some of its benefits over WEP are:

- Usage of temporal key integrity protocol (TKIP) for providing confidential services.
- More secure user authentication mechanism.
- Proper use of the RC4 algorithm makes networks more secure.
- Use of more complex and secure hash functions.
- Avoids re-use of the initialization vector.

There exist two modes of WPA, namely enterprise WPA; personal/WPA-PSK (pre-shared key). In enterprise mode, there is a centralized network entity called RADIUS server which provides services related to authentication, authorization and access control, while in personal mode there is no such concept of the RADIUS server and the client needs to know the WPA shared key generated by the AP and SSID of the network to be connected.

3.5 *Working of WEP*

WEP was the very first technique to provide security in WLAN by use of the RC4 encryption algorithm [1, 2]. Its working at sender and receiver side can be explained as follows:

3.5.1 At Sender Side

As shown in Fig. 1 at the start, both the sender and receiver share a secret key K_S. Assume S to be the supplicant/client which sends M (message) to the receiver at the other side [1, 18]. S then also calculates checksum known as cyclic redundancy check, which is then appended or concatenated with message M. Let this be represented as X = (M, CRC). Then supplicant encrypts this X using the RC4 encryption algorithm that takes two inputs to generate a keystream KS. The two inputs are:

(1) Shared key K_S of length 40 bits.
(2) An initial seed, which is called initialization vector IV.

Now this keystream KS is XORed with X which in turn produces the desired ciphertext C. The major drawback is that IV is sent without using any encryption algorithm, i.e., clear text is communicated over the network. To re-produce the original keystream, the generated ciphertext is XORed with the same keystream KS, i.e.

$KS \oplus X = ((X \oplus KS) \oplus KS) = X \oplus (KS \oplus KS) = X.$

But in order for the receiver to reconstruct KS, IV should be known. Therefore, IV is appended to ciphertext before being sent over the network. The major drawback is

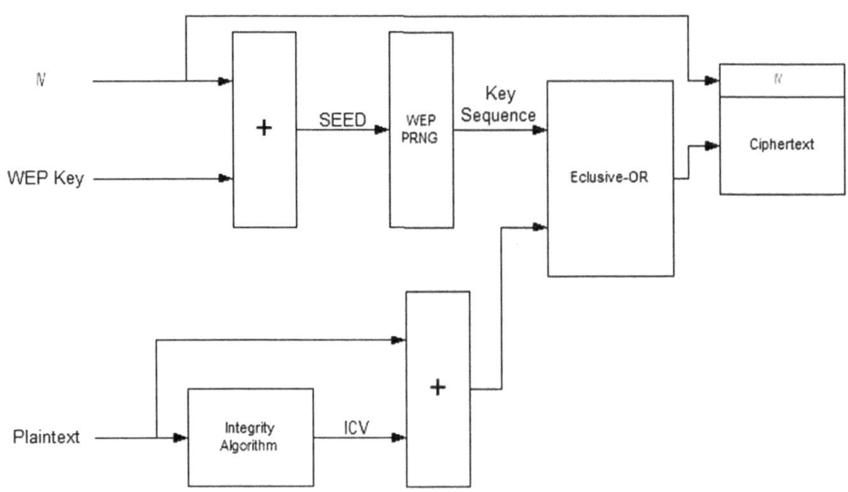

Fig. 1 WEP encryption algorithm (sender side)

that IV is sent without using any encryption algorithm, i.e., clear text is communicated over the network.

3.5.2 At the Recipient Side

As shown in Fig. 2, the WEP key and initialization vector is passed through the pseudorandom generator in order to obtain the keystream which is then XORed with the ciphertext to get the IV and the plaintext combination [2]. Now the plaintext is separated from the initialization vector and plaintext is passed through the integrity algorithm to get the new initialization vector IV_1, which is then compared with the received IV.

3.6 Flaws in WEP

WEP is considered very weak and it has been verified and justified that the WEP secret key can be broken within few minutes by the attacker. The major flaws [2–10] in WEP which make it insecure and vulnerable to various attacks are:

- Use of 24-bit initialization vector which exposes it to diverse attacks since it is of very short length and is appended with ciphertext as it is without using any encryption technique.
- No mechanism to prevent replay attacks.
- No support for key management and mutual authentication.
- Improper use of RC4 algorithm for providing privacy and authentication services since at every stage of RC4 encryption the same keystream is being used for encryption.
- Use of a 40-bit WEP key for encryption has been proven to be insecure as the key can be broken within few minutes. Therefore, a larger key of 128 bits is suggested.

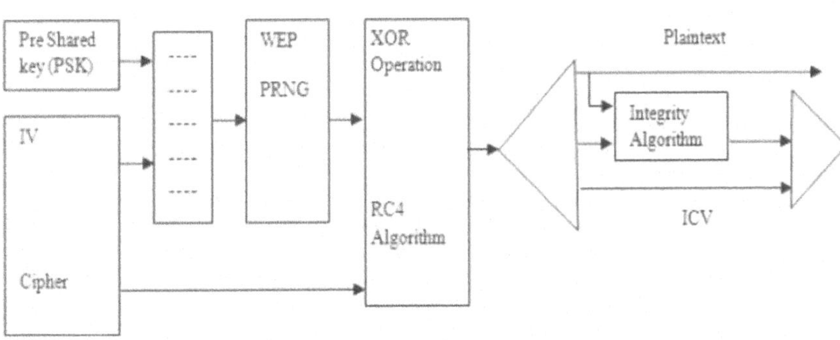

Fig. 2 WEP decryption algorithm (recipient side)

- Data source authentication: There is no mechanism for the source of data being authenticated. The use of CRCs permits attackers to frame their fake messages which have the same CRC as of original message and impersonate them as they are being originated and arrived from an authentic or known entity. Using MAC can be a very good measure in order to prevent this type of attack as they are used for data source authentication. Other measures can be to make CRC inaccessible to attackers by encrypting it or applying some kind of technique, but WEP failed to achieve this.
- Use of only one mechanism to implement all security services which are presently based on data privacy service in case of WEP which is also a major drawback of any security service.

3.7 Enhancements Over WEP

In order to counter the inherent flaws in WEP, a new algorithm was developed which was more secure and is interoperable with wired equivalent privacy (WEP), i.e., no extra hardware required for its implementation.

3.7.1 Enhanced WEP (eWEP)

eWEP [14] is one of the leading accomplishments in securing the wireless network. Its applications are analogous to that of WEP except it tries to probe the mechanisms to protect the initialization vector which is dispatched and relayed in plaintext over the network, thus providing one of the solutions for securing the network from attacks.

3.7.2 Working

To start or begin the process of encryption, sender S and receiver R mutually agree on some initial IV (IV_1) [14]. Then a new random IV, i.e., IV_2 is generated by S. Now sender S with the help of key K_s and IV_1 generates a keystream KS by using RC4 as encryption algorithm. Then CRC is calculated and succeeded or attached to M_1 which in turn is equivalent to $X_1 = (M1, CRC)$, IV_2 is appended to X_1. Then this whole message is XORed with previously generated keystream KS_1. The process continues this way for all the fragments M_1, M_2 … M_n as shown in Fig. 3. This whole message is then sent over the network to receiver R.

The process is almost similar in comparison to that of WEP. The major difference or change is that here we will encrypt X = (M, CRC) and IV (initialization vector) with RC4 encryption algorithm in turn to hide IV from an attacker. In this sender S encrypts X_i appended to IV_{i+1} with the help of an IV_i from the previous step. Therefore, the receiver needs to know only the initial IV, i.e., IV_1 is required to decrypt the first frame, which in turn reveals IV_2 used for the decryption of the

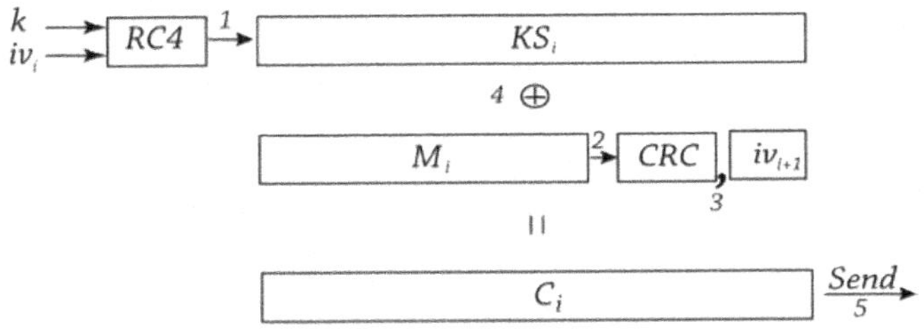

Fig. 3 Encryption process in eWEP

second frame and the process continues in the same manner. At the receiver end, R first decrypts the message by knowing IV_1 and then stores the appended IV_2 with it, which in turn is then used to decrypt the next frame being received from S and this process continues, which ultimately leads to the decryption of all successive frames being received by receiver R.

4 Conclusions

The major contribution of this paper is to analyze the major issues on security in WLAN. IEEE802.11i is the latest standard being used to provide security in WLANs. It specifies two frameworks for being used in 802.11 WLANs, one being the robust security network (RSN) and the other being the pre-RSN. A network entity is assumed to being RSN-capable if it is able to create the RSN associations between the communicating entities, otherwise, it is assumed as pre-RSN entity. Any network is termed as an RSN security framework if it allows robust security network associations with RSN-capable network equipments. Similarly, any network that is able to allow only pre-RSN association between the network entities is termed as pre-RSN framework for network security. The main point of difference between these two frameworks is that of four-way handshake procedure, depending on whether it is included in the authentication and association process.

References

1. Khan MA, Hasan A (2008) Pseudo random number based authentication to counter denial of service attacks on 802.11. In: 5th IFIP international conference on wireless and optical communications networks, WOCN '08, pp 1–5
2. Yao Y, Chong J, Xingwei W (2010) Enhancing RC4 algorithm for WLAN WEP protocol. In: The control and decision conference (CCDC), pp 3623–3627
3. Chen J-C, Wang Y-P (2005) Extensible authentication protocol (EAP) and IEEE 802.1x: tutorial and empirical experience. IEEE Commun Mag 43:supl.26–supl.32
4. Gast M (2005) 802.11 wireless networks: the definitive guide. O'Reilly Publication

5. Mohapatra H, Rath S, Panda S, Kumar R (2020) Handling of man-in-the-middle attack in wsn through intrusion detection system. Int J 8(5):1503–1510
6. Mohapatra H, Rath AK, Landge PB, Bhise D, Panda S, Gayen SA (2020) A comparative analysis of clustering protocols of wireless sensor network. Int J Mech Prod Eng Res Dev (IJMPERD) ISSN (P) (2020):2249-6890
7. Abbas K, Afaq M, Khan TA, Rafiq A, Song W-C (2020) Slicing the core network and radio access network domains through intent-based networking for 5G networks. Electronics 9(10):1710
8. Saqib M, Mehmood A, Rafiq A, Muhammad A, Song W-C (2020) Distributed SDN based network state aware architecture for flying ad-hoc network. In: 2020 21st Asia-Pacific network operations and management symposium (APNOMS). IEEE, pp 25–30
9. Abbas K, Afaq M, Khan TA, Mehmood A, Song W-C (2020) IBNSlicing: intent-based network slicing framework for 5G networks using deep learning. In: 2020 21st Asia-Pacific network operations and management symposium (APNOMS). IEEE, pp 19–24
10. Dowling B, Fischlin M, Günther F, Stebila D (2015) A cryptographic analysis of the TLS 1.3 handshake protocol candidates. In: Proceedings of the 22nd ACM SIGSAC conference on computer and communications security, pp 1197–1210
11. Diáz G, Cuartero F, Valero V, Pelayo F (2004) Automatic verification of the TLS handshake protocol. In: Proceedings of the 2004 ACM symposium on applied computing, pp 789–794
12. Ma Y, Yan L, Huang X, Ma M, Li D (2020) DTLShps: SDN-based DTLS handshake protocol simplification for IoT. IEEE Internet Things J 7(4):3349–3362
13. Cai J, Huang X, Zhang J, Zhao J, Lei Y, Liu D, Ma X (2018) A handshake protocol with unbalanced cost for wireless updating. IEEE Access 6:18570–18581
14. Wagner D, Schneier B (1996) Analysis of the SSL 3.0 protocol. In: The second USENIX workshop on electronic commerce proceedings, vol 1, no 1, pp 29–40
15. Han S-W, Kwon H, Hahn C, Koo D, Hur J (2016) A survey on MITM and its countermeasures in the TLS handshake protocol. In: 2016 eighth international conference on ubiquitous and future networks (ICUFN). IEEE, pp 724–729
16. Van Berkel K, Bink A (1996) Single-track handshake signaling with application to micropipelines and handshake circuits. In: Proceedings second international symposium on advanced research in asynchronous circuits and systems. IEEE, pp 122–133
17. Yin Z, Leung VCM (2005) Third-party handshake protocol for efficient peer discovery in IEEE 802.15. 3 WPANs. In: 2nd international conference on broadband networks. IEEE, pp 840–849
18. Mzid R, Boujelben M, Youssef H, Abid M (2010) Adapting TLS handshake protocol for heterogenous IP-based WSN using identity based cryptography. In: 2010 international conference on wireless and ubiquitous systems. IEEE, pp 1–8
19. Dowling B, Fischlin M, Günther F, Stebila D (2016) A cryptographic analysis of the TLS 1.3 draft-10 full and pre-shared key handshake protocol. IACR Cryptol. ePrint Arch. p 81
20. Liu Yi, Wang H, Li T, Li P, Ling J (2018) Attribute-based handshake protocol for mobile healthcare social networks. Futur Gener Comput Syst 86:873–880
21. Petridou S, Basagiannis S (2012) Towards energy consumption evaluation of the SSL handshake protocol in mobile communications. In: 2012 9th annual conference on wireless on-demand network systems and services (WONS). IEEE, pp 135–138
22. Du X, Li K, Liu X, Su Y (2016) RLT code based handshake-free reliable MAC protocol for underwater sensor networks. J Sens
23. Mao J, Zhu H, Liu YL, Liu YJ, Qian W, Zhang J, Huang X (2018) RSA-based handshake protocol in internet of things. In: 2018 9th international conference on information technology in medicine and education (ITME). IEEE, pp 989–993
24. Zhang J, Yang L, Gao X, Tang G, Zhang J, Wang Q (2021) Formal analysis of QUIC handshake protocol using symbolic model checking. IEEE Access
25. Park J, Kang N (2014) Lightweight secure communication for CoAP-enabled internet of things using delegated DTLS handshake. In: 2014 international conference on information and communication technology convergence (ICTC). IEEE, pp 28–33

26. Qing L, Yaping L (2009) Analysis and comparison of several algorithms in SSL/TLS handshake protocol. In: 2009 international conference on information technology and computer science, vol 2. IEEE, pp 613–617
27. Sharma P, Lal N, Diwakar M (2013) Text security using 2d cellular automata rules. In: Proceedings of the conference on advances in communication and control systems-2013. Atlantis Press, pp 363–368
28. Diwakar M, Patel PK, Gupta K, Chauhan C (2013) Object tracking using joint enhanced color-texture histogram. In: 2013 IEEE second international conference on image information processing (ICIIP-2013). IEEE, pp 160–165
29. Kumar P, Sehgal V, Chauhan DS, Diwakar M (2011) Clouds: concept to optimize the quality of service (QOS) for clusters. In: 2011 world congress on information and communication technologies. IEEE, pp 816–821

Secure Authentication in WLAN Using Modified Four-Way Handshake Protocol

Manoj Diwakar, Prabhishek Singh, Pramod Kumar, Kartikay Tiwari, Shashi Bhushan, and Manju Kaushik

Abstract This paper presents a brief description of the wireless networks, their advantages over wired networks and security issues requiring immediate attention. It is then followed by 802.11 architecture and various services offered by it and then the motivation for conducting the research. Thereafter, a problem statement outlining the inherent flaws in the present IEEE 802.11i standard is presented. Hence, a method is proposed to overcome the denial of service using the handshaking approach. Finally, the results are discussed and concluded.

Keywords Handshake protocol · Network security · Denial of service · WLAN

M. Diwakar
Graphic Era Deemed to be University, Dehradun, India
e-mail: manoj.diwakar@gmail.com

P. Singh (✉)
Amity School of Engineering and Technology, Amity University, Noida, Uttar Pradesh, India

P. Kumar
Krishna Engineering College, Ghaziabad, Uttar Pradesh, India
e-mail: pramodkumar.hod@krishnacollege.ac.in

K. Tiwari
Thapar University, Patiala, Punjab, India
e-mail: ktiwari_be18@thapar.edu

S. Bhushan
UPES, Dehradun, India
e-mail: sbhushan@ddn.upes.ac.in

M. Kaushik
Amity University, Jaipur, Rajasthan, India

A. Tomar et al. (eds.), *Machine Learning, Advances in Computing, Renewable Energy and Communication*, Lecture Notes in Electrical Engineering 768,
https://doi.org/10.1007/978-981-16-2354-7_56

Table 1 Security services w.r.t security threats

Security services	Security threats				
	Traffic analysis	Eavesdropping	Masquerade	DoS	Authorization
Privacy	✓	✓	✓		✓
Authentication			✓		✓
Integrity			✓		✓
Access control			✓		✓
Non-repudiation			✓		✓

1 Introduction

WLAN is a wireless technology that is gaining universal acceptance due to its low cost, scalability and ease of installment, and the major advantage is support for mobility. In today's world, the major requirement of any user is of having continuous connectivity while roaming [1–5]. Despite providing these features it also suffers from issues like security and low quality of service, including lower data rates and frequent disconnections. Now five security services need to be achieved in order to provide security in any wireless network, namely privacy, authentication, integrity, access control and non-repudiation [6–10]. Table 1 shows which security services are required in order to prevent security threat and secure the wireless network. Therefore, in order to provide these security services and to prevent these attacks, various techniques and standards were defined [11–15]. Wired equivalent privacy (WEP) was the first major breakthrough in providing security to wireless networks, but it suffered from various types of attacks, and the secret key used was not secure enough and was proved to be breakable within few minutes. The latest standard which is being used presently to secure the wireless network is IEEE802.11i, which uses the best of earlier techniques and mechanisms to give a better and secure solution [4, 16–19].

2 Wired Equivalent Privacy (WEP)

WEP [20] was the first security algorithm that was originally included as a privacy mechanism in IEEE802.11 standard and was later ratified in Sep 1999 to incorporate more security features. The main aim of WEP was to provide some level of security to wireless networks. Various services [21–24] being offered by it include:

- Data secrecy/privacy: It is simply the prevention of data from the attackers so that it can only be read by authentic members. This comprises various encryption algorithms which are used to encrypt data being communicated over the network.
- Data integrity: It simply guarantees that data being communicated over the network has not been altered and is genuine.

- Access control: It basically depends on the integrity of data. A message which is corrupted is rejected and deemed as non-authenticated.

  WEP defines two methods of authentication for WLAN:

(1) **Open system authentication**: It is one of the convenient and feasible authentication algorithms [25–27] and is used as a default algorithm for providing authentication services as shown in Fig. 1. The steps followed are:

- Client sends an authentication request frame to the access point.
- AP in response sends the authentication response frame which specifies approval or disapproval of the authentication request.

(2) **Shared key authentication**: As shown in Fig. 2, it is a four-step process and begins by sending the authentication request message by the client to an access point which in response generates a challenge text and sends it to the client in other authentication of this frame [28–32]. The client after receiving this frame extracts the challenge text, encrypts it and reverts back the message frame to AP. AP then decrypts the message and tries to correlate it with the

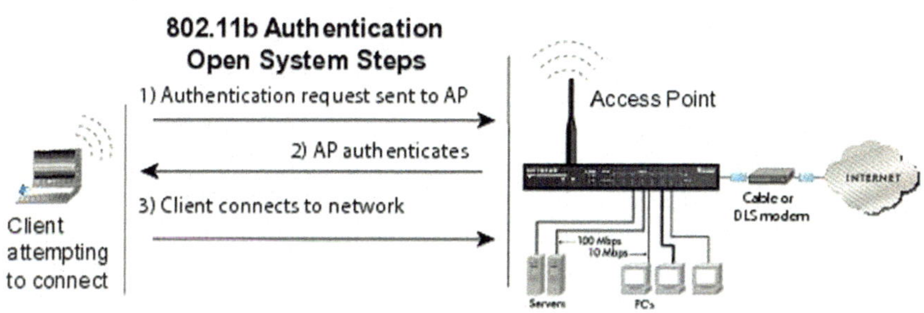

Fig. 1 Open system authentication

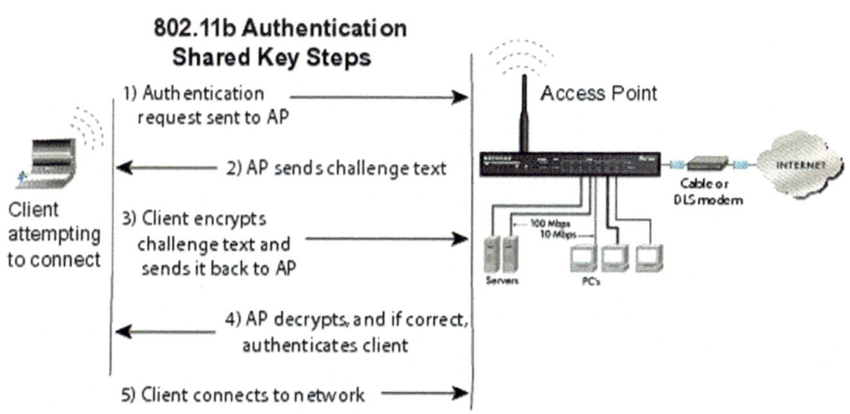

Fig. 2 Shared key authentication

original authentic challenge text being sent by it. If it matches then it justifies the identity and status of the client and it responds accordingly.

3 Proposed Methodology (Modified Four-Way Handshake)

This segment proposes an option changed four-way handshake answer for forestall DoS and DoS flooding assaults more than four-way handshake period of IEEE802.11i standard as appeared in Fig. 3. The message streams are:

(i) Mail 1: [A, Nonce1, Sequence, Msg1];
(ii) Mail 2: [B, Nonce2, Nonce1, Sequence + 1, Msg2, MIC_{PTK} ([Nonce2 XORNonce1], Sequence + 1, Msg2)];
(iii) Mail 3: [A, Nonce2, Seq + 1, Msg3, MIC_{PTK} (Nonce1, Sequence + 1, Msg3)];
(iv) Mail 4: [B, Sequence + 1, Msg4, MIC_{PTK} (Sequence + 1, Msg4)];
Here, PTK = PRF (PMK, Nonce2, A, B).

Here the significant change over standard four-way which has been done to forestall DoS assaults is that Nonce1 esteem communicated by the authenticator in Mail 1 which is the significant wellspring of propelling DoS assaults is never put away at

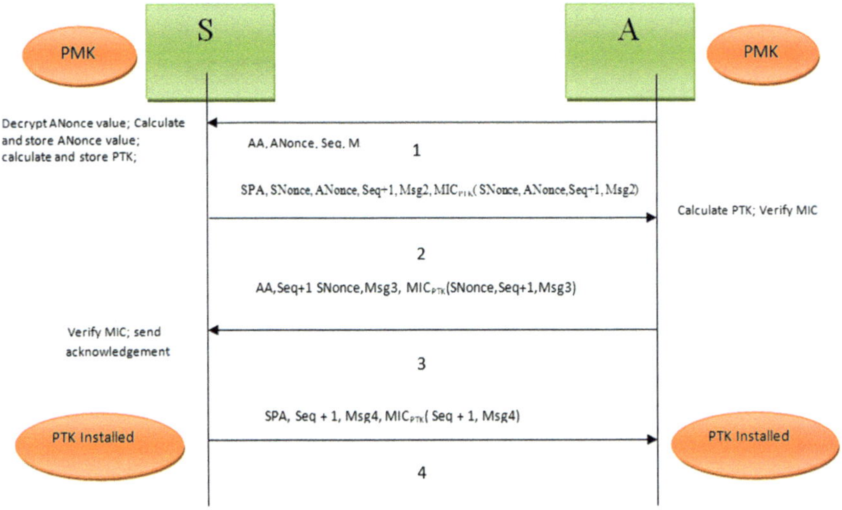

Fig. 3 Modified four-way handshake process

the petitioner side and rather the public transient key (PTK) is determined uniquely as the capacity of SNonce esteem which is the pseudorandom number created at the petitioner site. Additionally, the determined PTK is put away just a single time at the petitioner side in the wake of accepting the message1.

Steps to be followed are:

(i) AP builds message1 and sends it to petitioner S.
(ii) When Supplicant gets Message1:

- Generate and store SNonce esteem.
- Calculate PTK as the capacity of SNonce (ANonce esteem isn't thought of).
- Store PTK and transfer message.

(iii) After getting Message2 by AP:

- Calculate PTK by a similar instrument.
- Verify MIC.
- Construct and send Message3.
- Acknowledge receipt of Message3, construct and send Message4.

4 Results and Discussion

The proposed solution was implemented by developing a program using Java as a language on the Windows32 platform. The program is having three entities, namely a supplicant, access point and attacker. First, the authentication process in order to validate the PMK and to generate other succession keys begins between the supplicant and the authenticator, i.e., AP. The attacker is the entity that tries to forge messages (Message1) being sent over the network and thus tries to mount DoS and DoS flooding attacks. There is a separate message program file imported by every entity of the program, and it consists of the general frame fields with respective get and set methods that are used to access the data fields of the frames.

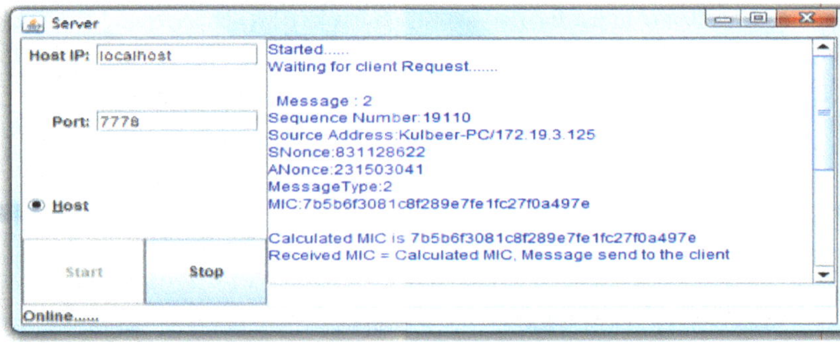

Access Point/Authenticator

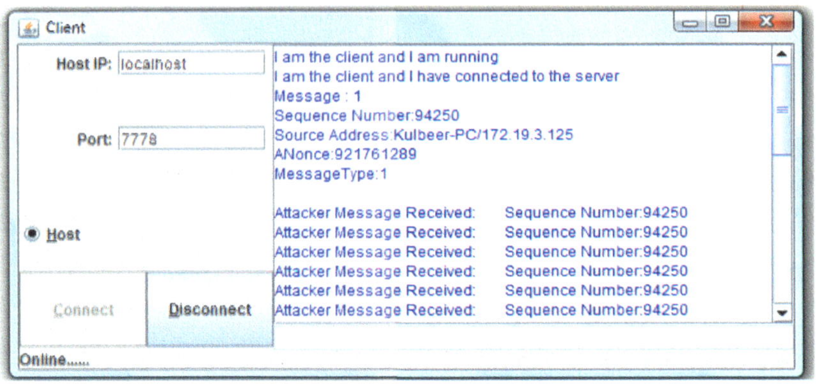

Supplicant/Client

Attacker

When a standard four-way handshake process is implemented, it has been noticed that as the fake packet rate, i.e., Message1 is increased, the time at which memory gets exhausted decreases rapidly. The memory size of the client was fixed to 5 Mb. As Table 2 shows when Message1 frames were sent by the attacker at the specified rate after sending Message2 by supplicant and before receiving Message3 frames from the authenticator.

Table 2 Memory DoS flooding attack on the proposed solution

Delay between the packets (ms)	Average time until memory exhaustion	Packets sent by the attacker
50	–	20
30	–	35
10	–	50

Table 3 Comparison of four-way handshake versus proposed solution

Parameters	Four-way handshake	Modified four-way handshake
DoS attack	Possible	Secure
Memory exhaustion	Occurs	Never occurs
Security	Less	More
Overhead	More	Least
Protection of management and control frames	Not protected	Not protected

Table 3 compares the four-way handshake process with the proposed solution. The overhead involved in the modified four-way handshake process is the least since PTK is calculated only once at the client-side for the whole process and the SNonce value is also generated and stored once in the whole session.

5 Conclusions

The most helpless piece is the sending of message1 which is sent decoded over the system. Likewise, it gives a way to making sure about control and the executive outlines which are additionally handed off in plain content over the system and can along these lines effectively be manufactured by the aggressor. It is likewise protected from memory depletion assaults since Nonce1 is never put away at the petitioner side since it very well may be unscrambled distinctly by the petitioner. Presently so as to make management and control outlines secure, the FCS field of these casings is adjusted to clear a path for the message integrity check field (MIC) without annexing any additional field for it in this manner making it perfect with prior control and the executive outlines.

References

1. Lashkari AH, Towhidi F, Hosseini RS (2009) In: The international conference on future computer and communication wired equivalent privacy (WEP), ICFCC 2009, pp 492–495
2. Lashkari AH, Mansoori M, Danesh AS (2009) Wired equivalent privacy (WEP) versus Wi-fi protected access. In: The ICCDA Singapore conference
3. NIST Special Publication 800-97 (2007) Establishing wireless robust security networks: a guide to IEEE 802.11i
4. Shiyang D (2010) Compare of new security strategy with several others in WLAN. In: The 2nd international conference on computer engineering and technology (ICCET)
5. He C, Mitchell JC (2004) Analysis of the 802.11i 4-way handshake. In: Proceedings of the 3rd ACM workshop on wireless security, Philadelphia, PA, USA, pp 43–50
6. Benefits and vulnerabilities of Wi-Fi protected access 2 (WPA 2). http://cs.gmu.edu/~yhwang1/INFS612/Sample_Projects/Fall_06_GPN_6_Final_Report.pdf
7. Rango FD, Lentini DC, Marano S (2006) Static and dynamic 4-way handshake solutions to avoid denial of service attack in Wi-Fi protected access and IEEE 802.11i. EURASIP J Wirel Commun Netw 2006:1–19
8. He C, Mitchell JC (2004) Analysis of the 802.111 4-way handshake. In: Proceedings of the ACM workshop on wireless security (WiSe '04), Philadelphia, PA, USA, Oct 2004, pp 43–50
9. Faria DB, Cheriton DR (2002) DoS and authentication in wireless public access networks. In: Proceedings of the ACM workshop on wireless security (WiSe '02), Atlanta, GA, USA, Sept 2002, pp 47–56
10. Edney J, Arbaugh WA (2003) Real 802.11 security: WiFi-protected access and 802.11i. Addison Wesley, New York, NY, USA (2003)
11. Xing X, Shakshuki E, Benoit D, Sheltami T (2008) Security analysis and authentication improvement for IEEE 802.11i specification. In: The IEEE global telecommunications conference, GLOBECOM 2008
12. Han S-J, Oh H-S, Park J (1996) The improved data encryption standard (DES) algorithm. In: Spread spectrum techniques and applications proceedings
13. Hassan HR, Challal Y (2005) Enhanced WEP: an efficient solution to WEP threats. In: The second IFIP international conference on wireless and optical communications networks, pp 594–599
14. Lashkari AH, Danesh MMS, Samadi B (2009) A survey on wireless security protocols (WEP, WPA and WPA2/802.11i). In: The 2nd IEEE international conference on computer science and information technology, pp 48–52
15. Chen J-C, Jiang M-C, Liu Y-W (2005) Wireless LAN security and IEEE 802.11i. Wirel Commun. 12. IEEE
16. Eum S-H, Cho S-J, Choi H-K, Choo H (2008) A robust session key distribution in 802.11i. In: The international conference on computational sciences and its applications ICCSA 2008
17. Ali KM, Owens TJ (2010) Access mechanisms in Wi-Fi networks state of art, flaws and proposed solutions. In: The IEEE 17th international conference on telecommunications (ICT)
18. Odhiambo ON, Biermann E, Noel G (2009) An integrated security model for WLAN. In: The AFRICON 2009, pp 1–6
19. Zhao S, Shoniregun CA, Imafidon C (2008) Addressing the vulnerability of the 4-way handshake of 802.11i. In: The third international conference on digital information management, ICDIM 2008, pp 351–356
20. Myneni S, Huang D (2010) IEEE 802.11 wireless LAN control frame protection. In: The 7th IEEE consumer communications and networking conference (CCNC)
21. Khan MA, Hasan A (2008) Pseudo random number based authentication to counter denial of service attacks on 802.11. In: The 5th IFIP international conference on wireless and optical communications networks, WOCN '08, pp 1–5
22. Yao Y, Chong J, Xingwei W (2010) Enhancing RC4 algorithm for WLAN WEP protocol. In: The control and decision conference (CCDC), pp 3623–3627

23. Chen J-C, Wang Y-P (2005) Extensible authentication protocol (EAP) and IEEE 802.1x: tutorial and empirical experience. IEEE Commun Mag 43:supl.26–supl.32
24. Gast M (2005) 802.11 wireless networks: the definitive guide. O'Reilly Publication
25. Liu T, Liu K, Cheng Y, Cai LX (2019) Security analysis of camera file transfer over Wi-Fi. In: ICC 2019-2019 IEEE international conference on communications (ICC). IEEE, pp 1–6
26. Sarjana FW, Arif TY, Adriman R, Munadi R (2019) Simple prevention of advanced stealth man-in-the-middle attack in WPA2 Wi-Fi networks. In: 2019 international conference on electrical engineering and computer science (ICECOS). IEEE, pp 349–353
27. Nandi S (2019) Elliptic curve cryptography based mechanism for secure Wi-Fi connectivity. In: 15th international conference on distributed computing and internet technology, ICDCIT 2019, Bhubaneswar, India, 10–13 Jan 2019, vol 11319, p 422. Springer
28. Singh GM, Kohli MS, Diwakar M (2013) A review of image enhancement techniques in image processing. Int J Technol Innov Res (IJTIR) 5
29. Diwakar M, Singh P (2020) CT image denoising using multivariate model and its method noise thresholding in non-subsampled shearlet domain. Biomed Signal Process Control 57:101754
30. Diwakar M, Kumar M (2015) CT image denoising based on complex wavelet transform using local adaptive thresholding and bilateral filtering. In: Proceedings of the third international symposium on women in computing and informatics, pp 297–302
31. Sharma P, Lal N, Diwakar M (2013) Text security using 2d cellular automata rules. In: Proceedings of the conference on advances in communication and control systems-2013. Atlantis Press, pp 363–368
32. Kumar P, Sehgal V, Chauhan DS, Diwakar M (2011) Clouds: concept to optimize the quality of service (QOS) for clusters. In: 2011 world congress on information and communication technologies. IEEE, pp 816–821

Teaching Bot to Play Thousand Schnapsen

Andżelika Domańska, Maria Ganzha, and Marcin Paprzycki

Abstract During recent years, AI found its ways to games. Here, imperfect information games, such as Thousand Schnapsen, bring about major challenges. In this work, the rules and characteristics of this game have been described. Next, an overview of existing literature, focusing on similar problems, has been presented, followed by a summary of selected methods for finding an optimal strategy along with applied modifications. Finally, the results of the experiments are discussed.

Keywords Card games · Thousand Schnapsen · Machine learning

1 Introduction

Thousand Schnapsen (also known as Russian Schnapsen) is one of the most popular card games in Poland. It is also played in countries such as Russia, Belarus, and Ukraine. The goal of the game is to score a total of at least 1000 points. The first player to do this wins. The game can be played by 2, 3, or 4 people. In this work, a 4-player version is considered. Local variants of the game introduce additional rules, e.g., in different regions of Poland, different scoring for "marriages" is applied. Hence, let us summarize the rules used in this contribution.

A. Domańska · M. Ganzha
Warsaw University of Technology, Warsaw, Poland
e-mail: M.Ganzha@mini.pw.edu.pl

M. Paprzycki (✉)
Systems Research Institute Polish Academy of Sciences, Warsaw, Poland
e-mail: marcin.paprzycki@ibspan.waw.pl

A. Tomar et al. (eds.), *Machine Learning, Advances in Computing, Renewable Energy and Communication*, Lecture Notes in Electrical Engineering 768,
https://doi.org/10.1007/978-981-16-2354-7_57

1.1 Game Rules and Characteristics

Consecutive games are played until one of the players scores 1000 points. The player who collected at least 800 points can get more if and only if (s)he wins an auction. In the 4-player version, the dealer is not an "active player". Still, (s)he can score points (see, **Declaration**). Each game consists of the following 5 stages.

Dealing A standard deck of 24 cards (9—Ace) is used. The dealer deals 7 cards to each player and the remaining 3 constitute the stock (hidden cards).

Bidding Dealer does not participate in this stage. Bidding starts from the person sitting to his left, who must bid at least 100 points. The next player (to the left) may raise the declared number, by a multiple of 10, or pass. Bidding continues until all but one player have passed. Since the maximum number of points that can be scored in a single game is 360, this is the maximum value that can be declared. A player can score 360 points when (s)he has three strongest marriages (valued at 240 points) and an ace of their suit. With an extremely "bad hands" of all opponents and their suboptimal game, it is possible to collect all cards (scoring 120 points), for a total of 360 points.

Declaration Next, the stock is revealed and the dealer receives points "found there" (Ace gives 10 points; marriages give values presented in Table 1).

Stock cards are taken by the highest bidder, who gives one card to each opponent (from that game). Before the game starts, (s)he must declare how many points (s)he plans to score (number not lower than that from the bidding).

Gameplay The game consists of placing one card on the pile by successive players. It consists of 8 rounds (one for each card in hand). The winner of the bidding starts the game. The winner of each round is the player who played the highest card. The winner of the round starts the next one. Here, two rules apply.

1. **The same suit**—if possible, the player must place a card with the same suit as the first card on the pile.
2. **Card with higher value**—if possible, the player must place a card with a higher value than the strongest card on the pile.

When the pile is empty, the player may check-in (place queen or king of the same suit, provided that (s)he has both). Here, point bonuses are awarded as in Table 1.

Each card has a rank and color. Here, T denotes the "10" card in the standard deck. Note that, T is "stronger" than J, Q, and K. Overall, set of cards is defined as $\mathcal{D} = \mathcal{R} \times \mathcal{C}$, where $\mathcal{R} = \{9, J, Q, K, T, A\}$ is set of ranks and $\mathcal{C} = \{C, D, H, S\}$ is set of colors. Each card has "value" depending on rank, defined by function $f : \mathcal{R} \longrightarrow \{0, 2, 3, 4, 10, 11\}$ represented in Table 2.

Table 1 Values of marriages

Suit	♠	♣	♢	♡
Value	40	60	80	100

Table 2 Rank values

r	9	J	Q	K	T	A
$f(r)$	0	2	3	4	10	11

However, the "strongest" card in the pile depends on the current state of the game. Here, relevant are: (1) color of the first card on the pile $F \in \mathcal{C}$ and (2) current marriage color (if any marriage was checked-in) $M \in \mathcal{C} \cup \{\emptyset\}$. Thus, the *contextual value* of the game is defined by $h : \mathcal{D} \longrightarrow \mathbb{R}$. This function must meet the following conditions: (a) card of the same color with a higher value has higher contextual value; (b) card of color, other than the color of the first card, and the current marriage has a lower contextual value than any card of the color of the first card or of the current marriage; and (c) any card of the color of the first card has a lower value than any card of the color of the current marriage (if they differ).

Counting points At the end of the game, players count rank values of collected cards and bonuses for checked-in marriages. If the winner of the auction has scored at least the number of declared points, this number is added to her total score otherwise, it is subtracted. The remaining players receive the scored points unless they have exceeded the threshold of 800 points.

Thousand Schnapsen is a **trick-taking** card game. Players try to collect as many opponents' cards as possible. As the opponents' cards are unknown, it is an **imperfect information** game. Moreover, it is **not** a **zero-sum** game. The value of all cards, in a single game, is 120, while the number of checked-in marriages depends both on the card split and the implemented strategies. This complicates the development of game-playing strategies (and training of bots).

Another difficulty is the large number of states. Before the game starts, each player has 8 cards (9,464,816,790 initial hands). The bidding winner knows her cards and one card from each opponent. Thus, the number of initial opponents' hands equals 3,432. The remaining players know their cards (they do not know what happened with two cards from the stock). Thus, the number of possible initial hands equals 12,870. Overall, the lower bound on the size of the game tree is 328,801, while the upper bound is 323,593,859,345,481. Obviously, these numbers are recalled only to illustrate the complexity of the game.

2 Related Work

Thousand Schnapsen is not a widely known game. Hence, instead, we discuss card games with incomplete information. Here, authors, typically, focus on variants of Poker or Bridge. However, attempts to apply AI in Austrian Schnapsen [10, 11] or the Swiss game Jass [4] can be found. In the latter article, an overview of methods that can be used for this type of game can be found. Overall, in the literature, algorithms based on expert knowledge (see, [1, 5, 7, 9]), reinforcement learning (e.g., *Temporal Difference Learning*, *Policy Gradient*, *First-Order Methods*, *Neural Fictious Self Play* [3], and *Counterfactual Regret Minimization* discussed in Sect. 2.1), Monte Carlo methods [11], and evolutionary algorithms are considered. Interested readers should consult references for pertinent details. Upon careful examination of the literature, we have decided to experiment with CFR and its modification that uses neural networks for function approximation—Deep CFR [2]—which showed the most promise in similar games.

2.1 Counterfactual Regret Minimization (CFR)

The CFR algorithm was proposed in 2007, for imperfect information games with large state spaces [12]. It uses iterative generation of new strategies (strategy profiles) that should converge to the Nash equilibrium. Its aim is to minimize regret value, introducing a new concept of counterfactual regret. Let u_i be the payoff function for player i, Σ_i set of his strategies, σ^t the strategy profile at time t, and σ^t_{-i} strategies of opponents at time t. Then regret can be defined as follows:

$$R_i^T = \frac{1}{T} \max_{\sigma_i^* \in \Sigma_i} \sum_{t=1}^{T} \left(u_i(\sigma_i^*, \sigma_{-i}^t) - u_i(\sigma^t) \right) \tag{1}$$

The concept introduced by the authors of the CFR algorithm is to define regret $R^T_{i,imm}(I)$ separately for each information set $I \in \mathcal{I}$. This approach is appropriate if and only if minimizing $R^T_{i,imm}(I)$ for each $I \in \mathcal{I}$ means minimizing R_i^T. On the basis of the calculated regret, it is possible to determine a new strategy using, for example, *regret matching*. The basic version of CFR requires the value of $R_i^T(I, a)$ to be stored for each possible state $I \in \mathcal{I}$ and action $a \in A(I)$. If $|\mathcal{I}|$ is very large, this is not feasible, due to limited RAM size. Two solutions are then proposed. (1) To use expert knowledge to create a state abstraction using advanced domain knowledge. (2) To use neural networks to approximate the function (see [8]). This modification is called Deep CFR and in application to Limit Texas Hold'em, it outperforms the NFSP [2].

Theoretical convergence of the CFR to the Nash equilibrium is ensured only for 2-player games [2]. Nevertheless, attempts to use it for 3-player games have shown its potential [6]. Even if, despite the lack of theoretical foundations, as a result of the CFR algorithm, a good approximation of the Nash equilibrium is calculated, it does not mean that an optimal strategy is found. Moreover, even finding the exact Nash equilibrium also does not, in theory, guarantee the calculation of the optimal strategy. This shows that games for more than 2 players are problematic for all algorithms based on Nash equilibrium calculation.

3 Applying Selected Approaches to the Game Playing

Let us elaborate on the scope of our work. First, note that, during the single game, player can make non-optimal decisions that lead to worse result, but also prevent him from immediate loss in the whole game. Considering it while training the bot is non-trivial and therefore it has been omitted. It allowed to consider each game completely independently of the previous ones. Finally, the 4-player version does not differ from the 3-player one, in the *game-playing phase*. Therefore, in what follows, the 3-player version is considered.

In the game, two problems can be distinguished. (1) Optimal bidding, taking into account the points, cards in hand, and the state of the auction. (2) Finding the optimal strategy for playing cards, based on the history of the game, cards in hand, cards in the pile, and the color of the last checked-in marriage. Since these are two independent problems, the latter is the focus of this work.

3.1 *Reasoning*

As the game progresses, players reveal information about their cards, through successive moves. Therefore, it is important to efficiently reason and remember which cards opponents definitely do/do not have. Based on the rules of the game, basic rules for inferring the status of opponents' cards can be defined.

If the opponent checked-in a "marriage" using Q or K, then (s)he is sure to have a second card of the pair. If, as the first card on the pile, the opponent chose a Q or K and didn't check-in, then (s)he does not have the second card of the pair. The application of above-stated rules, concerning cards that should be added to the pile, allows us to determine which cards opponents do not have.

3.2 *Payoff Function*

To implement the game, it is necessary to define the payoff function. Let H be the set of all states, $Z \subset H$ the set of final states, and $\mathcal{P} = \{1, 2, 3\}$ the set of players. Also, let the function $p_i : Z \longrightarrow \mathbb{Z} \cap [0, 360]$ return the number of points earned by i-th player in state $z \in Z$. Simple reward function defined as $u_i(z) = p_i(z)$ doesn't meet fixed-sum condition 2.

$$\forall_{z \in Z} \sum_{p \in \mathcal{P}} u_p(z) = 400 \tag{2}$$

Let $m : \mathcal{C} \longrightarrow \{40, 60, 80, 100\}$ be a function, which returns bonus for a marriage in a given color. Moreover let $C_{\text{unused}} : Z \longrightarrow 2^{\mathcal{C}}$ be a function that returns the set of colors of not checked-in marriages, for the final state. Modified function u_p can be defined as follows:

$$u_i(z) = p_i(z) + \frac{1}{|\mathcal{P}|} \sum_{c \in C_{\text{unused}}(z)} m(c) \tag{3}$$

Proof that function 3 meets the condition 2 is trivial.

3.3 Implementing CFR

The CFR requires a full traversal of the game tree, which can take a very long time (over 40 min on a moderate computer). Therefore, the External-sampling MCCFR variant, which does not require a full traversal, was chosen. Here, full tree traversal is replaced with the Monte Carlo sampling. Here, in the nodes where the movement belongs to the highlighted player, a full traversal is performed, and in the remaining nodes, a random action is selected. The probability of selection is defined by the current strategy.

Since the number of game states is very large, it was necessary to use abstraction. Moreover, proper encoding of the card set was needed. Since the used deck consists of 24 cards, it can be represented as a 24-bit number. If the i-th bit is set, it means that a card of index i is in the given set. For example set $\{(9, S), (Q, D), (A, H)\}$ is encoded by 8 404 993, with binary representation:

```
 1 0 0 0 0 0 0 0 0 1 0 0 0 0 0 0 0 0 0 0 0 0 0 1
 | | | | | | | | | | | | | | | | | | | | | | | |
AH                QD                            9S
```

The proposed representation saves memory, but requires additional computations to encode and decode the cards.

In addition to encoding the cards, reasoning, described in Sect. 3.1, was used. Ultimately, the state was defined using:

- encoded set of cards that can be used by the player,
- encoded set of rest player's cards,
- encoded set of cards possibly owned by opponents,
- encoded sets of cards that are for sure owned by each opponent.

To reduce the state set, it is possible to represent similar situations the same. Hence, it was proposed to "unify" the sets of cards by "shifting" the colors. This operation, shown in Example 1, consists of "cutting out" the encoding fragments corresponding to the colors, from which there is no card in any of the given sets, and completing the encoding, on the right, with zeros to the length of 24.

Example 1 There are two sets of card sets:

(a) {(9, S), (Q, D), (A, H)}
{(T, S), (K, D)}

(b) {(9, S), (Q, C), (A, H)}
{(T, S), (K, C)}

These sets are originally encoded as follows:

(a) `100000000100000000000001`
`000000001000000000010000`

(b) `100000000000000100000001`
`000000000000001000010000`

The code fragments, corresponding to the colors that are empty, in each encoding from a given set, have been marked in blue. After performing the unification, the obtained encodings are as follows:

(a) `100000000100000001000000`
`000000001000010000000000`

(b) `100000000100000001000000`
`000000001000010000000000`

3.4 Deep CFR

Deep CFR [2] is a variant of CFR, based on neural network function approximation. The method scheme is very similar to that for tabular CFR. Calculation of the current strategy, in a given state, is done using the *advantage network*. During subsequent traversals of the game tree, new training data is obtained for the network and stored in buffer(s) with a limited capacity. These buffers use the *reservoir sampling* method. At the end of each iteration, the *policy network* is also trained, which "stores" the calculated strategy. Both networks may or may not have the same architecture.

Encoding a game state, for use in Deep CFR, should meet different requirements than encoding for standard CFR. The use of neural networks eliminates the need to code sets of cards with a single integer. Moreover, unification described in Sect. 3.3 is not advisable, as this extra work may slow the learning process.

To reduce the number of states, without losing information that can be useful in the decision-making process, is to use inference about opponent's cards (see, Sect. 3.1). Remembering game history is redundant. Information about the course of the game can be represented by: (a) set of cards that can be owned by both opponents, (b) set of cards that can be owned by the first opponent, and (c) set of cards that can be owned by the second opponent. Moreover, encoding of checked-in marriages can be achieved by coding with the set of colors and the designation of the last checked-in.

Note that, while the first round is started by the winner of the bidding, the next is started by the winner of the previous round. However, based on the information which player starts which turn, it is possible to clearly determine which player will be next (clockwise). This allows to replace the original opponent IDs with IDs in the context of the current sequence. After all simplifications, the full game state consists of: (1) set of player's cards, (2) list of cards in the pile, (3) set of cards that can

Table 3 Game state encoding

Position	Data
0–23	Set of player's cards
24–71	List of cards in the pile
72–95	Set of cards that can be owned by both opponents
96–119	Set of cards that are for sure owned by the first opponent
120–143	Set of cards that are for sure owned by the second opponent
144–148	Set of checked-in marriages
149–151	Trump

be owned by both opponents, (4) set of cards that are for sure owned by the first opponent, (5) set of cards that are for sure owned by the second opponent, (6) set of checked-in marriages, and (7) trump.

Card-sets were encoded as binary vectors of length 24. Since there can be 0, 1, or 2 cards in the pile, it is encoded by 2 binary vectors (of length 24), corresponding to the first and the second card. When there is no second or, even, first card in the pile, the corresponding vectors are all zeros. Otherwise, in the position corresponding to the index of the card it encodes, there is one.

Colors of checked-in marriages are encoded by a binary vector of length 4. The trump is also coded by a binary vector of length 4. When no check-in has occurred yet, the vector consists of all zeros. All codes are combined into one binary vector of length 152, summarized in Table 3.

In Sect. 3.2, an approach to modifying the reward function has been proposed so that the game can be classified as a fixed-sum game. This solution was also used for the Deep CFR algorithm. Additionally, the standardization of the values expected on the *advantage network* output has been introduced. The payoff values are in the range $[0, 400]$. Their weighted average also is in this range. Due to the regret definition, it is in the range $[-400, 400]$. Standardizing by dividing by 400 reduces this range to $[-1, 1]$.

4 Experimental Results

Let us now discuss experimental results obtained for the CFR and the Deep CFR. Obviously, since there are no known results of applying AI to Thousand Schnapsen, we had to implement a "random strategy" player, performance of which will be used as a "baseline". This player selected a random action, from the set of available actions (where the probability of choosing each action is equal).

Overall, 10,000 games were played, all using random strategy, to establish performance, depending on players' order in the first round. It showed that the first player wins, on average, 38.48% of the games, the second 29.85%, and the third 31.67%.

Hence, the player that starts the game has an advantage. Interestingly, the second best result does not belong to the second player, but to the third one. The reason for this may be that the third player, often, has a smaller number of possible actions (due to the re-raise requirement). Therefore, the probability of picking the "wrong move" is smaller than for the second player.

Interestingly, the experiment showed that 1,000 games give a good enough estimate of effectiveness because afterward the number of wins for each player is close to that obtained after 10,000 games. This observation was used to determine the number of simulations to be performed to estimate the effectiveness of training.

4.1 CFR Results

After establishing the baseline, let us now look into the results obtained by the training approaches. Despite the use of minimalist representation of the game state, the number of possible states remained too large to save them in memory of a computer with 20 GB RAM and the swap memory limit set at 15 GB. After about 70 million states were saved, training stopped due to lack of memory.

Therefore, an even more limited state representation was tested. It contained: (1) set of cards that can be used by the player, (2) set of cards that are for sure owned by the first opponent, (3) set of cards that are for sure owned by the second opponent, and (4) information if the pile is empty. On the one hand, the number of states was still very large, and on the other, available information was very limited. This led to situations where two states, in which completely different decisions should be made, were treated the same. The performance of the obtained strategy was exactly the same as the random one.

A potential solution to these problems may be to develop a better abstraction of the state, both reducing the number of possible states and the loss of significant information. It could be also possible to use a much more powerful computer. However, the latter was not an objective of the research. It just illustrates one of the important technical issues related to the use of the CFR-based approach.

4.2 Deep CFR Results

The problems that made use of the tabular CFR unsuccessful, i.e., too many states and too much information limitation, did not occur in the Deep CFR method. Both *advantage network* and *policy network* are MLP networks consisting of 4 hidden layers with 192, 96, 48, and 24 neurons. The activation function in each of the hidden layers was $TanH$. The remaining parameters were set to: (a) number of traversals per iteration—10, (b) number of epochs—30,000, (c) batch size—100, (d) learning rate—$1e-5$, (e) *advantage network* buffer's size—$1e6$, and (f) *policy network* buffer's size—$2e6$. After each learning iteration, the algorithm's effectiveness was

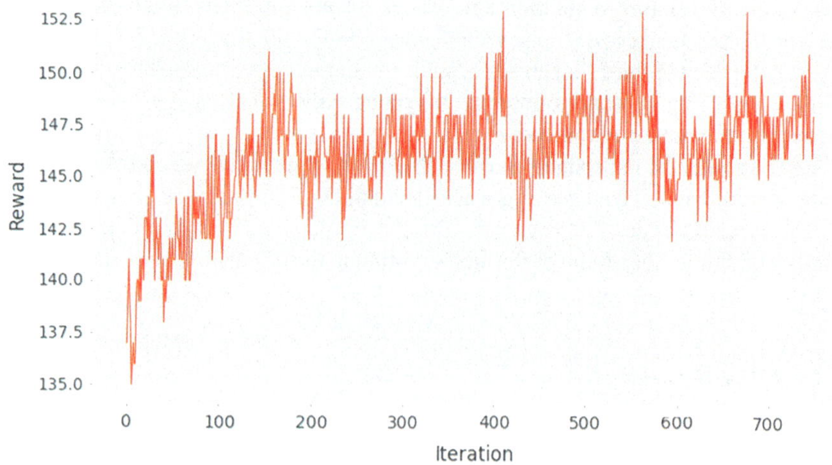

Fig. 1 Deep CFR algorithm performance achieved in the first 750 training iterations

evaluated. It consisted in conducting 1,000 games between the Deep CFR agent and the agents implementing the random strategy. The results are presented in Fig. 1.

After a total of 750 iterations, the maximum average reward achieved is 153. A trained agent's percentage of success depends on which player plays the card on the first turn. It was calculated on the basis of 10,000 games played against agents implementing a random strategy, for each of the situations. Somewhat similarly to the random case, the player that plays the first hand has the advantage. Specifically, the first player won 51,5% of games, the second 39,4% of games, and the third 41,5% of games. Here, again, we see that the best "positions" to be in is the first and the last player of the first round.

Overall, trained agents definitely outperform these using random strategy. The greatest "progress" can be observed for the player that starts the game, it is about 13% points.

The results of 10,000 simulations in which *each* player implements a strategy calculated using the Deep CFR algorithm are similar to the baseline. The first player wins about 40.00% of games, the second one 29.14%, the third one 30.86%.

5 Conclusions

Obtained results differ from other card game focused AI in at least two aspects:

- Thousand Schnapsen is not popular all over the world, so no algorithm has been found in the literature that works at the level of a human expert,
- the considered algorithms, in theory, do not work for more than 2-player games.

Therefore, developing a solution that is more effective than the random strategy can be considered a success.

As part of the project, an abstraction of the game state was also developed, which can undoubtedly be used to implement other algorithms that deal with other card games, not explored in this work. Obviously, the obtained results open the way for further research in multiple directions.

References

1. Bergh MVD, Hommelberg A, Kosters W, Spieksma F (2016) Aspects of the cooperative card game hanabi. In: BNCAI
2. Brown N, Lerer A, Gross S, Sandholm T (2019) Deep counterfactual regret minimization. In: Proceedings of the 36th international conference on machine learning, pp 793–802
3. Heinrich J, Silver D (2016) Deep reinforcement learning from self-play in imperfect-information games (03)
4. Niklaus J, Alberti M, Pondenkandath V, Ingold R, Liwicki M (2019) Survey of artificial intelligence for card games and its application to the SWISS game JASS (06)
5. Osawa H (2015) Solving hanabi: estimating hands by opponent's actions in cooperative game with incomplete information. In: Ganzfried S (ed) Computer poker and imperfect information, papers from the 2015 AAAI workshop, Austin, Texas, USA. 26 Jan 2015
6. Risk N, Szafron D (2010) Using counterfactual regret minimization to create competitive multiplayer poker agents (01), 159–166
7. Robilliard D, Fonlupt C, Teytaud F (2014) Monte-Carlo tree search for the game of "7 wonders" (08), 64–77
8. Sutton RS, Barto AG (2018) Reinforcement learning: an introduction. The MIT Press, 2nd edn. http://incompleteideas.net/book/the-book-2nd.html
9. Ward C, Cowling P (2009) Monte Carlo search applied to card selection in magic: the gathering (10), 9 – 16. https://doi.org/10.1109/CIG.2009.5286501
10. Wisser F (2010) Creating possible worlds using sims tables for the imperfect information card game schnapsen 2(11), 7–10. https://doi.org/10.1109/ICTAI.2010.76
11. Wisser F (2015) An expert-level card playing agent based on a variant of perfect information monte carlo sampling. In: Proceedings of the 24th international conference on artificial intelligence, pp 125–131. IJCAI'15, AAAI Press
12. Zinkevich M, Johanson M, Bowling M, Piccione C (2007) Regret minimization in games with incomplete information. In: Proceedings of the 20th international conference on neural information processing systems, pp 1729–1736. NIPS'07, Curran Associates Inc., Red Hook, NY, USA

Correction to: Analysis of Algorithms in Medical Image Processing

Tina, Sanjay Kumar Dubey, Ashutosh Kumar Bhatt, and Mamta Mittal

Correction to: Chapter "Analysis of Algorithms in Medical Image Processing" in: A. Tomar et al. (eds.), *Machine Learning, Advances in Computing, Renewable Energy and Communication*, Lecture Notes in Electrical Engineering 768, https://doi.org/10.1007/978-981-16-2354-7_10

In the original version of the book, the following belated corrections have been incorporated: The author name "Mamata Mittal" has been changed to "Mamta Mittal" in the Frontmatter, Backmatter and in Chapter "Analysis of Algorithms in Medical Image Processing".

The chapter and book have been updated with the changes.

The updated version of this chapter can be found at
https://doi.org/10.1007/978-981-16-2354-7_10

A. Tomar et al. (eds.), *Machine Learning, Advances in Computing, Renewable Energy and Communication*, Lecture Notes in Electrical Engineering 768,
https://doi.org/10.1007/978-981-16-2354-7_58

Author Index

A. Tomar et al. (eds.), *Machine Learning, Advances in Computing, Renewable Energy and Communication*, Lecture Notes in Electrical Engineering 768,
https://doi.org/10.1007/978-981-16-2354-7

Printed by Printforce, the Netherlands